A Guide to Polymeric Geomembranes

Wiley Series in Polymer Science

Series Editor
John Scheirs
Excelplas Geomembrane Testing Services
PO Box 2080
Edithvale, VIC 3196
Australia
scheirs.john@pacific.net.au

Modern Fluoropolymers
High Performance Polymers for Diverse Applications

Polymer Recycling
Science, Technology and Applications

Metallocene-based Polyolefins
Preparations, Properties and Technologies

Polymer-Clay Nanocomposites

Dendrimers and Other Dendritic Polymers

Modern Styrenic Polymers
Polystyrenes and Styrenic Copolymers

Modern Polyesters
Chemistry and Technology of Polyesters and Copolymers

Feedstock Recycling and Pyrolysis of Waste Plastics
Converting Waste Plastics into Diesel and Other Fuels

A Guide to Polymeric Geomembranes

JOHN SCHEIRS

Excelplas Geomembrane Testing Services, Edithvale, Australia

Wiley Series in Polymer Science

A John Wiley and Sons, Ltd., Publication

This edition first published 2009
© 2009 John Wiley and Sons Ltd

Registered office
John Wiley & Sons Ltd, The Atrium, Southern Gate, Chichester, West Sussex, PO19 8SQ, United Kingdom

For details of our global editorial offices, for customer services and for information about how to apply for permission to reuse the copyright material in this book please see our website at www.wiley.com.

Library of Congress Cataloging-in-Publication Data

Scheirs, John.
  A guide to polymeric geomembranes : a practical approach / John Scheirs.
     p. cm.
  ISBN 978-0-470-51920-2 (cloth)
 1. Geosynthetics. 2. Polymers. I. Title.
  TA455.G44S34 2009
  624.1'8923–dc22

                                       2009012002

A catalogue record for this book is available from the British Library.
ISBN 978-0-470-51920-2 (H/B)

Set in 10/12 Times by Laserwords Private Limited, Chennai, India.
Printed and bound in Great Britain by CPI Antony Rowe, Chippenham, Wiltshire

Contents

Series Preface

The Wiley Series in Polymer Science aims to cover topics in polymer science where significant advances have been made over the past decade. Key features of the series will be developing areas and new frontiers in polymer science and technology. Emerging fields with strong growth potential for the twenty-first century such as nanotechnology, photopolymers, electro-optic polymers etc. will be covered. Additionally, those polymer classes in which important new members have appeared in recent years will be revisited to provide a comprehensive update.

Written by foremost experts in the field from industry and academia, these books have particular emphasis on structure-property relationships of polymers and manufacturing technologies as well as their practical and novel applications. The aim of each book in the series is to provide readers with an in-depth treatment of the state-of-the-art in that field of polymer technology. Collectively, the series will provide a definitive library of the latest advances in the major polymer families as well as significant new fields of development in polymer science.

This approach will lead to a better understanding and improve the cross fertilization of ideas between scientists and engineers of many disciplines. The series will be of interest to all polymer scientists and engineers, providing excellent up-to-date coverage of diverse topics in polymer science, and thus will serve as an invaluable ongoing reference collection for any technical library.

John Scheirs

Preface

Geomembranes are flexible polymeric sheets mainly employed as liquid and/or vapour barriers. Polymeric (and elastomeric) geomembranes are designed as relatively impermeable liners for use in a variety of containment situations (e.g. to contain water, process fluids, leachates, mine liquors and contaminated industrial liquid effluents), in applications where natural clay or other containment options are not possible or viable.

Geomembranes are used extensively in a broad array of industries such as water conservation, mining, construction, waste management, agriculture, aquaculture, and wine making. Their diverse applications include: water protection; conveyance and storage; basins and ponds; municipal solid waste (MSW) and hazardous waste (HSW) landfills; process water ponds and leachate collection ponds; storm water collection ponds; evaporation aprons; private and commercial water features; floating covers and other containment facilities. In all of these applications geomembranes intercept the flow path of liquid through soil, performing fluid barrier functions in the containment system.

Geosynthetic engineers, specifiers, designers, facility owners and operators are presented with a diverse range of geomembrane materials which all appear to provide similar benefits, based on basic mechanical properties. The key factor to consider when assessing the suitability of a given geosynthetic product is its performance when it comes to installation, welding, chemical resistance and environmental durability.

This book covers the various types of materials used for geomembranes, their attributes and shortcomings. Each type of geomembrane material has different characteristics which affect installation procedures, durability, lifespan and performance. There are a number of geomembrane types available so it is often difficult to select the geomembrane with the right combination of properties required for a given application. Geomembrane materials are selected for their overall chemical resistance, mechanical properties (elastic modulus, yield strength, puncture/tear resistance), and weathering resistance. Good material selection coupled with excellent design and construction methods could yield a 'theoretically' flawless liner. In practice, however, some degree of installation-related imperfections and applied stress/strain on the liner system is inevitable.

This book is intended to assist project specifiers, engineers and purchasers in their understanding and evaluation of polymeric geomembranes. All aspects of polymeric geomembrane materials, performance, testing, design, engineering, installation considerations, welding practices, case histories and field failures are included in the book. An overview of the manufacture, structure-property relationships, material properties and quality control of geomembrane materials is provided. The material properties covered which relate to the manufacture and quality are referred to as index properties as well as those related to the design and in-service properties are referred to as design or performance properties. Both the index and performance properties are important in specifying geomembranes for containment, liner and cover applications.

Geomembranes have become critical components in the design and environmental performance of mining facilities. The mining industries extensively utilize geomembranes in heap leach pads, solution ponds and evaporation ponds. Due to the enormous size of many of these mining applications, mines have come to represent a significant percentage of geomembrane consumption.

Mining companies stretch the capabilities of geomembranes to their limits and sometimes to the point of failure. For example leach pads are heaps of rocks/ore up to 120–180 m high piled on a geomembrane pad. In addition, there is traffic on the liner leach pad in the form of truck hauls or dozer pushes. Furthermore, the liquors used are highly acidic with pH values of around 1. Thus leach pads present one of the most aggressive service environments for geomembranes.

Due to their functionality geosynthetic liners and membranes are utilized in some of the most demanding applications that synthetic materials have been called upon to perform in. Service environments often combine extreme heat and UV exposure together with high mechanical loading and exposure to aggressive solutions and slurries. In addition, geosynthetic liners are expected to exhibit long-term durability with service lives being measured in decades rather than years. Expected service lives generally range from 20 to >100 years.

Geosynthetic liners are used extensively in critical applications such as protecting the water table from toxic landfill leachates or preventing corrosive mine process solutions from contaminating soil and aquifers. The failure of such geosynthetic barriers can have devastating environmental consequences. Given that they must withstand the extremes of weather and be laid over coarse and soft subgrades it is important that a generous safety factor is employed in their installation designs. Geosynthetic liners such as landfill caps need to have excellent longevity since they are intended to become permanent features of the landscape.

This book discusses the structure-property relationship of various geomembrane materials and compares and contrasts their individual advantages and shortcomings. The geotechnical designer needs to understand the limitations of various geomembrane products – relatively thin materials (0.5–0.75 mm) can be damaged by abrasion, for instance, and the texturing on spray-on geomembranes can be scratched off.

In landfill design, geomembranes are typically used as base liners (or basal liners) which are placed below waste to minimise seepage of leachate into the underlying soil and into the water table. Geomembrane covers are placed over the final waste to keep surface water and rain water from infiltrating the waste and adding to the volume of leachate solution. The geomembrane cover also serves to capture the landfill gas preventing release of methane which is a potent greenhouse gas. Base liners are typically HDPE because of its inherently good chemical resistance and strength, whereas the covers are generally LLDPE, VLDPE or fPP since they are more flexible than HDPE and hence conform better to the underlying decomposing waste and are better able to tolerate the strains associated with waste settlement.

About the Author

Dr. John Scheirs is a polymer technologist with ExcelPlas Geomembrane Testing Services. He specializes in geomembrane properties, polymer selection, failure analysis and testing and evaluation of geomembrane materials. He has extensive experience in durability testing of many polymeric geomembranes, in addition to routine mechanical analysis, forensic investigation and weld testing of geomembranes and polymeric liners. He has worked widely with HDPE, fPP and PVC membranes. Previously he worked with Exxon-Mobil on the stabilization and long-term durability of HDPE polymers.

Symbols Used

Throughout this book a number of symbols are used in the text as an aid to the reader for passages in italics that warrant particular attention because of their importance or relevance to the subject.

The symbols used in this book are:

 A drawing pin to highlight interesting or significant facts

 A clipboard to indicate "TAKE NOTE" for important notes to avoid pitfalls

 An eye to indicate "AT A GLANCE" for summarized information

A thumbs up to indicate "RULES OF THUMB" for general (simplified) rules of science or engineering

Acknowledgments

The following persons are acknowledged for their generous assistance in proof-reading selected chapters:

Rick Thomas, Texas Research International (TRI), Texas, USA

Dr. Ian D. Peggs, I-CORP INTERNATIONAL, Florida, USA

Prof. Kerry Rowe, Queen's University, Canada

David Bishop & Vera Olischlaeger from NAUE GmbH & Co. KG, Germany

Felon Wilson, Sue Uhler and Bill Shehane from Seaman Corporation, USA

Prof. Stephen W. Bigger, Faculty of Health, Engineering and Science at Victoria University, Australia

Fred Gassner, Golder Associates, Melbourne, Australia

Liza Du Preez, Golder Associates, Perth, Australia

Lance St Hill, Fabric Solutions, Queensland, Australia

Simon Hsu, Nylex, Victoria, Australia

Rod Parry, ExcelPlas Geomembrane Testing, Melbourne, Australia

Phil Bennet, GEOTEST, Adelaide, Australia

Andrew Mills, Layfield, Canada

1

Introduction to Polymeric Geomembranes

1.1 INTRODUCTION

The large number of commercially available geomembranes (or polymeric geosynthetic barriers) can make it challenging to select which geomembrane has the most appropriate combination of performance properties for a given application. Each type of geomembrane material has different characteristics that affect its installation procedures, durability, lifespan and overall performance. It is therefore necessary to match the project performance criteria with the right combination of properties of a particular geomembrane. Geomembrane materials are generally selected for their overall performance in key areas of chemical resistance, mechanical properties (elastic modulus, yield strength, puncture/tear resistance), weathering resistance, product life expectancy, installation factors and cost effectiveness.

The properties of polymeric geomembranes are determined mainly by their polymer structure (architecture of the chains), molecular weight (i.e. the length of the chains) and the crystallinity (packing density of the chains). Polymer crystallinity is one of the important properties of all polymers. Polymers exist both in crystalline and amorphous forms.

Common geomembranes can be classified into two broad categories depending on whether they are thermoplastics (i.e. can be remelted) or thermoset (i.e. crosslinked or cured and hence cannot be remelted without degradation) (see Table 1.1). Since thermoset geomembranes are crosslinked, they can exhibit excellent long-term durability.

When selecting a geomembrane for a particular application the following aspects need to be considered:

- choice of polymer;
- type of fabric reinforcement;
- colour of upper ply (e.g. white to maintain lower temperatures for sun exposed applications);
- thickness;

A Guide to Polymeric Geomembranes: A Practical Approach   J. Scheirs
© 2009 John Wiley & Sons, Ltd

Table 1.1   Main plastic classifications for common geomembrane types

Thermoplastic geomembranes	Thermoset geomembranes	Combinations of thermoplastic and thermoset
HDPE, LLDPE	CSPE (crosslinks over time)	PE-EPDM
fPP	EPDM rubber	PVC-nitrile rubber
PVC	Nitrile rubber	EPDM/TPE (Trelleborg)
EIA	Butyl rubber	Polymer-modified bitumen
TPU, PVDF	Polychloroprene (Neoprene)	

- texture (e.g. smooth or textured for improved friction angles);
- product life expectancy;
- mechanical properties;
- chemical resistance;
- ease of installation.

Table 1.2 lists various advantages and disadvantages of common geomembrane types.

Firstly, geomembrane quality begins with base polymer resin selection. It is important to select or specify high-grade polymer resins that have been manufactured to meet the specific, unique demands encountered by geomembranes. Polymeric geomembrane properties are a function of the chemical structure of the base polymer resin, the molecular weight, the molecular weight distribution and the polymer morphology (e.g. the crystallinity). Next it is necessary to select the right combination of additives to protect the geomembrane, such as premium carbon black as well as antioxidant additives and stabilizers to ensure long life even in exposed conditions. Finally, it is necessary to select the most appropriate geomembrane manufacturing method.

1.2 VISCOELASTIC BEHAVIOUR

Polymers exhibit both viscous and elastic characteristics when undergoing deformation and hence are termed viscoelastic. Viscous materials (like honey), resist shear flow and strain linearly with time when a stress (e.g. in-service loading) is applied. Elastic materials strain (i.e. elongate) instantaneously when stretched and quickly return to their original state once the stress is removed (e.g. as in the case of EPDM liners). Viscoelastic materials have elements of both of these properties and, as such, exhibit time dependent strain.

Some phenomena in viscoelastic materials are:

1. If the strain is held constant, the stress decreases with time (this is called relaxation).
2. If the stress is held constant, the strain increases with time (this is called creep, as can be observed with HDPE liners).

Viscoelastic behavior comprised of elastic and viscous components is modelled as linear combinations of springs and dashpots, respectively.

The Maxwell model for viscoelastic behaviour can be represented by a viscous dashpot (a piston in oil) and an elastic spring connected in series, as shown in Figure 1.1(a). In

Table 1.2 Advantages and disadvantages of commonly used synthetic geomembranes

Geomembrane	Advantages	Disadvantages
HDPE	Broad chemical resistance	Potential for stress cracking
	Good weld strength	High degree of thermal expansion
	Good low temperature properties	Poor puncture resistance
	Relatively inexpensive	Poor multiaxial strain properties
LLDPE	Better flexibility than HDPE	Inferior UV resistance to HDPE
	Better layflat than HDPE	Inferior chemical resistance to HDPE
	Good multiaxial strain properties	
fPP	Can be factory fabricated and folded so fewer field fabricated seams	Limited resistance to hydrocarbons and chlorinated water
	Excellent multiaxial properties	
	Good conformability	
	Broad seaming temperature window	
PVC	Good workability and layflat behaviour	Poor resistance to UV and ozone unless specially formulated
	Easy to seam	Poor resistance to weathering
	Can be folded so fewer field fabricated seams	Poor performance at high and low temperatures
CSPE	Outstanding resistance to UV and ozone	Cannot be thermally welded after ageing
	Good performance at low temperatures	
	Good resistance to chemicals, acids and oils	
EPDM	Good resistance to UV and ozone	Low resistance to oils, hydrocarbons and solvents
	High strength characteristics	Poor seam quality
	Good low temperature performance	
	Excellent layflat behaviour	
Butyl rubber	Good resistance to UV and weathering	Relatively low mechanical properties
	Good resistance to ozone	Low tear strength
		Low resistance to hydrocarbons
		Difficult to seam
Nitrile rubber	Good resistance to oils and fuels (but not biodiesel)	Poor ozone resistance unless properly formulated
		Poor tear strength

this model if the polymer is put under a constant strain, the stresses gradually relax. That is, the tension in the spring (the stress) is gradually reduced by movement of the piston in the dashpot after a strong elongation (or displacement). Stress relaxation describes how polymers relieve stress under constant strain.

The Kelvin–Voigt model for viscoelastic behaviour also known as the Voigt model, consists of a viscous dashpot and Hookean elastic spring connected in parallel, as shown in Figure 1.1(b). It is used to explain the creep behaviors of polymers. When subjected

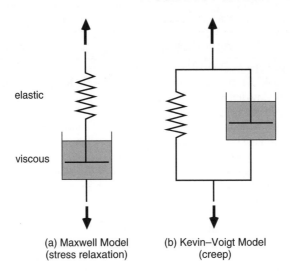

(a) Maxwell Model   (b) Kevin–Voigt Model
(stress relaxation)   (creep)

Figure 1.1 Polymers are viscoelastic materials having the properties of both viscous and elastic materials and can be modelled by combining elements that represent these characteristics. One viscoelastic model, called the Maxwell model, predicts behavior akin to a spring (elastic element) being in series with a dashpot (viscous element), while the Kelvin–Voigt model places these elements in parallel. Stress relaxation describes how polymers relieve stress under constant strain. Because they are viscoelastic, polymers behave in a nonlinear, non-Hookean fashion. This nonlinearity is described by both stress relaxation and a phenomenon known as creep, which describes how polymers strain under constant stress

to a constant stress, viscoelastic materials experience a time-dependent increase in strain (i.e. change in length). This phenomenon is known as viscoelastic creep. In this model on the application of a force, the spring gradually expands until the spring force equals the applied stress. Creep describes how polymers strain under constant stress.

The temperature dependence of strain in polymers can also be predicted using this model. An increase in temperature correlates to a logarithmic decrease in the time required to impart equal strain under a constant stress. In other words, it takes less energy to stretch a viscoelastic material an equal distance at a higher temperature than it does at a lower temperature.

1.3 POLYMER STRUCTURE

Polymer structure describes the chemical makeup of the polymer chains. HDPE, for example, is comprised of linear molecules of repeating CH_2 groups as shown in Figure 1.2.

Chemical structures of the main classes of geomembranes are shown in Figure 1.3. Note that HDPE due to its regular, symmetrical structure is crystalline and quite stiff but by substituting one of the hydrogen atoms (shaded) with a bulky methyl group (as in flexible polypropylene) or an even more bulky chlorine atom (as in PVC) the crystallinity of the polymer is disrupted and the material becomes more flexible.

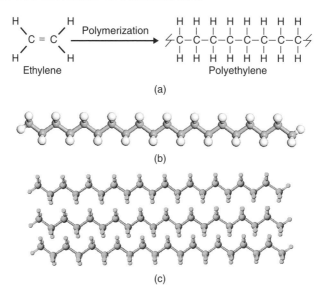

Figure 1.2 Schematic of (a) the chemical structure of ethylene gas and polyethylene, (b) the molecular structure of a single polyethylene chain or 'backbone' and (c) multiple HDPE chains showing the close packing behaviour of polyethylene chains which gives high-density polyethylene a semi-crystalline morphology and its high density.

The degree of incorporation of chlorine in the polymer structure also has a large bearing on the final properties. Chlorinated polyethylene (CPE) for instance contains between 36–42 wt% chlorine while PVC has 57 wt% chlorine. The low crystallinity of CPE allows high plasticizer and filler loadings and gives it rubbery elastomer properties. In the chlorination process the larger diameter chlorine atoms randomly replace the far smaller hydrogen atoms. The random substitution and the size discrepancy effectively disrupts the crystalline structure. Furthermore the incorporation of a polar plasticizer in both CPE and PVC destroys the dipole attraction between the chains and these polymers become very flexible and rubbery. The use of a ketone ethylene ester (KEE) polymeric plasticizer for PVC (in the case of EIA geomembranes) creates a material that is permanently plasticized since the plasticizer cannot be extracted or lost.

The incorporation of chemically active cure sites in the polymer structure such as diene in EPDM and the sulfonyl chloride group in the case of CSPE allows these materials to be crosslinked or cured to give thermoset elastomers. Where the partially (surface) fluorinated HDPE gives it increased chemical resistance, the polymerized polyvinyliene fluoride is a thermoplastic liner material that has outstanding chemical resistance to all those chemicals that can swell or oxidize HDPE (such as aromatic solvents and oxidizing acids).

1.4 MOLECULAR WEIGHT

Molecular weight (Mw) is basically the length of the polymer chains. Polymer chains are very long (made up of thousands of carbon atoms linked in series) and hence are

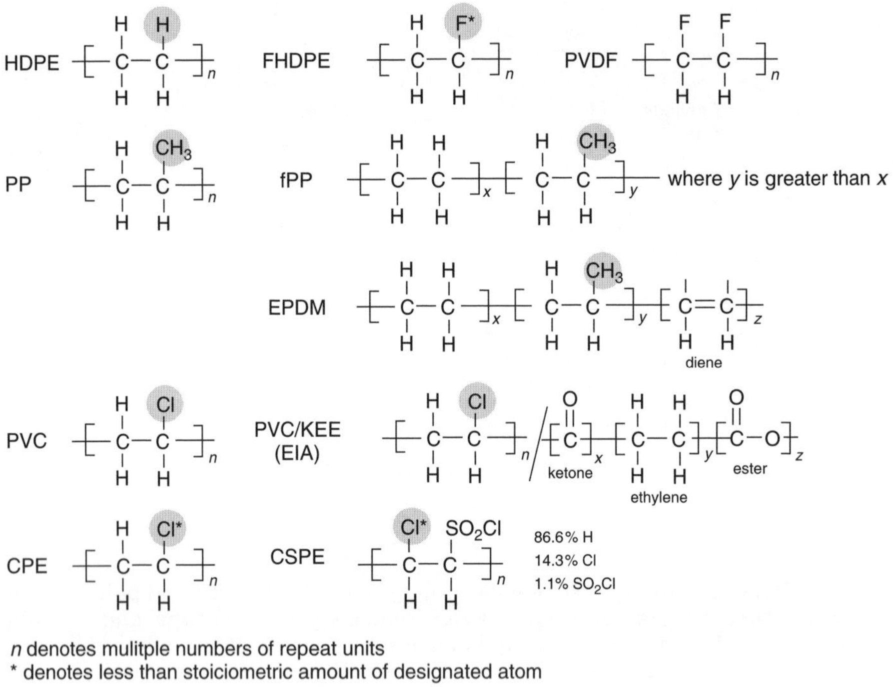

Figure 1.3 Chemical structures and repeat units of various geomembrane polymers. Note the most basic repeat unit is that of HDPE. Substitution of a hydrogen atom in the HDPE structure confers properties such as greater flexibility, greater polarity, greater solvent resistance and the ability to undergo crosslinking

also referred to as macromolecules. In general terms, as the polymer molecular weight increases, the geomembrane strength increases.

The molecular weight of the polymer can affect physical properties such as the tensile strength and modulus, impact strength, puncture resistance, flexibility and heat resistance as well as its long-term durability properties.

It is difficult to measure the molecular weight directly so generally a simpler way of expressing molecular weight is by the melt index (MI) (also referred to as melt flow index (MFI) or melt flow rate (MFR)). The melt index is inversely proportional to the polymer's molecular weight. For example, a low melt index value indicates higher molecular weight and stiffer melt flow behavior (i.e. higher melt viscosity) while a high melt index value indicates a lower molecular weight and easier melt flow (i.e. low melt viscosity) (Scheirs, 2000). Note: MFI is not applicable to PVC polymers.

Table 1.3 shows the effect of molecular weight and melt index on polymer properties.

HDPE geomembrane resins are generally high MW resins and therefore have low melt flow index values (see Figure 1.4). For this reason they are referred to as 'fractional melt' and 'HLMI' (high load melt index) resins. The term 'HLMI' HDPE refers to those polyethylene resins that should really be called High Molecular Weight resins with an HLMI of less than 15 g/10 min using ASTM D1238, Condition F (21.6 kg load).

Table 1.3 Effect of molecular weight and melt index on polymer properties

Property	As Molecular weight increases	As melt index increases
Molecular weight (chain length)	Increases	Decreases
Tensile strength (at yield)	Increases	Decreases
Tensile elongation	Increases	Decreases
Stiffness	Increases	Decreases
Impact strength	Increases	Decreases
Stress crack resistance	Increases	Decreases
Permeability	Decreases	Increases
Chemical resistance	Increases	Decreases
Abrasion resistance	Increases	Decreases
Processability	Decreases	Increases

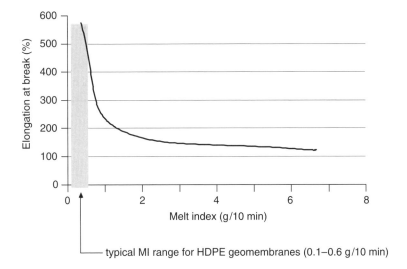

Figure 1.4 Relationship between melt flow index and % elongation at break for MDPE. Note the typical MI range for HDPE geomembranes (when tested with a 2.16 kg weight at 190 °C.)

The relationship between polymer molecular weight and melt index is summarized in Table 1.4

In addition to the length of the polymer chains (i.e. the molecular weight) the mechanical and physical properties of the plastics are also influenced by the bonds within and between chains, chain branching and the degree of crystallinity.

1.5 MOLECULAR WEIGHT DISTRIBUTION

The molecular weight distribution (MWD) is a fundamental polymer property which determines the processability and the end use properties of the polymer. Since an increase in the molecular weight of a polymer improves the physical properties, there is a strong

Table 1.4   Molecular weight and melt index relationship

Classification	Number of carbon atoms	Molecular (Mw) weight	MI (2.16 kg) standard melt index	MI (21.6 kg) high load melt index
Medium MW	7500–12 000	100 000–180 000	0.6–10	25–50
High MW	18 000–56 000	250 000–750 000	0.06–0.15	7–25

demand for polymers having high molecular weights. However, it is the high molecular weight molecules that render the polymer more difficult to process. A broadening in the molecular weight distribution tends to improve the flow of the polymer when it is being processed at high rates of shear as the low molecular weight tail acts as a "processing aid" for the higher MW chains.

Thus due to the high viscosity of higher molecular weight resins such as low HLMI HDPE used for geomembranes, the molecular weight distribution becomes a very important consideration in the processability of these HDPE resins. Resin manufacturers can tailor the molecular weight distribution (MWD) by catalyst and process selection. Geomembrane resins benefit from a broad to very broad distribution. While narrow distribution resins are tougher than broad distribution resins (at equivalent molecular weights), processability becomes easier as the MWD broadens. Table 1.5 shows the effect of molecular weight distribution on polymer geomembrane properties.

1.6   CRYSTALLINITY

In addition to the chemical structure, the properties of polymers are very dependent on the polymer morphology – particularly crystallinity.

The term crystallinity refers to the presence of crystalline regions where the polymer chains pack efficiently into dense regions that are impervious to both oxygen and chemicals (see Figures 1.5). Hence highly crystalline polyethylene has excellent chemical resistance and oxidative stability.

The ordered and aligned portions of the polymer chain form small regions that are called crystallites. The non-ordered regions are called amorphous. These amorphous regions that

Table 1.5   Effect of molecular weight distribution (MWD) on polymer geomembrane properties

Property	As molecular weight distribution broadens
Stiffness	Decreases
Impact strength	Decreases
Stress crack resistance	Increases
Melt strength	Increases
Processability	Increases

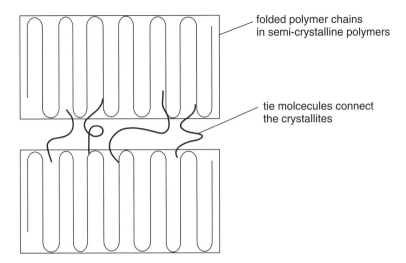

folded polymer chains
in semi-crystalline polymers

tie molcecules connect
the crystallites

Figure 1.5 Schematic of crystallites in semi-crystalline polymers. The polymer chains fold tightly in densely packed crystallites which are impervious to oxygen and chemicals. The crystallites (or lamellae) are interconnected by tie molecules which span the amorphous regions

are not crystalline contain more random orientation of the polymer chains. The proportion of crystalline (ordered and tightly packed) regions to the amorphous (disordered) regions is expressed as the degree of crystallinity of the polymer.

Polymer chains can fold and pack together to form ordered (crystalline) regions. These regions where parts of the polymer molecules are arranged in regular order are called crystallites. In between these ordered regions molecules are arranged in a random disorganized state and these are called amorphous regions.

Amorphous regions of a polymer are made up of randomly coiled and entangled chains. Amorphous polymers have lower softening points and are penetrated more by solvents than are their crystalline counterparts. PVC represents a typical amorphous polymer.

The role of crystallinity is very important in explaining the behavior of polymeric geomembranes. The amount of crystallinity of geomembrane materials varies from nil in the case of PVC, to as high as 55–65% for HDPE (see Table 1.6).

The degree of crystallinity has a pronounced effect on the performance properties of the geomembrane, especially the mechanical properties and chemical resistance. The tightly packed molecules within the crystallites of HDPE, for example, creates dense regions with high intermolecular cohesion and these areas are resistant to penetration by chemical, gases and vapours. In constrast the complete lack of crystallinity of PVC geomembranes makes them susceptible to permeation and solvation by small solvent molecules.

Table 1.6 Typical % crystallinity values for various
geomembrane polymers

Polymer	Crystallinity (%) (average values)
HDPE	55
MDPE	40
LLDPE	15
VLDPE	10
FPP	5
CPE	1–2
PVC	0

🖈 *The highly crystalline nature of HDPE is responsible for its higher density and stiffness, as well as its low permeability and high chemical resistance.*

HDPE is semi-crystalline but introducing an alkene comonomer (e.g. butene or hexene) into the polymer backbone gives side chains that reduces the crystallinity. This in turn has a dramatic effect on polymer performance, which improves significantly as the side-chain branch length increases up to hexene, and becomes less significant with octene and longer chains. It is by manipulating this side branching that various grades of polyethylene varying in crystallinity are produced.

The greater number of crystalline regions is what differentiates HDPE from its lower density cousins such as LLDPE, MDPE, LDPE and VLPE. This semi-crystalline microstructure of HDPE imparts excellent chemical resistance and high strength; however it also makes HDPE susceptible to environmental stress cracking (ESC). fPP, CPE and PVC owing to their low crystallinities are more flexible and not susceptible to ESC (see Figure 1.6).

Polymer chains with side branches (e.g. LLDPE) or irregular pendant groups (e.g. PVC, CSPE) cannot pack together regularly enough to form crystallites. This is the reason why LLDPE and VLDPE, which have a controlled number of side branches, have much lower crystallinities than HDPE.

HDPE crystallizes from the melt under typical conditions as densely packed morphological structures known as spherulites. These are small spherical objects (usually from 1 to 10 μm) in diameter composed of even smaller structural subunits: rod-like fibrils that spread in all directions from the spherulite centres, occupying the spherulite volume. These fibrils, in turn, are made up of the smallest morphological structures distinguishable, small planar crystallites called lamellae. These crystallites contain folded polymer chains that are perpendicular to the lamella plane and tightly bend every 5 to 15 nm (see Figure 1.7).

Lamellae are interconnected by a few polymer chains, which pass from one lamella, through a small amorphous region, to another. These connecting chains, or tie molecules, are ultimately responsible for mechanical integrity and strength of all semi-crystalline polymer materials. Crystalline lamellae offer the spherulites rigidity and account for their high softening temperatures, whereas the amorphous regions between lamellae provide flexibility and high impact strength to HDPE products.

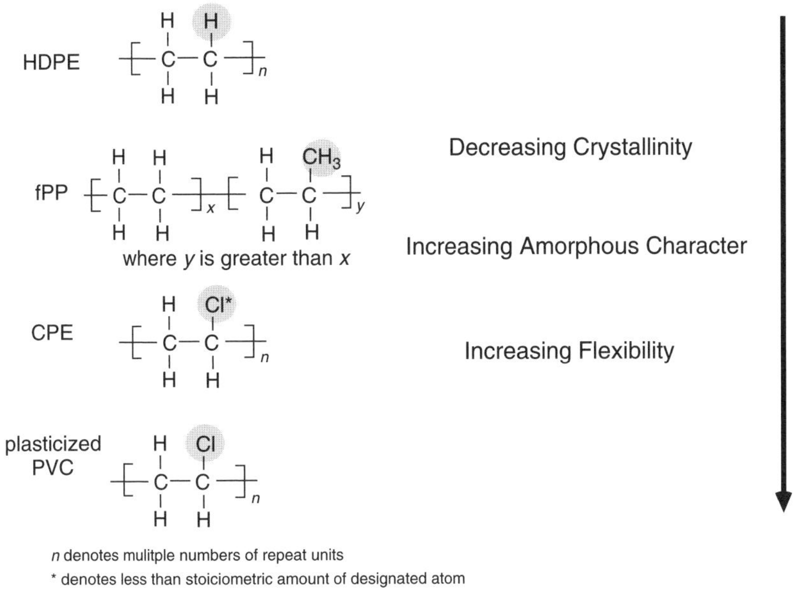

Figure 1.6 Effect that the substitution of a hydrogen atom in HDPE by substituents of increasing size (e.g. methyl group, chlorine atom) has on the crystallinity and the flexibility of the polymer

Highly crystalline polymers are rigid, high melting and less affected by solvent penetration. Hence HDPE geomembranes which have some 55–60% crystallinity exhibit excellent solvent resistance. Crystallinity makes polymers strong, but also lowers their impact resistance. For instance, samples of HDPE prepared with crystallinities of 95% are extremely brittle.

An increase in the degree of crystallinity leads to a direct increase in rigidity and tensile strength at yield point, hardness and softening point and to a reduction in diffusion and permeability. However increasing crystallinity also means a reduction in the number of 'tie' molecules in the amorphous regions which are susceptible to chemical attack (e.g. oxidation) and tie chain pullout from the crystallites (i.e. stress cracking) (see Figure 1.8).

Increasing crystallinity results in the following property attributes: increased tensile strength, increased stiffness or hardness, increased chemical resistance, decreased diffusive permeability (or vapour transmission), decreased elongation or strain at failure and decreased stress crack resistance.

In semi-crystalline polymers, the antioxidants reside in the amorphous regions which fortuitously are the same regions where oxygen can diffuse into cause oxidation. In contrast, the crystallites are too dense for either oxygen or antioxidant and diffuse into. The more amorphous polyolefins are more prone to oxidative degradation since oxygen can

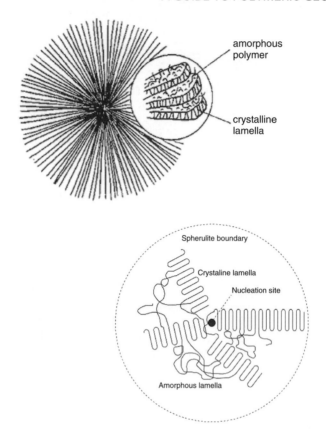

Figure 1.7 Schematic of spherulites in semi-crystalline polymers

diffuse more freely throughout their entire structure and there is a greater volume of polymer that must be protected by the antioxidant. In addition, antioxidants can also diffuse more readily through and migrate more easily from amorphous polymers compared to their more crystalline forms (Scheirs, 2000).

Crystalline thermoplastics (also called semi-crystalline) include HDPE, LLDPE and polypropylene. In these materials the polymeric chains are folded in a crystal lattice. The folded chains form lamellae (plate-like crystals).

Whilst the crystallites (i.e. tightly packed crystalline regions) are impervious to both oxygen and chemical ingress, the 'tie' molecules which interconnect the crystallites are susceptible to oxidation and chemical attack. The area in which the tie molecules reside is termed the amorphous region (i.e. disordered region) and these areas have lower density than the crystallites and so oxygen and opportunistic chemical can diffuse into these areas.

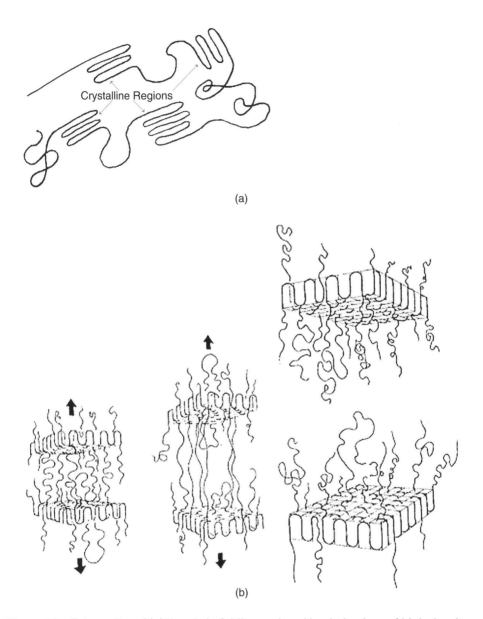

Figure 1.8 Schematics of (a) the chain folding and packing behaviour of high-density polyethylene chains to form crystalline regions and (b) the crystalline regions or 'crystallites' in HDPE (which are interconnected by 'tie molecules') which can be pulled apart under the combined action of stress and a chemical agent (referred to as environmental stress cracking). Reprinted with permission from *Polymer*, Importance of tie molecules in preventing polyethylene fracture under long-term loading conditions by A. Lustiger and R. L. Markham, **24**(12), 1647. Copyright (1983) Elsevier

1.7 PROPERTIES OF POLYETHYLENES

Polyethylene is by far the most widely used polymer to manufacture geomembranes. Polyethylene resins are manufactured in very-low-density, low-density, linear low-density, medium-density and high-density varieties. The density range for all polyethylene geomembrane polymers falls within the general limits of 0.85 to 0.960 g/cm^3.

Molecular weight, molecular weight distribution and crystallinity (i.e. density) are the three most important characteristics of polyethylene resins and play a major role in determining the durability and end-use performance properties of HDPE and LLDPE geomembranes.

Polyethylene is classified into several categories based on its density and branching. HDPE has little branching, giving it stronger intermolecular forces and higher tensile strength than lower density polyethylene, thereby making it ideal for geomembrane applications. HDPE is defined as having a density of equal to or greater than 0.941 g/cc.

The density of polyethylene is primarily controlled by the frequency and length of the side branches (which in turn are determined by the type and level of comonomer). The side branches prevent the PE chains from packing closely together, so the longer the side branches, the more open the structure and hence the lower the density. Homopolymer HDPE has a density greater than 0.960 g/cc while copolymers have densities less than 0.960 g/cc.

Note that true homopolymer HDPE is not used for geomembranes due to its tendency to undergo environmental stress cracking.

Typical comonomers are butene, hexene and octene which are carbon chains with 4, 6 and 8 carbons respectively. These comonomers are denoted as C4, C6 and C8 for simplicity. They all have a reactive double bond at the end of the chain and are referred to as alpha olefins. The 'olefin' indicates they contain a C=C bond in their structure while alpha indicates the double bond is between the first and second carbon atoms.

The type of comonomer used determines the end-use performance characteristics of the resin. Hexene and octene copolymers are tougher and more flexible; however butene copolymers are typically less expensive.

It is important to emphasize that HDPE geomembranes are actually manufactured using a polyethylene resin with a density 0.932–0.940 g/cm^3 which falls into the MDPE category as defined in ASTM D-883. It is the addition of carbon black that pushes the final density of the geomembrane up into the density range between 0.941 and 0.950 g/cm^3 which corresponds to a HDPE as defined in ASTM D-883. For this reason the 'HDPE' nomenclature is used to describe most black polyethylene geomembranes.

Note that 'HDPE' geomembrane resins are in fact MDPE base polymer with the addition of 2% carbon black, which raises its density into the classification range of HDPE.

The effect of increasing density on various PE geomembrane properties is shown in Table 1.7.

Table 1.8 lists the density classifications for polyethylene resins.

Table 1.7 Effect of density on PE geomembrane properties

Property	As density increases
Crystallinity	Increases
Tensile strength (at yield)	Increases
Stiffness	Increases
Impact strength	Decreases
Stress crack resistance	Decreases
Permeability	Decreases
Chemical resistance	Increases
Abrasion resistance	Increases
Processability	Decreases

Table 1.8 Density classifications for polyethylene resins

Polyethylene type	Defined density range (g/cc)
HDPE	0.941–0.965
MDPE	0.926–0.940
LLDPE	0.915–0.925
LDPE	0.910–0.915
VLDPE	0.880–0.910

HDPE is the most common field-fabricated geomembrane material primarily due to its low material cost, broad chemical resistance and excellent mechanical properties.

MDPE is a substantially linear polymer, with high levels of short-chain branches, commonly made by copolymerization of ethylene with short-chain alpha-olefins (e.g. 1-butene, 1-hexene and 1-octene).

LLDPE is a substantially linear polymer, with significant numbers of short branches, commonly made by copolymerization of ethylene with short-chain alpha-olefins (e.g. 1-butene, 1-hexene and 1-octene). As its name implies, Linear Low Density Polyethylene is a lower density polymer (<0.939 g/cm^3), with increased material flexibility. LLDPE is mainly used for liners where large settlements are anticipated for long term consolidation, such as for landfill covers. Capping contaminants with LLDPE geomembranes, not only makes it possible to control the release of carbon dioxide and methane (by-products of the decomposition of organic matter), but allows their capture and reuse. The flexibility of the LLDPE is also useful for geomembrane liners that are installed on subgrades prone to differential settlement.

LLDPE has a higher tensile strength and higher impact and puncture resistance than LDPE. It is very flexible and elongates under stress. It can be used to make thinner sheets, with better environmental stress cracking resistance. It has good resistance to chemicals and to ultraviolet radiation (if properly stabilized). However it is not as easy to process as LDPE, has lower gloss and a narrower operating range for heat sealing. Hence it finds application in plastic sheets (where it permits use of lower thickness profile than comparable LDPE), coverings of cables, geomembranes and flexible tubing.

LLDPE geomembranes are available in a smooth, textured or single textured finish. The comonomers used to produce the resin can include hexene or octene.

Low-density polyethylene (LDPE) has very poor environmental stress crack resistance and rather poor mechanical properties and so it does not find application as a geomembrane. LDPE has a high degree of short- and long-chain branching, which means that the chains do not pack to form a dense crystal structure as well. It has therefore less strong intermolecular forces, as the instantaneous-dipole induced-dipole attraction is less. This results in a lower tensile strength and increased ductility.

VLDPE is most commonly produced using metallocene catalysts and is a highly flexible and ductile material.

1.8   STRESS–STRAIN BEHAVIOUR OF POLYMERS

The stress–strain behaviour of polymers used to manufacture geomembranes is largely determined by the properties discussed above, namely the molecular weight, molecular weight distribution and crystallinity or density.

Figure 1.9 shows a typical stress–strain curve for HDPE which identifies the following:

- the linear elastic region (where it obeys Hooke's Law);
- the plastic region where the polymer draws and extends;
- the yield stress;
- the ultimate strength (or tensile strength at break);
- the modulus of elasticity (i.e. the gradient of the initial linear slope).

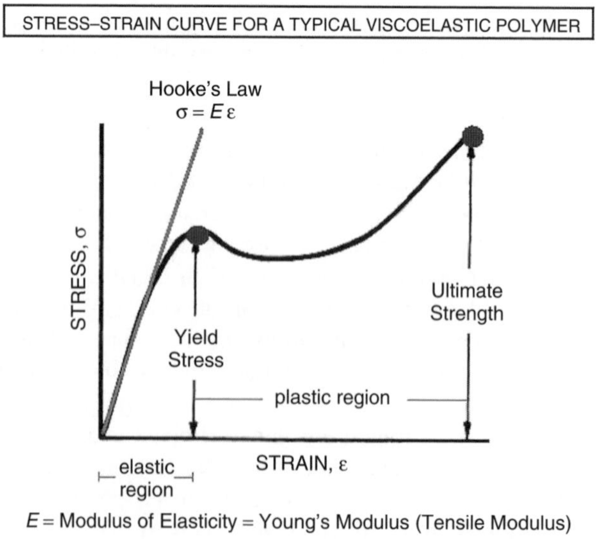

STRESS–STRAIN CURVE FOR A TYPICAL VISCOELASTIC POLYMER

Hooke's Law
$\sigma = E\varepsilon$

STRESS, σ

Yield
Stress

Ultimate
Strength

plastic region

elastic
region

STRAIN, ε

E = Modulus of Elasticity = Young's Modulus (Tensile Modulus)

Figure 1.9   Stress–strain curve showing elastic and plastic regions

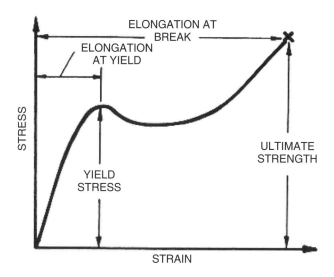

Figure 1.10 Stress–strain curve for HDPE showing main tensile test value parameters

Figure 1.10 is a similar stress–strain curve which also identifies the elongation at yield (also know as the yield strain) and the elongation at break (also known as the breaking strain).

Figure 1.11 shows the stress–strain curves for various polymer types. If the load rises linearly to fracture with no plastic deformation then the material is said to be brittle as is the case for PVC liners where the plasticizers have been extracted or, for HDPE geomembranes after extensive oxidation. More commonly though the behaviour of geomembranes is ductile but may exhibit brittle behaviour depending on the temperature. The brittle transition occurs at sub-zero temperatures for common geomembranes (see section on low temperature properties) (Scheirs, 2000). For geomembranes applications either hard and tough (e.g. HDPE) or soft and tough (e.g. CSPE, fPP, EIA) polymers are the most suitable.

1.8.1 YIELD BEHAVIOUR

'Yield' is defined as the onset of plastic deformation in a polymer under an applied load. This is an important parameter because it represents the practical limit of use more than does ultimate break or rupture. The yield properties depend on the polymer crystallinity and the polymer morphology. The yield behaviour also depends on the test conditions used. The yield properties vary with both the test temperature and the speed of the test. For this reason it is very important that tensile testing of polymer geomembranes be conducted at 23 °C where possible. This may therefore cast doubt on field tensiometer measurements where higher or lower temperatures might be encountered. Since the speed of the tensile test is also critical, the tensile test speed (also known as the *crosshead speed* and determined by the strain rate) must be standardized and defined (Scheirs, 2000).

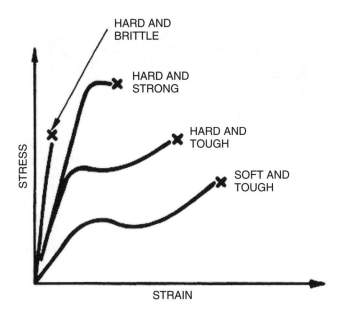

Figure 1.11 Stress–strain curves for various polymer types

1.8.2 PLASTIC DEFORMATION

'Plastic deformation' is the deformation that remains after a load is removed from a polymer sample. It is also called permanent deformation or non-recoverable deformation. Under small enough loads less than the yield stress the deformation is elastic and is recovered after the load is removed (i.e. the specimen returns to its original length). Yielding thus represents the transition from elastic to plastic behaviour. Consider a HDPE geomembrane sample under an applied tensile load. The length of the specimen will increase (as measured by the elongation). As the elongation increases, the load at first increases linearly but then increases more slowly and eventually passes through a maximum where the elongation increases without any increase in load (as in Figure 1.10). This peak in the stress–strain curve (i.e. the load–elongation curve) is the point at which plastic flow (permanent deformation) becomes dominant and is defined as the yield point. Not all polymers exhibit a defined yield point such as that exhibited by HDPE. PVC, for example, shows no obvious yield point in the stress–strain curve.

1.8.3 STRESS

The shape and magnitude of the load–elongation curve depends on the particular polymeric geomembrane being tested. Rather than load, the properly normalized variable is stress which is defined as the load per unit cross-sectional area of the test specimen. Stress therefore has units of pressure (1 MPa = 1 MN/m^2 = 145 psi).

1.8.4 STRAIN

Rather than quoting elongation, the proper normalized variable is strain which is the extension divided by the initial length. Strain is therefore dimensionless whereas elongation is expressed as a percentage.

1.8.5 TYPES OF LOADING

The most common type of loading used for testing polymeric geomembranes is *uniaxial tension* but other types of loading are arguably more important such as compression, hydrostatic compression and uniaxial (i.e. multiaxial tensile) loading. The simplest variation of the tensile test is the uniaxial compression test which should not be confused with hydrostatic compression in which the load is applied from all sides. It has been found that compressive stresses are higher than tensile stresses for a given strain value.

1.8.6 TEMPERATURE EFFECTS

The shape and magnitude of the stress–strain curve is very dependent on temperature. As the temperature increases, the yield stress, elastic modulus (i.e. stiffness) and yield energy all decrease while the yield strain (elongation at yield) increases (see Figure 1.12).

1.8.7 STRAIN RATE EFFECTS

Strain rate determines to the speed of the application of force on the material being tested. High strain rates (i.e. high testing speed) have the effect of making the polymer behave in a more brittle fashion – in the same way that reducing the temperature makes the polymer stiffer and more brittle (Figure 1.13).

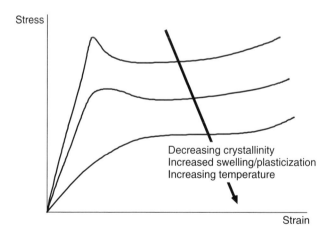

Figure 1.12 Effect of swelling and increased temperature on the stress strain properties of HDPE geomembranes. The material becomes softer and more rubbery but loses its tensile strength

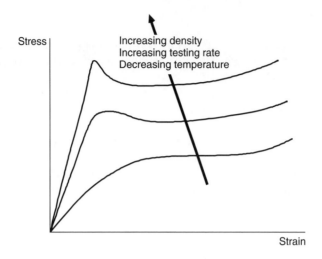

Figure 1.13   Effect of increasing density or increasing testing rate (i.e. strain rate) or decreasing temperature on the stress–strain properties of HDPE geomembranes. The material becomes stiffer and stronger as the density or testing speed increase or as temperature is decreased

Table 1.9   Melting points of various polymer resins

Polymer	Melting point (°C)
Poly(ethylene vinyl acetate) (EVA)	85
Metallocene polyethylene (mPE)	90–100
Low-density polyethylene (LDPE)	108
Linear low-density polyethylene (LLDPE)	125
High-density polyethylene (HDPE)	130
Flexible polypropylene (fPP)	150

1.9   MELTING POINTS

Polymer geomembrane resins have very different melting points as shown in Table 1.9. The polymer melting point (or more correctly the melting range) is of importance during thermal welding; particularly when welding different geomembrane materials to each other.

REFERENCES

Scheirs, J., *Compositional and Failure Analysis of Polymers: A Practical Approach*, John Wiley & Sons, Ltd, Chichester, UK, 2000, 766 pp.

2

Geomembrane Manufacturing Methods

Polymeric geomembranes can be manufactured by a number of different techniques and the nature of the specific manufacturing methods can impart various characteristics to the final product. The two main manufacturing methods are the extrusion and the calendering operations (see Table 2.1). Extrusion can further be divided into three sub-methods namely, blown extrusion, cast extrusion and extrusion coating.

Polyolefin geomembranes (i.e. HDPE, LLDPE and fPP) are all manufactured by an extrusion method where the polymer resin in pelletized form is mixed with a pelletized concentrate called a masterbatch. The masterbatch comprises the additive formulants such as carbon black (or titanium dioxide in the case of white membranes), antioxidants and stabilizers. The mixture is then fed to an extruder where the materials are heated, intimately mixed and sheared using a special tapered flighted screw. The melt is both distributively mixed and dispersively mixed to ensure homogeneous distribution and dispersion of the additives (see Figure 2.1). The melt is then forced through a die – either a flat die (in the cast sheet extrusion method) or an annular die (in the blown film method).

The cast extrusion is alternatively called 'Flat die', 'Flat bed' or 'Slot die' extrusion. Flat-die geomembrane manufacturing delivers greater thickness and gauge control than round die blown-film. Blown film on the other hand provides certain advantages of polymer orientation not present in flat-die produced material. For instance, the vertical bubble provides biaxial orientation of the film to give it improved tear resistance that would not be possible on cast-film liners.

On account of the weight of the vertical bubble of film, the blown film process tends to introduce a degree of balanced orientation to the liner. Polymer orientation can improve the mechanical properties of the liner. Therefore for the same starting resin, a blown film geomembrane has the potential to produce a higher performance geomembrane than a flat-die manufactured liner.

A Guide to Polymeric Geomembranes: A Practical Approach   J. Scheirs
© 2009 John Wiley & Sons, Ltd

Table 2.1   Some of the geomembrane types manufactured by extrusion and calendering methods respectively

Extrusion (by blown film or flat die)	Calendering (counter-rotating rollers)
HDPE	FPP and fPP-R
LLDPE	PVC
fPP	CSPE-R
VLDPE	EPDM and EPDM-R

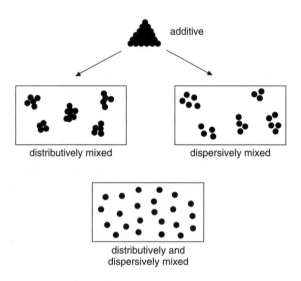

distributively mixed   dispersively mixed

distributively and
dispersively mixed

Figure 2.1   Schematic showing the importance of both distributive and dispersive mixing on the dispersion of an additive (e.g. carbon black) in the polymer geomembrane. Reproduced by permission of NAUE

2.1   BLOWN FILM (ROUND DIE)

In the blown film method the molten plastic is extruded through a vertical orientated annular die to produce an inflated bubble (see Figure 2.2) that is hauled off vertically and slit to produce a flat sheet.

Blown film lines can produce geomembranes 7.0 m wide in thicknesses of 0.5 to 2.5 mm at rates of more than 1200 kg per hour. Blown film lines use complex three-layer dies to optimize melt flow, eliminate dead spots and prevent overheating of the resin. Typically three extruders feed the die. The B extruder is used for the core of the sheet while the A and C extruders supply the polymer to the inner and outer skins of the geomembrane. The multilayer can be textured for greater friction and traction by pumping nitrogen gas into the A and C extruders. When the nitrogen gas exits the die in the outer and inner layers, it disrupts the flow of resin and creates a controlled texturing of the skins. The largest roll

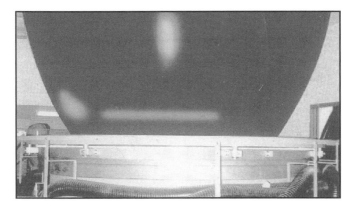

Figure 2.2 Geomembrane bubble tube produced by the blown film method. Note that all the nip rollers when the bubble is collapsed 'crease' lines can form on the inside of the geomembrane which is a typical characteristic of blown sheets. Reproduced by permission of NAUE

that can be wound is 900 mm in diameter, although the geomembrane is generally wound into rolls 400 to 500 mm in diameter. According to Battenfeld Gloucester the blown film method is the process of choice for geomembranes, with about 80% of products extruded in this way (Battenfeld Gloucester, 2006).

The blown film manufacturing method is less costly than the cast method. Geomembranes made by the blown film method can contain fold lines from collapsing of the bubble and these can give the final geomembrane sheet different strength characteristics (Figure 2.3). Figures 2.4 to 2.6 show these fold lines.

Geomembrane made by the blown film method can also have tears and pin holes due to various defects known as unmelts, gels and die build-up. These are areas of poorly fused and/or crosslinked polymer that have smeared along the polymer surface when the bubble exits the annular die, creating small holes or tears (see Figure 2.7).

2.2 FLAT SHEET EXTRUSION (FLAT DIE)

In the flat sheet extrusion method the molten plastic is forced through a flat die and then between polished chill rollers to produce the flat geomembrane sheet (see Figure 2.8). The role of the flat die is to uniformly distribute the molten plastic along the width of the die and to develop uniform flow patterns. Flat die extrusion of geomembranes has greater thickness control than the blown film method; however advances in gauge control for the blown film method is closing this gap.

Table 2.2 summarizes some of the advantages of the flat sheet extrusion method for manufacturing HDPE geomembranes.

Table 2.3 summarizes the differences in the properties of blown and flat sheet extruded HDPE geomembrane.

Figure 2.3   Photograph (taken looking upwards) of an HDPE geomembrane made by
the blown film method. Note that the bubble narrows at the top of the tower due to the
converging collapsing frame assembly. Reproduced by permission of Battenfeld
Gloucester

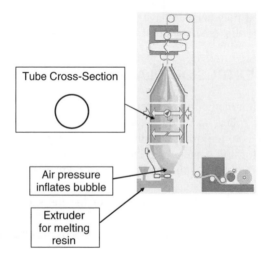

Figure 2.4   Schematic of the blown film geomembrane manufacturing method.
Reproduced by permission of NAUE

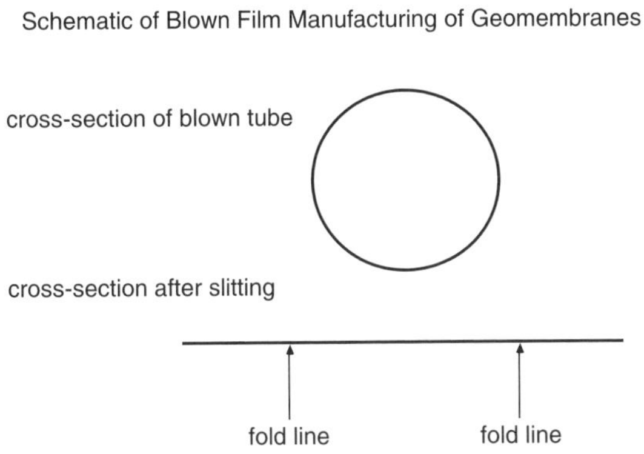

Schematic of Blown Film Manufacturing of Geomembranes

cross-section of blown tube

cross-section after slitting

fold line   fold line

Figure 2.5 Schematic showing that a geomembrane produced by the blown film method has foldlines in its cross-section after slitting

Figure 2.6 Photograph of fold lines in a deployed HDPE geomembrane. The fold lines are a legacy of the blown film manufacturing process where the blown tube or bubble is collapsed at the top of the blown film tower. Reproduced by permission of NAUE

2.3 COEXTRUSION

Coextrusion uses two or more extruders to feed one die to give a product composed of various layers (see Figure 2.9). In coextrusion the layers of molten polymer simultaneously exit the die so that they form molecular entanglements with each other. This interpenetration gives a monolithic structure if the same polymer formulation is used for all three layers. Coextrusion is fundamentally different from lamination since no plane of weakness exists with coextrusion.

Coextrusion provides for novel combinations and structures by simply changing one or more of the polymer types in the A:B:A or A:B:C construction. For example, in order to

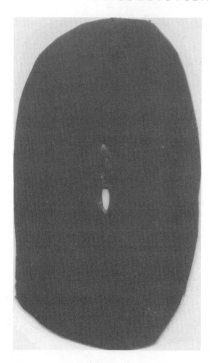

Figure 2.7 Photograph of a typical manufacturing defect in a HDPE geomembrane sheet made by the blown film process. The hole defect is caused by a cold slug of extrudate in the melt forming an annular slit as it passes through the die lips. Reproduced by permission of NAUE

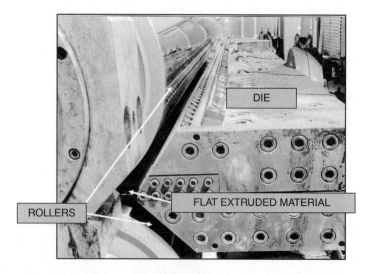

Figure 2.8 Photograph of a flat sheet extrusion die. The extruded sheet exits the gap in the die block on the left and then passes between the polished nip rollers. Reproduced by permission of NAUE

Table 2.2 Summary of advantages of the flat sheet extrusion method for manufacturing HDPE geomembranes. Reproduced by permission of NAUE

Attribute	Stated advantages
Higher MFI resins can be used for flat sheet extrusion	This translates to higher sheet flexibility and less energy to weld (due to the higher MFI)
Greater welding speed	Typically 2.4 m/min for 1 mm HDPE
No fold lines	Homogeneous sheets
Good thickness control	Thickness variation <5%
No blown film defects	Defects such as 'cold slugs' and 'gels' which can lead to holes in the blown geomembrane do not occur in cast liners
Good dimensional stability	Shrinkage <1% at 100 °C for 1 h
Shiny/glossy surface finish	Easy damage recognition

Table 2.3 Comparison of blown and flat sheet extruded HDPE geomembranes. Reproduced by permission of NAUE

Property	Blown film	Flat sheet (cast)
Stiffness	Low melt index 0.6 g/10 min (190 °C/5 kg) and therefore higher stiffness	Melt index >2.0 g/10 min (190 °C/5 kg) and therefore lower stiffness
Fold lines	Fold lines present can give variations in modulus	No fold lines
Thickness variation	7–15%	<5% (therefore better weld quality)
Dimensional stability	<2% (100 °C for 1 h) and therefore more folds in sunlight	<1% (100 °C for 1 h) and therefore less folds in sunlight
Welding speed	1.9 m/min for 1 mm sheet	2.4 m/min for 1 mm sheet

combine the flexibility and out-of-plane elongational properties of VLDPE with the chemical resistance of HDPE, a coextruded HDPE/VLDPE/HDPE three-layer (20%:60%:20%) geomembrane can be manufactured (see Figure 2.10).

For textured geomembranes a blowing agent can be added to one or both of the outer layers to give either singly or doubly textured sheet. The purpose of such texturing being to give improved friction angles.

Other coextrusion variations that are employed are:

- The top layer may be white-surfaced to reduce heat build-up and therefore extend the geomembrane lifetime and reduce desiccation (i.e. drying out) of the underlying clay.
- The middle layer could have high levels of electrically conductive carbon to facilitate spark testing.
- The top layers can be heavily UV stabilized for long-term UV exposure applications.

A point that is little discussed is that, three-layer coextrusion capability also provides the geomembrane manufacturer the option of introducing regrind or recycle into the geomembrane. A lower quality/off spec/recycled material can be used in the middle layer

Figure 2.9 Photograph of a top view of a flat sheet (cast) extrusion line for manufacturing HDPE geomembranes. Note that the two extruder barrels on the left feed the slot die right of middle. The two extruders are required to produce a multilayer sheet. Reproduced by permission of NAUE

accounting for 75% of the structure; these materials may be purchased for two-thirds of the cost of virgin geomembrane resins.

2.4 CALENDERING

PVC, CSPE and scrim-reinforced geomembranes including CSPE-R and fPP-R are not produced by conventional extrusion methods and are manufactured instead by calendering. In this process the polymer resin, carbon black, fillers and various additive are mixed either in a heated batch mixer (such as a Banbury) or a heated continuous mixer (such as a Farrel FCM). The mixture is then masticated using mixing units customary in the rubber processing industry such as a roll mill where it is homogenized. The mixed mass is then passed through a set of calender counter-rotating rollers to form the final sheet. Most scrim-reinforced geomembranes (SRG) are highly flexible liners manufactured by this type of calendering method.

Typical SRG base polymers are PVC, CSPE and fPP. In the calendering process the polymer formulation is intensively mixed in a special Banbury mixer and then passed through a two-roll mill where it is flattened. The dough-like material is then passed through a set of counter-rotating rollers which comprise the calender to produce the final sheet. The sheet is made up of polymer plies laminated to the scrim support (which is an

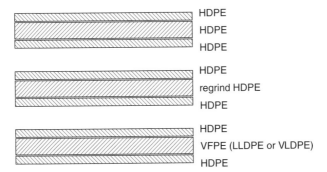

Figure 2.10 Some possible three-layer coextruded geomembrane structures possible with modern coextrusion equipment

open weave fabric generally made from polyester). The adhesion of the plies is due to 'strike-through' through the open weave fabric. The openings in the scrim (i.e. the apertures) need to be large enough to enable good adhesion between the plies. Inadequate ply adhesion can lead to delamination which is one of the potential shortcomings of scrim-reinforced geomembranes. The scrim-reinforced geomembranes can either have one central scrim and two outer plies (hence are three-ply geomembranes) or, comprise two scrim layers on each side of a central polymer layer and with two polymer plies on each outer face (hence are five-ply geomembranes).

It is important that the apertures (i.e. the openings) in the scrim are sufficiently large to enable the polymer plies to adhere to one another. This gives 'key and lock' bonding that is necessary to achieve the required ply adhesion and prevent delamination failure. These reinforced geomembranes are generally slightly thicker than the unit dimensions due to the additional thickness of the scrim. For instance, a 1.21 mm fPP-R geomembrane consists of thin plys of 0.5 mm fPP each side of a scrim. Reinforced geomembranes are often only available in limited widths (e.g. 2–3 m) owning to width limitations of the calendering equipment.

2.5   SPREAD COATING

Spread coating (or solvent coating) is a process by which the polymer is dissolved in a solvent which is then spread onto the textile and heated to evaporate the solvent and leave the coating.

Reinforced ethylene interpolymer alloy (EIA-R) type geomembranes (e.g. XR-5) are generally produced by this manufacturing process. In the spread coating process the molten polymer is spread in a relatively thin coating over a dense fabric substrate (i.e. tightly woven fabric or non-woven fabric) using a spreading knife. The coating knife runs parallel to the weft.

Penetration of the viscous polymer to the opposite side of the fabric is limited due to the dense weave of the fabric and so the material is turned over and the process repeated on the other side of the fabric. Since the polymer coating intimately encapsulates the fabric there is no tendency for delamination.

2.6 EXTRUSION COATED GEOMEMBRANES

Extrusion coated geomembranes are made by a specialized technology where fabrics or scrims are coated with polymer. It is a process by which the molten polymer is extruded as a flat sheet film and pressed into the fabric surface, adhering to and coating the surface (e.g. EIA-R geomembranes from Cooley).

The extrusion coating provides a number of purported advantages over spread coating, solvent coating and lamination such as:

- Extrusion coated fabric achieves all material properties immediately upon cooling. Non-extrusion coatings may undergo changes post-production as the adhesives and solvent-borne coatings continue to cure.
- Extrusion coating provides intimate contact between membrane and fabric resulting in a stronger bond than the abrupt glue line that separates coating and fabric in the lamination process.
- Extrusion coating does not involve volatile organic constituents (VOCs) typical in solvent coating processes. Therefore, off-gassing and unpleasant odours are not an issue with extrusion coating.
- Extrusion coated fabrics have high wear and abrasion resistance unlike laminated products in which the adhesive provides a failure plane for delamination.
- Extrusion coating is monolithic with none of the micropores typically found in solvent coated fabrics as a result of the solvent evaporating during processing (Cooley, 2006).

2.7 PIN-HOLE DETECTION

The geomembrane production line should be fitted with an in-line pin-hole detection system. This normally comprises a spark testing bar fitted to the geomembrane production line. This works by holding a charge in the bar at the location where the geomembrane passes over a metal roller. If there are any pin-holes in the geomembrane, then the charge passes through the hole and makes contact with the metal roller to complete a circuit, which sets off an alarm. The spark tester is generally capable of detecting defects or pin-holes less than 0.25 mm in diameter.

2.8 TEXTURING

Texturing is a randomized surface roughness technique intended to enhance friction to prevent geomembrane liners from sliding down slopes. Textured geomembranes have roughened surfaces in order to increase the friction angle in contact with soil or geosynthetic layers. They can also be patterned or embossed with structured profiles to give engineered liners that physically interact and engage with various mating surfaces. The surface of a textured geomembrane significantly increases the interfacial friction with adjacent materials as compared to the same geomembrane with smooth surfaces.

Textured surfaces give increased friction and shear stability for applications on steep slopes.

Sliding failures or slope failures with geosynthetic interfaces have been well documented over recent years. Usually these failures have occurred at the geomembrane/geotextile interface or the geomembrane/soil interface. The advent of textured and/or structured geomembranes however have significantly reduced the likelihood of sliding failures.

Textured geomembranes are thus much less prone to sliding on steeper slopes. Moreover, in the case where the textured side faces upward, nor will the material (for example MSW in a landfill) slip on top of the textured geomembrane surface. This however may lead to failure due to high shear forces and high tensile forces being applied to the membrane. In such cases it is important to use a geomembrane with a smooth top surface and so any material on top slides on the liner rather than inducing damaging stresses into the liner.

The friction angle between a textured geomembrane and the soil layer (or a geotextile) is measured by short-term friction tests in the laboratory. Since these tests are relatively short term they only reflect the friction properties of the new sheet. The long-term friction angle (and hence the slope stability) however is dependent on the ability of the texturing particles to sustain shear stress. Accordingly, in order to ensure the friction is maintained at a constant level, the particles at the sheet surface must not yield under the applied shear stress. Slow but steady loss of adhesion of sheet surface particles (ultimately leading to catastrophic loss of stability and slope failure) may occur due to creep, stress cracking and oxidative ageing.

Textured or structured sheet made by flat-die extrusion and embossed calendars have been touted as premium products over the texturing produced by blown film coextrusion processes (see Table 2.4). The latter technique has limitations with respect to variable quality and lower-than-expected asperity height and cross-roll friction values. This has led to some documented slope failures (Sieracke, 2005).

 The texturing processes for HDPE differs between manufacturers.

The flat-die extrusion method enables the formation of a textured or embossed surface that does not affect the core thickness. This overcomes some of the limitations associated with blown film coextrusion texturing such as non-uniformity, variable peak height, variable area coverage and reduced mechanical properties. The reduction in

Table 2.4 Techniques used to produce textured geomembranes to enhance their frictional properties

Texturing method	Process description	Comments
Blown film coextrusion with blowing agent	In-line with primary bonding	Used extensively in North America
Hot particle impingement	Secondary process with secondary bonding	Resembles small thread-like nodules
Hot foam laminated	Secondary process with secondary bonding	Not widely used
Patterned (or structured) roller	In-line with primary bonding	Used extensively in Europe

mechanical properties associated with blown film coextrusion texturing is of particular concern in long-term applications where differential settlement (e.g. due to localized settlement) can cause damaging multiaxial (i.e. out-of-plane) stresses to develop.

Geomembrane liners can be textured on either one side or on both sides. In order to prevent soil and other cover materials from slipping along the top of smooth geomembrane sheet, the upper layer can also be textured or structured.

It should be noted however that the interface shear strength of the upper interface should not be higher than that of the lower interface otherwise potentially damaging tensile stresses can be induced in the geomembrane. Such tensile stresses can be particularly detrimental for HDPE liners if residing over the long term, due to likelihood for stress rupture and brittle cracking.

📌 *It is very difficult to remove dust from the edges of textured geomembranes. For this reason it is necessary to specify that the geomembrane has a non-textured flat edge for convenient welding.*

2.8.1   COEXTRUSION TEXTURING

The blown textured surface is produced using coextrusion processing equipment. The texture can be applied to one or two sides of the liner and is an integral component of a three-layer coextruded geomembrane. Texturing combines the durability of HDPE liners with a roughened surface that provides very good frictional characteristics against a variety of soils and geosynthetic surfaces. This increase in friction helps keep cover soil in place and improves the overall liner stability on slopes.

In the blown film coextrusion texturing method nitrogen gas is injected into the polymer melt and when the extruded polymer exits the die the nitrogen bubbles rupture (see Figures 2.11 and 2.12). As the bubbles of nitrogen gas expand they give a burst bubble effect, producing a roughened textured surface (see Figure 2.13). This method of texturing is however highly variable within a single roll or across the roll width. In addition, it is difficult to standardize this method of texturing as it is highly variable from one manufacturer to another.

📌 *Since nitrogen blown texturing occurs in a relatively uncontrolled fashion it is difficult to produce a consistent asperity height across the roll width of the geomembrane. Individual asperity height readings can vary from 0.225 to 0.8 mm (Ivy, 2003).*

Blown-film textured geomembranes which have smooth edges are covered by US Patent Nos 5 763 047 and 5 804 112. The blown-film texturing provides increased friction angles for higher stability on steep slope applications while the smooth edges result in easier, more cost-effective, more consistent and better welding.

2.8.2   IMPINGEMENT TEXTURING (ALSO KNOWN AS SPRAY-ON TEXTURING)

Impingement texturing involves a secondary process where granulated LDPE (which has a lower melting point) is dropped (or sprayed) onto the hot surface of an extruded geomembrane. The lower melting point particles fuse to the surface of the HDPE geomembrane as

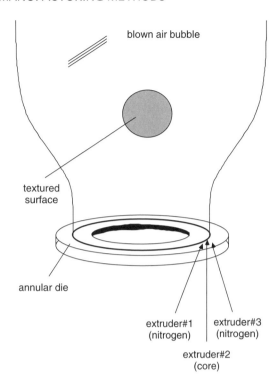

Figure 2.11 Schematic showing the method for double texturing of polyolefin geomembranes by the nitrogen gas blown film method

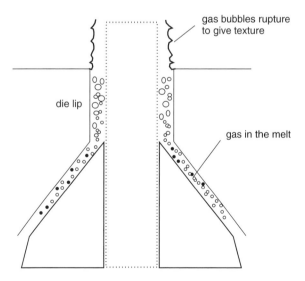

Figure 2.12 Schematic showing the mechanism for texturing of blown geomembranes by injecting nitrogen gas into the melt

Figure 2.13   Photograph of a HDPE geomembrane with *nitrogen blown* texturing where a blowing agent (based on nitrogen gas) is injected into the melt during coextrusion manufacture of the GM sheet. The limitations of this type of texturing over other forms of texturing are variable core thickness, non-uniform area coverage, inconsistent asperity heights on the surface and generally reduced tensile, elongational and stress crack resistance

small nodules (Figure 2.14). It is important to note that the minimum thickness of these impingement textured geomembranes remains the same as if the material were smooth sheets.

The polymer used for the textured particles that are deposited on the geomembrane surface should of the same type (i.e. compatible) as the parent geomembrane so that good adhesion is achieved and the texturing is not the weak link.

With impingement texturing care needs to be taken with the distribution of the thread-like particles and the homogeneity of the coverage (see Figure 2.15). The uniformity of the impingement texturing can vary across (and along) the geomembrane roll and this can lead to differences in the frictional properties of the geomembrane.

The surface texturing on geomembranes can vary to the extent that the surface roughness can vary from one location to the next. Significant surface agglomeration of texturing is present if it can be seen as obvious shade variation from a distance of 15 m from the surface being inspected and if the patches are larger than 20 mm in diameter.

The other important factor to consider with impingement texturing is the level of adhesion between the particle and the sheet. Clearly the level of adhesion between the particles and the base sheet needs to exceed the shear forces applied in a shear box test or else the texturing will simply scrape off in service.

It is possible that if the sprayed-on texturing is removed it can leave depressions/nicks on the geomembrane sheet that may be defect sites due to localized reductions in the geomembrane strength and thickness. The angular nature of texturing can cause notches, which are stress concentrators and precursors to cracks.

impingement texturing
with thread-like particles

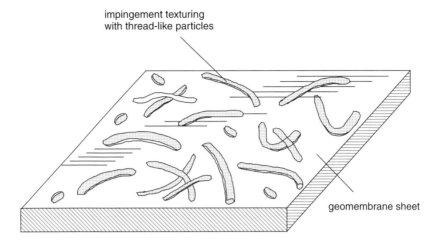

geomembrane sheet

Figure 2.14 Schematic showing the thread-like appearance of impingement textured geomembranes

Figure 2.15 Photograph of an HDPE geomembrane with *impingement* texturing where hot polyethylene particles are sprayed onto previously manufactured GM sheet. The advantages of this type of texturing over nitrogen blown texturing is consistent core thickness and a more homogeneous structure; hence the texturing has little effect on tensile and elongation properties

2.8.3 STRUCTURING

Structuring or patterning is a method of texturing whereby the smooth sheet made by the flat bed cast method passes between two counter-rotating patterned rollers immediately upon leaving the die lips (see Figure 2.16). These rollers have patterned or embossed

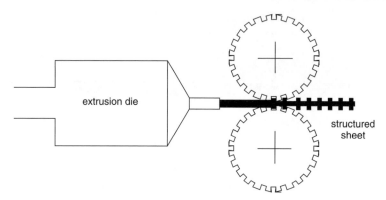

Figure 2.16   Schematic showing the process for structuring HDPE geomembranes between embossed rollers to imprint a friction-enhanced surface pattern

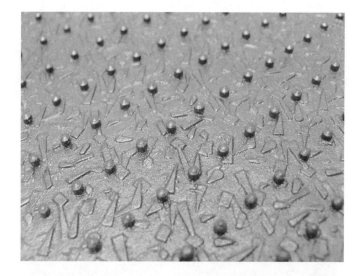

Figure 2.17   Flat-die moulded textured surface (surface-friction profile). This profile is known as microspikes since the spikes are 1 mm high and spaced 5 mm apart

surfaces and this pattern is transferred to the hot sheet. The result is that the final sheet has raised surface patterns on its surface (see Figures 2.17 and 2.18). The edges of the sheet are left untextured up to 15 cm from the edge so that field welding can be facilitated. Structured or patterned geomembranes are considered as being less critical than nitrogen blown or impingement textured sheets from a long-term performance perspective because they are composed of the same resin and formed in an integrated processing step.

Structured or embossed geomembrane (textured) surfaces exhibit both consistent core thickness and consistent asperity height by virtue of the manufacturing process (see Table 2.5). This allows QC and CQA checks to be decreased since multiple measurements

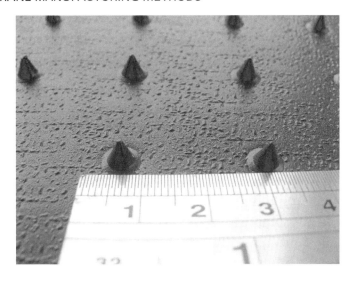

Figure 2.18 Photograph of an HDPE geomembrane with *spikes* on its surface to improve friction angles and grip with very steep soil slopes such as clay or coarse clay (Agru liner). The spikes are 6 mm high and spaced 25 mm apart

Table 2.5 Comparison scorecard of attributes of coextruded (textured) geomembranes and embossed (structured) geomembranes. Adapted with permission from R. K. Frobel, *Geosynthetics*, **25**, 12 (2007)

Design consideration	Coextruded textured	Embossed structured
Consistent thickness (cross roll)	✗	✔
Consistent texture (cross roll)	✗	✔
Consistent asperity heights	✗	✔
Asperity heights >0.375 mm	✗	✔
Texturing does not reduce mechanical properties	✗	✔
Resists texture comb over under shear	✗	✔

to determine the average or minimum values are not required. Structured geomembranes with sharp spikes have excellent steep slope potential.

📌 *Embossed (structured) flat-die geomembranes have some advantages over textured blown film (round die) geomembranes such as consistent core thickness, consistent texture from roll to roll and a completely homogeneous structure, hence and so the texture has little effect on the tensile and elongational properties of the geomembrane. In contrast, blown film (nitrogen textured) geomembranes generally exhibit variable core thickness, non-uniform area coverage, inconsistent asperity heights on the surface, inconsistent/variable texture from roll to roll and generally reduced tensile, elongational and stress crack resistance properties. Horizontal flat-die sheet that is pattern calendered also*

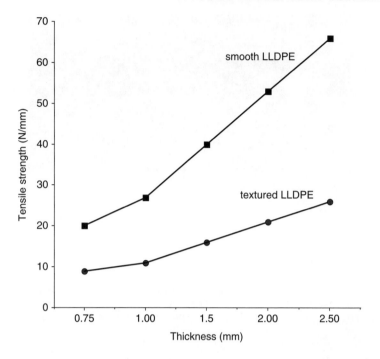

Figure 2.19   Tensile strength at break (N/mm) for smooth and textured LLDPE geomembranes

enables a smooth region on the edge of the geomembrane for easy field seaming as well as excellent control over the final thickness.

2.8.4   DEFECTS CREATED BY TEXTURING

Surface texturing creates notches at the edges of a test specimen. As the specimen is stretched during a tensile test, stresses can concentrate in the newly created grooves. These will be the areas for cracks initiate and to propagate during the tensile test and can eventually cause the material to prematurely break at a lower than expected stress. This is why the tensile strength at break is generally significantly reduced by texturing of geomembranes (see Figure 2.19). This is especially true for HDPE geomembranes where the predominant mode of premature failure is a quasi-brittle fracture initiated at stress concentrating surface notches. Texturing also leads to a reduction in puncture strength of geomembranes as shown in Figure 2.20.

Care should be taken to ensure that the texturing process does not lead to 'notch effects' since the high shear forces on slopes can translate into strong local stress concentrations. Such stress concentrating forces can propagate notches into cracks.

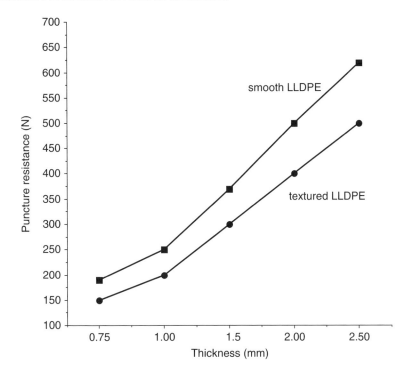

Figure 2.20 Puncture resistance (N) for smooth and textured LLDPE geomembranes

2.8.5 MEASURING THE THICKNESS OF TEXTURED GEOMEMBRANES

The thickness of textured geomembrane sheets is expressed as the minimum core thickness between the roughened peaks (referred to as 'asperities'). The core thickness is measured using a tapered-point micrometer as described in ASTM D5994 and where the tapered-points have a 60° angle and a 0.08 mm diameter flat tip. Since the tapered points can penetrated into the surface of the geomembrane giving underestimates of the thickness it is necessary to limit the amount of force applied to typically 0.56 N.

Only one tapered point is needed for a single-sided textured sheet while a micrometer with two opposing tapered tips is required for a double-sided textured geomembrane sheet. The test requires that ten measurements are taken across the roll width with the average core thickness calculated.

2.8.6 ANCHOR SHEET

Acidic corrosion is the primary durability concern in concrete sewerage systems (pipes, detention tanks, treatment facilities). Mechanically anchored HDPE corrosion protection linings, also known as 'anchor sheets' or Corrosion Protection Liners (CPLs) are widely used to protect such concrete sewerage structures which are potentially susceptible to attack by sulfuric acid. HDPE anchor sheets can be incorporated into both precast and *in situ* reinforced concrete structures and have service lives in excess of 100 years.

Anchor liners (or stud liners) have an array of studs or knobs on their surface. These liners provide protection to concrete structures protecting them against chemical attack such as sulfide attack. They can be cast into new structures or grouted into existing structures. Anchor liners are commonly used in aggressive sewers and pipelines, sumps, manholes, tanks, bund areas, channels and chutes, as well as applications where acid spillages can occur.

Anchor Lining Systems (Milnerton, South Africa) manufacture a cast-in liner, called an Anchor Knob Sheet (AKS) designed to attach to concrete structures through the embedding of its 1230 anchor knobs/studs per square metre into the concrete. The optimal size, shape and number of knobs ensures an extremely high pull-off resistance from the concrete (i.e. 78 t/m^2). The studs are produced simultaneously with the HDPE liner during the extrusion manufacturing process, thereby ensuring total homogeneity with the base liner material. This gives a sheet which is referred to as 'monolithic' since the knobs are an integral part of the sheet and not welded-on in a secondary step (i.e. unlikely to separate under load or in use).

The high chemical resistance of HDPE, along with proven welding and testing systems, ensures that sewers, manholes, pump stations and other concrete structures, which operate in chemically aggressive environments, are provided with long-term protection against corrosion. Additional benefits, like a raised carrying capacity through a low friction coefficient, when used for liquid conveyance, makes this a cost-effective solution for a wide variety of lining applications.

2.9 ADDITIVES FOR GEOMEMBRANES

Additives are critical for the proper performance of geomembranes and geosynthetics in general.

2.9.1 PIGMENTS

Pigments are used in geomembranes to provide the colour to control the surface temperature of the geomembrane but more importantly they act as UV stabilizers. Most geomembranes are either black, beige or shades of grey. Carbon black and titanium dioxide are the main two pigments used in geomembranes. Carbon black is an excellent UV screen, absorbing most of the UV radiation that strikes the geomembrane and converting it to heat. Titanium dioxide, on the other hand, reflects almost all UV radiation. Both these pigments offer excellent UV protection. Geomembranes can be exposed for years with minimal UV degradation when using these pigments together with special stabilizer additives.

2.9.2 CARBON BLACK

Carbon black not only acts as a UV screening agent (absorbing damaging UV light) but also performs a radical trapping function by binding up damaging free radicals. Carbon black only reaches its maximum effectiveness when a certain fine particle size is achieved (e.g. 20 nm) and the particle-to-particle distance is minimized in order to provide dense coverage (this is a function of proper distribution and dispersion of the carbon black aggregates). The carbon black used in HDPE geomembranes is generally required to conform to category Group 3 and/or lower as defined in ASTM D-1765.

Levels of carbon black only up to 2.5–3.0% can be used in HDPE since larger amounts can detract from the mechanical properties of HDPE simply because the high crystallinity of HDPE leaves little free volume (i.e. space) to accommodate additives.

Carbon black pigments come in different particle sizes. The more expensive carbon blacks have a smaller average particle size (i.e. 18–22 nm) while the cheaper carbon blacks are coarser (e.g. 60–150 nm). The smaller particle size carbon blacks are more efficient at blocking UV light and preventing ultraviolet degradation of polymeric geomembranes (see Figure 2.21). The level of carbon black in the geomembrane referred to as the carbon black content (CBC) is determined according to ASTM D-1603, ASTM D-4218 or ISO 6964. Not only are the amount and particle size of the carbon black important but also is its uniform dispersion in the geomembrane. The test method for determining the extent of dispersion of carbon black is ASTM D-5596 or ISO 11 420.

To check that the right amount of carbon black is added and that it is dispersed thoroughly within the geomembrane sheet, both carbon black content and carbon black dispersion tests are performed. Carbon black content tests measure the percentage of carbon black added to the geomembrane by heating the geomembrane until only the carbon black remains. For HDPE geomembranes, typically 2 to 3% carbon black content is within acceptable limits.

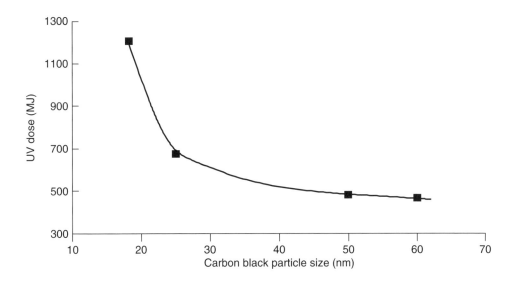

Figure 2.21 Failure point of a HDPE geomembrane under UV irradiation as a function of carbon black particle size. Note that the smaller the carbon black particle size, the greater the UV stability of the HDPE geomembrane. For this reason it is important that the correct carbon black particle size is used

Carbon black dispersion tests estimate dispersion of carbon black within a geomembrane sheet by observing a thin film of the geomembrane under a microscope. The dispersion observed is qualitatively compared to a chart containing several degrees of carbon black dispersion. Most geomembrane specifications allow a dispersion rated as A-1, A-2 or B-1. For HDPE and LLDPE, the carbon black agglomerates need to fall mainly (i.e. 9 out of 10 views) into Category 1 (see Figure 2.22) or Category 2 (Figure 2.23) and only 1 is allowed in Category 3 (see Figure 2.24). No carbon black agglomerates in Category 4 or Category 5 (see Figure 2.25) are acceptable. Poor carbon black dispersion may result in reduced UV and environmental stress crack resistance and may reduce tensile properties.

Note that the LLDPE carrier resin from black masterbatches can preferentially reside in the centre of the GM and can cause delamination problems. This is often identified as black streaks when performing carbon black dispersion tests.

2.9.3 STABILIZER PACKAGE

Antioxidants and stabilizers are added to polymeric materials to inhibit oxidation and extend the induction period to onset of degradation. Because geosynthetics are manufactured at high temperatures (200 to 220 °C), antioxidants are needed that function at the high temperatures associated with manufacturing as well as the lower temperatures associated with in-service applications. Consequently, manufacturers generally use a combination of two or more types of antioxidants and stabilizers to provide overall stability. Figure 2.26 shows the autooxidation cycle for the oxidiation of polyolefins. Cycle-breaking antioxidants and stabilizers function at steps (a), (b) and (c) to scavenge the damaging free radicals.

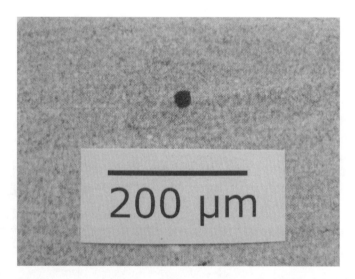

Figure 2.22   Photograph of a carbon black agglomerate in a HDPE geomembrane as viewed under magnification on a microtomed slice. The agglomerate particle has a diameter of approximately 25 µm and so falls into the Category 1 classification (i.e. diameter <35 µm; total area of <960 µm²)

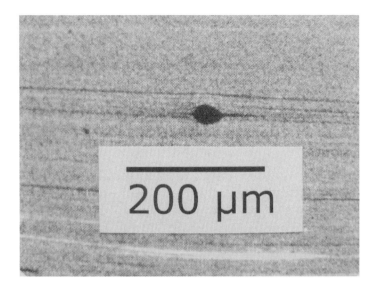

Figure 2.23 Photograph of a carbon black agglomerate in a HDPE geomembrane as viewed under magnification on a microtomed slice. The agglomerate particle has a diameter of approximately 40 µm and falls into the Category 2 classification (i.e. diameter 35–75 µm; total area of >960–4390 µm²)

Figure 2.24 Photograph of a relatively large carbon black agglomerate in a HDPE geomembrane as viewed under magnification on a microtomed slice. The agglomerate particle has a diameter of approximately 150–200 µm but with an area of approximately 17 500 µm² it just falls into the Category 3 classification (i.e. diameter 75–175 µm; total area of >4390–24 053 µm²)

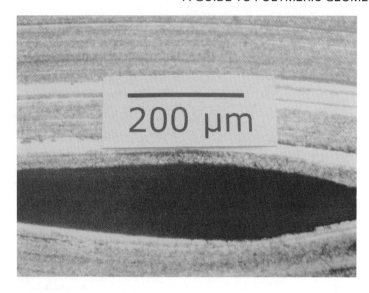

Figure 2.25   Photograph of an elongated carbon black agglomerate in a HDPE geomembrane as viewed under magnification on a microtomed slice. The agglomerate particle has a diameter of approximately >200 μm and hence falls into the Category 4 or even 5 classification (i.e. diameter 175 to >300 μm)

The additive package and specifically the stabilizer package, needs to be based on best practice principles which means that it should comprise a high performance stabilizer package formulated for potentially aggressive and exposed environments.

It is quite difficult to write a specification for polyolefin geomembranes around a particular additive, or group of additives, because they are generally proprietary. Furthermore, there is ongoing research and development in the stabilization area and thus additives are subject to changes over time. If additives are included in a specification, the description must be very general as to the type and amount.

A best practice stabilization package for polyolefin geomembranes can comprise at least three antioxidant/stabilizer additives:

- Hindered Phenolic Antioxidant (HPA).
- Hindered Phosphite Processing Stabilizer (HPPS).
- Hindered Amine Stabilizer (of high molecular weight, i.e. oligomeric or polymeric) (non-migratory) (HALS-HMW).
- Hindered Amine Stabilizer (of low molecular weight) (migratory) (HALS-LMW).

Antioxidants

Typical antioxidants used in polyolefin geomembranes are shown in Table 2.6.

Hindered Amine Stabilizers

Hindered amine light stabilizers (HALS) are widely used in polymeric geomembranes to impart both UV and thermooxidative (i.e. heat) protection. HALS are the stabilizer product

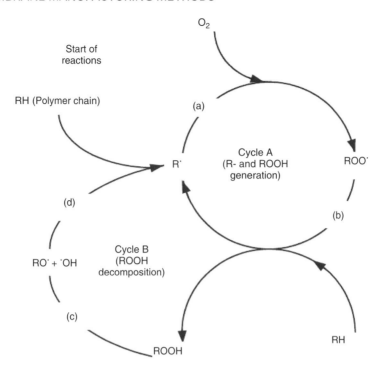

Figure 2.26 Autooxidation cycle for the oxidation of polyethylene. The oxidation chain mechanism comprises two interacting cyclical processes. In cycle A, free radicals and hydroperoxide groups are produced. In cycle B, hydroperoxides are decomposed to form more free radicals. Chain breaking antioxidants function at steps (a), (b) and (c) to scavenge free radicals. Reprinted with permission from S. B. Gulec, T.B. Edil and C. H. Benson, *Geosynthetics International*, **11**, 60, Copyright (2004) Thomas Telford Publishing

of choice for polyolefins because they exhibit multi-functional stabilization activity. HALS is multifunctional in its capacity by: trapping free radicals, decomposing alkyl hydroperoxides (which are unstable intermediates in the oxidation cycle), quenching excited states and regenerating its active species. Commercially available HALS contain the tetramethyl piperidinyl moiety which is the precursor to the 'active site' for stabilization activity. Once activated, the nitroxyl active site undergoes many cycles until it is ultimately deactivated. Although nitroxyl deactivation is a major pathway by which basic HALS lose effectiveness, there are also other causes such as chemical side reactions with acidic compounds.

HALS are the stabilizers of choice for polyolefin geomembranes because they exhibit multi-functional stabilization action and can protect the geomembranes against both heat and UV light. HALS are capable of decomposing radical precursors, trapping free radicals and also can potentially quench excited states, all of this at the same time as regenerating the active species.

Table 2.6 Typical antioxidants for polyolefin geomembranes

Stabilizer (Manufacturer)	Comments
Irganox 1010 (Ciba)	This hindered phenolic antioxidant is the 'workhorse' antioxidant for HDPE, LLDPE and fPP geomembranes and is almost universally used
Irganox 1076 (Ciba)	Hindered phenolic antioxidant used for many years now, in polyethylenes and polypropylenes
Hostanox 03 (Clariant)	A hindered phenolic antioxidant claimed to be superior to Irganox 1010 types in imparting long-term heat stability. It also boasts very low migration from polyolefins and much lower water extraction than 1010 types, as well as lower adsorption onto fillers
Irganox B225 (Ciba)	Synergistic 1:1 blend of Irgafos 168 (hindered phosphite) and Irganox 1010 (hindered phenolic)
Irganox B215 (Ciba)	Synergistic 2:1 blend of Irgafos 168 (hindered phosphite) and Irganox 1010 (hindered phenolic)

The tetramethyl piperidinyl group of HALS stabilizers is the precursor for the active site for stabilization activity and once activated, the resultant nitroxyl active site undergoes many regeneration cycles until it is ultimately deactivated. Typical HALS used in polyolefin geomembranes are shown in Table 2.7.

HALS stabilizers can lose their effectiveness because of the following reasons:

- nitroxyl deactivation (which is the major pathway);
- acid attack (protonation of the active nitroxyl site);

Table 2.7 Typical hindered amine stabilizers for polyolefin geomembranes

Stabilizer (Manufacturer)	Comments
Tinuvin 783 (Ciba)	Tinuvin 783 is a 1:1 blend of a workhorse HALS (Chimassorb 944) and a tertiary HMW HALS (Tinuvin 622) and is recommended for polyolefin-based geomembranes. It is a common HALS stabilizer system used in polyolefin geomembranes. Addition rate of 5000 ppm is recommend for a 20 year life
Hostavin N-30 (Clariant)	This stabilizer fits in the category of a 'HATS' (hindered amine thermal stabilizer). It is a low-basicity, polymeric, HMW HALS used in geomembranes where it can serve as both an antioxidant and UV stabilizer
Uvinul 4050H (BASF)	Uvinul® 4050H is a UV stabilizer that belongs to the group of low molecular-mass sterically hindered amines (monomeric HALS). It has FDA food contact approval
Uvasil 2007 (Great Lakes Chemical)	This is a synergistic blend of LMW and HMW HALS said to result in a broad distribution of stabilizer mobility within a polyolefin matrix. The HMW HALS provides long-term permanence while the LMW HALS imparts high activity during outdoor exposure

- extraction and loss;
- chemical side reactions.

The total concentration of the HALS additives should be equal to or greater than 5000 ppm (i.e. 0.5%) for 15–20 year exposed lifetimes in HDPE and fPP.

Initially only low-molecular weight HALS additives were developed such as Tinuvin 770. Whilst they were effective light stabilizers for polyolefins they had shortcomings such as a high migration rate and only moderate resistance to extraction. To overcome these disadvantages high-molecular weight (or polymeric) HALS were developed. These additives show remarkable effectiveness for stabilizing polyolefins against not only UV degradation but also thermal oxidation.

Tinuvin 770 and Tinuvin 622 are HALS additives based on a polyester structure. Unfortunately the polyester backbone is prone to hydrolytic- and photolytic-cleavage reactions which can lead to additive loss. The hydrolysis reactions can be accelerated by the presence of acids and typical formulation components such as stearates. Chimassorb 944 is a polymeric hindered amine stabilizer which has stable triazine backbones in its structure. Due to the absence of ester groups, Chimassorb 944 is not prone to hydrolytic breakdown and furthermore the triazine rings in the backbones impart superior thermal and photochemical stability as well as low volatility.

Note that combinations of the various antioxidants and stabilizers listed in Table 2.8 are often used to provide optimum stabilization; for instance low and high molecular weight HALS are often used in combination, such as Tinuvin 770/Chimassorb 944 or Uvinul 4050/Uvinul 5050.

Figure 2.27 shows the effective temperature ranges of different antioxidant and stabilizer types used in polymeric geomembranes. Note: High molecular weight HALS are also

Table 2.8 Common antioxidants/stabilizers used in polymeric geomembranes and their molecular weights and effective temperature ranges

Antioxidant/Stabilizer[a]	Molecular weight	Effective temperature range ($^\circ$C)[b]
Irgafos 168 (phosphite)	647	150–300
Irganox 1010 (phenolic)	1178	Up to 300
Irganox 1330 (phenolic)	775	Up to 300
Irganox B225 (blend)	1:1 phosphite/phenol	Up to 300
Irganox B215 (blend)	2:1 phosphite/phenol	Up to 300
Tinuvin 770 (low MW HALS)	481	Up to 150
Uvinul 4050 (low MW HALS)	450	Up to 150
Chimassorb 944 (high MW HALS)	2580	Up to 150
Tinuvin 622 (high MW HALS)	3100–4000	Up to 150
Tinuvin 783 (high MW HALS blend)	1:1 C–944/T-622	Up to 150
Cyasorb 3346 (high MW HALS)	1500–1800	Up to 150
Hostavin N30 (high MW HALS)	>1500	Up to 150
Uvinul 5050 (high MW HALS)	3500	Up to 150

[a]Irgafos™, Irganox™, Chimassorb™, Tinuvin™ are trademarks of Ciba; Cyasorb™ is a trademark of Cytec; Hostavin™ is a trademark of Hoechst; Uvinul™ is a trademark of BASF.
[b]Fay and King (1994).

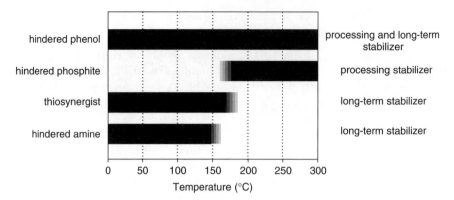

Figure 2.27   Effective temperature ranges of different antioxidant and stabilizer types used in polymeric geomembranes

commonly referred to as oligomeric or polymeric stabilizers and they are difficult to extract from the geomembrane polymer matrix.

Commercial examples.
5000 ppm Tinuvin 783 (Ciba) comprises Chimassorb 944 (high molecular weight hindered amine) and Tinuvin 622 (a low molecular weight hindered amine stabilizer) in a 1:1 ratio.

5000 ppm of a 1:1 mixture of a high molecular weight hindered amine (Cyasorb UV-3529 from Cytec) in combination with a low molecular weight hindered amine stabilizer (Cyasorb UV-3853 from Cytec) which is highly mobile, to protect the surface of the geomembrane but also has high solubility in the polymer to resist extraction.

Typical Stabilization Formulations

Polyolefin stabilization packages have as a rule of thumb at least 5000 ppm (i.e. 0.5 wt%) of HALS stabilization to give greater than 20 years UV stability in combination with 2.5% wt. carbon black.

Typical polyolefin geomembrane additive packages are shown below as examples.

For polyolefin geomembranes in potable water contact:

- 3000 ppm Irganox 1010 (hindered phenolic);
- 1500 ppm Irgafos 168 (hindered phosphite);
- 5000 ppm Chimassorb 944 (hindered amine).

For non-potable applications:

- 3000 ppm Irganox 1010 (hindered phenolic);
- 1500 ppm Irgafos 168 (hindered phosphite);
- 3000 ppm Chimassorb 944 (hindered amine);
- 2000 ppm Tinuvin 770 (hindered amine).

Note for non-potable applications, Tinuvin 770 is used to replace part of the Chimassorb 944 for the purpose of reducing cost, since Tinuvin 770 is not approved for contact with potable water.

For improved antioxidant leaching resistance in water contact applications, the following mixture has been used:

- 3500 ppm of Irganox 1330 (hindered phenolic);
- 5000 ppm of Chimassorb 944 (hindered amine);
- 2000 ppm of Tinuvin 770 (hindered amine).

The last formulation uses Irganox 1330 in place of Irganox 1010. Irganox 1330 is a trifunctional hindered-phenolic antioxidant whereas Irganox 1010 is tetrafunctional. The molecular weight of Irganox 1330 is actually lower than that of Irganox 1010 (cf. 775 *versus* 1178). The extraction and leaching resistance of Irganox 1330 however, is purported (by Ciba) to be higher due to the molecule being more rigid and planar (due to its central phenyl core), while the Irganox 1010 molecule is more pliant and mobile.

Irganox 1330 is also preferred over Irganox 1010 in terms of its resistance to alkaline hydrolysis and its low adsorption potential onto carbon black.

Irgastab FS811 at levels of 1 wt% (10 000 ppm) has been used for the stabilization of fPP geomembranes as well as fPP roofing products. This additive package at 1 wt% loading is expected to give the 0.75 mm FPP product a 10 year lifetime. Note Irgastab FS 811 is a 5:3:2 mixture of Chimassorb 944, Tinuvin 770 and Irgastab FS 042, and so the formulation is as follows:

- 5000 ppm of Chimassorb 944 (a high molecular weight hindered amine);
- 3000 ppm of Tinuvin 770 (a low molecular weight hindered amine);
- 2000 ppm of Irgastab FS 042 (a hydroxylamine stabilizer).

The hydroxyl amine (Irgastab FS 042) has good extraction resistance and is often used in applications for instance, where there is a high surface area to volume ratio such as in fibres.

Non-Reactive HALS

In recent years it has been recognized that strong acids can interact with basic hindered amine stabilizers and significantly reduce their effectiveness. The reason is that standard HALS stabilizers are basic in nature (i.e. alkaline) and therefore acidic compounds can deactivate them by a neutralization reaction to form a non-active salt. Acidic species (e.g. in some mining leach solutions) can deactivate hindered amines not having the N-OR group, and this interferes with the activity of 'normal' hindered amine stabilizers.

A solution to this problem has been the development of methylated HALS. The low basicity of methylated HALS is of particular value in the stabilization of polyolefins where the activity of the more basic hindered amine stabilizes is significantly reduced because of interaction with the acid species.

2.9.4 BLOOMING OF ADDITIVES

Note if the total stabilizer package exceeds or approaches 0.5–0.6% (i.e. 5000–6000 ppm), then waxy 'blooms' can form on the liner surface due to exudation of the additives by migration/diffusion processes to the liner–air interface. These additives can interfere with weldability of the HDPE as they act as weak boundary layers. Before wedge welding, this waxy bloom needs to be removed with a polar solvent such as acetone or limonene.

2.9.5 INTERACTION OF ADDITIVES

The typical pigment used for black geomembranes is N110 carbon black. If the carbon black loading exceeds 3 wt% and in particular falls in the range 3–5 wt%, then this can reduce the stability of the overall formulation since antioxidants can become absorbed and immobilized on the surface of the carbon black. Carbon black has a very high surface area and it can adsorb and immobilize antioxidants and reduce their effectiveness (Peña et al., 2002).

Small particle size blacks such as the 17 nm N110 type often possess high surface areas and therefore exhibit greater degrees of jetness (or 'blackness') than larger particle size blacks. Small particle size blacks are highly effective UV light absorbers but they are difficult to disperse at high loadings and often result in a high viscosity concentrate.

Another commonly used carbon black is Vulcan P (from Cabot) (17 nm particle size) at levels of 2.75–3.25 wt%. This is approaching the loading range where antioxidant adsorption onto carbon black may become an issue to consider.

The degree of steric hindrance of phenolic hydroxyl groups by alkyl groups in *ortho*-positions has been found to be the major factor affecting the adsorption activity of phenolic antioxidants onto carbon black. Irganox 1330 shows negligible adsorption. Therefore it may be concluded that the 2,6-ditert-butyl phenyl group has insignificant interaction with the carbon black due to the steric hindrance (Peña et al., 2002).

It is known that different grades of carbon black adsorb Irganox 1010 to different extents. The ester groups of Irganox 1010 act as sites for adsorption on the carbon black. Therefore owing to the presence of the same phenolic component and ester linkage, Irganox 1010 and Irganox 1076 show broadly similar responses. Despite having a smaller number of functional groups per molecule, Irganox 1076 has a greater adsorption/ desorption energy onto carbon blacks due to sterically more exposed ester groups.

Peña et al. (2002) concluded that stabilizer adsorption activity is mainly governed by the number of active functional groups per molecule, together with the steric accessibility of the functional groups to the carbon black surface. Steric hindrance of the adsorption active functional groups (H–N< and Ar–O–H) by alkyl groups has been shown to significantly reduce adsorption activity. Carbonyl groups have also been shown to contribute to additive adsorption activity. Therefore Irganox 1010 is more adsorbed than Irganox 1330 due to the four ester groups per molecule in the case of Irganox 1010. The nature of the carbon black substrate also plays an important part in adsorption. It has been shown that highly oxidized carbon blacks, bearing plentiful carboxylic and phenolic functionality, are more effective at adsorbing stabilizers.

The use of Irganox 1010 in combination with the wrong type of carbon black can therefore result in a lower degree of stabilizer performance due to the high surface area per unit mass of the carbon black coupled with some surface porosity. This finding supports the move away from Irganox 1010 to Irganox 1330.

REFERENCES

Battenfeld Gloucester [http://www.strictly-extrusion.com/2006/03/09/battenfeld-gloucester-one-of-worlds-largest-blown-film-lines-for-geomembranes/] (2006).

Cooley, 'Tech Note 043-R0', The Cooley Group [www.cooleygroup.com] (February 2006).

Fay, J. J. and King III, R. E., Antioxidants for geosynthetic resins and applications, in Hsuan, G. and Koerner, R. M. (Eds), *Proceedings of the 8th GRI Conference on Geosynthetic Resins, Formulations and Manufacturing*, Industrial Fabrics Association International (IFAI), St Paul, MN, USA, pp. 74–92 (1994).

Frobel, R. K., Using structured geomembranes in final solid-waste landfill closure designs, *Geosynthetics*, **25**, 12 (March 2007).

Gulec, S. B., Edil, T. B. and Benson, C. H., Effect of acidic mine drainage on the polymer properties of an HDPE geomembrane, *Geosynthetics International*, **11**, 60 (2004).

Ivy, N., Asperity height variability and effects, *GFR Magazine*, **21**(8), 28 (October/November 2003).

Peña, J. M., Allen, N. S., Liauw, C. M., Edge, M. and Valange, B., Factors affecting the adsorption of stabilizers onto carbon black, in Proceedings of EuroFillers'99 Conference, Lyon, France, 1999. Available at: [http://www.chem-mats.mmu.ac.uk/Research/EuroFillers/s1066.pdf].

Sieracke, M. D., Geosynthetic Manufacturing Concerns from a Consultant's Perspective, in *Proceedings of the GRI/NAGS Conference*, Las Vegas, NV, USA, 2005.

3

HDPE Geomembranes

3.1 INTRODUCTION

The most commonly used polymer worldwide, to manufacture geomembranes is high-density polyethylene (HDPE) (or more specifically medium-density polyethylene (MDPE)). HDPE exhibits high strength and excellent chemical resistance to a wide range of chemicals due mainly to its crystalline microstructure (it has a crystallinity of 40–60%). The molecules (or molecular chains) of HDPE are quite linear and this enables them to efficiently pack into crystallites (i.e. highly crystalline domains). Due to the close packing of the polymer chains in HDPE there is little room (or free volume) for chemicals to permeate the polymer, hence giving it outstanding chemical resistance. It is the most chemically resistant member of the polyolefin family on account of its dense molecular configuration (density >0.94 g/cm^3). Although less flexible than its LLDPE counterpart, it still offers good elongation properties allowing up to 12% deformation at its yield point.

✎ *Regular HDPE cannot be used as a geomembrane because it is too stiff and prone to environmental stress cracking. Instead a lower density version of HDPE, namely MDPE, is used for geomembranes. The industry however has commonly adopted the term 'HDPE' (as is also the case in this book) when it in fact refers to MDPE geomembrane liners. The apparent misnomer is due to the fact that the addition of carbon black to the MDPE resin raises the final product density into the realm of HDPE.*

HDPE geomembranes gained popularity over the traditional PVC geomembranes in the early 1980s due to their broad spectrum chemical resistance and their ability to be thermally seamed rather than using solvent cements and glues. In those days the seams of PVC geomembranes were peelable whereas the thermal fusion seams of HDPE were not.

Then the period from 1985–1995 saw a major swing from rubber liners (e.g. butyl, nitrile and CSPE) to HDPE geomembranes for use in the municipal and hazardous waste

A Guide to Polymeric Geomembranes: A Practical Approach   J. Scheirs
© 2009 John Wiley & Sons, Ltd

industry due to its combination of superior overall chemical resistance, high strength and low cost.

📌 *HDPE is an enormously popular product due to its UV resistance, low cost, versatile mechanical properties and very good overall chemical resistance.*

The low material cost of HDPE resins allows for thicker lining cross-sections to be used compared to other more expensive geomembrane materials, thus providing increased resistance to puncture and abrasion.

HDPE is typically manufactured as a blown film or flat sheet product. It is the cheapest (per mm thickness) industrial geomembrane commercially available, Therefore it is generally the first geomembrane material to be considered for a geomembrane application and is only rejected in those cases where there are sound scientific reasons for not using it. In practice at present, HDPE is the most selected geomembrane material, being chosen in some 95% of geomembrane applications. Figure 3.1 shows a HDPE geomembrane being installed to line a large water storage.

📌 *The low initial material cost of HDPE allows thicker sections to be used compared to other geomembrane materials. A thick, durable, HDPE liner can be placed in exposed applications where the cost of other materials would be prohibitive. The excellent chemical resistance of HDPE is often the driving force behind its selection. However, since HDPE is a field fabricated material, welding and testing need to be completed with great care.*

Some shortcomings of HDPE are:

- HDPE is almost never pre-fabricated in a controlled factory setting and therefore all welds on environmental grade HDPE geomembranes are field welds which greatly increase installation and third party field quality control costs.

Figure 3.1   Photograph of a HDPE liner being installed in a large impoundment

- HDPE is a very stiff liner with a high coefficient of thermal expansion, often requiring special design considerations.
- Particular care must be taken against mechanical damage while installing any HDPE geomembrane as it is quite notch sensitive and any surface notches can propagate into cracks under applied stress/strain.
- HDPE is prone to environmental stress cracking (ESC) due to a crystalline lattice structure. As indicated above true HDPE resins are not suitable for the manufacture of geomembranes due to their pronounced sensitivity for stress cracking.
- HDPE's puncture resistance is lower than most other competing materials. Puncture resistance properties require that a minimum of 1.5 mm of thickness be specified for most exposed applications.

Table 3.1 lists the limitations of HDPE geomembranes and gives some practical examples of how these shortcoming influence installation or long-term performance.

Since HDPE is a stiff material it cannot be prefabricated into panels. Rather the HDPE geomembrane is delivered to site in rolls (between 6–9 m in width) and all the welding is performed on site.

HDPE exhibits a distinct (i.e. well-defined) yield point in its uniaxial tensile stress/strain curve at around 12% strain (i.e. elongation) (see Figure 3.2). Up to this point HDPE behaves elastically without damage to the polymer's microstructure structure. Above this point further deformation (e.g. thinning) can occur without additional load. This arguably represents one of the main deficiencies of HDPE geomembranes. In contrast LLDPE has a far less pronounced yield point at about 40% strain whereas PVC has no defined yield point.

The yield stress values for various thickness HDPE geomembranes are shown in Figure 3.3 and the break stress values are shown in Figure 3.4.

At greater strains HDPE yields and the deformation is plastic up to the failure point. The design basis for HDPE geomembranes generally dictates that strains in service should not exceed about 3% in order to reduce the likelihood of liner failure by creep rupture and environmental stress cracking phenomena. This level of strain however can easily occur as a result of factors such as temperature-induced stresses, subgrade settlements and indentation by underlying or overlying materials.

HDPE therefore often performs in service, in practice, at the limit of acceptable design strain. Geomembranes such as LLDPE and fPP on the other hand can operate safely at much greater strains without risk of failure.

Arguably the most limiting material property for HDPE geomembranes is its stress cracking resistance (SCR). The SCR is the single most important property of HDPE that determines its long-term mechanical durability and its resistance to mechanical damage due to stresses imposed during welding and installation.

As mentioned above although HDPE has excellent broad spectrum chemical resistance and is of relatively low cost it does suffer from a number of shortcomings which are listed in Tables 3.1 and 3.2.

Table 3.1 Some limitations of HDPE geomembranes (Peggs, 1997)

Limitation	Comments and examples
Lack of flexibility	HDPE can be difficult to install due to its lack of flexibility and its rigid nature especially in the higher thicknesses (2.0–2.5 mm)
	Flexibility is an important factor for fabrication and ease of installation. HDPE is quite stiff and this makes prefabrication (i.e. factory seams) not feasible
	HDPE geomembranes provide less conformance to uneven subgrades due to their stiffness
	Flexibility also determines how a geomembrane can be affected by folds, wrinkles or waves in the material. HDPE develops less wrinkles than fPP and PVC during placement which can be an advantage
High coefficient of thermal expansion[a]	HDPE geomembranes have a high thermal expansion coefficient that causes them to expand and form 'waves' (wrinkles) when exposed to the sun (see Figure 3.5) and conversely also to pull tight (and develop tensile stresses) at low temperatures
	The high thermal expansion of HDPE can preclude the installation of these geomembranes at low temperatures since when they are covered at significantly higher temperatures then wrinkles and waves will develop
	PVC has a 60% lower coefficient of thermal expansion than HDPE
HDPE welding requires great care and special attention	The extrusion fillet welding method creates a large heat affected zone (HAZ) compared with hot wedge welding. Therefore care needs to be exercised in conducting extrusion fillet welding for detailed work around pipes, connections and at T-junctions
HDPE has low deflection capabilities	HDPE geomembranes have low deformation capabilities. As a result differential settlement (for example, as can occur on a landfill cap) can induce stresses in HDPE geomembranes and this can promote ESC
HDPE is easily scratched	When deploying an HDPE geomembrane it should not be dragged on rough subgrades since this will lead to scratches and scorelines that are precursors to stress cracking. It should be deployed with care on a smooth subgrade without protruding stones and gravel

[a]The amount of expansion or contraction that occurs when a geomembrane is heated or cooled is expressed as its coefficient of linear thermal expansion. It is determined by accurately measuring the dimensions of a test section at two different temperatures and calculating the percent change/$°$C. HDPE has a high coefficient of thermal expansion which can cause stress in the seamed areas. HDPE has a coefficient of thermal expansion of 0.018%/$°$C. This means for example, that a liner on a 150 m beam will expand more than 1.8 m on a day when the temperature starts at 0 $°$C and the black liner reaches up to 70 $°$C in the early afternoon sun. To accommodate this much expansion, the liner develops large undulations or waves on the beams and exposed side slopes. This high linear expansion also places limitations on when seaming can be done. Seaming on hot sunny days can result in split seams or the liner lifting away from the corners at the top of the side slopes as the liner contracts during cooling in the evening.

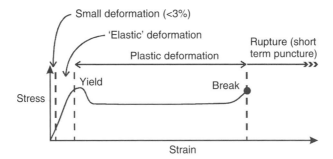

Figure 3.2 The stress–strain properties for a typical HDPE geomembrane. Note that although angular particles may not puncture the liner they can cause localized elongation (i.e. localized strain <3%) that can lead to the development of stress cracking. Hence the need to prevent localized strain of the geomembrane by using a cushioning layer. Reprinted with permission from the Proceedings of 56th Canadian Geotechnical Conference, Selection of Protective Cushions for Geomembrane Puncture Protection by Eric Blond, Martin Bouthot, Oliver Vermeersch and Jacek Mlynarek Copyright (2003) Eric Blond

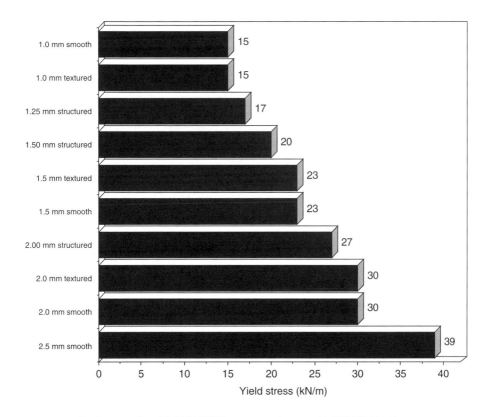

Figure 3.3 Yield stress for AGRU HDPE geomembranes (ASTM D6693)

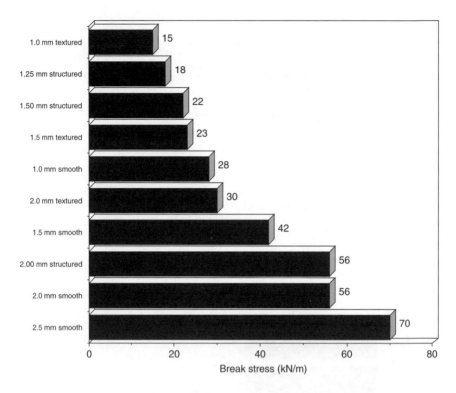

Figure 3.4 Break stress for AGRU HDPE geomembranes (ASTM D6693)

Figure 3.5 Photograph of 'waves' in an HDPE geomembrane resulting from high temperatures causing thermal expansion. Reproduced by permission of NAUE

Table 3.2   Summary table of some common limitations of HDPE geomembranes and recommended mitigation strategies

Limitation	Mitigation strategy
The high crystallinity of HDPE makes it sensitive to stress cracking	Only use those HDPE geomembranes that have been tested according to SP-NCTL and exceed 300 hrs
The relatively high coefficient of thermal expansion can lead to standing 'waves' and wrinkles that make it difficult to achieve intimate contact with the subgrade[a]	Use German beams installation method (this can also lead to tensile stresses so it must be done carefully)
The relatively high stiffness of HDPE can make good conformance to certain subgrades difficult to achieve	Select a LLDPE or HDPE/VLDPE/HDPE geomembrane which is intrinsically less stiff
HDPE geomembranes have poor axisymmetric tensile elongational properties	Select membranes with excellent axisymmetric tensile elongational properties such as fPP or, for added chemical resistance select coextruded HDPE/LLDPE/HDPE geomembranes where the LLDPE core imparts high flexibility
Smooth HDPE geomembranes have a very low friction coefficient which can lead to slope instability problems	Select a textured HDPE to give enhanced interface friction

[a]Note that it is a requirement of most specifications that a composite liner has intimate contact between the geomembrane and underlying CCL or GCL; hence 'waves' and wrinkles are highly undesirable.

📌 *A thicker HDPE geomembrane is generally preferred for a number of reasons, including the ability to weld without damage to the liner, increased tensile strength, greater stress crack resistance and less susceptibility to folding which can lead to stress cracking.*

3.2   STRUCTURE–PROPERTY RELATIONSHIPS

In HDPE the polymer chains can fold and pack efficiently together to form ordered (crystalline) regions while those areas where the chains are more random and less ordered are called the amorphous regions. Crystallinity is therefore a measure of the amount of crystalline regions in the polymer relative to its amorphous content.

Crystallinity positively influences many of the HDPE's properties, some of which are:

- hardness;
- modulus (i.e. stiffness);
- tensile strength;
- melting point;
- chemical resistance.

3.2.1 CHEMICAL RESISTANCE OF HDPE GEOMEMBRANES

Highly crystalline polymers are rigid, high melting and less affected by solvent penetration. Hence HDPE geomembranes which have some 40–60% crystallinity have excellent solvent and chemical resistance. Crystallinity makes a polymer strong, but also lowers its impact resistance and increases its susceptibility to environmental stress cracking; whereas, a decrease in polymer crystallinity (as in the case of LLDPE relative to HDPE), is associated with decreasing mechanical stiffness and chemical resistance. HDPE does not posses any functional groups in its structure which favour potential chemical attack. For polyethylene geomembranes the chemical resistance is linked with the polymer density; hence for chemicals where HDPE is considered to be of 'limited resistance', the alternative of a lower density material (e.g. LLDPE, fPP) is not recommended. The resistances of HDPE geomembranes to various chemicals are shown in Tables 3.3 and 3.4.

HDPE geomembranes often compete with PVC and CSPE geomembranes for similar end applications. Table 3.5 compares the various attributes of these three geomembrane options and shows that HDPE performs well when measured against the various selection criteria.

3.3 COMPARISON OF HDPE GEOMEMBRANES WITH OTHER GEOMEMBRANES

HDPE geomembranes offer an impressive property profile relative to other polymeric geomembranes such as PVC and CSPE (see Table 3.5).

The main advantage of HDPE geomembranes compared to other geomembranes is their better chemical resistance to hydrocarbons and solvents. However, chemical resistance data tables do not consider chemical resistance under the stresses that will be imposed on the liner system. The semi-crystalline nature of HDPE may make it more susceptible to stress cracking when tested under stress in the presence of chemicals.

3.4 DURABILITY AND SURVIVABILITY OF HDPE

The long term durability and survivability of HDPE geomembranes is a function of the environmental stress cracking resistance (ESCR) of the specific grade of HDPE resin used, the efficacy and permanence of stabilization packages (i.e. antioxidants and stabilizers),

Table 3.3 HDPE chemical resistance quick facts

Category of chemical	Chemical resistance
Concentrated acids, alcohols and bases	Excellent resistance (no attack)
Aldehydes, esters, aliphatic and aromatic hydrocarbons, ketones, mineral and vegetable oils	Good resistance (minor attack) (possibility of swelling)
Halogenated hydrocarbons and oxidizing agents	Limited resistance (moderate attack) and suitable for short-term use only

Table 3.4 Chemical resistances of HDPE geomembranes for various classes of chemicals, tested in accordance to the NRW (Northrhine–Westphalia) Guidelines (Darmstadt Report (1998))

| Chemical type | Test liquid | Test liquid run number | Relative change in mean value (%)[c] | | | | | | |
| | | | Weight | | Yield stress | | Yield strain | |
			28 days	90 days	28 days	90 days	28 days	90 days
Benzenes (Otto fuels) and aromatic hydrocarbons	40 vol% trimethylpentane 20 vol% benzene 20 vol% methylbenzene 10 vol% dimethylbenzene 10 vol% methylnaphthalene	1[a]	+1.9	+2.1	−7.2	−6.7	−19.4	+15.5
Fuel oil, diesel fuel, paraffin oil, lubricating oil	30 vol% fuel oil 35 vol% paraffin oil 35 vol% lubricating oil HD 30	2[a]	+1.9	+3.5	−3.3	−6.2	+10.6	+26.8
Amines	Dimethylamine	3[a]	+0.1	+0.1	+1.0	+1.0	+1.7	0
Alcohols	30 vol% methanol 30 vol% isopropanol 40 vol% ethandiol	4[a]	0	0	+0.8	+3.1	−2.0	+1.6
All hydrocarbons as well as used motors and gear oils	60 vol% toluene 30 vol% xylene 10 vol% methyl naphthalene	5[a]	+1.3	+1.4	−4.9	−4.6	+16.0	+20.0
Benzene and mixtures containing benzene	30 vol% benzene 30 vol.% toluene 30 vol% xylene 10 vol% methyl naphthalene	6[a]	+1.2	+1.4	−5.7	−3.6	−11.9	+17.1
All esters and ketones	50 vol% ethyl acetate 50 vol.% methyl isobutylketone	7[a]	+0.9	+0.9	−3.3	−1.6	+11.4	+14.6

(continued overleaf)

Table 3.4 *(continued)*

Chemical type	Test liquid	Test liquid run number	Relative change in mean value (%)[c]					
			Weight		Yield stress		Yield strain	
			28 days	90 days	28 days	90 days	28 days	90 days
Aromatic esters and ketones	50 vol% salolmethylester 50 vol% acetophenone	8[a]	+1.3	+1.5	−3.3	−1.6	+6.4	+8.9
Aliphatic aldehydes	35–40% commercial formaldehyde solution	9[a]	+0.1	+0.1	+0.8	+4.1	−0.2	−4.9
Organic mineral acids	50 vol% acetic acid 50 vol% propionic acid	10[a]	+0.5	+0.6	−1.2	+1.0	+1.1	+4.9
Inorganic mineral acids	50 vol% sulfuric acid 50 vol% nitric acid	11[a]	+0.6	+0.6	+0.3	−9.8	−0.3	−13.0
Inorganic alkalis	60% caustic soda	12[a]	0	0	−0.5	+4.1	−0.3	−0.8
Inorganic salt liquids	Saturated aqueous solution NaCl/Na$_2$SO$_4$ liquid (1:1)	13[a]	0	0	+1.0	+3.6	−0.6	−4.1
Leachates	Mixtures according to NRW guidelines	14[b]	0	0	−0.1	+3.1	+1.7	+3.3

[a]Highly concentrated liquids. After exposure to these liquids, the geomembrane complies with the NRW guidelines if the relative change in weight is ≤15 %, yield stress is ≤25% and yield strain is ≤25%.
[b]Aqueous solutions (leachates). After exposure to these liquids, the geomembrane complies with the NRW guidelines if the relative change in weight is ≤10%, yield stress is ≤20% and yield strain is ≤20%.
[c]Note that the original membrane properties (i.e. prior to liquid exposure) after conditioning for 7 days in a standard atmosphere (DIN 50 014–23/50-2) were as follows: yield stress – range of 18.8–19.8 MPa, mean value 19.4 MPa; yield strain – range of 10.8–13.1%, mean value 12.3%.

Table 3.5 Comparison of HDPE geomembrane properties with those of PVC and CSPE

Property	HDPE	PVC	CSPE
Tensile strength	✔✔✔✔	✔✔✔	✔✔
Puncture resistance	✔✔	✔✔✔	✔✔
Chemical resistance	✔✔✔✔	✔✔ (attacked by polar solvents)	✔✔✔
UV resistance	✔✔✔	✔	✔✔✔✔
Heat resistance	✔✔✔✔	✔	✔✔✔✔
Microbial resistance	✔✔✔	✔ (needs biocides)	✔✔
Ease of placement	✔	✔✔✔✔	✔✔
Cold weather problems	✔✔	✔ (brittle)	✔✔
Installed cost	Moderate	Low	High

the magnitude of stresses generated in the geomembrane during installation and operation as well as the stress relaxation rate. The lifetime of the HDPE geomembrane depends on the above factors both individually and in concert.

There are three basic categories of desirable properties for survivability of geomembranes:

- properties required for installation (e.g. conformability, weldability);
- properties required for backfilled geomembranes (e.g. puncture resistance);
- properties required for exposed geomembranes (e.g. UV resistance, oxidative resistance).

There are a range of different commercially available HDPE resins, each with differing mechanical properties and long-term durability. The variations are a function of the HDPE copolymer used during polymerization, coextruded structures used during liner manufacturing and overall additive packages employed which are often proprietary to the resin and geomembrane manufacturer respectively.

The stress cracking resistance of HDPE is a function of the providence of the resin used and thus varies widely from one geomembrane manufacturer to another. Most geomembrane resins now have excellent ESCR properties; however, stress cracking of HDPE liners can still proceed by other mechanisms such as:

- re-crystallization;
- oxidative embrittlement;
- stress rupture.

3.5 SELECTION OF QUALITY HDPE GEOMEMBRANES

The minimum specifications for a quality HDPE geomembrane product are contained in GRI Test Method GM-13 Standard Specification for 'Test Properties, Testing Frequency and Recommended Warranty for High Density Polyethylene (HDPE) Smooth and Textured Geomembranes'. This specification was developed by the Geosynthetic Research Institute (GRI), with the co-operation of HDPE geomembrane manufacturers. This specification sets forth a set of minimum, physical, mechanical and chemical properties that must be met, or exceeded by the geomembrane being manufactured (see Table 3.7). This standard specification is intended to ensure good quality and performance of HDPE geomembranes in general applications. If the HDPE GML product meets the requirements of the GRI GM-13 specification then it is regarded as a best practice product.

Black HDPE geomembranes should be manufactured with the critical constituents shown in Table 3.6.

The tear strength values for various thickness HDPE geomembranes are shown in Figure 3.6 and the puncture resistance values are shown in Figure 3.7.

Note that selection of quality HDPE liners is dependent on the proper choice of HDPE resin, proper formulation and the method of liner manufacture.

3.5.1 HDPE RESIN SELECTION

A major issue with selecting an HDPE geomembrane is that each supplier can choose from a number of different HDPE resins available on the market and therefore the end performance properties (especially stress crack resistance and oxidative resistance) can vary widely from one HDPE geomembrane to the next. Geomembrane quality therefore begins with base resin selection. It is important to select or specify high-grade polyethylene resins that have been specially formulated to meet the specific, unique demands encountered by geomembranes.

HDPE resin manufacturers have tried to develop HDPE resins that are more resistant to stress cracking, chemical attack, oxidation, as well as being more cost effective. This allows downstream HDPE geomembrane manufacturers to differentiate their products. It should also be noted that 'HDPE' is a misnomer in the industry because the density of the HDPE geomembrane resins ranges from 0.934 to 0.938 g/cm^3 which is actually in the medium-density range (MDPE). Of course, geomembrane durability increases with increasing resin density; however increased resin density also results in increased stress cracking. In fact, the major reason for using MDPE instead of HDPE for geomembranes is increased stress crack resistance.

HDPE Resins for Geomembrane Manufacture

There are a range of commercially available polyethylene resins for geomembrane applications (see Table 3.8).

Table 3.6 Typical formulation of an HDPE geomembrane

Ingredients	Critical aspects	Specifications to meet
97.5% polyolefin polymer	Must have the right stress crack resistance	NCTL >300 h
>2% carbon black	Must be of the right particle size	Particle size ~20 nm
At least 0.5% anti-oxidants and HALS stabilizers	Must have high effectiveness and permanence	HP-OIT >400 min

Table 3.7 Physical properties of smooth sheet HDPE geomembranes as per the GRI GM-13 Specification

Property	Test method	1.00 mm values	1.5 mm values	2.0 mm values	2.5 mm values
Sheet thickness	ASTM D 5199	nom. – 10%	nom. – 10%	nom. – 10%	nom. – 10%
Specific gravity or density	ASTM D 792 or ASTM D 1505	>0.940 g/cm³	>0.940 g/cm³	>0.940 g/cm³	>0.940 g/cm³
Melt index	ASTM D 1238, at 190/2.16	<1.0 g/10 min	<1.0 g/10 min	<1.0 g/10 min	<1.0 g/10 min
Carbon black content	ASTM D 1603 (3)	2.0–3.0%	2.0–3.0%	2.0–3.0%	2.0–3.0%
Carbon black dispersion[a]	ASTM D 5596	Note 4	Note 4	Note 4	Note 4
Tensile strength at Yield (1)	ASTM D, 6693 Type IV	>15 kN/m	>22 kN/m	>29 kN/m	>37 kN/m
Tensile Strength at break (1)	ASTM D, 6693 Type IV	>27 kN/m	>40 kN/m	>53 kN/m	>67 kN/m
Elongation at yield (1)	ASTM D, 6693 Type IV	>12%	>12%	>12%	>12%
Elongation at break (1)	ASTM D, 6693 Type IV	>700%	>700%	>700%	>700%
Tear resistance	ASTM D 1004	>125 N	>187 N	>249 N	>311 N
Puncture resistance	ASTM D 4833	>320 N	>480 N	>640 N	>800 N
Dimensional stability	ASTM D 1204, 100 °C for 1 h	± 2%	± 2%	± 2%	± 2%
Oxidative induction time (min, average) (5)	ASTM D 3895	>100 min	>100 min	>100 min	>100 min
Standard OIT or high pressure OIT	ASTM D 5885	>400 min	>400 min	>400 min	>400 min
Oven Ageing @ 85 °C (5), (6)	ASTM D 5721				
(a) Standard OIT (min, average), % retained after 90 days	ASTM D 3895	55%	55%	55%	55%
(b) HP-OIT (min, average), % retained after 90 days	ASTM D 885	80%	80%	80%	80%

(continued overleaf)

Table 3.7 *(continued)*

Property	Test method	1.00 mm values	1.5 mm values	2.0 mm values	2.5 mm values
Stress crack resistance (2) Notched constant tensile load stress crack resistance (single point test at 30% of yield stress)	ASTM D 5397 (SP-NCTL)	>300 h	>300 h	>300 h	>300 h
UV resistance (7)					
(a) Standard OIT (min, average) or	ASTM D 3895	N.R. (8) 50%	N.R. (8) 50%	N.R. (8) 50%	N.R. (8) 50%
(b) HP-OIT (min, average), % retained after 1600 h (9)	ASTM D 5885				

[a]Note 1: Dispersion only applies to near spherical agglomerates. 9 of 10 views shall be Category 1 or 2, No more than 1 view from Category 3.

1. Machine direction (MD) and transverse direction (TD) average values should be on the basis of 5 specimens each direction. A rate of strain of 51 mm/min in both machine and transverse directions.
   Yield elongation is calculated using a gauge length of 33 mm.
   Break elongation calculated using a gauge length of 50 mm

2. The yield stress used to calculate the applied load for the SP-NCTL test should be manufacturer's mean value via MQC testing.

3. Other methods such as D-4218 (muffle furnace) or microwave methods are acceptable if an appropriate correlation to D-1603 (tube furnace) can be established.

4. Carbon black dispersion (only near spherical agglomerates) for 10 different views:
   9 in Categories 1 or 2 and 1 in Category 3.

5. The manufacturer should ideally provide results of both the OIT and HPOIT test methods.

6. It is also recommended to evaluate samples at 30 and 60 days to compare with the 90 day response.

7. The condition of the test should be an 20 h UV cycle at 75 °C followed by a 4 h condensation at 60 °C.

8. Not recommended since the high temperature of the standard OIT test produces an unrealistic result for some of the antioxidants in the UV exposed samples.

9. UV resistance is based on percent retained value regardless of the original HP-OIT value.

10. The density of the uncoloured base resin shall be greater than 0.940 g/cm^3.

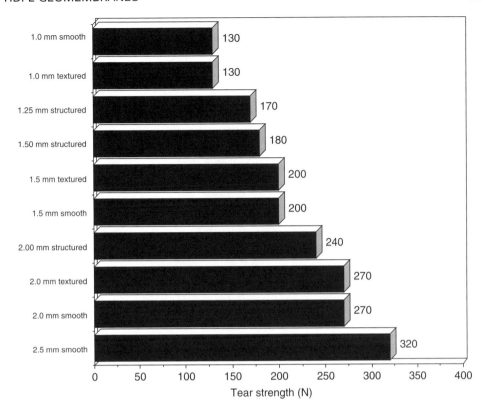

Figure 3.6 Tear strength for AGRU HDPE geomembranes (ASTM D1004)

High-density PE grades used for the extrusion of geomembranes generally have a high MW, high melt strength and broad molecular weight distribution. Many are copolymers with hexene to lower the density and thereby improve the ESCR.

Two of the most widely used HDPE geomembrane resin grades are Marlex K306 and K307 HDPE resins from Chevron Phillips. Marlex K306 and K307 resins have very different molecular weights and are tailored for different geomembrane production methods. For instance, K306 is for round dies (blown film method) and K307 for flat dies (cast extrusion method). An earlier related resin TR-400 is an MDPE used for pipe and membrane production. It comes with a bare minimum antioxidant package of 500 ppm or about 45 min of S-OIT. TR-400 was the 1980s version of geomembrane resin; in the 1990s it was renamed TR 400G and then in 2001 the K306/307 fully formulated resins were introduced. Therefore, the TR-400 (medium density, linear hexene copolymer pipe grade) is the predecessor resin to K306.

The main difference between TR 400G and K306/307 is the HALS stabilizer package. The fully formulated K307 resin has a typical HP-OIT value of 823 min. So the geomembrane-grade HDPE is clearly well stabilized with HALS. The only reason liner manufacturers may still use TR-400 over K307 is if it is available at a lower pricing.

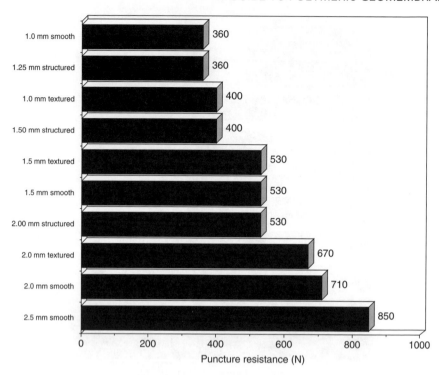

Figure 3.7   Puncture resistance of AGRU HDPE geomembranes (ASTM D4833)

Table 3.8   Common commercially available HDPE (MDPE) geomembrane resins

Manufacturer	Grade	MFI	HLMI (21 kg)	Density (g/cm^3)
Atofina	Finacene M3302 (metallocene)	0.15–0.30 (2.16 kg)	–	0.935
Borealis	Borstar FB 4370 G	2.04 (5 kg)	–	0.937
Borealis	Borstar FB1350	0.70 (5 kg)	15	0.935
BP-Solvay	Fortiflex G36-10-150	–	10	0.937
Chevron-Phillips	Marlex K306	0.11 (2.16 kg)	12	0.937
Chevron-Phillips	Marlex K307	–	21	0.937
Dow	50045E	0.50 (5 kg)	–	0.950 (black)
Dow	2342 M (flat die)	0.85 (2.16 kg) 2.6 (5 kg)	–	0.932 (resin) 0.944 (sheet)
Mobil	HAD601	0.55 (5 kg)		0.933
Petromont	S7000	0.13 (2.16 kg)	16	0.937
Petromont	S7001	0.20 (2.16 kg)	20	0.938
Petromont	S7002	0.23 (2.16 kg)	24	0.937
Repsol (Spain)	Alcudia 3802-N	0.80 (5 kg)	18	0.949 (black)
Sabic	0132HS00	0.8 (5 kg)		0.932
Solvay (US)	Fortiflex G36-10-150	–	10	0.937
Total Petrochemical	3802B	0.9 (5 kg)	18	0.948 (black)
Total Petrochemical	mPE*3509	0.9 (2.16 kg)	25	0.935

*mPE=metallocene-based polyethylene

Table 3.9 HDPE geomembrane properties made from Chevron Phillips resins[a]

Property (Test method)	Typical values	Comments
Density (ASTM D1505)	0.937 g/cm^3 (resin)	
	0.947 g/cm^3 (sheet)	Minimum 0.940 g/cm^3 (sheet)
Melt index (ASTM D1238)	0.11 g/10 min (resin)	Maximum 1.0 g/10 min (resin)
2.16 kg and 190 °C	0.15 g/10 min (sheet)	
High load melt index rate	12.0 (resin)	–
Carbon black content (ASTM D1603)	2.4% (sheet)	Minimum >2%
DSC melting point	126 °C	–

[a]The resins are natural (that is, without carbon black added).

It is important to note that K307 is a natural (unpigmented) fully formulated resin and therefore carbon black needs to be added.[1] The masterbatch to be used with Marlex K307 resin generally does not contain HALS and should not since it is usually only carbon black and phenolic antioxidants required for process stabilization (i.e. to prevent oxidation during extrusion). Table 3.9 shows the various resin and sheet properties for HDPE geomembrane made with a Chevron Phillips resins.

The Chevron Phillips HDPE geomembrane resin grades are widely used by a number of high volume HDPE geomembrane liner manufacturers (notably GSE (Rayong, Thailand) and Hui-Kwang Chemical Company (Tainan, Taiwan)).

Petromont HDPE resins are produced in Canada and were used by Solmax for the manufacture of HDPE geomembranes. For instance, Solmax 480T-2000 is a commonly used HDPE GML which was made from Petromont S7000. The properties of Petromont geomembrane grade resins are listed in Table 3.10.

The lower MI(21 kg) resins have higher MW and these are tailored for blown film (i.e. round die) processing. The higher MI(21 kg) resins have lower MW and these are tailored for cast extrusion (flat bed) processing.

Dow have introduced a new bimodal HDPE resin for geomembranes, 50045E (0.950 g/cc, 0.50 HLMI), that has high molecular weight and high melt strength for thick blown films. It contains carbon black for UV stability.

Table 3.10 Petromont geomembrane grades

Petromont resin grade	MI(21 kg)	Density (g/cc)	Comments
Petromont 7000	16	0.9375	Blown film GML production
Petromont 7001	20	0.9385	Cast extrusion (flat bed) GML production
Petromont 7002	24	0.9375	Cast extrusion (flat bed) GML production

---

[1] Marlex M248 is a black masterbatch from Chevron Phillips Chemical Company containing carbon black (N550) and antioxidants in an LLDPE carrier designed to provide sheet oxidative induction time (OIT) values of over 100 min when combined with HDPE GML resins such as Marlex K306/307, HHM TR-305 or HHM TR-400G. The Marlex M248 masterbatch is normally added to the base geomembrane resin at a 9:1 letdown ratio to achieve a 2.5% carbon black concentration in the sheet.

Finacene M3302 from Fina is a metallocene HDPE (MI_2 range for this polymer is 0.15 to 0.30 g/min) and density in the range 0.9315 to 0.9350 g/cm^3. The copolymer resin contains about 2000 ppm of Irganox 1010, 1200 ppm of Tinuvin 622, approximately 200 ppm of zinc oxide, about 600 ppm of Viton™ Z100 and some 150 ppm of CARBOWAX™ PEG 300 (US Patent, 2006).

Alcudia 3802-N HDPE from Repsol (Spain) is used by Sottrafa Geomembranes (Spain) for their Alvatech 5002 HDPE GML. Alcudia 3802-N has a MI(5) of 0.80 and a MI(21) of 18 g/10 min.

Borstar FB 4370 G is a geomembrane resin used by Agru for their HDPE geomembranes

Resin Substitution

Due to difficulties with obtaining their regular resins, some geomembrane liner (GML) manufactures may on occasion resort to using so-called 'equivalent' resins. Whilst such resins may be equivalent in terms of MFI and density they can have other variations, for instance, widely different resistance to stress cracking.

There have been well-publicized cases of HDPE manufacturers who used a substitute HDPE resin and suffered numerous failures since the HDPE GML had an extremely low stress cracking resistance. In this case, the additional heat of welding was sufficient to initiate stress cracks that propagated through the sheet resulting in extensive cracking and leaking of the installed product.

 Ensure the SCR values of any new HDPE resin are checked and validated.

3.5.2   ADDITIVE FORMULATION SELECTION

It is necessary to select the right combination of additives to protect the geomembrane, such as premium carbon black and antioxidant additives and stabilizers to ensure long life even in exposed conditions. Formulation related measurements such as OIT and HP-OIT values for different HDPE geomembranes vary widely, demonstrating that large differences exist in durability between geomembranes from different manufacturers (Table 3.11).

Table 3.11   HDPE resins used by major geomembrane manufacturers

Manufacturer (Country)	HDPE resin raw material used
Agru (Austria)	Borstar FB 4370 G
Daelim (Korea)[a]	Marlex TR-418BL
GSE (Thailand)	Marlex K307 (cast), Marlex K306 (blown)
Hui-Kwang (Taiwan)	Marlex K306 (blown)
Solmax (Canada)	Petromont 7000 (up to 2007)
Sotrafa (Spain)	Alcudia 3802-N HDPE

[a]Daelim (Korea) manufacture HDPE geomembrane under the 'Geosheet' trademark in thicknesses from 1 mm to 3 mm in thickness and a maximum width of 6.5 m. It is produced only in flat-die extrusion system using TR-418BL compounded HDPE resin.

The GRI specification GM-13 is a generic HDPE geomembrane specification. Whilst GM-13 is an excellent general-purpose specification it is inadequate in some respects. In the case of long-term durability, the accelerated testing sections which cover QUV exposure and heat aging (oven aging) do not replicate immersion of the geomembrane in process liquids, nor do they consider chemical resistance testing (which are application-specific).

Given the aggressive and exposed service environment of some HDPE geomembrane applications, incubation tests or 'pot testing' in specific chemicals or mixtures followed by assessment of mass change, swelling and secondary effects such as change in OIT and change in SCR therefore need to be also considered. Furthermore, chemical resistance should also be considered under strain.

HDPE resin used to manufacture the HDPE geomembranes should have a minimum density of 0.934 g/cc and the geomembrane should contain approximately 2.5–3.0% carbon black, antioxidants, hindered amine light stabilizers (HALS) and no other additives, fillers or extenders. The geomembrane should be manufactured from virgin resin, with no more than 3% regrind material. If regrind material is used, it must be of the same formulation as the virgin resin. No post-consumer resin of any type is allowed in the formulation.

The formulated coloured resin shall have properties equivalent to ASTM D-1248, Type III, Category 4 or 5, and Grade P34. The resin shall contain not less than 97% of the base polymer, and not less than 2% carbon black as defined in ASTM D-1248, Class C, to impart maximum weather resistance. The dispersion shall be even and uniform such that the variation in carbon black content is no greater than ±5% from the average carbon black content.

3.5.3 THE GEOMEMBRANE MANUFACTURING METHOD

HDPE geomembranes are manufactured by extrusion methods that involve melting and mixing of polymer resin in an extruder. The thickness of geomembrane sheets typically varies between 0.75 to 3.0 mm with widths varying from 1.8 to a maximum width of 9.5 m.

HDPE liner products can be made using either blown film method or by the cast extrusion method. Table 3.12 summarizes some of the interchangeable terms used to describe these processing techniques.

Table 3.12 Terms used to describe the two manufacturing techniques for HDPE geomembranes

Manufacturing technique	Synonymous terminology/usage
Blown film	Round die
	Round bubble
	Annular die
Cast extrusion	Flat bed
	Slot die
	Flat die

The type of manufacturing method used for the production of the HDPE geomembrane material should be documented in the project specifications, as the technique used confers some performance differences in the end product. For example, flat-die geomembrane manufacturing delivers greater thickness and gauge control than round die blown film. Blown film on the other hand provides certain advantages of polymer orientation not attained in flat-die produced material. For instance, the vertical bubble provides opportunity for biaxial orientation of the film to give it tear resistance that would not be achievable on cast-film lines.

Due to the weight of the vertical column of film, the blown film process tends to introduce a degree of balanced orientation to the liner. Molecular orientation can improve the mechanical properties of the liner. Therefore if the same resin is employed, the blown-film method can generate a higher performance geomembrane than flat-die manufactured liner.

Note however that different HDPE resins are required for the different processing methods. Generally the one resin manufacturer will have different HDPE resins tailored specifically for the different extrusion processes. See, for instance, Table 3.14 which shows that Chevron Phillips have four main HDPE grades for geomembranes.

The lower MI (i.e. based on 21 kg load) resins have higher molecular weight (MW) and these are tailored for blown-film (i.e. round die) processing. The higher MI resins have lower MW and these are tailored for cast extrusion (flat bed) processing. It is obviously important to ensure that the wrong grade of resin is not selected for a particular manufacturing process.

3.5.4 STANDARD CRITERIA FOR HDPE GEOMEMBRANE LINERS

The standard performance criteria for HDPE geomembrane can be summarized as follows:

1. The geomembrane must be made of high-density polyethylene (HDPE) and not contain any recycled polymer.
2. The oxidative induction time of the geomembrane must exceed (a) 100 min, as determined by ASTM D-3895-95 (Test Method for Oxidative-Induction Time of Polyolefins by Differential Scanning Calorimetry) or (b) 400 min, as determined by ASTM D-5885-95 (American Society for Testing and Materials Standard Test Method for Oxidative-Induction Time of Polyolefin Geosynthetics by High-Pressure Differential Scanning Calorimetry).
3. The oxidative induction time of the geomembrane after oven ageing at 85 °C for 90 days, as described in ASTM D-5721-95 (Standard Practice for Air-Oven Ageing of Polyolefin Geomembranes), must exceed (a) 55% of the value for the original geomembrane, as determined by ASTM D-3895-95 (Standard Test Method for Oxidative-Induction Time of Polyolefins by Differential Scanning Calorimetry), or (b) 80% of the value for the original geomembrane, as determined by ASTM D-5885-95 (Standard Test Method for Oxidative-Induction Time of Polyolefin Geosynthetics by High-Pressure Differential Scanning Calorimetry).

3.5.5 ACCEPTABLE HP-OIT RESULTS

The OIT and HP-OIT values for different HDPE geomembranes vary widely, demonstrating that large differences exist in durability between geomembranes from different manufacturers.

HP-OIT measurements have been used effectively to monitor depletion of HALS stabilizers (Li and Hsuan, 2004). HDPE with a specific antioxidant package consisting of hindered amine light stabilizers was studied using HP-OIT. Oxidative resistance was based on percent retained value regardless of the original HP-OIT value. Conclusions from this work are:

- 'for hindered amines the high-pressure OIT is the appropriate test' (Hsuan and Guan, 1997);
- '150 °C is the maximum effective temperature for HALS';
- '150 °C is the highest temperature one should consider (as a HP-OIT test temperature) to avoid degradation of the HALS'.

Several investigators have found that the combination of hindered phenolic antioxidants and hindered phosphites do not give HP-OIT values higher than about 370 min (Thomas and Ancelet, 1993; Hsuan and Guan, 1997). Therefore HP-OIT values greater than 400 min indicate the HDPE is stabilized with hindered amines stabilizers. Whilst the GM-13 specification lists the minimum acceptable HP-OIT value as >400 min this should only be considered the bare minimum and HALS levels that give HP-OIT values of greater than 700 min should be specified for critical long-term exposed installations (see Table 3.13).

A recently published figure for HP-OIT (ASTM D5885) of a GSE HDPE liner is 660 min (Rowe and Rimal, 2008).

Note that OIT results on geomembranes with modern stabilization packages need careful interpretation since some antioxidants/stabilizers may not give long OIT values but nevertheless may provide good service life.

Table 3.13 Typical HP-OIT values for common HDPE GML materials

HDPE GML material	HP-OIT value at 150 °C (min)
Polytex GEO HDPE (ex Chile)	1091
Solmax HDPE (ex Canada)	926
GSE HDPE (ex Thailand)	750–850
HDPE (ex Germany)	280
Huikwang Huitex™ HD (ex Taiwan)	1142

Table 3.14   Marlex™ geomembrane grades from Chevron-Phillips (USA)

Marlex™ resin grade	MI(21 kg)	Density (g/cc)	Comments
K306	12.0	0.937	Blown film GML production
TR-400G	12.0	0.937	Predecessor to K306
K307	21.0	0.937	Cast extrusion (flat bed) GML production
TR-305	21.0	0.937	Predecessor to K307

3.5.6 ACCEPTABLE SCR PERFORMANCE

The GM-13 specification value for an SCR of >300 min should be regarded as a minimum acceptable value only.

For example, Chevron-Phillips have four main HDPE grades for geomembranes, as shown in Table 3.14.

These HDPE resins have outstanding ESCR values with test times of >900 h in the SP-NCTL stress cracking test (while the GRI GM-13 specification only calls for SCR values of greater than 300 h). Therefore the Chevron-Phillips HDPE liner grades have some 300% better stress crack resistance than the minimum specification value.

For critical long-term installations target SCR value of >600 h could therefore be specified, hence providing a higher safety factor.

3.5.7 FACTORY SAMPLING AND TESTING

Factory sampling and testing of raw materials shall be in conformance with ASTM D-4354 (Standard Practice for Sampling of Geosynthetics for Testing). The vendor provides standard factory testing reports for raw materials including chemical composition (proportions of base polymer and additives), molecular weight (of virgin base polymer), melt index, specific gravity, low temperature brittleness and oxidative induction time.

The tests on the HDPE geomembrane resin shown in Table 3.15 are required once per batch of resin.

The tests on the HDPE geomembrane liner shown in Table 3.16 are required for every roll of geomembrane.

Table 3.15   HDPE resin testing frequency

Resin property	Test method	Frequency
Density	ASTM D792	Once per batch of resin
Melt flow rate	ASTM D1238, 190 °C/2.16 and 21.6 kg	Once per batch of resin
Low temperature brittleness	ASTM D746 Procedure B	Once per batch of resin
Oxidative induction time	ASTM D3895	Once per batch of resin
High pressure oxidative induction time	ASTM D5885	Once per batch of resin

Table 3.16 HDPE liner manufacturing testing frequency

HDPE GML property	Test method	Frequency
Thickness	ASTM D5199	Every roll
Tensile strength at yield[a]	ASTM D6693	Every roll
Tensile strength at break[a]	ASTM D6693	Every roll
Elongation at yield[a]	ASTM D6693	Every roll
Elongation at break[a]	ASTM D6693	Every roll
Tear resistance[a]	ASTM D1004	Every roll
Puncture resistance	ASTM D4833	Every roll
Specific gravity	ASTM D1505 or D792	Every 10th roll
Carbon black content	ASTM D1603 or D4218	Every 10th roll
Carbon black dispersion	ASTM D5596	Every 10th roll
Dimensional stability	ASTM D1204, 100 °C for 1 h	Every 20th roll
Stress crack resistance (SP-NCTL)	ASTM D5397	Every 20th roll

[a]Tensile tests (ASTM D 6693) and tear resistance (ASTM D 1004) are performed at the frequency reported above in both the machine and transverse directions.

For smooth GML, tensile properties (ASTM D 6693), tear resistance (ASTM D 1004) and puncture resistance (ASTM D4833) tests are to be performed on at least one test specimen from each roll (in each direction, where applicable) and the moving average as well as the minimum value from each five consecutive rolls are to be reported.

The finished geomembrane should be free from surface blemishes, scratches and other defects (additive agglomerates, visually discernible rework such as gels and unmelts, etc.).

The HDPE geomembrane manufacturer should identify all rolls with the manufacturer's name, product code, thickness, roll number, roll dimensions and date of manufacture. The cores on which the geomembrane is wound should be at least 150 mm in diameter (OD) and should be sufficiently strong that the roll can be lifted by a forklift stinger or with slings without excessively deflecting or structurally buckling the roll.

The stacking of rolls at the manufacturing facility should not cause buckling of the cores or flattening of the rolls. Rolls are stacked in such a way that the upper roll is cradled into the valley of the two underlying rolls. Normally the maximum stacking limit is five rolls high. The geomembrane must be protected against puncturing and load-induced damage at all times, including during storage and transportation.

3.6 COMMON FAILURE MODES OF HDPE GEOMEMBRANES

HDPE geomembranes which are widely used as landfill liners and as liners for mine tailings and solution dams are unlikely to fail due to chemical degradation by contact with MSW leachates or mine liquors since HDPE has such excellent broad chemical resistance. HDPE is more likely to fail due to other factors first.

For instance, the predominant cause of failures in HDPE geomembrane is due to man-induced damage during construction, such as bulldozer damage, stone punctures and puncturing by depth measuring stakes. If HDPE geomembranes are not damaged during the installation or construction phase then the next most likely cause of premature failure

Figure 3.8   Photograph of brittle failure in a 1 mm HDPE geomembrane

is by environmental stress cracking at folds or areas of elevated stress (e.g. seams, angular stone protrusions, etc.) (see Figure 3.8) or by oxidation which leads to embrittlement.

The creation of wrinkles is detrimental to HDPE liners especially if they are compressed into sharp crease marks when hydrostatically loaded. It is noteworthy that waves and wrinkles in HDPE liners do not lend themselves to being smoothed out by top loading. This is due to the practically incompressible (Poisson ratio: 0.49)[2] nature of HDPE. Hence compressive forces acting on long, high waves transforms them to smaller but very steep waves, causing high, localized flexural strain in the geomembrane (BAM, 1999; Soong and Koerner, 1999).

When using HDPE in critical mining applications it should be noted that if leachates or mine liquors contain high levels of aromatic solvents (e.g. kerosene), oxidizing acids (e.g. sulfuric acid) or chlorinated solvents, then swelling and chemical attack of HDPE can occur. Although HDPE is recognized for its broad resistance to chemicals, exposure to specific chemicals can also lead to environmental stress cracking of HDPE geomembranes. The propensity of HDPE geomembranes to chemical attack varies depending on the type of HDPE resin used to manufacture the HDPE geomembrane. The susceptibility of the liner to long-term stresses will be a function of the SCR of the specific resin used, and such resistances presently vary by a factor of about 10–500.

---

[2] The Poisson ratio is a measure of the tendency of a polymer to expand laterally when compressed uniaxially – for instance when a cube of material is compressed in one direction, it tends to expand in the other two directions.

3.7 MULTILAYER HDPE GEOMEMBRANES

Three-layer extrusion capability provides substantial benefits to the geomembrane manufacturer, such as the ability to tailor the composition of each layer to achieve specific
performance benefits such as the use of HDPE skins on a more flexible LLDPE core
to give both improved flexibility/conformability whilst still having the good chemical
resistance of HDPE.

The layer composition could be as follows, for example:

- 10% HDPE (top layer);
- 80% LLDPE (middle layer);
- 10% HDPE (bottom layer).

For textured HDPE a blowing agent can be added to one or both of the outer layers to
give either singly or doubly textured HDPE sheets to provide improved friction angles.

Other variations that are used are:

- Top layer may be white-surfaced to reduce heat build-up and therefore extend the
  geomembrane lifetime and reduce desiccation (i.e. drying out) of the underlying clay.
- Bottom layer could have high level of electrically conductive carbon for spark testing.
- Top layers can be heavily UV stabilized for long-term UV exposure applications.

Co-extruded HDPE can also be produced with a top layer made with an additive
package specifically formulated for extreme environments such as heap leach pads in
contact with concentrated sulfuric acid. The top layer of coextruded geomembrane can
be heavily fortified with expensive antioxidants and stabilizers.

A point that is little discussed is that three-layer coextrusion capability also provides
the geomembrane manufacturer the option of introducing regrind or recyclate into the
geomembrane. A lower quality/off specification/recycled material can be used in the
middle layer accounting for 75% of the structure; these materials can be purchased for
two-thirds the cost of virgin GM resins.

Most conventional HDPE geomembrane liners are in fact HDPE/HDPE/HDPE and
appear monolithic in nature (i.e. the individual plies are so well bonded that no discernable
interface exists). The reason for this is that modern geomembrane production lines all have
three-layer capability but often only standard HDPE geomembrane resins run in all three
layers. In some circumstances there can be 'delamination' of HDPE geomembrane layers
and this is a phenomena termed 'separation in plane'.

3.7.1 WHITE-SURFACED HDPE GEOMEMBRANES

The black colour of most geomembranes is not favoured by some authorities and facility
owners due to difficulties in identifying potential defects during maintenance inspections.
White on black geomembranes on the other hand readily show damage to the top white
surface. Black geomembranes also have a tendency to heat up significantly in the sun due
to the 'black body' effect. White-surfaced HDPE geomembranes have a white-pigmented
light-reflective surface layer approximately 0.13 mm thick on top of a regular black HDPE
geomembrane.

HDPE with a white surface layer reflects sunlight and results in the temperature of the liner being about 20–35 °C cooler than black geomembranes. As a result thermal expansion and hence wrinkling and wave formation during installation are minimized. This allows installers to achieve intimate contact between the liner and the subgrade. This is especially important in the case of a clay subgrade since if the geomembrane and clay are not in complete intimate contact, then a hole in the liner or in a faulty weld, will not be effectively sealed by the adjacent clay. This will allow water to pool between the liner and the clay. Moreover, as large folds and wrinkles in the liner can be compressed flat into crease lines, the stresses generated by folding may subsequently promote stress cracking. Another major benefit of white liners is that the thin white layer facilitates easy detection of any geomembrane damage. Since the bulk of a white liner is black, any surface damage will cause the underlying black HDPE to show through and the colour contrast will be readily apparent.

Advantages of white coextruded geomembranes include:

- less wrinkles caused by thermal expansion;
- lower risk of damage to liner which can result from the interaction of wrinkles and earthmoving equipment;
- improves damage detection due to contrast caused by exposing black underlying layer;
- minimizes radiant heat buildup and reduces worker fatigue;
- improves visibility in low light applications.

A black liner can reach temperatures of 72 °C or more, whereas a white layered geomembrane reduces heat buildup on the liner by as much as 50% (see Figure 3.9). The white surface increases the reflection of radiant energy and thus reduces thermally induced wrinkles by 50% or more (GSE, 2008a).

Depending upon the temperature differences during the day, an HDPE geomembrane may exhibit thermal expansion of 1% or more. The installer must allow an extra length of geomembrane during hot installations to account for subsequent geomembrane contraction as the liner cools. A geomembrane that is installed without sufficient slack may experience bridging at the toe of the slopes, at tank corners, at sumps and along ditches. Experienced installers will allow for wrinkles in the liner to avoid overtensioning the liner when it cools. Unfortunately subsequent earthmoving contractors may only have had infrequent experience with geomembranes. A wrinkled liner surface can be distorted by bulldozers pushing soil cover. Industry experts attribute most liner leakage to post-installation damage which occurs after the geomembrane installer has completed the specified work. White geomembranes can be installed with fewer, smaller wrinkles because the liner will exhibit less expansion and contraction.

Another benefit of lower liner temperature is decreased moisture evaporation from underlying soil layers. High liner temperatures can cause moisture in the underlying soil layer to condense on the underneath side of the geomembrane. This evaporation can lead to dessicated clay layers, and to condensation buildup and migration under the liner with a potential for liquid build-up and saturation at the base of slopes.

As discussed above the light-reflective surface layer (see Figures 3.10 and 3.11) also acts as a damage detection layer. Should any damage occur the black primary layer of the geomembrane will be exposed, thus making visual inspection more reliable. The improved

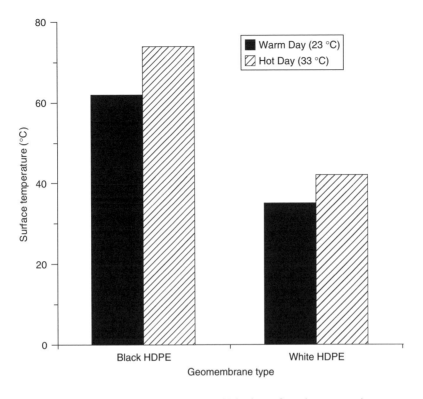

Figure 3.9 Surface temperature of white- and black-surfaced geomembranes

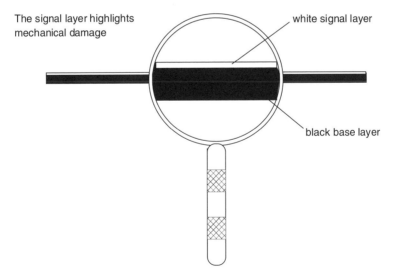

Figure 3.10 Schematic showing the white signal layer on a black geomembrane which serves to highlight geomembrane damage during installation

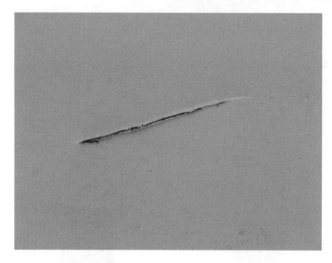

Figure 3.11   Photograph of installation damage to a white/black HDPE geomembrane. Note the thin top layer of white polyethylene (known as the 'signal' layer) readily highlights damage by providing contrast enhancement

visual inspection capability is especially important for exposed applications such as tank liners and secondary containments which do not have the protective benefit of cover soil.

The white surface is part of the extrusion process and not a laminated layer; it is molecularly bonded and cannot be separated from the black portion. White/black HDPE is thus a coextruded material making the white surface an integral part of the membrane which cannot delaminate or detach from the black base sheet.

White geomembranes also have reduced movement of panel overlaps and minimal 'fishmouths' along the seams. Additionally, where black welding rod is used to seam patches, it is much easier to determine if all required destructive seam tests have been performed because each location is clearly visible (GSE, 2008a).

The white layer is formulated to be highly resistant to UV degradation and can be used on exposed applications. While black geomembrane will have a longer life than white, accelerated laboratory tests indicate an estimated lifetime of 18–50 years for the white surface in exposed conditions (GSE, 2008a).

3.7.2   CONDUCTIVE HDPE GEOMEMBRANES

In conductive HDPE the underside of a liner has a coextruded layer containing a higher proportion of carbon black which renders it the electrically conductive layer. This permits the use of electrical testing for pinholes in the geomembrane. This is performed by applying an electric potential between a brass brush that is dragged over the surface of the liner and the conductive coating on the underside of the liner (see Figure 3.12).

At a pinhole or puncture there is a spark discharge (which is both visible and audible) since the current carried by the high carbon layer on the underside of the liner makes contact with the brass brush. This detection method allows holes and breaches in the liner to be located and be repaired before any cover or liquid is placed on the liner.

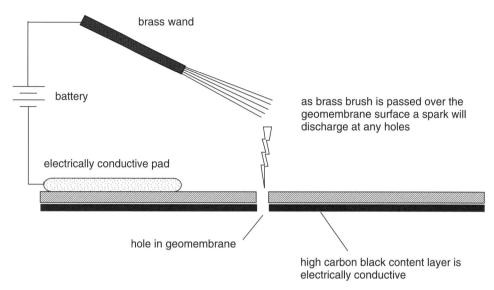

Figure 3.12   Schematic showing spark testing of a conductive geomembrane that incorporates an electrically conductive layer containing high-purity carbon black, approximately 0.08–0.13 mm thick, coextruded as the geomembrane's underside and this enable spark testing over the entire installed liner surface. Spark testing is performed by charging the conductive layer through electrical induction using a conductive contact pad attached to a battery. An operator then sweeps the conductive brass wand just above the geomembrane's nonconductive top surface. Any holes or penetration in the geomembrane will automatically transmit a visible spark from the charged undersurface as the wand passes over and set off an audible alarm

Conductive HDPE incorporates an electrically conductive layer containing high-purity carbon black approximately 0.08–0.13 mm thick, coextruded as the liner's underside. Due to surface effects caused by the conductive layer the tensile properties of conductive HDPE often only achieve the minimum average values. Conductive HDPE generally has an overall carbon black percentage above 3% (as opposed to 2.0–2.5% for regular HDPE liners) due to the high carbon black loadings required for a conductive layer.

A high level of carbon black can be incorporated into one layer of the HDPE geomembranes to give an electrically conductive layer. This enables the electrical detection of pinholes in a HDPE geomembranes. This is done by applying an electric potential between the conductive layer on the underside of the liner and a brass brush which is swept over the surface of the HDPE liner. Directly over a pinhole, a spark discharge can be recorded (as well as seen and heard) since the upper HDPE layer provides no electrical insulating effect. Thus allows pinhole leaks and thin areas to be detected before any liquid/soil/waste is actually placed on the liner, thus making repair much easier.

Conductive HDPE geomembranes also increase the effectiveness of 'water-lance' leak surveys since these normally rely on the subgrade soil to act as the negatively charged layer. However in some cases the geomembrane is wrinkled and not in intimate contact with the subgrade; this makes the water lance-survey technique problematic. The highly carbon filled coextruded geomembrane with the conductive layer overcomes this problem since the liner and the conductive layer are fully bonded.

In order to assure a continuous conductive surface during surface testing, the surface must be smooth. Therefore only single-sided conductive textured geomembranes are available. In order to conduct a spark test, the smooth side must be down and the textured side up. For this reason there are no conductive double-sided textured geomembranes.

A small amount of dirt or water on the liner surface generally has no effect on the spark test. However, when the spark test electrode comes in contact with a conductor on top of the liner (such as water or dirt) with a sufficient surface area, that conductor can behave in the same manner as the spark test grounding pad and discharge through the sheet. As a result, there will be a false indication that there is a hole. In that case, simply remove the water or dirt and retest the area.

It is important to note that unlike traditional electric leak surveys, spark testing of conductive geomembranes does not require water flooding with its added expense and time-consuming delays.

For spark testing large areas an all-terrain vehicle (ATV) can be used to pull a spark testing unit mounted on a trailer. The unit, which exerts less pressure on the liner than normal foot traffic includes a power supply which charges a sliding neoprene contact pad and a 1.8 m wide brass brush. It also contains voltage and sensitivity controls used to minimize any interference or noise created by moisture or humidity. As the ATV moves forward at 3 to 5 km/h covering 4000 to 12 000 m^2/day, the brush sweeps over the geomembrane surface to detect penetrating damage (GSE, 2008b).

3.7.3   HDPE WITH INTERNAL ALUMINIUM BARRIER LAYER (HDPE–Al)

HDPE geomembranes are commercially available with a layer of aluminium sandwiched in the membrane. The aluminium barrier liner (e.g. Agrusafe™ from Agru) can be used as a liner for collection basins for containing media that may endanger ground water – particularly hydrocarbons which may diffuse through conventional HDPE. The Agru HDPE-Al geomembrane is manufactured by laminating a 0.15 mm aluminium layer between two HDPE layers. HDPE-Al liners prevent diffusion of hydrocarbons (HCs), chlorinated hydrocarbons (CHCs) and chlorofluorocarbons (CFCs). HDPE-Al liners are ideal as GMLs for spill containment for petrochemicals or for hazardous wastes.

HDPE-Al geomembranes provides a total diffusion barrier for hydrocarbons and chlorofluorocarbons.

3.8   FLUORINATED HDPE (F-HDPE)

Fluorinated HDPE (F-HDPE) is a modified version of regular HDPE where the surface has been partially fluorinated (see Figure 3.13). The surface fluorination of HDPE geomembranes is performed by exposing the HDPE sheet to elemental fluorine gas which exchanges with hydrogen along polyethylene chains at the surface of the

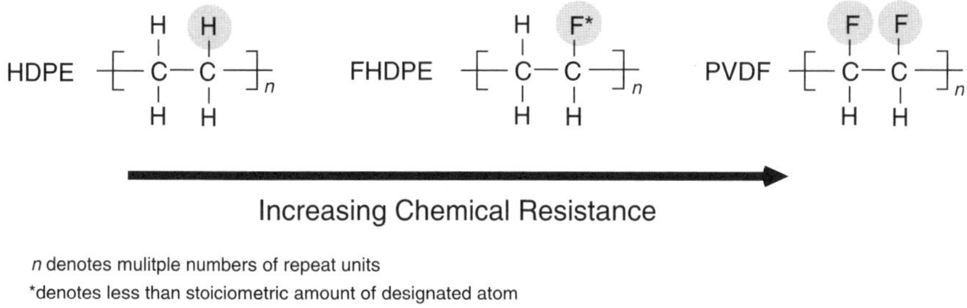

Increasing Chemical Resistance

n denotes mulitple numbers of repeat units
*denotes less than stoiciometric amount of designated atom

Figure 3.13 Effect of increasing fluorine substitution on the chemical resistance (solvent resistance) of HDPE geomembranes

polymer. A fluorination surface treatment on HDPE geomembranes allows them to be used in hydrocarbon liquid environments that would otherwise be absorbed into the surface layers and eventually permeate through the HDPE. Aromatic solvents such as benzene and toluene are particularly good at permeating into HDPE liners. Due to their similar chemical structure, chemicals such as petrol and kerosene can diffuse right though a HDPE liner and condense on the opposing surface giving an 'apparent leakage' rate.

Thus F-HDPE is a minimal cost approach to upgrade the resistance of HDPE geomembranes to hydrocarbon liquids and gases by the fluorinated layer functioning as a barrier to hydrocarbon liquid absorption and subsequent diffusion. The barrier properties of HDPE and LLDPE geomembranes to aggressive landfill leachates and potential problems due to its seepage into groundwater can be significantly improved by this simple surface treatment.

The lifetimes of HDPE and LLDPE geomembranes are essentially dependant on the time it takes for the antioxidant/stabilizer system to be depleted. The principal factors affecting the loss of antioxidants due to their extraction at the geomembrane surface are: the diffusivity of the antioxidants in the HDPE and, swelling of the geomembrane which enhances the molecular mobility of the chemicals. When the geomembrane is in contact with liquid hydrocarbons for an extended period of time, swelling can occur due to sorption and the protective additives can be extracted. One means of improving the effectiveness of a geomembrane with regard to the permeation of organic contaminants is to treat the material in order to reduce its affinity to these chemicals by fluorinating the surface of the geomembrane.

The technique of fluorinating HDPE consists of applying elemental fluorine, which substitutes the hydrogen along the polymer chains at the surface of a polyolefin substrate. The reactive gas mixture that the HDPE geomembrane is exposed to is a combination of fluorine (F_2), nitrogen (N_2) and oxygen (O_2). This combination of gases actually increases the surface energy of the HDPE (making it more wettable by polar liquids).

The resulting fluorinated layer is thus a molecular layer of both carbon–fluorine (C–F) bonds and carbon–oxygen (C–O) bonds. F-HDPE has improved resistance to hydrocarbons (the rate of diffusion of hydrocarbons into the sheet drops by 1.5–4.5 times), and thus the rate of antioxidant loss is significantly decreased and hence the predicted lifetime of the geomembrane is extended (Sangam and Rowe, 2005).

A study by Rimal *et al*. compared the change in oxidative induction time (OIT), tensile properties and crystallinities of both an untreated geomembrane and a fluorinated geomembrane over time. The results indicated that immersion in jet fuel accelerates the antioxidant depletion rate relative to that observed in water or municipal solid waste leachates (Rimal *et al*., 2004). It was found that antioxidant depleted at a much faster rate from untreated geomembrane than fluorinated geomembrane showing that fluorination had beneficial effects. Tensile test results indicated that immersion in jet fuel decreases the yield stress and increases the strain at yield of both the unfluorinated and fluorinated samples. Overall it was found that fluorination of HDPE geomembrane provided significant beneficial effects (Rimal *et al*., 2004).

An added benefit of the fluorination of HDPE is that it also allows the geomembrane to be adhesively bonded to itself and to concrete and metal surfaces using a two-part epoxy adhesive.

✍ *F-HDPE geomembranes have increased resistance to organics, increased oxidation resistance, and increased resistance to stress cracking on the surface. F-HDPE can also be adhesively bonded to concrete and itself (this is not possible with regular HDPE).*

The improvement of the diffusive barrier function of HDPE geomembranes to volatile organic compounds when subjected to surface fluorination has been studied by Sangam *et al*. (2005). Sorption and diffusion tests were performed on both traditional 'untreated' and 'fluorinated' 1.5 mm HDPE geomembranes using dilute aqueous organic contaminants commonly found in municipal solid waste leachates. Surface fluorination resulted in a reduction in both the diffusion and the permeation coefficients by factors ranging between 1.5 and 4.5, depending on the hydrocarbon examined. To achieve the same level of protection as provided by the fluorinated geomembrane underlain by 0.60 m of compacted clay, one would need an additional 0.4–0.9 m of compacted clay in conjunction with a conventional (untreated) geomembrane. The importance of the thickness of the fluorinated layer was also highlighted (Sangam and Rowe, 2005).

Sangam *et al*. (2001) also found that fluorinating HDPE geomembranes decreases the permeability of aromatic solvents (which are found in municipal sold waste landfills) such as benzene, ethyl benzene, toluene and xylene by factors between 2.6 and 4.8 (Sangam and Rowe, 2001).

3.8.1 DEGREE OF FLUORINATION

Fluorinated polyethylene (F-HDPE) can be classified according to the degree of fluorination:

• 1F HDPE – 20–28% of available hydrogen sites on the surface replaced with fluorine.
• 2F HDPE – 28–32% of available hydrogen sites on the surface replaced with fluorine.
• 3F HDPE – 33–40% of available hydrogen sites on the surface replaced with fluorine.

An F-HDPE geomembrane is fluorinated in the rolled state which requires the reactive gas to flow from the outer edge to the centre resulting in a progression of level of fluorination from the centre of the web to the edge. For commercial F-HDPE geomembranes the centre areas are fluorinated to a level slightly higher than the 2F quoted above

Table 3.17 Solvent compatibility with polymers and elastomeric materials[a,b]

Liner materials	Aromatic 150 fluid	Aromatic 200 fluid	Aromatic 200 ND fluid
Thermoplastics			
PVC	U	U	U
HDPE	U	U	U
1F HDPE	PS	PS	PS
2F HDPE	PS	R	R
3F HDPE	PS	R	R
PVDF	R	R	R
Elastomers			
EPDM rubber	U	U	U
Neoprene rubber	U	U	U
Nitrile rubber	U	U	U
Viton™ rubber	PS	PS	PS

Polymers	Exxsol™ D80 fluid	Isopar™ M fluid	Orchex™ 796 fluid
Thermoplastics			
PVC	PS	PS	PS
HDPE	M	PS	PS
1F HDPE	PS	PS	PS
2F HDPE	PS	R	R
3F HDPE	PS	R	R
PVDF	R	R	R
Elastomers			
EPDM rubber	U	U	U
Neoprene	M	PS	PS
Nitrile rubber	PS	R	R
Viton™	PS	PS	PS

[a]Where: R, recommended; PS, probably satisfactory; M, marginal; U, unsatisfactory/not recommended.
[b]F HDPE – 20–28% of available hydrogen sites on the surface replaced with fluorine; 2 F HDPE – 28–32% of available hydrogen sites on the surface replaced with fluorine; 3 F HDPE – 33–40% of available hydrogen sites on the surface replaced with fluorine.

and the outer edges will be toward the upper limit of the 3F range. The chemical resistance of F-HDPE with varying levels of fluorination towards aromatic fluids is shown in Table 3.17.

3.8.2 FLUORINATION PROCESS DETAILS

The preparation of fluorinated geomembranes is performed by post-manufacture surface fluorination. The process of fluorinating polyethylene was developed over 20 years ago by Air Products in the USA to reduce the permeability of HDPE automotive petrol tanks. By using a mixture of fluorine in nitrogen and oxygen, the fluorination causes a chemical modification of the surface of the polyolefin, thereby enhancing the solvent barrier properties of many polyolefins such as HDPE.

The fluorination benefit is based on the formation of a fluorocarbon barrier layer on the polymer surface, by the replacement of hydrogen atoms on the carbon chains with fluorine. The fluorocarbon barrier changes the surface characteristics of the polymer in terms of polarity, cohesive energy density and surface tension. This in turn has a major effect in reducing the wetting, dissolution and diffusion of non-polar solvents relative to the polymer. Thus fluorination is effective in reducing the permeability of non-polar compounds, including solvents through a polymer surface. Post manufacture surface fluorination modifies only those polymer molecules near the surface, and so there is no significant change to the bulk properties of the geomembrane such as tensile strength and elongation. The depth of surface fluorination is related to the time exposed to the fluorination process, which is a function of the diffusion rate of the gaseous reactants of the process.

To enable the gaseous reactants to have full contact with the geomembrane surface, two strategies can be employed:

- Embossed geomembrane – a geomembrane with microspikes on one side of the sheet[3] is used as the spikes keep the adjacent surfaces of the geomembrane in the rolls apart and create a pathway for gas flow through the space to react with the surface of the geomembrane (Gassner *et al.*, 2008).
- A geonet is incorporated within the roll of smooth geomembrane to create a pathway for gas flow through the geonet and react with the surface of the smooth geomembrane. This type of geonet has a thickness of approximately 5 mm.

The fluorine treatment process creates a highly oxidative environment. Particles of materials that are reactive in strong oxidizing environments, such as organic matter or moisture, can potentially cause combustion of the geomembrane during treatment. Significant care is required during the incorporation of the geonet to minimize the presence of foreign particles. If great care is not taken during the treatment process the rolls of geomembrane can be damaged by combustion.

Fluoroseal, a toll fluorination company, have disclosed that it uses three times the normal concentration of fluorine gas for HDPE geomembranes as compared to that for increasing the barrier properties of blow moulded drums and containers. This is partially a function of achieving effective fluorination over the entire surface area of the geomembrane roll. The concentration of fluorine gas for treatment of the smooth geomembrane with a geonet was found to be 1.5 to 2 times greater than the concentration required to treat the embossed geomembrane. This is likely due to the increase in HDPE surface area created by the inclusion of the geonet (Gassner *et al.*, 2008).

The surface energy (i.e. wetting tension) of the polyethylene generally increases from 35 dyne (for the untreated geomembrane) into a range of 55 to 80 dyne.

3.8.3 SUPERFICIAL LAYER

The fluorination of the surface of the geomembrane occurs at a molecular level and the resultant layer of fluorinated polyethylene is very thin (approximately 3 to 5 μm).

---

[3] e.g. a 'structure spike' HDPE geomembrane (AGRU (Austria) GmbH)) which has 5 mm high spikes at 25 mm centres on the surface.

For this reason extra care is required to minimize the risk of damage to the fluorinated surface by scratching, scuffing or abrasion. Construction specifications should highlight the requirement that the geomembrane must be carefully handled to prevent damage and that there is no option for re-fluorination of surface damaged areas, due to the safety risk of working with fluorine gas. For this reason access and traffic across the surface of the geomembrane should be restricted and the deployment of the geomembrane needs to be limited to pulling the geomembrane over the underlying GCL only.

It is also recognized that extrusion welds would be most vulnerable to any hydrocarbon permeation/swelling effects, due to the use of conventional HDPE extrudates. The use of extrusion welds with F-HDPE should therefore be minimized, and no extrusion welds are permitted on the side of the geomembrane where it is important to preserve the fluorinated layer. Any significant damage to the surface of the geomembrane is required to be removed and a new panel installed and joined by fusion welding (Gassner *et al.*, 2008).

Both fusion and extrusion welding methodologies can be used to join F-HDPE geomembrane panels. The welding parameters are similar to those used for a non-fluorinated HDPE geomembrane with the fluorination treatment having no perceived impact upon the weldability of the geomembrane. Strength testing undertaken on seams formed using fusion and extrusion welding methods exhibited weld strengths for both peel and shear failure modes similar to non-fluorinated HDPE geomembrane seams (Gassner *et al.*, 2008).

3.8.4 ASSESSING THE LEVEL OF SUBSTITUTION

The level of substitution (or fluorination) of the HDPE can be assessed in the laboratory using infrared spectroscopy by determining the concentration of the fluorine absorption (C–F peak) (1320 to 930 cm^{-1}) relative to a reference (C–H peak) (1482 to 1442 cm^{-1}). The greater the C–F peak intensity relative to the C–H reference peak, the higher the level of fluorination.

Since the fluorinated geomembrane appears identical (by eye) to the unfluorinated membrane, a quick and reliable field test has been developed to validate that the geomembrane is actually fluorinated.

Due to the fact that the fluorinated process is known to increase the surface energy of the polymer from about 35 dyne to up to 55–80 dyne (i.e. it becomes more wettable) a dyne pen or marker can be conveniently used to test for increased wettability.

For example, a 64 dyne pen offers a quick and reliable way for installers to validate that the HDPE sheet has been fluorinated. Since fluorination doubles the surface energy of HDPE, a '64 dyne pen' marking will bead-up on untreated HDPE but wet-out on fluorinated HDPE.

3.8.5 ADHESIVE BONDING

The surface modification treatment also allows an adhesive bond to be used with the F-HDPE geomembrane instead of traditional heat seaming. The seams between F-HDPE liner panels are overlapped by 150 mm and glued together using a two-part epoxy adhesive recommended by Huntsman. Duct tape and sandbags can be used to apply pressure to the seams during adhesive curing (Bathurst *et al.*, 2006).

3.8.6 APPLICATIONS

Surface-fluorinated HDPE and LLDPE geomembranes have been used to line a prescribed industrial waste landfill designed to contain solid tars, gas-work waste, foundry waste and hydrocarbon contaminated soil from remediated sites.

REFERENCES

BAM Document, 'Certification Guidelines for Plastic Geomembranes Used to Line Landfills and Contaminated Sites', Federal Institute for Materials Research and Testing (1999).

Bathurst, R. J., Rowe, R. K., Zeeb, B. and Reimer, K., A geocomposite barrier for hydrocarbon containment in the Arctic, *International Journal of Geoengineering Case Histories*, **1**(1), 18 (2006).

Blond, E., Bouthot, M. and Mlynarek, J., 'Selection of Protective cushions for geomembrane puncture protection', CTT Group/SAGEOS, St-Hyacinthe, Quebec, Canada.

Darmstadt Report, Staaliche Materialprüfungsanstalt Darmstadt Test Report K980777, 17/04/98 (from NAUE) (1998).

Gassner F. W., Jackson E. D. and Scheirs J., Fluorinated Geomembrane for Hydrocarbon Site, in *Proceedings of EuroGeo4*, Paper Number 232 (2008).

Geosynthetics Specifier's Guide 2008, IFAI, 2008.

'GSE Technical Bulletin on GSE White™ HDPE' [www.gseworld.com] (2008a).

'GSE Technical Bulletin on GSE Conductive™ HDPE' [www.gseworld.com] (2008b).

Hsuan, Y. G. and Guan, Z., Evaluation of the Oxidation Behavior of Polyethylene Geomembranes Using Oxidative Induction Time Testing, in Riga, A. T. and G. H. Patterson (Eds), *Oxidative Behaviour of Materials by Thermal Analytical Techniques*, ASTM STP1326, American Society for Testing and Materials (1997).

Li, M. und Hsuan, Y. G., Temperature and pressure effects on the degradation of polypropylene tape yarns – depletion of antioxidants, *Geotextiles and Geomembranes*, **22**, 511–530 (2004).

Peggs, I. D., Stress cracking in HDPE geomembranes: what it is and how to avoid it, in *Proceedings of Geosynthetics Asia '97*, Bangalore, India, pp. 409–416 [http://www. geosynthetica.net/tech_docs/hdpe_sc_paper.asp] (1997).

Rimal, S., Rowe, R. K. and Hansen, S., Durability of Geomembrane Exposed to Jet Fuel A-1, in *Proceedings of the 57th Canadian Geotechnical Conference*, Session 5 D (2004).

Rowe, R. K and Rimal, S., Depletion of Antioxidants from an HDPE Geomembrane in a Composite Liner, *Journal of Geotechnical and Geoenvironmental Engineering*, **134**, 1, (2008).

Sangam, H. P. and Rowe, R. K., Effect of Surface Fluorination on Diffusion through a High Density Polyethylene Geomembrane, *Journal of Geotechnical and Geoenvironmental Engineering*, **131** (6), 694–704 (2005).

Sangam, H. P., Rowe, R. K., Cadwallader, M. W. and Kastelic, J. R., Effects of HDPE Geomembrane Fluorination on the Diffusive Migration of MSW Organic Contaminants, in *Proceedings of the Geosynthetics Conference 2001*, IFAI, Roseville, MN, USA, pp. 163–176 (2001).

Soong, T.-Y. and Koerner, R., Behavior of Waves in High Density Polyethylene Geomembranes: a Laboratory Study, *Geotextiles and Geomembranes*, **178**, 81–104 (1999).

Thomas, R. W. and Ancelet, C. R., The Effect of Temperature, Pressure and Oven Ageing on the High Pressure Oxidative Induction Time of Different Types of Stabilizers, in *Proceedings of the Geosynthetics '93 Conference*, IFAI, St. Paul, MN, USA, pp. 915–924 (1993).

US Patent 7081285, to Fina Technology (2006).

4

Linear Low-Density Polyethylene Geomembranes

4.1 INTRODUCTION

Linear low-density polyethylene (LLDPE) geomembranes were introduced to address the principle shortcoming of HDPE which is its relative lack of flexibility. LLDPE polymers are less crystalline forms of polyethylene and as such are more flexible and less prone to brittle stress cracking. The trade-off however is that the degree of chemical and UV exposure resistance of LLDPE (and VLDPE) are lower than those of HDPE. Although more flexible than HDPE, LLDPE is not as flexible as PVC. Common LLDPE applications include landfill caps, pond and channel liners, tank liners and sewage processing ponds.

Linear low-density polyethylene geomembranes are typically produced by the blown sheet method. LLDPE closely resembles HDPE albeit having much more flexibility, offering better multiaxial stress resistance, higher puncture resistance (in large scale hydrostatic puncture), better elongational properties (both uniaxial and multiaxial) and possessing far superior stress crack resistance. It is lower in density (and therefore crystallinity) than HDPE liners, allowing flexibility and softness without the use of migrating plasticizers or other plasticizing additives.

LLDPE geomembranes are produced with resins that encompass a relatively wide density range. For instance the higher density grades exhibit mechanical properties approaching those of MDPE and HDPE whereas the lower density products show properties closer to those of VLDPE. The specification GRI GM-17 gives typical properties of the full range of LLDPEs. LLDPE geomembranes are approximately 15%–20% crystalline compared with 40–50% crystallinity in the case of HDPE geomembranes. As a consequence LLDPEs do not possess the chemical resistance or tensile strength of HDPE.

The intrinsic flexibility, toughness and good resistance to environmental stress cracking of LLDPE and VLDPE geomembranes make them well suited for:

- Sites with difficult access and site conditions (since they can install large factory-fabricated panels which reduce installation time).
- Sites with either coarse or soft subgrades.

A Guide to Polymeric Geomembranes: A Practical Approach   J. Scheirs
© 2009 John Wiley & Sons, Ltd

- Landfill caps and closures of landfills, leach pads and other installations that require high puncture resistance and where a significant amount of subsidence may occur (i.e. they can more easily accommodate differential settlement and localized strain while maintaining liner integrity). Differential or localized subgrade settlements can also occur in such applications as water reservoirs, artificial lakes and canals.

4.2   ATTRIBUTES OF LLDPE GEOMEMBRANES

Attributes of LLDPE geomembranes include:

- They are more flexible than HDPE (one quarter of the stiffness of HDPE).
- Excellent large scale puncture resistance compared with HDPE.
- Good resistance to stress cracking.
- LLDPE is readily available in potable water grades and is relatively inexpensive.
- LLDPE can be easily and simply welded utilizing standard fusion or extrusion welding equipment and technologies.

4.3   LIMITATIONS OF LLDPE GEOMEMBRANES

Some of the principal shortcomings of LLDPE include:

- LLDPE has only moderate weathering/UV resistance qualities and therefore, must be heavily stabilized or buried/covered.
- LLDPE has only moderate chemical resistance to hydrocarbons and poor dimensional stability.
- LLDPE is susceptible to oxidation which is accelerated by the catalytic effects of multivalent transition metal ions in a chemically activated state. Of these the ferrous/ferric (Fe^{2+}/Fe^{3+}) ion is the most common but copper and manganese ions are effective oxidation catalysts.
- Surface friction properties are poor unless the geomembrane is heavily textured.
- LLDPE geomembranes used in exposed installation may require fencing to prevent damage by animals.
- LLDPE can be problematic to repair since this requires the use of a specialized extrusion gun which requires considerable skill to operate properly.

4.4   MECHANICAL PROPERTIES

The uniaxial tensile and elongational properties of LLDPE are similar to those of HDPE with the % elongation at break for LLDPE being slightly higher (800% vs. 700%).

The tensile strength at break values for various thickness LLDPE geomembranes are shown in Figure 4.1 and the % elongation at break values are shown in Figure 4.2.

Textured LLDPE geomembranes typically have only half the elongation at break values of smooth LLDPE (800% compared to 300–400%). The texturing also tends to reduce the puncture resistance of textured LLDPE relative to smooth LLDPE (see Figure 4.3). Note that the AGRU micro-spike texturing process is different to most other texturing

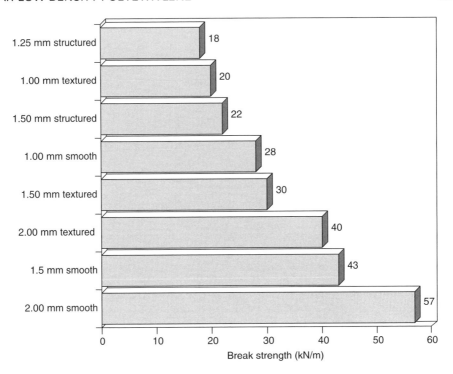

Figure 4.1 Break strength of Agru LLDPE geomembranes (ASTM D-6693)

processes. Nitrogen gas textured geomembranes suffer even greater losses in elongation properties than the textured and structured geomembranes from AGRU.

The index puncture resistance of LLDPE versus HDPE geomembranes is shown in Figure 4.4. LLDPE geomembranes have consistently lower rod puncture resistance (ASTM D-4833) than HDPE geomembranes.

The puncture resistance of LLDPE (as measured by the index puncture test ASTM D-4833) increases fairly linearly with sheet thickness from 200 N for 0.75 mm to over 600 N for 2.5 mm LLDPE (see Figure 4.5). While LLDPE has a somewhat lower puncture resistance in the index puncture test than HDPE geomembranes, LLDPE shows markedly better performance in the large-scale puncture test (ASTM D-5514) using the truncated cone. LLDPE has a critical cone height of 7.5 cm compared with only 1.5 cm for HDPE. In fact the large-scale puncture resistance of LLDPE geomembranes is even better than that of PVC.

The tear strength values for various thickness LLDPE geomembranes are shown in Figure 4.5.

The tear strength of LLDPE is somewhat lower than that of HDPE (220 N vs. 249 N for 2 mm sheet).

Due to the lower chemical resistance of LLDPE vs. HDPE and also the higher flexibility of LLPDE vs. HDPE, triple layer coextrusions (i.e. laminates) of HDPE and LLDPE are produced with LLDPE as the middle highly flexible layer and HDPE as the outer chemically resistant layers. The chemical resistance of various polyolefin geomembranes

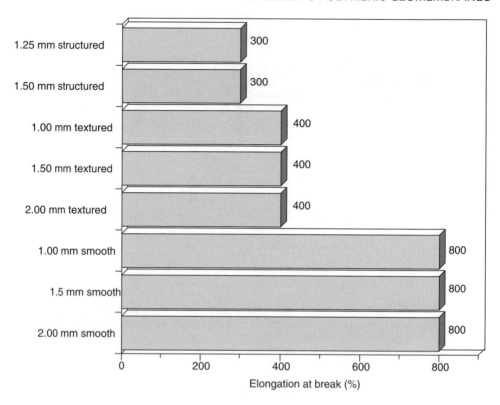

Figure 4.2   % Elongation at break for Agru LLDPE geomembranes (ASTM D-6693)

in terms of absorption of hydrocarbons and chlorohydrocarbons follows the following trend from greatest to least absorption: fPP > VLDPE > LLDPE > HDPE (Aminabhavi and Naik, 1999).

4.5   LLDPE GEOMEMBRANE RESINS

There are a number of LLDPE resins produced for geomembrane applications.

Chevron-Phillips produces a LLDPE geomembrane grade resin, namely CP Marlex K203, which is a high molecular weight hexene copolymer tailored for sheet and geomembrane applications such as landfill covers and caps that require outstanding ESCR. It has a broad fusion range for good weldability. It is customized to give improved output and texturing and can also be used as a coextruded cap layer for HDPE geomembranes.

Chevron-Phillips also offers a range of resins containing mPACT metallocene catalyzed LLDPE (Grade D139, density 0.918 g/cm^3) blended with the Enhanced LLDPE Grade K203 (density 0.922–0.924 g/cm^3) geomembrane resin. The target application for these resins is to produce geomembrane liners that provide an alternative to PVC and PP for landfill caps and other applications where high ductility and elongation is required.

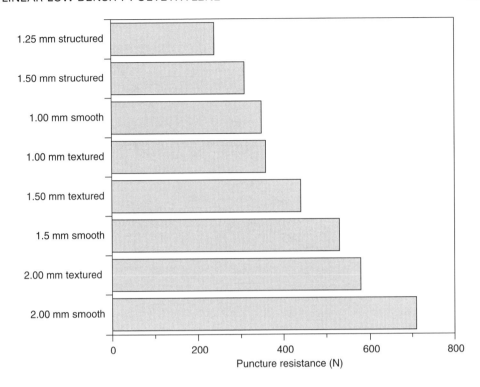

Figure 4.3   Puncture resistance of Agru LLDPE geomembranes (ASTM D-4833)

4.6   SPECIALITY FLEXIBLE POLYETHYLENE GEOMEMBRANES

4.6.1   PROPRIETARY FLEXIBLE POLYETHYLENE GEOMEMBRANES

While there are many generic flexible polyethylene geomembranes there are some with a proprietary formulations such as the Enviro Liner™ from Layfield (Edmonton, Alberta). These have a flexibility similar to conventional LLDPE but are made from a proprietary polymer formulation and are referred to as a 'fabric supported polyolefin geomembrane'. The properties of these geomembranes are shown in Tables 4.1 and 4.2. The Enviro liner™ products are heavily fortified with stabilizers for outstanding heat and UV stability.

Note the mechanical properties for the Enviro Liner 6000 geomembranes closely match those for unreinforced LLDPE as listed in GRI™ GM-17.

4.6.2   TRI-DIRECTIONAL REINFORCED POLYETHYLENE (RPE)

RPE geomembranes typically consist of polyethylene ply layers encapsulating a scrim reinforcement. For example, Dura-Skrim™ polyethylene geomembranes (manufactured by Raven, Sioux Falls, SD) are reinforced with a heavy encapsulated 1300 Denier tri-directional polyester reinforcement. In addition to excellent dimensional stability the

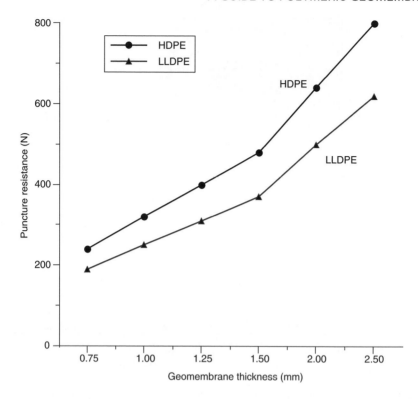

Figure 4.4   Puncture resistance (N) of HDPE and LLDPE geomembranes (ASTM D-4833)

tri-directional reinforcement provides exceptional tear and tensile strength. Dura-Skrim™ J-Series membranes are formulated with thermal and UV stabilizers to assure a long service life. The outer layers consist of a high-strength polyethylene film manufactured using virgin grade resins and stabilizers for UV resistance in exposed applications.

Reinforced polyethylene geomembranes are manufactured by interlaminating a layer of woven polyester scrim between two or more sheets of polyethylene film. The scrim is first bonded into a 'molten polyethylene bonding layer' and then encapsulated between two plies of 'high strength' outer PE film. Such geomembranes are highly tear resistant (see Table 4.3). The scrim reinforcement responds effectively to tears in the surrounding liner matrix and arrests the tear.

4.6.3   REINFORCED POLYETHYLENE (RPE)

Reinforced polyethylene (RPE) (produced by Layfield and others) is used for canal liners, drilling sump liners, soil remediation liners, tailings dam liners and interim landfill caps. RPEs are flexible in extremely low temperatures and have very good chemical resistance; however they have only moderate UV resistance and are suited for short-term exposure applications only.

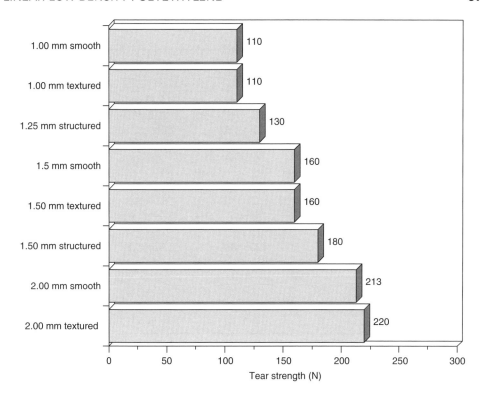

Figure 4.5   Tear strength of Agru LLDPE (ASTM D-1004)

Table 4.1   Properties of Enviro Liner™ unsupported polyolefin geomembranes from Layfield [www.layfieldgroup.com][a]

Properties	Enviro Liner™ 6020	Enviro Liner™ 6030	Enviro Liner™ 6040
Thickness (ASTM D-5199)	0.5 mm	0.75 mm	1.0 mm
Tensile strength at break (ASTM D-638)	13.5 N/mm	20.1 N/mm	26 N/mm
Tensile elongation at break (%) (ASTM D-638)	800%	800%	800%
Tear resistance (ASTM D-1004)	49 N	71 N	98 N
Puncture resistance (ASTM D-4833)	142 N	213 N	271 N
Low-temperature impact resistance (ASTM D-1790)	$-70\,°C$	$-70\,°C$	$-70\,°C$
High-pressure OIT (ASTM D-5885)	>2000 min	>2000 min	>2000 min
	2500 min typical	2500 min typical	2500 min typical

[a]Note also that Enviro Liner 6000 has exceptional UV resistance according to ASTM D-4329 with >90% strength retained after 30 000 h at an irradiance of 0.90 W/m^2 at a wavelength of 310 nm and under the following testing conditions: 10 h UV cycle (at 60 °C) and a 2 h condensation cycle (at 50 °C).

Table 4.2 Properties of Enviro Liner™ polyolefin geomembranes (fabric[a]-supported polyolefin geomembrane) from Layfield[b]

Properties	Enviro Liner™ 8024	Enviro Liner™ 8030	Enviro Liner™ 8036
Thickness (ASTM D-5199)	0.61 mm	0.75 mm	0.91 mm
Coated weight (ASTM D-5261)	540 g/m^2	680 g/m^2	800 g/m^2
Tensile strength at break (ASTM D-751)	890 N/mm	1000 N/mm	1100 N/mm
Tensile elongation at break (ASTM D-751)	30–40%	30–40%	30–40%
Trap tear resistance (ASTM D-4533)	440 N	440 N	440 N
Puncture resistance (ASTM D-4833)	270 N	270 N	270 N
Hydrostatic resistance (ASTM D-751)	1200 kPa	1400 kPa	1700 kPa
Low-temperature flex resistance (ASTM D-2136)	−40 °C	−40 °C	−40 °C
High-pressure OIT (ASTM D-5885)	>2000 min 2500 min typical	>2000 min 2500 min typical	>2000 min 2500 min typical

[a]The Enviro Liner 8000™ material is coated onto a standard 9 × 9, 1000 denier polyester fabric scrim.
[b]Note that presence of the scrim reinforcement increases the tear strength some 4–10 fold from 49–98 N to 440 N.

UV resistance (by the standard ASTM G-53-84) shows that RPE has >80% retained tensile strength after 2000 h of QUV exposure. Note that these UV results are for the black product only and colours other than black may have a lower UV resistance.

RPE can be regarded as a low cost, lightweight polytarp material which is available in thickness from 0.3 to 0.6 mm while the actual coating thickness is only 44–63 μm on each side. The tear strength (by the tongue tear method ASTM D-2261) is between 311 and 400 N.

There are no theoretical limits to the size of liner panel that can be factory fabricated however there are some practical handling weights. Panel weight is important because of the limitations of handling equipment available in the field. If a forklift or other piece of equipment is available for placement and deployment, then a panel of up to 1800 kg can be accommodated. If only manual handling is available then a panel weight of 1400 kg should be regarded as the practical maximum weight.

GRI GM22 'Standard Specification for Test Methods, Required Properties and Testing Frequencies for Scrim Reinforced Polyethylene Geomembranes Used in Exposed Temporary Applications', covers scrim-reinforced polyethylene geomembranes in thicknesses of 0.50 mm, 0.30 mm and 0.20 mm used in exposed temporary applications.

Table 4.3 Properties of Dura-Skrim™ Geomembranes manufactured by Raven Industries (these are four-layer reinforced laminates containing no adhesives). The outer layer consists of a high-strength polyethylene film (loaded with UV stabilizers) reinforced with a 1300 denier encapsulated tri-directional scrim reinforcement [www.ravenind.com][a,b]

Grade	Thickness (mm)	Ply adhesion ASTM D-413	Tensile strength 25 mm width ASTM D-7003	Grab tensile strength ASTM D-7004	Tongue tear strength ASTM D-5884	Trap tear strength ASTM D-4533
J30BB	0.75	9 kgf	50 kgf (MD) 36 kgf (TD)	99 kgf (MD) 95 kgf (TD)	44 kgf (MD) 41 kgf (TD)	66 kgf (MD) 64 kgf (TD)
J36BB	0.9	11 kgf	51 kgf (MD) 39 kgf (TD)	101 kgf (MD) 101 kgf (TD)	47 kgf (MD) 42 kgf (TD)	86 kgf (MD) 78 kgf (TD)
J45BB	1.125	14 kgf	63 kgf (MD) 48 kgf (TD)	117 kgf (MD) 117 kgf (TD)	53 kgf (MD) 54 kgf (TD)	88 kgf (MD) 87 kgf (TD)

[a]Notes:

- To convert kg(f) to Newtons(N) multiply by 9.8.
- All values are typical roll averages.
- Tear strength values are dominated by the scrim properties and so there is only a marginal increase in tear strength with membrane thickness.
- 25 mm strip tensiles are highly dependent on sample direction whereas wide width grab tensiles give a more useful representation of tear strength of the scrim-reinforced laminates since it is less dependent on scrim directional properties and 'scrim pull-out' end effects.
- Tensile elongation (film break) for all thickness is ~750% while the tensile elongation (scrim break) for all thicknesses is in the range 30–36%.

[b]Data from Raven Industries (Sioux Falls, SD, USA).

4.6.4 REINFORCED METALLOCENE POLYETHYLENES

A specialty RPE is made by Hiraoka (Japan) called WTF-1000D. It consists of metallocene polyethylene plies encapsulating a 1100 dtex woven polyester scrim. The membrane is made in a thickness of 0.9 mm and has excellent tear strength 530 N (MD) and 400 N

Table 4.4 Properties of the Hiraoka WTF-1000D membrane liner[a,b]

Property	Value
Scrim type	Woven polyester (1100 dtex)
Coating	Linear low-density polyethylene (metallocene) (0.4 mm each side)
Unit mass	890 g/m²
Thickness	0.9 mm
Tensile strength (N/50 mm)	1850 (MD); 1810 (TD)
Trapezoidal tear (N)	530 (MD); 400 (TD)
Temperature range	−40 to +70 °C

[a]Note: to convent Newtons (N) to kgf divide by 9.8.
[b]Data from Hiraoka (Japan) [www.tarpo-hiraoka.com].

(TD)) and tensile strength (on a 50 mm strip specimen) of 1850 (MD) and 1810 N (TD) (see Table 4.4). The membrane is used for water tank liners and it has high strength, high abrasion resistance, excellent coating adhesion and excellent water head resistance.

4.6.5 POLYTARP LINERS

Polytarp liners consist of polyethylene film coextruded over an HDPE tape reinforcement layer. These polytarp products are not normally utilized in geomembrane applications as they do not have the required mechanical properties of tensile strength, tear strength and puncture resistance. Furthermore, UV stability and delamination are potential problems.

Polytarp liners have been used in channel lining applications. One needs to ensure the seams are parallel with the length of the channel to ensure flowing water does not peel open the seams. Weld strength is low since fusion depth is limited by the trilaminate construction.

4.7 VERY LOW-DENSITY POLYETHYLENE (VLDPE) GEOMEMBRANES

Very low-density polyethylenes (VLDPE) are a form of linear-low density polyethylenes that have increased chain branching which results in lower density and higher flexibility. Very flexible polyethylene (VFPE) is a classification that includes LLDPE and very low-density polyethylene (VLDPE). VLDPE geomembranes were developed to compete directly with PVC geomembranes and liners.

Advantages of LLDPE/VLDPE over PVC geomembranes include:

- good weldability without hydrochloric acid fumes;
- high flexibility without the need for plasticizers;
- maintains its mechanical properties without plasticizer migration;
- generally better chemical resistance;
- 1/3 lower density.

Unimodal VLDPE geomembranes that are based on a vanadium catalyst system, exhibit lower UV resistance than unimodal VLDPEs made with titanium based catalysts.

Some membranes are coextruded structures with three layers comprising HDPE/VFPE/HDPE to optimize both high flexibility and high chemical resistance.

VLDPE is manufactured by Dow but is not available in North America.

REFERENCES

AGRU [www.agruamerica.com].
Aminabhavi, T. M. and Naik, H.G., Chemical compatibility of geomembranes – sorption, diffusion and swelling phenomena, *Journal of Plastic Film and Sheeting*, 15(1), 47–56 (1999).
Chevron-Phillips Marlex LLDPE [www.cpchem.com].
Hiraoka (Japan) [www.tarpo-hiraoka.com].
Layfield Enviro Liner™ Polyolefin Geomembranes [www.layfieldgroup.com].
Raven Dura-Skrim™ Geomembranes [www.ravenind.com].

5

Flexible Polypropylene (fPP) Geomembranes

5.1 INTRODUCTION

Flexible polypropylene (fPP) geomembrane resins are produced by the incorporation of high levels of ethylene propylene rubber (EPR) into the semicrystalline polypropylene (PP) matrix directly in the polymerization reactor. The inclusion of ethylene segments in the polymer backbone breaks up the crystallinity of regular PP and confers elastomeric properties.

fPP resins such as Astryn CA743GA are made by Basell using the Catalloy™ process which produces a reactor blended elastomeric alloy of PP and EP rubber. Unlike PVCs and PVC blends, fPP polymers contain no plasticizers that can migrate out of the membrane and thus are intrinsically flexible. The Catalloy™ process enables a high percentage of ethylene–propylene rubber to be copolymerized with polypropylene in the polymerization reactor. fPP is a very flexible material which bears little resemblance to regular polypropylene sheet which is far more rigid. fPP is even more flexible than VLDPE; however not as flexible as PVC.

🖈 *fPP geomembranes are manufactured from Astryn™ resin which is a reactor alloy of ethylene–propylene rubber particles dispersed on a molecular level in a polypropylene matrix. Rather than discrete EPR domains within the PP matrix the morphological structure of Astryn™ resins is best described as a molecularly blended alloy. Astryn™ CA743GA, for example, is believed to have an EPR content of 65%. The remaining polypropylene phase has a heat of fusion of 13.5 J/g which translates to a crystallinity of approximately 6% (based on a heat of fusion value of 207 J/g for 100% crystalline PP).*

fPP is thus different from other PP-based olefinic polymers in that it is not a physical blend but rather a reactor blend on a molecular level. The fPP polymer structure is therefore completely homogeneous even at very high EPR levels. In addition to its high degree of flexibility, fPP polymers have a broad melting transition, which allows them to be thermally seamed with a wide range of seaming equipment. fPP nevertheless maintains many

A Guide to Polymeric Geomembranes: A Practical Approach   J. Scheirs
© 2009 John Wiley & Sons, Ltd

of the inherent characteristics of polypropylene. fPP resins can be extruded, calendered or blown into geomembrane sheet which exhibits high flexibility (for convenient handling), high toughness and puncture resistance and excellent topographical conformability.

5.2   ATTRIBUTES OF fPP GEOMEMBRANES

fPP polymers possess a range of favourable attributes which makes them well suited as geomembrane liners. For instance, fPP geomembranes offer high resistance in the following areas: weathering/UV, chemical, puncture, tear, abrasion and cold weather. fPP liners are very flexible and have excellent dimensional stability. fPP geomembranes have low crystallinity and are not affected by traditional environmental stress cracking.

fPP geomembranes can be conveniently wedge- or extrusion-welded and can be patched using a heat gun.

fPP is a chemically inert and intrinsically flexible polymer that does not require plasticizers.

The excellent conformability of fPP liners enables extra shear interface points for maximum holding power on side slopes.

fPP geomembranes exhibit excellent multiaxial strain at rupture performance with the strain values being many times higher than that of HDPE geomembranes (see Figure 5.1).

The ability to be factory fabricated and folded for transport to the installation site means a significant reduction in field seaming compared to the more rigid HDPE geomembranes (see Figure 5.2).

Attributes of fPP Geomembranes (Montell, 1998):

- *large scale multiaxial elongation of greater than 150%;*
- *critical cone heights in puncture resistance of greater than 10 cm;*
- *uniaxial elongation of greater than 800%;*
- *wide heat seaming temperature windows;*
- *a low coefficient of thermal expansion (half to one fourth compared to HDPE);*
- *a high level of flexibility enabling factory fabrication of large panels that can be folded and transport for deployment at the installation site.*

5.3   SHORTCOMINGS OF fPP GEOMEMBRANES

Whilst possessing an almost ideal combination of properties for geomembranes and liners, fPP polymers do suffer from some important limitations.

Flexible polypropylene geomembrane liners can be susceptible to oxidative stress cracking particularly along folds and creases.

Polypropylene is susceptible to oxidation which is accelerated by the catalytic effects of multivalent transition metal ions in a chemically activated state. Of these the ferrous/ferric

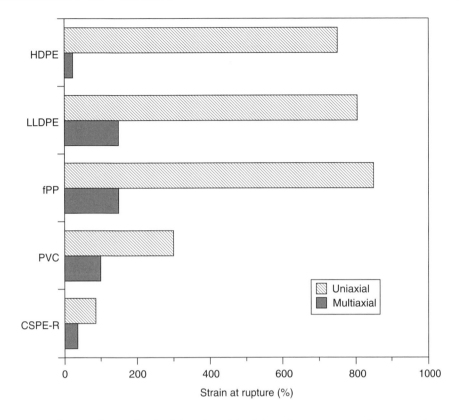

Figure 5.1   Comparison of uniaxial and multiaxial strain at rupture of geomembranes

(Fe^{2+}/Fe^{3+}) redox couple is the most common but copper and manganese ions have also been shown to be damaging.

fPP geomembranes can be affected by halogenated aliphatic hydrocarbons (e.g. trichloroethylene, methylene chloride, chloroform, chlorinated solvents), aromatics (e.g. dichlorobenzene and chlorinated solvents), aliphatic hydrocarbons (butane, pentane, hexane, light esters) and aromatics (benzene, toluene, xylene). Care must be exercised when in long term contact with the following chemicals: organic acids (acetic, stearic); volatile organics (ketones, aldehydes, esters, amides); oils and waxes as well as strong oxidants (e.g. potassium permanganate, potassium dichromate, chlorine, perchloric acid, peroxides).

In particular, fPP swells tremendously in chlorinated hydrocarbons, for instance in trichloroethylene fPP swells 500% compared with 395% and 21% swelling for PVC and HDPE, respectively (Montell, 1998). fPP also exhibits high swelling when immersed in cyclohexane with swelling of 356% compared with 4% and 25% for PVC and HDPE, respectively.

Panel configuration: 8 panels of 20 m x 95 m x 1 mm (thickness)
Panel weight: 1680 kg
756 m of field seams at 3 m/min = 4.2 h welding time

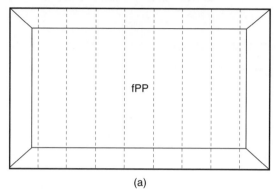

(a)

Panel configuration: 24 panels of 6.8 m x 95 m x 1 mm (thickness)
2268 m of field seams at 2.5 m/min = 15 h welding time

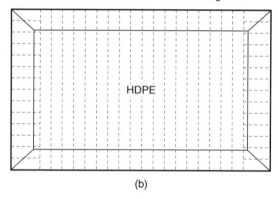

(b)

Figure 5.2 Schematics illustrating the reduction in field seaming possible by using flexible liners such as fPP relative to HDPE

5.4 PERFORMANCE PROPETIES OF fPP GEOMEMBRANES

Flexible polypropylene liners possess a unique combination of mechanical toughness, flexibility and resistance to environmental factors. Their intrinsic flexibility is a result of their molecular structure which has high levels of ethylene propylene rubber incorporated into the otherwise semicrystalline polypropylene matrix. As a result, the crystallinity of fPP polymers is amongst the lowest of all common geomembrane materials and this provides outstanding resistance to environmental stress cracking. However, owing to its low crystallinity the chemical resistance of fPP is less than that for HDPE.

The tear strengths of various thickness fPP geomembranes are shown in Figure 5.3. For applications where additional tensile strength and tear resistance are required, fPP liners are available in textile scrim reinforced versions known as fPP-R (or R-fPP). The tear

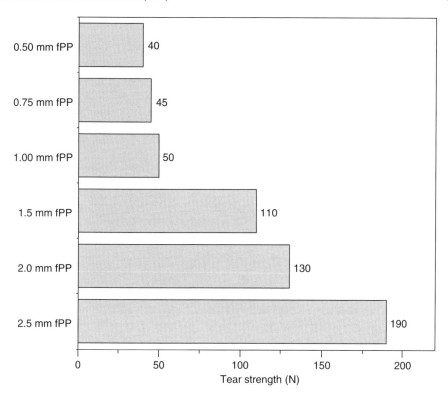

Figure 5.3   Tear strength versus thickness for fPP geomembranes (ASTM D-1004) (data from Geosynthetic Specifier's Guide, IFAI publication, 2009)

strength of scrim-reinforced fPP is some five times higher than the unreinforced material (see Figure 5.4). fPP has better seaming behavior than HDPE, for instance it can easily be seamed by hot air equipment at low ambient temperature (fPP has been successfully seamed even at a temperature of $-9\,^\circ$C in strong wind and snow (Montell, 1998).

fPP geomembranes exhibit outstanding resistance to multiaxial strain ($>120\%$ strain) and are therefore the liners/covers of choice along with LLDPE for sites with difficult conditions such as unstable site subgrade, or with out of plane subsidence such as a landfill cover or cap. fPP geomembranes provide excellent conformability to uneven or unstable subgrades. These geomembranes have stiffness (i.e. modulus) values of around 100 MPa which is some eight times lower than those of HDPE (830 MPa) and about half that of LLDPE and CSPE geomembranes (which are in the range 200–210 MPa).

Thermally generated wrinkles (and resultant stresses) are minimized using fPP liners by virtue of its relatively low coefficient of thermal expansion. This allows large panels in the field to remain dimensionally stable over a wide temperature range (e.g. from cool night temperatures to full exposure in the midday sun).

fPP geomembranes demonstrate outstanding puncture resistance in large-scale hydrostatic puncture testing (ASTM D-5514) where they produce critical cone heights of 10 cm – some ten times greater than those for HDPE and 20–30% greater than those

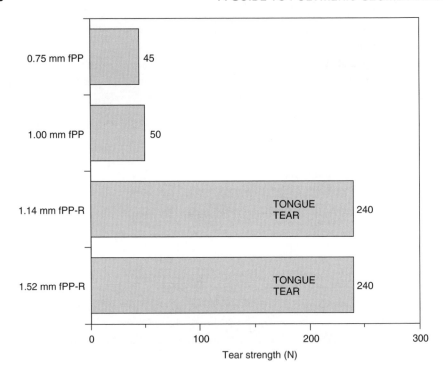

Figure 5.4 Tear strength for unreinforced and reinforced fPP geomembranes (ASTM D-1004 and D-5884 (data from Geosynthetic Specifier's Guide, IFAI publication, 2009)

obtained by other flexible geomembranes such as LLDPE and PVC (see Figure 5.5). Even if fPP geomembranes are punctured, tears do not propagate readily, even when subjected to tensile stresses such as on a side slope. For added tear strength the scrim reinforced fPP (R-fPP) can be used since the scrim acts as a 'ripstop' and arrests tear propagation.

R-fPP is typically manufactured by either calandering (rolling of molten resin with rollers in series) or extrusion operations into plies that are usually combined over a reinforcing polyester scrim fabric during the manufacturing process. Scrims are fully contained within two layers of polypropylene to eliminate 'wicking' (that is, fluid conduction into the liner that could delaminate plies).

Good low temperature liner properties are required if geomembranes are to be deployed in cold climates. fPP membranes can be unfolded and installed at temperatures of $-30\,^{\circ}$C and the material is capable of remaining ductile until $-50\,^{\circ}$C. The brittle temperature of fPP is $-50\,^{\circ}$C as tested by ASTM D-1790.

fPP has superior low-temperature properties to CSPE geomembranes particularly with regard to easy installation in cold weather. fPP also has the advantage of being easy to repair when aged whereas CSPE geomembranes cross-link over time and become more difficult to repair. fPP also has a broad melting range that allows it to be thermally seamed with a wide range of seaming equipment.

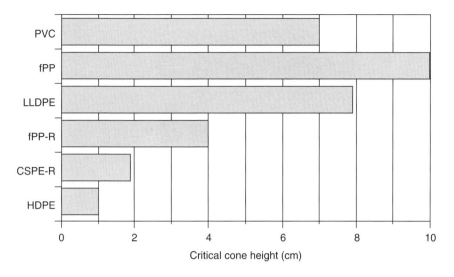

Figure 5.5   Critical cone height for various geomembranes (ASTM D-5514)

The low-temperature impact performance of fPP liners is far superior to that of PVC, with PVC geomembranes experiencing brittle failure at −20 and −40 °C while fPP retains full ductility (see Figure 5.6).

The increased surface friction of fPP geomembranes increases friction angles and thus side slope stability concerns are reduced. When smooth geomembranes are pushed against other geosynthetics or soil, the degree of friction can be measured. Hard and smooth geomembranes such as HDPE possess little friction; however softer materials such as LLDPE, fPP, EPDM, and PVC have greater friction under the application of an applied normal stress.

The performance properties and test values (minimum averages) of non-reinforced and reinforced fPP geomembranes are listed in Table 5.1.

5.4.1   STRESS CRACK RESISTANCE

When fPP is tested in the constant load stress cracking test (ASTM D-5397) using a 5% soap solution, 50 °C solution temperature and a notch depth of 20% of the sheet thickness, it elongates but no samples are observed to fail in a brittle manner. In other words, as shown in Figure 5.7, there is no 'knee' in the curve, indicating that no ductile to brittle transition occurs. The samples will elongate over time but none of the samples show signs of cracks even when viewed under 200× magnification (Montell, 1998). In contrast, some grades of HDPE are notorious for displaying ductile to brittle transition behaviour in surfactant (i.e. soapy) solutions.

Samples of fPP were tested using the 'Notched Constant Load Environmental Stress Crack Resistance Test', with no failures reported in an aggressive soap solution (5% soap

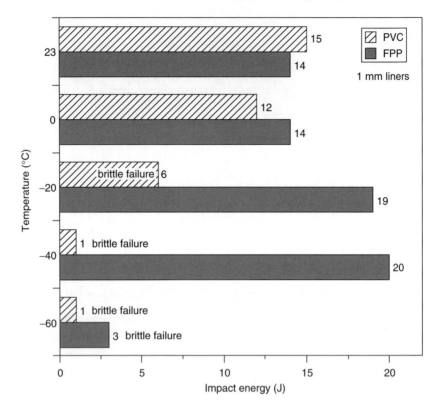

Figure 5.6 Low-temperature impact performance for PVC and fPP geomembranes (from Montell)

solution skimmed from the surface of a 52% black liquor effluent obtained from a paper processing facility).

Since fPP is not susceptible to stress cracking that can affect HDPE it has been used for lining lagoons for holding black liquor at pulp mills.

5.4.2 UV STABILITY

Black fPP geomembranes exhibit outstanding weatherability. Basell, the manufacturer of the base resin for all fPP membranes, have shown that 1 mm fPP geomembranes made with Astryn™ fPP fully formulated resin resist greater than 3×10^6 langleys[1] with only negligible effects on the mechanical properties (Montell, 1998).

EMMAQUA accelerated UV exposure test results can be correlated with real-time weathering for specific geographical regions. For instance, 3×10^6 langleys of sun

---

[1] A langley (Ly) is a unit of measure for radiation power distribution over area and is used to measure solar radiation. 1 langley (Ly) is 41 840.00 J/m^2 = 0.04184 MJ/m^2 while 1 kLy = 1 kcal/cm^2 = 41.84 MJ/m^2 and 1 kLy/year = 1.33 W/m^2.

Table 5.1 Performance properties of non-reinforced and reinforced fPP geomembranes (adapted from the GRI GM-18 specification)

Property[a]	0.75 mm fPP	1.0 mm fPP	0.91 mm fPP-R	1.14 mm fPP-R
Thickness (min. ave.), ASTM D-5199	0.75 mm	1.00 mm	0.91 mm	1.14 mm
Thickness (lowest individual specimen), ASTM D-5199	0.68 mm	0.90 mm	0.82 mm	1.03 mm
Tensile strength (min. ave.) (dumbell specimen), ASTM D-638-IV	11 kN/m	13 kN/m	–	–
Tensile strength (min. ave.) (grab specimen), ASTM D751-A	–	–	890 N	1100 N
Tensile elongation (min. ave.) (dumbell specimen), ASTM D-638-IV	700%	700%	–	–
Tensile elongation (min. ave.) (grab specimen), ASTM D751-A	–	–	22%	22%
Multiaxial elongation, ASTM D-5617	120%	120%	–	–
Tear resistance (non-reinforced), ASTM D-1004	45 N	50 N	–	–
Tear resistance (reinforced), ASTM D-5884	–	–	245 N	245 N
Puncture resistance, ASTM D-4833	110 N	130 N	330 N	380 N
Ply adhesion, ASTM D-6636	–	–	65 N	65 N
Low-temperature flexibility, ASTM D-2136	$-40\,^{\circ}$C	$-40\,^{\circ}$C	$-40\,^{\circ}$C	$-40\,^{\circ}$C

[a]Note that the tensile elongation and the tear resistance (reinforced) are dominated by the scrim properties; thus the results are invariant with thickness.

exposure equates to $125\,520$ MJ/m^2. From annual solar radiant exposure tables it is recorded that Phoenix, Arizona receives about 8000 MJ/m^2 per year, and so 3×10^6 langleys is equivalent to 15.7 years of real-time exposure in Phoenix. Miami, Florida receives 6500 MJ/m^2 per year of solar radiation; hence this equates to about 19.3 years of real-time exposure in Miami.

The tensile strength and elongation of black fPP remain virtually constant even after an accelerated dose of UV equivalent to 15 years in the Arizona desert; alternatively, over 19 years in a hot and moist climate or 24 years in a typical North American city (e.g. Akron, Ohio which receives 5110 MJ/m^2 of UV radiation per year) (Montell, 1998).

Furthermore, a study by Stevens found that an fPP geomembrane had passed exposure to 4000 h of 'xenon weather-o-meter testing' followed by severe bend stress testing (Montell, 1998).

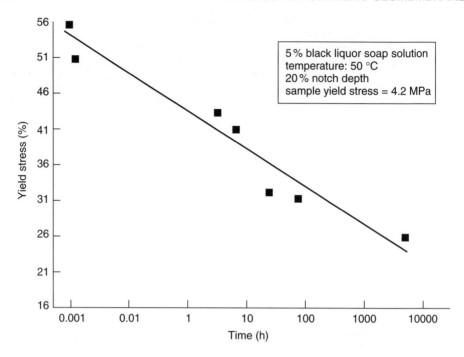

Figure 5.7 Flexible polypropylene (fPP) geomembranes are not susceptible to environmental stress cracking as is evidenced by the absence of a 'knee' in the NCTL plot (as per ASTM D-5397) for specimens in black liquor soap solution

5.4.3   CHEMICAL RESISTANCE OF fPP GEOMEMBRANES

Owing to their lack of crystallinity and strongly amorphous nature, fPP geomembranes have lower chemical resistance than HDPE and are susceptible to attack by specific chemical groups (see Table 5.2).

FPP geomembranes can be adversely affected (e.g. swelled, softened or plasticized) by:

- aliphatic hydrocarbons (e.g. alkanes such, pentane, hexane, cyclohexane);
- halogenated aliphatic hydrocarbons (e.g. trichloroethylene, methylene chloride, chloroform, chlorinated solvents);
- chlorinated aromatics (dichlorobenzene);
- aromatics (e.g. benzene, toluene, xylene) (BTX).

Care must also be exercised when fPP geomembranes are in long-term contact with the following chemicals groups:

- fatty acids (e.g. oleic acid, stearic acid);
- volatile carbonyl-containing organics (e.g. ketones, aldehydes, esters, amides);
- oils and waxes (animal fats and dripping);
- strong oxidants (e.g. potassium permanganate, potassium dichromate, perchloric acid, peroxides, aqueous chlorine solutions).

Table 5.2   fPP Geomembrane chemical resistance quick facts (Montell, 1998)

Category of chemical	Chemical resistance
Inorganic acids (hydrochloric acid, nitric acid, sulfuric acid), organic bases (amines), inorganic bases[a] (sodium hydroxide, ammonium hydroxide), alcohols (methanol, propanol, ethylene glycol), salts (sodium chloride, potassium bromide)	Good resistance (no significant attack)
Organic acids (acetic acid, stearic acid), volatile/semi-volatile organics (ketones, aldehydes, esters, amides, esters), oil and greases (fats, oils and greases), strong oxidizers (chlorine, peroxides, potassium permanganate)	Marginal resistance[b] (minor attack). Possibility of swelling and property reduction. Suitable for short-term use only
Aliphatic halogenated hydrocarbons (trichloroethylene, methylene chloride, chloroform or chlorinated solvents), aromatic halogenated hydrocarbons (dichlorobenzene and other chlorinated solvents), aliphatic hydrocarbons (butane, pentane, hexane, light petroleum ethers), aromatic hydrocarbons (benzene, toluene, xylene)	Poor resistance (moderate to major attack). Not recommended

[a] Although inorganic acids and inorganic bases have little effect on the base polymer these chemicals can decompose the stabilizers and antioxidants respectively that are present in the fPP geomembrane.
[b] fPP is affected to some degree and therefore the final geomembrane should be tested with the actual chemicals under actual or simulated service conditions.

Table 5.3   Chemical resistance of fPP when immersed in specific chemicals (% swelling in 100% concentration of the specific chemical) compared with HDPE and PVC (Montell, 1998)

Chemical	fPP	HDPE	PVC
Methanol	0.44	1.45	−8
Water	0.5	1.6	3
1-propanol	1	1	−4
Acetone	4	4.5	83
Tetralin	92	12.3	60
Iso-octane	137	11.5	1
Ortho xylene	182	20	8
Indolene	198	–	–
Cyclohexane	356	25	4
Trichloroethylene	500	21	395

The chemical resistance of fPP when immersed in specific chemicals (% swelling in 100% concentration of the specific chemical), compared with HDPE and PVC, is shown in Table 5.3.

Note the material cost for fPP-R is about double that of HDPE geomembranes. Suppliers of fPP Geomembranes include Cooley Engineered Membranes, Carlisle and Raven.

A summary of the attributes and benefits of fPP geomembrane liners is shown in Table 5.4.

Table 5.4   Summary of attributes and benefits of fPP geomembrane liners

Attribute	Specific benefits
High resistance to tearing	If fPP geomembranes are punctured, tears do not propagate easily, even when subjected to tensile stresses such as on a side slope. For enhanced tear resistance the scrim reinforced fPP can be utilized since the scrim acts as a 'ripstop' and inhibits tear propagation
Outstanding resistance to multiaxial strain (>120% strain)	Low risk of geomembrane failure under conditions of unstable site subgrade or out of plane subsidence such as a landfill cover or cap
High hydrostatic puncture resistance	Less chance of failure over cracks and protrusion in the subgrade
Low coefficient of thermal expansion (CTE)	Thermally generated wrinkles (and resultant local stresses) are minimized by the low CTE of fPP permitting large panels in the field to remain dimensionally stable over a broad temperature range (e.g. from low night temperatures to full sun exposure at midday)
Excellent low temperature properties	fPP membranes can be deployed, unfolded and installed at temperatures of $-30\,^{\circ}$C with the benefit of the material remaining ductile until $-50\,^{\circ}$C
Outstanding resistance to environmental stress cracking	No stress cracking occurs in strong stress cracking agents such as black liquor from paper pulping operations
High friction angle	The increased surface friction of fPP increases friction angles and thus reduces side slope stability concerns

5.5   APPLICATIONS

Flexible polypropylene is commonly used in floating covers, baffles and primary containment applications. R-fPP (reinforced flexible polypropylene) is used widely in exposed wastewater treatment ponds, as landfill cover linings, in floating covers and baffle curtains, in lagoon linings and as a liner for ponds, lakes and water features.

5.6   fPP FIELD FAILURES

fPP geomembranes and liners have now been used for over ten years in a wide range of applications from mine tailing lagoons to solar ponds. In that time a number of failure mechanisms have emerged and these are discussed below.

One of the first publicized failures was the fPP liner of the El Paso Solar Pond. The US Bureau of Reclamation investigated the long-term properties of this fPP liner in 1989 (Comer, 1998). The performance of fPP geomembranes at two field test sites in which the geomembranes were either covered by soil or partially exposed to direct sunlight was studied. The buried fPP continued to have good mechanical properties after six years service. The exposed section of the fPP geomembrane at the El Paso Solar Pond showed cracking after 26 months. The particular stabilization package used in the exposed fPP was found to be inadequate to prevent UV degradation. Furthermore, the UV degradation was substantiated in a laboratory acceleration test using a 'UV weatherometer'. This illustrates the importance of proper stabilization and the need for UV weatherometer exposure and subsequent testing as a means for screening to evaluate the UV resistance of differently formulated fPP geomembranes.

Since then, other fPP failures have been reported for liners in contact with fats, oils and greases, as well as chlorinated water.

5.6.1 FIELD FAILURE CASE STUDY

A 1.14 mm reinforced polypropylene floating cover over animal fats from a chicken processing plant failed after just over five years. IR analysis showed the cover had undergone oxidation as evidenced by the presence of a significant carbonyl signal at 1725 cm^{-1} (see Table 5.5).

Animal protein is high in sulfur-containing amino acids and these sulfur-based molecules convert to hydrogen sulfide and other sulforous products during anaerobic digestion. The presence of these sulfur compounds leads to a acidic environment in animal fat processing lagoons. This type of acidic environment can attack and destroy (i.e. deactivate) conventional hindered amine stabilizers. That is, exposure of the membrane to acidic gases can cause the HALS stabilization system to become inactive. fPP, due to its amorphous nature, can have some of its additives modified and deactivated by acidic effluents. In particular, the hindered amine stabilizers are basic in nature and these can react with acidic containment effluents, leading to a loss of stabilizer activity. Once the HALS are neutralized by sulfur-derived acids they form a salt and it is not possible to regain any of the initial stabilizing ability. In other words, the stabilizer is 'dead'.

The ease to which acidic gases and sulfur-based acids can permeate the membrane increases with decreasing crystallinity in this order: HDPE < LLDPE < VLDPE < fPP. In addition, animal fats can solvate and plasticize these polyolefins, thus decreasing the resistance to permeation.

For installations which require maximum resistance to chemicals (especially fatty wastes) and installations requiring higher tensile strengths, HDPE is a preferred liner material. fPP has certain advantages for specific applications, such as flexibility and

Table 5.5 Infrared absorption bands of chemical groups detectable in oxidized fPP geomembranes

IR absorption band (cm^{-1})	Chemical group detected	Comments
1725	C=O	Carbonyl group – signature of oxidation
3130	–OH	By-product of oxidation
1100	C–O	By-product of oxidation
1460	–CH$_2$	Reference band (invariant with oxidation)

resistance of multiaxial strains, but its lower density and chemical resistance give it reduced lifetime compared to HDPE in certain applications.

Other limitations of fPP geomembranes that can lead to field failures are:

- Effluent that is rich in fats, oils and greases (FOG) can plasticize fPP (and certain other polymers) and can also extract additives from such liners. HDPE, due to its higher crystallinity, is quite resistant to the extractive effects of FOG-containing effluents. Fats, oils and greases can permeate flexible PP and cause a reduction in mechanical properties. fPP is amorphous (i.e. low crystallinity) and therefore small opportunistic molecules can permeate its structure.
- The polyester reinforcement in fPP is susceptible to acid-catalyzed hydrolysis. Water vapour can permeate fPP and initiate hydrolysis and breakdown of the polyester scrim reinforcement. Sewage is known to be acidic because of sulfurous compounds. The edges of fPP-R geomembranes must have no exposed scrim (except for roll ends) and they should have 10 mm edge encapsulation of the scrim on each side.
- fPP membranes are not susceptible to environmental stress cracking (that is, they are not susceptible to stress cracking in the traditional sense in their as-manufactured condition, to the presence of surfactants and detergents). However fPP is susceptible to oxidative stress cracking in a stressed configuration (where there are sharp folds leading to high surface strains), by various environmental factors such as chlorine, oxidizing acids, oxygenated water and in the presence of redox salts such as those based on transition metal ions.

fPP floating covers and liners particularly in potable water reservoirs have undergone oxidative stress-cracking failures along the tops of folds and at other locations of elevated stress (Figure 5.8). There appears to be synergism between stress and liquid environment,

Figure 5.8   Photograph of brittle cracking in a 0.75 mm fPP tank liner. The cracking was due to *oxidative stress cracking* on a fold in an oxidative environment containing chlorinated water

stress and thermal oxidation and also stress and UV radiation; all of these combinations can accelerate the loss of additives. However, the same fPP material that has undergone cracking in floating cover applications has performed very well in an exposed cap application.

Cracking in reinforced fPP (fPP-R) on the underside of floating covers at the bottom of drainage troughs has been reported by Peggs (2003).

The GRI GM-18 Standard Specification for 'Test Properties, Testing Frequency and Recommended Warrant for Flexible Polypropylene (fPP and fPP-R) Nonreinforced and Reinforced Geomembranes' was suspended as of May 3, 2004. The durability portion of this generic specification on flexible polypropylene geomembranes had come into question insofar as oxidation induction time values are concerned. OIT testing has proven to be an excellent predictor of polyethylene lifetime in that it is a relatively quick and standardized test. It is apparently more difficult for fPP due in part to its availability in colours (i.e. green, tan, blue) in addition to black, which complicate the situation making it difficult to set generic specification limits. The GM-18 specification was reintroduced on March 20, 2009 with 20,000 light hours of QUV exposure (2.3 years) and the use of OIT to access degradation was dropped.

Note new grades of fPP geomembranes have been developed by Firestone and Layfield that are specially stabilized to resist oxidative stress cracking even in strong chlorine solutions (e.g. 80 ppm free chlorine).

5.6.2   fPP TANK LINERS

It is important to note however that fPP tank liners should be installed without excessive wrinkles and that a pleated installation method (e.g. as in the case of a hung liner) may not be acceptable if it results in the formation of sharp creases in the material.

Basell, the manufacturer of the base resin for all fPP membranes, specifically advise the following (Motell, 1998):

> Stress concentrations, like sharp corners, wrinkles, etc. should be avoided, as in combination
> with an exposure to specific chemicals this can lead to premature failure of the liner.

REFERENCES

Comer, A. I., Hsuan, Y. G. and Konrath, L., 'Performance of flexible polypropylene geomembranes in covered and exposed environments'. US Bureau of Reclamation, Denver, CO, USA. In Rowe, R. K. (Ed.), *Proceedings of the 6th International Conference on Geosynthetics*, Atlanta, GA, USA, March 25–29, **1**, p. 359 (1998).

GRI GM-18 Standard Specification for 'Test Properties, Testing Frequency and Recommended Warrant for Flexible Polypropylene', 2004 [http://www.geosynthetic-institute.org/grispecs/gm18.pdf].

Montell, literature entitled 'Astryn Flexible fPP, High Performance Materials for Geomembranes' (1998).

Peggs, I. D., 'Geomembrane Liner Durability: Contributory Factors and the Status Quo', in *Proceedings of the UK IGS*, June 2003 [http://www.geosynthetica.net/tech_docs/IDPigsUKpaper.pdf] (2003).

6

CSPE Geomembranes

6.1 INTRODUCTION

CSPE is a thermally weldable geomembrane based on chlorosulfonated polyethylene (CSPE) synthetic rubber. CSPE geomembranes are available in a number of reinforced versions (CSPE-R). In its initial, manufactured form, CSPE is thermoplastic in nature, which permits seaming by chemical or thermal fusion methods. However, after installation and exposure to the environment, the membrane's individual polymer chains crosslink forming a highly stable, synthetic rubber material. Once cured to its ultimate strength, CSPE displays little change in extensibility and provides high resistance to a broad range of chemicals.

The basic polymer backbone of CSPE is essentially the same as polyethylene and because there are no double bonds, the long polymer chains are relatively impervious to attack from degrading agents such as oxygen, ozone or energy in the form of UV light. The chlorination of polyethylene via the introduction of chlorine atoms along with a controlled number of sulfonyl chloride (i.e. crosslinking sites) groups also introduced as side chains on the polyethylene backbone reduces the ability of the polymer to crystallize and the material becomes rubbery at a chlorine level of $>14\%$ (see Figure 6.1). The sulfonyl chloride groups are larger than the chloride atom; hence they are more efficient at breaking up the crystallinity and also provide chemically active cure sites.

Chlorosulfonated polyethylene geomembranes are generally referred to in the industry as Hypalon™. Hypalon™ is a trademark of DuPont who originally developed CSPE. There are various Hypalon™ elastomer grades which are used in geomembranes, industrial cable insulation, inflatable boats and roofing membranes. Hypalon™ geomembranes are widely used in the geosynthetic industry as liners, caps and floating covers.

6.2 GRADES OF HYPALON™

There are several grades of Hypalon™. The two most commonly used for liners are Hypalon™ 45 and Hypalon™ 48 which differ in their chlorine content:

- Hypalon™ 45: 24% chlorine (better mechanical properties);
- Hypalon™ 48: 43% chlorine (better chemical resistance).

A Guide to Polymeric Geomembranes: A Practical Approach   J. Scheirs
© 2009 John Wiley & Sons, Ltd

CSPE

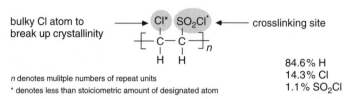

n denotes mulitple numbers of repeat units
* denotes less than stoiciometric amount of designated atom

84.6% H
14.3% Cl
1.1% SO₂Cl

Figure 6.1   Chemical structure of the repeat unit of CSPE geomembranes. The bulky chlorine atoms disrupt the crystallinity of the polyethylene while the sulfonyl chloride group provides an active site for crosslinking

Hypalon™ 45 gives higher modulus, tear strength, and hardness as compared with the Hypalon™ 48 types. Hypalon™ 45 is the grade most commonly used in Hypalon™ geomembranes. Hypalon™ 45 however has a lower chlorine content than does Hypalon™ 48; therefore, it is more flexible at low temperatures, but has reduced hydrocarbon and chemical resistance. It also provides better tear strength and slightly better heat resistance than the general purpose types.

Hypalon™ 45 can be compounded to give very good mechanical properties in uncured stocks. It is frequently used uncured in sheeting applications because of its strength and ease of seaming. Hypalon™ 45 was introduced in the early 1960s and has been widely used in the geosynthetic industry as liners, caps and floating covers since that time.

Hypalon™ 48 provides better hydrocarbon resistance, but poorer low temperature properties than Hypalon™ 45.

6.3   ATTRIBUTES OF CSPE GEOMEMBRANES

CSPE geomembranes exhibit an impressive array of properties that makes them well suited as geomembranes in exposed and aggressive applications. The main attributes of CSPE geomembranes are as follows:

- They have a proven long-term track record in exposed applications. They are dimensionally stable and offer excellent weathering and UV resistance. Pro-rated manufacturer warrantees are often available up to 20 years and longer.
- CSPE geomembranes (in their original as manufactured forms) can be both thermally and solvent weldable.
- They are available in potable water grades (Approved by the NSF).
- CSPE geomembranes are not subject to environmental stress cracking and display high resistance to a wide range of organic and inorganic chemicals.
- CSPE geomembranes retain flexibility in freeze/thaw conditions.
- CSPE geomembranes cure over time thereby increasing the liner's tensile strength, chemical resistance and UV resistance. Thus they have excellent UV resistance and outstanding weatherability in exposed applications (Schoenbeck, 1984).
- Due to the excellent long-term weathering resistance of CSPE membranes they are often selected for water and waste water applications and also are used to manufacture

floating covers. These membranes have been used as lining materials for more than 30 years now and have shown excellent durability as both a liner and a cover material.

No plasticizers are used in the formulation of CSPE geomembranes; hence there are no issues relating to plasticizers that can leach out and cause embrittlement or cracking of plasticized liners. Its inherent low coefficient of thermal expansion and contraction provides excellent dimensional stability and lay-flat characteristics. CSPE geomembranes are also highly flexible for easy conformance to earth contours.

CSPE has a very low thermal expansion coefficient compared to that of rigid membranes such as HDPE, and so expansion and contraction problems caused by temperature changes are virtually nonexistent. Elasticity is important in liners because of the settling that occurs after reservoirs or impoundments are filled. The elasticity of CSPE allows it to accommodate limited movement in the subgrade without cracking or developing leaks.

CSPE geomembranes can be pre-fabricated into large panels greatly minimizing field seaming requirements.

6.3.1 FACTORY PRE-FABRICATION

There are no theoretical limits to the size of the geomembrane liner panels that can be prefabricated in-house; however there are some practical constraints imposed by panel weight. The panel weight is important because of the limits of handling equipment that will be available in the field. The liners are normally limited to a maximum panel weight of 1200 kg; however if a skilled installation crew is available then panel weights of up to 1800 kg are possible. Note that these panels are folded in an 'accordion' fashion, that is, folded in one direction and then folded back on themselves.

An advantage of factory fabricated seams is that 100% of the welds are visually inspected by the welding operator and every fifth seam is destructively tested in peel and shear.

6.4 LIMITATIONS OF CSPE GEOMEMBRANES

CSPE does have its shortcomings and these include the following.

It is not as chemically resistant as some other specialty geomembranes. CSPE does not have good overall chemical resistance especially to hydrocarbons which readily swell and weaken it. In sewage applications, wind can push fatty acids onto the liner and under heating from the sun these can soften and penetrate the liner.

CSPE geomembranes are expensive compared to polyethylene and polypropylene and typically exhibit somewhat lower physical property characteristics.

Blistering problems have been reported, in particular associated with floating covers.

Solvent welding of CSPE becomes increasingly more difficult as the temperature drops.

CSPE geomembranes continue to cure or vulcanize over time and they crosslink into a cured or semi-cured rubber that looses its ability to be easily welded. Whilst this may be beneficial with regard to mechanical properties (for example, the tensile strength continues to increase), its ability to accept repairs decreases with age. For this reason special preparation and bonding agents are often required on older CSPE installations.

CSPE geomembranes have a shelf life and need to be fabricated and installed within 12 months in cool climates (e.g. the UK) and 6 months in hot and humid climates.

Table 6.1 Some limitations of CSPE geomembranes related to curing

Limitations of CSPE geomembranes	Comments
CSPE crosslinks (i.e. cures) in the field thus making thermal welding impossible after about 12 months	Repairs to CSPE geomembranes can be performed using two-part solvent-based adhesives. These however require extensive grinding and solvent scrubbing to activate the surface
While CSPE geomembranes can be solvent bonded in the field these are not always as strong as thermal seams	Solvent bonded field seams tend to experience a decline in strength after time due to polymer softening and scrim pullout.
CSPE geomembranes can partially cure on the roll making it difficult to unroll them	This 'blocking' can be difficult to detect until the material has been delivered to the field
CSPE needs to be activated with a xylene wipe	Xylene cannot be used in potable water reservoirs due to toxic residues

Some Limitations of CSPE geomembranes related to their curing behaviour are listed in Table 6.1.

6.5 GENERAL PROPERTIES OF CSPE GEOMEMBRANES

CSPE geomembranes offer excellent long term weathering properties, and are available with supporting scrims ranging from 530 to 2300 N values of grab tensile strength (see Figure 6.2).

CSPE liners are an excellent choice for water and waste water applications, and are commonly used to produce floating covers. CSPE geomembrane grades for drinking water contact are also available. Potable water grades of CSPE membranes have been approved by the American Water Works Association for storage of drinking water.

CSPE cures over time and becomes progressively more difficult to weld as it ages. Repairs on CSPE geomembranes materials can be made with an adhesive which often has a substantial component of xylene and other solvents. Note however that such solvents can leave residues that affect the contained fluid (e.g. potable water). A wipe with xylene can reactivate the surface of aged CSPE in most cases so that thermal welding can be accomplished. Chemical welding of aged CSPE is dependant on the material condition.

CSPE geomembranes exhibit their true advantage when exposed to high temperatures and oxidizing chemicals. They resist flex cracking and abrasion as well as the damaging effects of weather, UV/ozone, heat and chemicals. CSPE geomembranes exhibit a broad range of valuable properties which include ozone and abrasion resistance, very good chemical resistance for most chemicals, high tensile strength and tear strength.

To overcome the problems associated with the crystallinity of HDPE, the chlorine in CSPE interrupts or reduces the crystallinity and produces an amorphous, rubbery product. Since the crystallinity is the source of strength to HDPE, the addition of chlorine has reduced this inherent strength. A small amount of sulfur in the form of sulfonyl chloride is then added to obtain a stronger but much more flexible liner material. Compounded

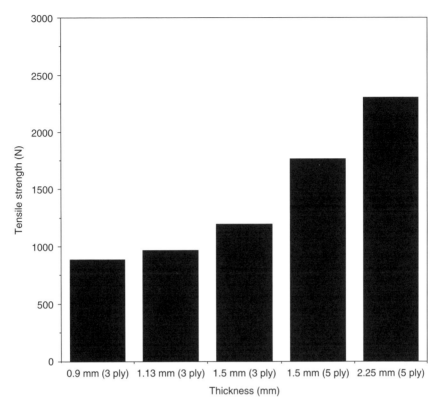

Figure 6.2 Tensile strengths of Hypalon™ geomembranes (ASTM D-751)

CSPE has achieved an excellent track record as a sheet membrane for roofs due to its excellent performance as an exposed membrane.

CSPE geomembranes contain no plasticizers and so have good resistance to growth of mold, mildew, fungus and bacteria.

CSPE geomembranes are sensitive to heat and atmospheric moisture (i.e. humidity) and can undergo premature crosslinking even when still on the roll. This causes the sheet to stick together (termed 'blocking') as a result of partial crosslinking reactions between adjacent sheet surfaces. Corn starch powder or talc can be used as an antiblocking agent.

Since notches and score lines in CSPE are potential defects and stress concentrations, equipment and other items should not be dragged across the surface of the CSPE geomembrane or be allowed to slide down slopes on the lining during installation. It is also important that all parties walking or working on the CSPE lining material wear soft-soled shoes.

6.5.1 CSPE FORMULATION

The carbon black content of CSPE geomembranes is typically quite high (e.g. 5–10%) as it also functions as a reinforcing filler and processing aid in addition to a UV blocking agent. The carbon blacks used in CSPE are usually N630 or N774 as per the ASTM D-1765–06 Standard Classification System for Carbon Blacks Used in Rubber Products.

The filler content of CSPE geomembrane is also quite high with between 20–40% of calcium carbonate being used for enhanced processability and 'body'. Potable grades of CSPE geomembranes also use magnesium oxide/hydroxide as an acid acceptor to assist with crosslinking.

CSPE geomembranes can be seamed with solvent cements or with heat. Field seaming is generally done with a bodied solvent consisting of 10% CSPE compound dissolved in a solvent (usually xylene), which evaporates after bonding leaving the composition of the seam to be the same as the remainder of the membrane.

CSPE was first manufactured by EI du Pont under the Hypalon™ trade name in 1951 and has been used extensively as a geomembrane. CSPE is often categorized as a non-vulcanized elastomer; however it does cure after exposure to the outside elements.

Despite a long history of service, CSPE membranes are notorious for the difficulty of repairing or re-welding of cured membrane. This can be performed, but it is a complicated procedure. There have also been reports of surface degradation, biological attack and some isolated failures.

Repairs to an installed CSPE liner can be made using scissors, brushes, bodied solvent and a hand roller.

It is important to note that the performance of CSPE-based sheets can also vary from manufacturer to manufacturer, as can the percentage of Hypalon™ found in each supplier's product. Overall however most CSPE-based membranes exhibit outstanding durability in a range of demanding environments.

6.5.2 CSPE-R

Scrim reinforced CSPE geomembranes are abbreviated as CSPE-R. CSPE-R is manufactured by a proprietary process in which two layers of CSPE compound are calendered to fully encapsulate a polyester reinforcing scrim. It is important that all exposed scrim edges shall be flood coated with CSPE high-solids chemical fusion solution along the edges. The CSPE encases the scrim by bonding to itself through the openings between the fibres. This double encapsulation avoids possible pinholes that can occur in unreinforced films. The scrim reinforces the CSPE and provides dimensional stability, increased tensile and tear strength (see Figure 6.3) and improved properties at elevated temperatures. A membrane with a single ply of scrim is generally used with a total thickness of 0.9 mm.

The reinforcement in CSPE geomembranes is generally a polyester scrim (i.e. woven polyester fabric in a standard basket weave). The woven fabric is usually a 10×10 scrim which indicates 10 strands per inch in the weft (i.e. cross machine) direction and 10 strands per inch in the waft (i.e. warp machine) direction. The strands of the weave are sufficiently far apart to give apertures through which the opposing polymer faces of the geomembrane can adhere to achieve what is termed in the industry as 'strike through'. Before the fabric scrim is encapsulated by the CSPE plies, an adhesion promoter or ply enhancer is coated onto the scrim. Scrim reinforced CSPE geomembranes are generally produced in widths of 1.52 m. The rolls are then fabricated in the factory under controlled conditions (without the typical field variables of air temperature, air velocity, relative humidity, equipment speed and contamination) into larger panels, custom sized to specific site specifications, drastically reducing the number of necessary field seams.

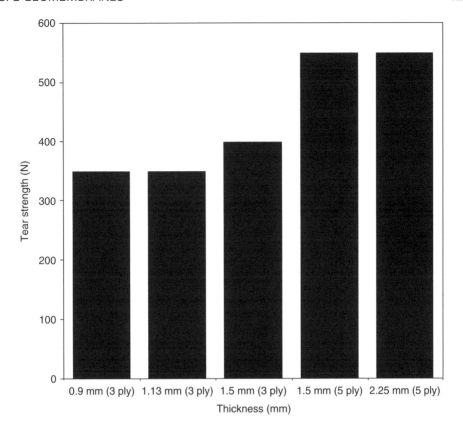

Figure 6.3 Tear strengths of Hypalon™ geomembranes (ASTM D-751)

The most commonly used scrim in CSPE-R geomembranes is a 10 × 10 × 1000 D which has 10 counts per inch in both directions.

Scrim reinforced CSPE has a coefficient of linear thermal expansion of 0.006%/°C (1/3 that of HDPE). With the greater flexibility of the CSPE liner, it can easily accommodate this much expansion without causing problems. A high coefficient of linear thermal expansion is undesirable as it can add tension to the seam areas causing failed seams.

The performance properties of CSPE-R geomembranes are shown in Table 6.2.

A comparison of the properties of 0.90 mm CSPE-R geomembranes from different manufacturers is shown in Table 6.3.

A comparison of the properties of 1.14 mm CSPE-R geomembranes from different manufacturers is shown in Table 6.4.

CSPE-R can be made in two main forms, either with 3 ply or 5 ply construction. The 3 ply construction is made up to a thickness of 1.5 mm while the 5 ply construction achieves a total thickness of 2.25 mm. Increasing the number of plies from 3 to 5 whilst maintaining a thickness of 1.5 mm increases the tensile strength by 48% (from 1200 to 1775 N) and increases the tear strength by 38% (from 400 to 550 N).

Table 6.2 Properties of CSPE-R geomembranes (polyester reinforcement) (Steven Geomembranes, 2008)

Properties	CSPE 0.9 mm	CSPE 1.14 mm	CSPE 1.52 mm
Tensile strength (ASTM D-751)	0.89 kN/m	1.11 kN/m	1.34 kN/m
Tear resistance (ASTM 5884 Method B Tongue tear)[a]	350 N	350 N	400 N
Puncture resistance (FTMS 101C Method 2031)	840 N	890 N	1100 N
Hydrostatic resistance (ASTM D-751 Method A)[b]	2.4 MPa	2.4 MPa	2.4 MPa
Specific gravity (ASTM D-792)	1.47	1.47	1.47
Dimensional stability (ASTM D-1204)	2%	2%	4%
Ply adhesion (ASTM D-413 MD)	1.2 kN/m	1.2 kN/m	1.2 kN/m
Indicative seam strength in shear (ASTM D-751)	0.71 kN/m	0.89 kN/m	1.07 kN/m
Indicative seam strength in peel (ASTM D-413)	1.75 kN/m	1.75 kN/m	1.75 kN/m

[a] Tear resistance of the liner is dominated by the tear resistance of the scrim and so the results do not vary with liner thickness.
[b] Hydrostatic resistance of the liner is dominated by the hydrostatic resistance of the scrim and hence the results do not vary with liner thickness.

Table 6.3 Comparison of 0.90 mm CSPE-R geomembranes from different manufacturers (Steven Geomembranes, 2008; Burke Industries, 2008)

Property	Steven Hypalon 36	Burke M283
Thickness	0.90 mm	0.90 mm
Tensile strength (ASTM D-751)	0.89 kN/m	1.11 kN/m
Tear resistance (ASTM D-5884) tongue tear	0.31 kN	0.44 kN
Puncture resistance (FTMS 101C Method 2031)	0.84 kN	1.06 kN
Hydrostatic resistance (ASTM D-751 Method A)	2.40 MPa	2.77 MPa
Ply adhesion (ASTM D-413 MD)	1.20 kN/m	1.37 kN/m

Table 6.4 Comparison of 1.14 mm CSPE-R geomembranes from different manufacturers (Steven Geomembranes, 2008; Burke Industries, 2008)

Property	Stevens Hypalon™ 45	Burke M284
Thickness	1.14 mm	1.14 mm
Tensile strength (ASTM D-751)	1.11 kN/m	1.24 kN/m
Tear resistance (ASTM D-5884) tongue tear	0.31 kN	0.46 kN
Puncture resistance (FTMS 101C Method 2031)	0.89 kN	1.11 kN
Hydrostatic resistance (ASTM D-751 Method A)	2.40 MPa	2.84 MPa
Ply adhesion (ASTM D-413 MD)	1.20 kN/m	1.71 kN/m

6.6   COMPARISONS OF CSPE WITH OTHER GEOMEMBRANES

CSPE like PVC is flexible and easy to solvent weld. However black CSPE geomembranes have a major advantage over PVC in that their excellent UV resistance makes them suitable for exposed applications.

A section of a CSPE-R cover of a municipal reservoir exposed to weathering for 15 years has been compared to a control sample of material stored indoors at the reservoir site for 15 years. The test results showed that after years of exposure to the sun, wind, snow, rain, pollutants and animal contaminants, the CSPE-R geomembrane had 92% of the strength of the control sample (Dow Geomembranes, 2008).

Unlike HDPE, CSPE-R can be factory fabricated into large panels. Care must be taken however with CSPE-R to prevent it from 'blocking' or sticking together on warm days. CSPE-R is more expensive than both HDPE and PVC geomembranes.

The basic polymer backbone of CSPE is the same as HDPE and since there are no double bonds,[1] the long polymer chains are relatively impervious to attack from degradation agents such as oxygen, ozone or chemical reagents.

CSPE is costly compared to flexible polypropylene and typically exhibits somewhat lower physical property characteristics. Furthermore, the density of CSPE is high at 1.47 g/cm^3 compared to say fPP (density 0.90 g/cm^3). The density of CSPE is thus 63% higher than that of polypropylene and this can add to freight costs.

6.7   APPLICATIONS OF CSPE GEOMEMBRANES

CSPE-based flexible membranes are used in a broad range of applications and weather conditions and have an extremely long service life such as:

- exposed lagoons;
- industrial waste ponds/lagoons;
- floating covers;
- ponds;
- water storages.

Reinforced CSPE liners gives them strength in high-stress applications such as steep slopes and floating covers.

The resistance of CSPE to oxidation and immunity to ozone and UV makes it a lining of choice in industrial waste applications, floating covers and exposed lagoons.

6.8   CHEMICAL RESISTANCE OF CSPE GEOMEMBRANES

The chemical resistance of CSPE geomembranes to various classes of chemicals is shown in Table 6.5.

---

[1] Note that double bonds refer to 'carbon–carbon double bonds' and are also called 'unsaturation'. Double bonds are sites for attack by ozone and other reactive chemicals.

Table 6.5   Chemical resistance of CSPE geomembranes to various classes of chemicals

Chemical Category	Rating	Rating Symbol
Alcohols	Excellent	◊
Aldehydes	Poor to fair	●
Alkalis, dilute	Good to excellent	◊
Alkalis, concentrated	Good to excellent	◊
Amines	Poor	●
Animal and vegetable oils	Good	◊
Brake fluids, non-petroleum based	Fair	●
Diester oils	Poor	●
Esters, alkyl phosphate	Poor	●
Esters, aryl phosphate	Fair	●
Ethers	Poor	●
Fuels, aliphatic hydrocarbons	Fair to good	● to ◊
Fuels, aromatic hydrocarbons	Fair	●
Fuels, extended (oxygenated)	Fair	●
Halogenated solvents	Poor	●
Hydrocarbons, halogenated	Poor	●
Ketones	Poor	●
Lacquer solvents	Poor	●
LP gases and fuel oils	Good	◊
Mineral oils	Good to very good	◊
Oil resistance	Fair to good	◊
Petroleum, aromatic	Poor	●
Petroleum, non-aromatic	Fair	●
Refrigerant ammonia	Good	◊
Refrigerant halofluorocarbons: R-11, R-12, R-13	Good	◊
Refrigerant halofluorocarbons w/oils: R-11, R-12, R-22	Good	◊
Silicone oils	Excellent	◊
Solvent resistance	Fair to good	● to ◊

REFERENCES

Burke Industries [www.burkeindustries.com] (2008).

Dow Geomembranes [http://www.dowgeomembrane.com/Corporate/Literature/HypSpec.pdf] [http://www.dowgeomembrane.com/Products/Hypalon/hypalon.htm] (2008).

Schoenbeck, M., Performance of Chlorosulfonated Polyethylene Geomembranes after Long Term Weathering Exposure, in Staff, C. E. (Ed.), *The Foundation and Growth of the Geomembrane Industry in the United States*, International Conference on Geomembranes, Denver, CO, USA, pp. 5–8 (1984).

Steven Geomembranes [www.stevensgeomembranes.com] (2008).

7

PVC Geomembranes

7.1 INTRODUCTION

PVC geomembranes are made from blends of rigid PVC (such as that used in PVC pipes) and softening agents called plasticizers. PVC resin must be mixed with plasticizers to impart flexibility before it can be manufactured into a geomembrane. These plasticizers make the PVC soft and supple so that it can be rolled, fabricated, and utilized as a liner material in complex installations. Plasticizers are used in quantities of up to 40% to create flexible PVC geomembranes.

Plasticizers function by disrupting and reducing the intermolecular attractions (secondary bonding forces) between adjacent PVC molecular chains. In other words, plasticizers act as lubricants between molecules, so that molecules can flex and slide past each other more freely, thus softening the otherwise hard and rigid PVC (Giroud and Tisinger, 1993). The typical plasticizers that are used to make PVC sheets flexible can also migrate out of the geomembrane over time, resulting in brittleness, shrinkage and even shattering in unreinforced membranes (see Figure 7.1).

Loss of plasticizers is such a major problem that some regions (e.g. the State of Florida) do not allow the use of plasticized PVC for liner materials. However, various studies comparing the chemical resistance of HDPE and PVC have shown that the landfill leachate has virtually no effect on PVC after 16 months (Artieres *et al.*, 1991).

Depending on the application, either low or high molecular weight plasticizers are used. Low molecular weight plasticizers are typically liquid esters such as phthalates. While these are excellent for increasing the flexibility of PVC they are extractable and can be chemically degraded by hydrolysis. Plasticizer loss by both water extraction and voltile loss can be limited by the use of high molecular weight plasticizers that have low migration rates (see Figures 7.2 and 7.3) (Stark *et al.*, 2005a,b). Such high molecular weight plasticizers are more chemically resistant and less susceptible to migration. However, PVC membranes containing these high molecular weight plasticizers typically do not have the level of cold temperature flexibility and low water absorption characteristics of monomeric-plasticized PVC geomembranes.

A Guide to Polymeric Geomembranes: A Practical Approach   J. Scheirs
© 2009 John Wiley & Sons, Ltd

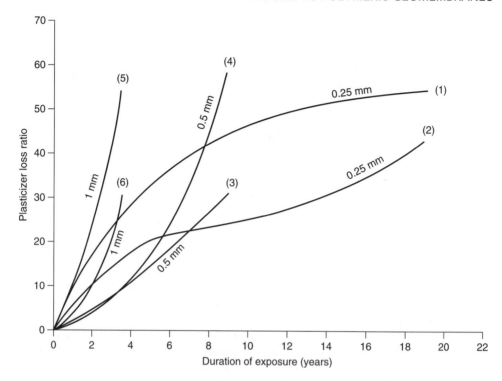

Figure 7.1 Plasticizer loss ratio from PVC geomembranes versus duration of exposure in various environments: (1) canals, Western USA, geomembrane protected by soil, above water level; (2) canals, Western USA, below water level; (3) landfill cover, Florida, geomembrane protected by soil, smooth bedding; (4) geomembrane protected by soil, rough bedding; (5) evaporation ponds, Sahara desert, geomembrane exposed just above sulfuric acid level; (6) evaporation ponds, Sahara desert, geomembrane immersed in sulfuric acid. Reprinted with permission from *GPR Magazine*, Influence of plasticizer molecular weight on plasticizer retention, by T. D. Stark, H. Choi and P. W. Diebel, **23**(2), 1, Copyright (2005) Industrial Fabrics Association

PVC geomembranes contain approximately 30% plasticizer designed to impart flexibility. Unfortunately these liquid plasticizers have a tendency to volatilize or be extracted out of the geomembrane over time, thus the requirement that PVC geomembranes should not be subjected to long-term UV exposure (i.e. they should be in covered applications). PVC can however be plasticized or alloyed with a polymeric plasticizer (namely KEE, ketone ethylene ester) to form ethylene interpolymer alloys (EIAs), the most well known of which is marketed under the trade name XR-5. These EIA geomembranes have outstanding weathering performance. For more information see Chapter 8 on 'EIA geomembranes'.

There are four key areas that determine the performance of a PVC geomembrane:

- formulation;
- thickness;

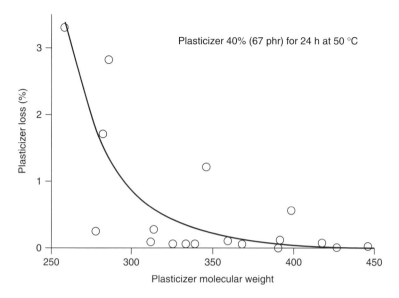

Figure 7.2 Relationship between plasticizer molecular weight and % plasticizer loss from PVC by water extraction for 24 h at 50 °C. Reprinted with permission from *GPR Magazine*, Influence of plasticizer molecular weight on plasticizer retention, by T. D. Stark, H. Choi and P. W. Diebel, **23**(2), 1, Copyright (2005) Industrial Fabrics Association

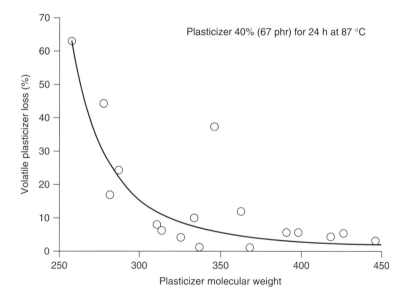

Figure 7.3 Relationship between plasticizer molecular weight and % volatile plasticizer loss from PVC during heat ageing in air for 24 h at 87 °C. Reprinted with permission from *GPR Magazine*, Influence of plasticizer molecular weight on plasticizer retention, by T. D. Stark, H. Choi and P. W. Diebel, **23**(2), 1, Copyright (2005) Industrial Fabrics Association

- reinforcement;
- the production technique (spread coating, calendering and extrusion).

PVC geomembranes are manufactured by either a calandering or an extrusion process. On account of their high flexibilities, PVC liners readily conform to subgrade contours and offer excellent interface friction without being textured.

For demanding applications, scrim-supported PVC membranes are recommended over unreinforced PVC. Reinforced geomembranes offer better tensile strength and tear/puncture resistance. Thicknesses of 0.75–1.00 mm are generally used. However, since the polymer typically represents 75% of the overall cost of the geomembrane, thicker gauges are more costly. The greater polymer thickness over the scrim does offer enhanced weatherability and abrasion resistance.

PVC offers a ductile geomembrane material at relatively low cost which can meet water potability requirements. PVC was one of the first materials used for geomembrane applications and remains an economical and versatile choice. PVC has been used for geomembrane applications for over 40 years.

PVC geomembrane liners have been widely employed, mainly due to their ease of installation compared with that of other geomembrane materials (notably HDPE which is far stiffer). However, they usually should be covered with soil to prevent excessive thermal/UV degradation. This limitation tends to offset somewhat its lower installation cost when compared with other liner materials that are not soil covered. Note that there are specially UV-stabilized PVC liners that can resist long-term UV exposure.

PVC is an elastic material that can conform readily to irregular shapes and differential settlement. PVC is one of the most versatile plastics available with formulations that can provide oil resistance, UV stability, low temperature resistance and other specific properties. Standard formulations of PVC geomembrane materials are blended for longevity and cost reduction. The regular industrial-grade PVC liners are not normally UV-stabilized.

PVC displays less chemical resistance than fPP, EIA and CSPE geomembranes but is usually less sensitive to animal fats and vegetable oils (FOGs) than EPDM geomembranes.

PVC liners are widely used for the conveyance of potable or agricultural water due to the ease of installation and fabrication. However unless PVC is specifically formulated with polymeric plasticizers (e.g. KEE plasticizers) or specialty UV stabilizers, the PVC should be covered with soil or protective layers.

7.2 ATTRIBUTES OF PVC GEOMEMBRANES

PVC liners are typically very flexible and readily conform to subgrade contours.

Due to this flexibility and conformance, PVC geomembranes offer excellent interface friction without being textured. In comparison, other geomembranes, especially HDPE, offer very little surface friction unless they are finished heavily textured.

PVC liners are solvent adhesive weldable, radio frequency (dielectric) weldable and thermal wedge weldable. Excellent seams can be achieved with these methods. PVC is typically prefabricated in a fabrication facility where high level quality control is easily achieved as opposed to field welding. This is an important advantage of PVC since it allows its fabrication into large sheets in a factory and requires less field seaming than

Figure 7.4 Photograph of a roll of a 0.75 mm fabricated PVC oil-resistant alloy geomembrane (HAZGARD 100) used for secondary containment as well as building underliners for contaminated sites. The material on the winder weighs 1800 kg and has a total area of 2000 m². Reproduced by permission of Layfield, Canada

HDPE membranes. Single PVC panels can be fabricated in sizes approaching one acre, minimizing the need for the more demanding field seaming operation (see Figure 7.4).

PVC offers excellent puncture resistance compared to HDPE. A 0.50 mm PVC membrane offers as much puncture resistance as a 1.5 mm HDPE membrane, according to an EPA report of March, 1988 titled 'Loading Point Puncturability Analysis of Geosynthetic Liner Materials'. In the truncated cone test, unsupported PVC achieves critical cone heights of 7 cm or more – some seven times that of HDPE (Stark *et al.*, 2008).

The main factors that make PVC a suitable geomembrane liner are as follows:

- *ease of installation and accommodation to factory fabrication as large sheets, thus significantly reducing field-seaming needs;*
- *superior frictional characteristics (compared to non-roughened geomembranes) in contact with both sand and filter fabrics (it is stable over slopes with angles up to 18% (3H: 1V; i.e. 3 horizontal to 1 vertical));*
- *acceptable uniaxial extension and excellent biaxial flexibility;*
- *easy and versatile weldability.*

7.3 SHORTCOMINGS OF PVC

PVC liners are typically not suitable for prolonged exposure to sunlight (UV) because they are not UV-stabillized. If specified for an exposed application the expected life span will generally be <10 years depending on specific climatic and chemical environment factors. PVC is not resistant to UV degradation unless specially formulated. PVC geomembranes therefore need to be buried by 30 cm of topsoil for UV screening protection. Most PVC manufacturers require this before warranting their installed products. Some manufactures offer UV stable PVC liners but the price is accordingly much higher.

PVC geomembranes have a high chemical resistance to the majority of acids, bases, salts and alcohols but the plasticizers can be affected by benzene, trichloroethylene, toluene and strong bases such as sodium hydroxide. Certain organic liquids and solvents can also extract the plasticizer and when the PVC 'dries out' it can crack. The PVC itself can be attacked by ketones, such as methyl ethyl ketone (MEK) and acetone.

PVC liners are sometimes formulated with biocides (i.e. in addition to plasticizers). Biocides protect PVC with inexpensive plasticizers (which are most commonly phthalate-based) from microbial attack. Without biocides, plasticizers will be consumed by soil borne microbes leaving the geomembrane brittle. Biocides found in PVC can be toxic to fish. Fish-grade PVC usually does not include a biocide since they use higher quality plasticizers that are not susceptible to biological attack and are usually more expensive.

Where liner seaming is concerned, PVC has inherently lower seam strength (e.g. than HDPE) which could cause issues in cases of excessive stresses on the welds.

Some of the common limitations of PVC are listed in Table 7.1.

The potential for loss of plasticizers from PVC excludes the use of PVC for liner material from many areas. Nonetheless, various studies comparing the chemical resistance of HDPE and PVC have shown that the landfill leachate has virtually no effect on PVC after 16 months (Artieres *et al.*, 1991)

Many geosynthetic engineers and designers broadly specify HDPE for its greater resistance to chemicals but neglect many of PVC's advantages and attributes. A major advantage of PVC geomembranes over HDPE is their ability to be fabricated into large sheets requiring less field welding than HDPE membranes. Compared to HDPE, PVC geomembranes have lower strength, but have high deformation capabilities and are not susceptible to environmental stress cracking.

Factory welds in PVC geomembranes and liners are generally of a high standard since they are made in a clean and controlled environment. The large prefabricated panels are shipped to the installation site on pallets and additional seaming is conducted in the field to produce their final configuration. In contrast, field welds can be problematic and lack integrity leading to leaks due to the uncontrolled nature of the outside environment with respect to temperature/wind and to the potential for contamination.

PVC is easy to install and easily factory fabricated into large panels up to 35 m wide which are then seamed together in the field. In many small applications, factory fabrication allows the panel to be made large enough so that no field seaming is required. The obvious advantage of factory fabrication is that the seaming is done under controlled conditions and the seaming equipment is preset and easily monitored. In addition, the quality control of the factory fabricated seams is conducted before the material is shipped to the job site.

Table 7.1 Some limitations of PVC geomembranes

Limitation	Comments
Plasticizers can be susceptible to migration and leaching (i.e. removal)	PVC can embrittle if the plasticizers are lost or extracted.
	This mainly presents a problem in installations where the PVC liner is in direct contact with substances that have an affinity for the plasticizers and 'strips' them at an accelerated rate. This stripping has been noted to happen with concrete, for example.
	This also makes PVC susceptible to contact with various chemicals and exposure to UV radiation and heat.
General PVC geomembranes are susceptible to UV degradation	Due to PVC's susceptibility to UV degradation it is often necessary to cover PVC geomembranes with soil or cover material in order to protect it from direct sunlight.
PVC is not resistant to burrowing or gnawing animals and can be susceptible to microbial degradation (unless formulated with biocides)	This means that PVC membranes require a protection layer to protect them from burrowing and gnawing animals.
PVC has relatively low tear strength (see Figure 7.5)	Scrim-reinforced PVC geomembranes are available that have enhanced tear strength.
PVC becomes brittle at low temperatures	This can limit the deployment of PVC geomembranes at cold temperatures.
PVC is susceptible to microbial degradation	This can be overcome by formulating with biocides in applications where this is appropriate (i.e. not for marine applications).
PVC has inherently lower seam strength compared to HDPE	This could cause issues in the case of situations where the welds are highly stressed.

7.4 PROPERTIES OF PVC GEOMEMBRANES

Typical performance properties of PVC geomembranes are listed in Table 7.2.

7.4.1 ADVANTAGES OF PVC GEOMEMBRANES OVER HDPE

The main structural difference between PVC and HDPE geomembranes is that PVC is amorphous (i.e. the polymer chains are randomly positioned and there are no ordered domains) while HDPE is semicrystalline (i.e. the polymer chains are tightly packed into ordered crystalline domains).

HDPE geomembranes exhibit a sharp and pronounced peak in their stress–strain curve (at the yield point) which is an indication that they tend to undergo a relatively abrupt change in their mechanical properties, whereas PVC undergoes a very large amount of elongation before property deterioration and failure. PVC has at least 100% available strain whereas HDPE has only 12% strain before yield.

Table 7.2 Typical PVC geomembrane properties (PVC Geomembranes Institute)

Properties	0.50 mm PVC	0.75 mm PVC	1.00 PVC	1.52 mm PVC
Tensile strength (MD) (ASTM D-882)	8.4 kN/m	12.8 kN/m	17.0 kN/m	24.0 kN/m
Tensile strength (XD)[a] (ASTM D-882)	8.0 kN/m	12.2 kN/m	16.2 kN/m	22.8 kN/m
Tear resistance (ASTM D-1004, Die C)	27 N	35 N	44 N	67 N
Hydrostatic resistance (ASTM D-751, Method A)	470 kPa	690 kPa	830 kPa	1240 kPa
Specific gravity (ASTM D-792)	1.20–1.30	1.20–1.30	1.20–1.30	1.20–1.30

[a]XD is the cross-direction, i.e. transverse direction.

PVC geomembranes are intrinsically flexible and easy to handle, while HDPE geomembranes (especially over 2 mm in thickness) are rigid, non-flexible and difficult to handle.

The intrinsic flexibility of PVC allows the fabrication of the majority of the seams to be performed under controlled factory conditions since the large fabricated panels can be folded easily. In contrast, HDPE geomembranes need to be seamed in the field where moisture, dirt, wind and temperature variations can affect the quality of the welding. It is noteworthy that a PVC liner requires only about 20% of the field seams that are required by an equivalently sized HDPE liner. This is an important advantage as it is well recognized that field seaming is potentially the most problematic aspect of liner construction.

PVC becomes easier to thermally weld with decreasing temperature and increasing thickness. On the other hand, chemical seaming is not generally recommended below 10 °C.

The stress–strain behavior of PVC is significantly different to that of HDPE. Even after reaching yield stress of the interface, PVC interfaces will not fail but maintain stability by stretching of the membrane material without loss of strength or material damage. In contrast, in the case of the HDPE and textured HDPE geomembranes, the stress-displacement response of the interface is such that after reaching peak stress, further shearing to a larger strain causes stabilization of the stress (Bhatia and Kasturi, 1996).

7.5 FAILURE MODES

The principal failure modes of PVC relate to its relatively low UV stability, its solvent susceptibility, its low tear strength (see Figure 7.5), as well as its reliance on plasticizing additives that can migrate or be extracted away or be consumed by microbes.

PVC absorbs certain organic liquids which can exert an 'over-plasticizing effect'. For example, PVC membranes used in sewerage ponds can unzip around the edge due to accumulation of fats/oils at the waterline and excessive plasticization of the liner.

PVC also tends to become brittle and darken when exposed to ultraviolet light or heat-induced degradation. Most industrial-grade PVC geomembranes can be degraded by sunlight and therefore it ideally requires being covered with a minimum of 30 cm of

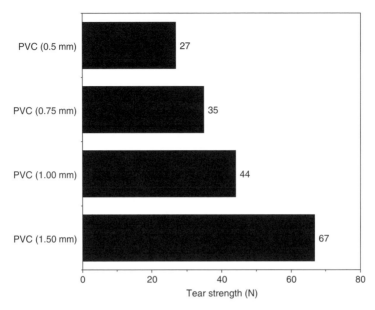

Figure 7.5 Tear strength for PVC geomembranes (ASTM D-1004) (data from Geosynthetics Specifier's Guide, IFAI publication, 2009)

clean fill to provide a service life of over 20 years. PVC geomembranes perform well when covered by at least 30 cm of soil. Note also that many PVC formulations, with quite different durability characteristics, are available on the market. Some premium PVC geomembranes are formulated for extended life (Newman *et al.*, 2004).

Damage to thin (0.25 and 0.5 mm) PVC accordion folded sheet has been observed where carton (cardboard) abrasion and long transport times result in holes due to abrasion. Long-term folding and creasing can also lead to permanent deformation where the elastic limit of the material is exceeded on the outer radius of the fold.

Despite its high innate friction angle, PVC geomembranes can nonetheless be prone to sliding failures. Slopes should be limited to a gradient of 3:1 to prevent sliding failures.

Due to its softness, abrasion and erosion of PVC geomembranes is a potential failure mode. For larger ponds and lakes, consideration for erosion and wave action should be taken into account with PVC liners. PVC liners are also not resistant to burrowing or gnawing animals and so biotic barriers should be considered.

Table 7.3 lists some of the potential failure modes of PVC geomembranes in common applications.

For the above applications PVC geomembranes have several attributes that make them very desirable as geomembranes and liners:

- very flexible which allows them to conform to irregular surfaces easily;
- fairly low modulus allowing them to stretch to conform to point stresses such as gravel;
- good puncture resistance and relatively good tear resistance when considering an unreinforced material.

Table 7.3 Common applications areas for PVC geomembranes and potential failure modes

Application areas	Comments
Decorative ponds	UV exposure may be an issue, with animal damage possible
Agriculture	Consider UV degradation and animal damage
Reservoirs	Must be potable water approved; UV exposure may be an issue
Tank linings	Abrasion and star creases may be an issue
Irrigation ponds	UV exposure may be an issue on exposed areas; also erosion arising from wave action should be considered
Golf course ponds, lakes and streams	UV exposure may be an issue
Mine tailings	PVC has good resistance to mine liquors although strong bases can degrade ester-based plasticizers
Secondary containment	Fuels and petroleum hydrocarbons can strip out plasticizers, leading to brittleness
Wastewater treatment facilities	Fats, oils and greases can plasticize and soften PVC
Landfill liners and caps	PVC has good multi-axial properties to cope with differential settlement of waste but UV exposure can cause issues
Canals and channels	UV exposure may be an issue; animal damage possible

7.6 FORMULATION OF PVC GEOMEMBRANES

There are four fundamental ingredients in modern PVC geomembranes:

- the PVC resin;
- plasticizers;
- stabilizers;
- pigments/fillers (note fillers can adversely effect cold temperature performance and elongation).

PVC geomembranes contain PVC resin at greater than 50% of the total polymer content suitably compounded with plasticizers, stabilizers, fillers, pigments and other ingredients to satisfy the physical property requirements and durability requirements. All four of these ingredients need to be optimized to achieve the right combination of flexibility, strength and weatherability of the geomembrane.

Dioctyl phthalate (DOP) has traditionally been the predominant plasticizer used in PVC, however there has been a shift towards using less migratory plasticizers such as diisodecyl phthalate (DIDP).

7.6.1 STABILIZATION OF PVC GEOMEMBRANES

Unfortunately conventional HALS stabilizers (which are commonly used in polyolefins) cannot be used in PVC. This is because as PVC degrades it releases hydrochloric acid which neutralizes and deactivates the HALS stabilizer, thus severely reducing the effectiveness of HALS as a light stabilizer. However the synthesis and development of a unique class of HALS light stabilizers known as NOR HALS (non-basic HALS or non-reactive

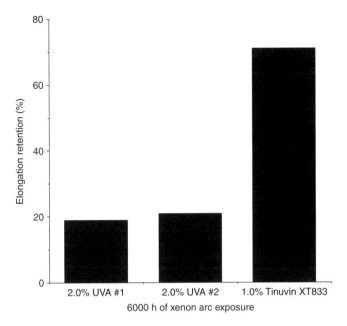

Figure 7.6 Elongation retention for flexible PVC geomembranes after 6000 h of xenon UV exposure

HALS) has overcome this problem and allows them to perform well in an acidic environment.

Tinuvin™ XT833 (Ciba, Basal, Switzerland) is an effective NOR HALS stabilizer for flexible PVC geomembranes. This additive can extend the life expectancy of flexible PVC beyond what can be achieved with conventional UV absorbers and pigments. Even at half the concentration levels, Tinuvin™ XT833 provides superior elongation retention in flexible PVC compared to two typical formulations containing ultraviolet absorbers (UVAs) (see Figure 7.6).

Thus Tinuvin™ XT833 can protect flexible geomembranes and liners from the harmful effects of light exposure and helps it maintain its initial tensile and elongation properties and physical integrity during long-term weathering. In PVC geomembranes, for example, it minimizes embrittlement and enables the membranes to retain their liquid barrier properties. In some cases, studies show, even in acidic environments, Tinuvin™ XT833 can double the expected lifetime of flexible PVC compared to conventional ultraviolet absorbers (UVAs).

7.6.2 NON-EXTRACTABLE PLASTICIZERS FOR PVC

PVC can be alloyed with ketone ethylene ester (KEE) (also known as Elvaloy™) which is essentially a polymeric non-liquid plasticizer or alloying agent. It is therefore much more tightly bound and will not volatilize or migrate out of the PVC. These ethylene interpolymer alloys (EIAs) geomembranes are more resistant to hydrocarbons than regular

PVC membranes and can therefore be left exposed (i.e. an installation advantage). This unique blend of resins has been optimized to produce a high performance geomembrane material which is not susceptible to plasticizer migration as in the case of conventional PVC geomembranes. EIA products have high temperature resistance as well as broad chemical resistance, including acids, oils and methane.

Typically EIA products are reinforced with a polyester scrim and are referred to as EIA-R or a scrim-reinforced EIA geomembrane. Such EIA-R geomembranes are manufactured by Cooley Engineered Membranes (Pawtucket, RI, USA) and Seaman Corporation (Wooster, OH, USA) amongst others. The scrim-reinforced geomembrane is usually manufactured by an spread-coating process (e.g. Seaman) or an extrusion coating process (Cooley). The finished membrane consists of two plies of a PVC-EIA terpolymer encapsulating one layer of reinforcing scrim. The reinforcing scrim is a rip-stop woven polyester to create an open-type weave that permits 'strike-through' (i.e. direct bonding of the surface layers of polymer through the apertures in the reinforcing scrim) in the case of extrusion coating while in the speed-coating process the fabric weave is more closed and the EIA coating covers scrim and gives a polymer coated fabric. Any exposed fabric from the scrim along the longitudinal edges of the rolls is not permitted unless a "non-wicking" fabric is used. See Chapter 8 for an in-depth treatment of EIA geomembranes.

REFERENCES

Artieres, O., Gousse, F. and Prigent, E., Laboratory Ageing and Geomembranes in Municipal Landfill Leachates, in *Proceedings of the 3rd International Landfill Symposium*, Sardinia, pp. 587–603 (1991).

Bhatia, S. K. and Kasturi, G., 'Comparison of PVC and HDPE Geomembranes – Interface Friction Performance', PGI Technical Bulletin. Available from the Geomembrane Institute, PGI – Technology Program, University of Illinois, 2215 Newmark Civil Engineering Laboratory, 205 North Matthews Avenue, Urbana, IL, USA [http://www.geomembrane.com/PGI%20Tech%20Bull/97-03.htm] (1996).

Giroud, J. P. and Tisinger, L. G., The Influence of Plasticizers on the Performance of PVC Geomembranes, in *Proceedings of Geosynthetic Liner Systems: Innovations, Concerns and Design*, IFAI, St Paul, MN, USA, pp. 169–196 (1993).

Layfield Canada [international@layfieldgroup.com].

Newman, E. J., Stark, T. D, Rohe, F. P. and Diebel, P., Thirty-year Durability of a 20 mil PVC Geomembrane, *Journal of Vinyl and Additive Technology*, **10**(4), 168 (2004).

PVC Geomembrane Institute [www.pvcgeomembrane.com].

Stark, T. D., Choi, H. and Diebel, P. W., Influence of Plasticizer Molecular Weight on Plasticizer Retention in PVC Geomembranes, *Geosynthetics International*, **12**(2), 99 (2005a).

Stark, T. D., Choi, H. and Diebel, P. W., Influence of Plasticizer Molecular Weight on Plasticizer Retention, *GFR Magazine*, **23**(2), 1 (2005b).

Stark, T. D., Boerman, T. R. and Connor, C. J., Puncture Resistance of PVC Geomembranes Using the Truncated Cone Test, *Geosynthetics International*, **15**(6), 480 (2008).

8

EIA Geomembranes

8.1 INTRODUCTION

Ethylene interpolymer alloy (EIA) geomembranes are PVC blends containing Elvaloy™ – a flexible terpolymer containing ketone, ethylene and ester monomers, all within the backbone of the polymer (see structure in Figure 8.1). EIA is a high molecular weight, solid and flexible thermoplastic polymer where the PVC resin is alloyed with the Elvaloy™ for improved strength and durability. The Elvaloy™ KEE modifier is a polymeric plasticizer (i.e. a high molecular weight plasticizer) that is more chemically resistant and has less susceptibility to plasticizer migration than monomeric plasticizers (i.e. low molecular weight plasticizers) used in most PVC membranes. The presence of the high molecular weight polymeric plasticizer enables the geomembrane to retain long term flexibility even in extreme temperature environments. However it should be noted that monomeric plasticizers generally offer greater low temperature flexibility and lower water absorption characteristics than EIA geomembranes.

Flexible PVC geomembranes attain their flexibility by the use of a lower molecular weight liquid plasticizer. However, the compromise with these lightweight plasticizers is that they can migrate away from the PVC polymer over time and the PVC can become brittle. Certain environments (e.g. high temperatures or exposure to certain chemicals) can accelerate this process, leaving a membrane prone to *in situ* shrinkage and stiffening. A minimum polymer thickness is required to function as a plasticizer reservoir to prolong the time it takes for liquid plasticizers to migrate to the surface and erode away.

Since PVC and Elvaloy™ are completely miscible they become a single-phase polymer when mixed. They disperse within each other and, as both polymers are high in molecular weight, they cannot migrate from each other when properly alloyed. Their affinity for each other ensures membrane flexibility in severe environments that would otherwise accelerate the loss/extraction of liquid plasticizers in conventional PVC geomembranes.

Chemically, Elvaloy™ is what's known as a 'polar polymer', which means its molecules have a natural affinity (miscibility) with other polar polymers such as PVC. When the Elvaloy™ and PVC molecules melt together into an alloy compound they remain together

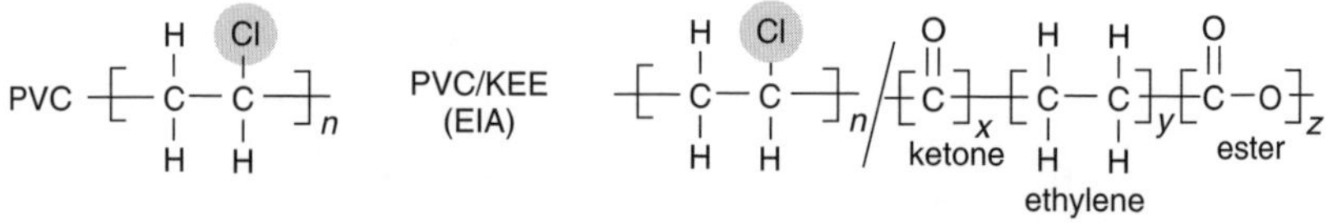

Figure 8.1 Chemical structures of PVC and EIA Polymers. EIA is a blend of PVC and a ketone–ethylene–ester (KEE) polymer. KEE acts as a non-extractable plasticizer for PVC geomembranes

naturally without phase separation problems. Since Elvaloy™ is a solid, high molecular weight polymer, it simply cannot migrate out or cause the PVC to become harder and more brittle over time even in the presence of good solvents for conventional plasticizers.

EIA geomembranes are generally reinforced with a denier weft reinforced polyester or nylon fabric (scrim) and are thus referred to as EIA-R. The reinforcement imparts high tensile strengths (up to of 65 MPa according to ASTM D-882). The weft reinforced polyester scrim is primed and coated with the EIA polymer formulation before the geomembrane is produced to make the sheet as monolithic as possible and reduce the potential for wicking and delamination. A proprietary process is used that impregnates the polyester fibres, promoting a molecular bond between the scrim and coating.

EIA-R geomembranes can be manufactured by a spread coating process where the heated PVC/KEE paste is applied to the scrim using a knife or blade in such a manner that it actually impregnates the PVC/KEE paste completely through the woven or weft inserted fibres, not just on the surface, to ensure maximum adhesion. Since the polymer coating intimately encapsulates the fabric there is (in theory) reduced tendency for delamination. Delamination of EIA-R geomembranes have been observed in the field however (see Figure 8.2).

The scrim reinforcement significantly reduces shrinkage and helps to dissipate stresses in the geomembrane by virtue of the proprietary cloth design. The reinforcement also makes the membrane highly puncture resistant. The coefficient of thermal expansion of EIA-R geomembranes is very low (1.4×10^{-5} cm/cm/$^\circ$C (maximum)) which ensures good dimensional stability and high degree of conformance even when deployed in hot weather.

It is important to note that the strength of EIA-R geomembranes is derived primarily from the fabric scrim and not from the thickness of the polymer plies or coating layers. The purpose of the polymer coating is to protect the fabric's inherent attributes. EIA-R geomembranes are thus also referred to as coated fabrics. They have been used in geomembranes applications since 1976. EIA-R geomembranes in particular XR-5® was developed to fulfil the demand for a geomembrane with better hydrocarbon and petroleum resistance than HDPE and CPE. EIA-R geomembranes have broad chemical compatibility and have been practically tested with many different petroleum products including diesel fuel, naphtha, kerosene and crude oil.

In addition moisture vapour or free water is not chemically reactive or degrading to either PVC or Elvaloy™. EIA geomembranes are however vulnerable to attack by liquid phthalate plasticizers. Although EIA formulations do not rely on liquid plasticizers for flexibility, some liquid plasticizers may be used during processing. High molecular weight KEE and PVC polymers have a natural affinity for lightweight plasticizers and exposing

Figure 8.2 Photograph of delamination of a 0.9 mm (36 mil) EIA-R geomembrane. The back ply has fully delaminated from the polyester scrim. Note this is a laminate and not a coated fabric. Reproduced by permission of ExcelPlas Geomembrane Testing

EIA to phthalates can promote 'over' plasticization and softening of the membrane under warm temperature conditions.

While seemingly benign substances such as fats, oils and greases (FOGs) can degrade the properties of PVC geomembranes by accelerating plasticizer loss, properly formulated EIA geomembranes are not prone to having their flexibility extracted by such chemicals.

8.2   ATTRIBUTES OF EIA-R GEOMEMBRANES

EIA-R Geomembranes offer a wide range of positive attributes.

EIA-R products have a wide spectrum of chemical resistances, including hydrocarbons. XR-5®[1] can be used for the primary and secondary containment of hydrocarbon liquids that may be chemically aggressive such as crude oil, fuel oils, aviation fuels, diesel, kerosene, refinery wastes, alcohols glycols, and many other substances (XR-5, 2008).

Certain EIA-R formulations are suitable for use with potable water and have excellent heat and chemical resistance, making them suitable for containment of a wide range of processing liquids.

EIA-R geomembranes also exhibits impressive dimensional stability, tensile strength and puncture resistance.

Elvaloy™ KEE flexibility is relatively constant over a wide temperature range, and this feature is inherited by EIA-R geomembranes. Thus the rolled and folded geomembrane remains workable even when cold.

EIA-R can be thermally or dielectrically seam welded.

EIA-R is flexible and can be pre-fabricated into large scale panels, minimizing required field seams.

---

[1] Note: XR-5® is a proprietary product from Seaman Corporation.

In many instances EIA-R can perform in service environments where no other geomembranes can survive.

8.3 LIMITATIONS OF EIA-R GEOMEMBRANES

Some limitations of EIA-R geomembranes include the following.

EIA-R geomembranes are relatively expensive but can prove to be the least expensive alternative in many containment applications.

They may exhibit poor flex cracking resistance (often attributed to a too high filler content or to creasing at low temperatures).

Floating covers can collapse due to cracking problems and scrim degradation by acid hydrolysis (so need to have encapsulated edges).

Heat degradation of the PVC based polymer can occur during overheating making the material susceptible to tearing (see Figure 8.3).

8.4 PERFORMANCE PROPERTIES OF EIA-R GEOMEMBRANES

The permanence of the phased polymer structure within the EIA coating, the coating's adhesion to the base fabric, and outstanding resistance to UV, chemical and microbiological attack are all attributes that have contributed to the acceptance and uptake of EIA-R geomembranes. EIA-R geomembranes are not subject to crazing or stress cracking from

Figure 8.3   Photograph of a tear failure adjacent to a weld in a reinforced 0.9 mm KEE-Elvaloy Terpolymer Sheet (EIA) reinforced with a polyester scrim reinforcement. The 'adjacent seam failure' is characteristic of heat affected material adjacent to the weld. Localized overheating of the weld has made the area directly along the weld susceptible to tearing. The fracture surface of the field failure exhibits a brittle crack edge with little or no signs of ductile tearing or deformation. Reproduced permission of ExcelPlas Geomembrane Testing

heating, high thermal expansion or localized stress concentrations (e.g. sharp folds or built-in notches).

Accelerated UV ageing tests using EMMAQUA (Equatorial Mount with Mirrors for Acceleration Plus Water Spray) have demonstrated that EIA-R geomembranes resist 'checking' (i.e. fine, shallow regular cracks appearing on the surface) and cracking after UV exposures equal to 20 years of normal weathering (Seaman Corporation, 2008).

EIA geomembranes combine good mechanical properties such as high tensile strength (e.g. 15.8 MPa minimum as per ASTM D-882) with the ability to conform to irregularities while still providing a tough and abrasion resistant membrane. This conformability property eliminates the need to regrade or reshape the existing subgrade.

Another important advantage of EIA-R geomembranes is their ability to be factory fabricated in one piece enabling quick installation and eliminating troublesome field seams. The best known commercially available EIA-R geomembrane is known as XR-5® and is manufactured by the Seaman Corporation. A less well known related product is XR-3®. Some performance properties and applications areas for XR-5® and XR-3® EIA-R geomembranes are shown in Table 8.1.

XR-5® is a versatile alloy which has been utilized for secondary containment of hydrocarbons such as diesel oils for over 30 years. XR-5® combines good resistance to combustible liquids (hydrocarbons) with a high performance polyester base fabric. XR-5® has also been widely used as an exposed lining material since it has outstanding long-term resistance to UV radiation. Testimony to its excellent UV resistance is that it has been used in exposed applications in equatorial climates where there are extreme UV exposures. The performance properties of XR-5® geomembranes are listed in Table 8.2.

The related polymer alloyed product, XR-3® has lower cost than XR-5® but shares many of its physical properties, however XR-3® is used in less demanding chemical applications where the superior chemical resistance of XR-5® is not warranted. XR-3® carries the NSF 61 certification for potable water applications.

XR-5® is flexible at temperatures above freezing, and is easy to install and repair, even in remote locations. Installation and repair instructions can be described to local contractors for efficient response to the client's containment needs. XR-5® is not normally used for installation in cold temperatures. When cold weather installation is required

Table 8.1 Product application chart for EIA geomembranes from Seaman Corporation (Qlining, 2008)

Properties	XR-5® 8130	XR-5® 8138	XR-5® 6730	XR-3® reinforced	XR-3® PW
High puncture resistance	✔	✔			✔
High UV resistance	✔	✔	✔	✔	✔
High strength applications	✔	✔	✔		✔
Floating covers (non-potable)	✔	✔	✔	✔	
Diesel and jet fuel containment	✔	✔	✔		
Chemical resistant applications	✔	✔	✔		*[a]
Potable water contact					✔

[a]*The XR-3 grade has only moderate chemical resistance for areas such as stormwater and domestic wastewater applications. Source: Seaman Corporation, Ohio, USA.

Table 8.2 Properties of XR-5® EIA-R geomembranes with PET reinforcement (Seaman Corporation, 2008)

Property	XR-5® 8130	XR-5® 8130 PW	XR-5® 8138
Thickness (ASTM D-5199)	0.75 mm	0.75 mm	1.00 mm
Weight (ASTM D-751)	1020 g/m^2	1020 g/m^2	1288 g/m^2
Tensile strength, (ASTM D-751)	2450 N	2450 N	2447 N
Tear strength (ASTM D-4533, Trap Tear)	155 N	155 N	155 N
Puncture resistance (ASTM D-4833)	1110 N	1110 N	1112 N
Hydrostatic resistance (ASTM D-751 A)	540 N/cm^2	540 N/cm^2	540 N/cm^2
Bursting strength (ASTM D-751, ball tip)	2890 N minimum	2890 N minimum	2892 N minimum
	3560 N typical	3560 N typical	3560 N typical
Ply adhesion (ASTM D-413)	65 N/2.5 cm	65 N/2.5 cm	65 N/2.5 cm
Bonded seam strength (ASTM D-751)	2450 N	2450 N	2447 N
Low-temperature brittleness (ASTM D-1790)	−35 °C	−35 °C	−35 °C

Table 8.3 Minimum property values for XPROgeo™ EIA-R membranes from LG Chem

Grade of XPROgeo™	Thickness (mm)	Tensile strength (N/m) (ASTM D-751 [2])	Tear strength (N) (ASTM D-5884, Method B)	Puncture resistance (N) (FTMS 101C, Method 2065)	Hydrostatic resistance kN/m^2 (ASTM D-751)
301E	0.75	2460	310	1560	5500
401E	1.00	2460	310	1560	5500
306E	0.75	2460	530	1560	5500
281CE	0.75	1100	130	930	2800
301PE	0.75	2460	310	1560	5500

two other materials are recommended – 8218 LTA and 8228 OR LTA from Seaman Corp. are excellent low-temperature materials, remaining flexible and durable for cold temperature installation and service. The 8218 LTA material is used for water and drilling mud containment while the 8228 OR LTA is used when hydrocarbon resistance is required or in municipal wastewater projects in extremely cold environments.

LG Chem manufacture reinforced EIA-R membranes made with Elvaloy (DuPont) under the tradename XPROgeo™ (introduced in 2007). They produce fives grades in thicknesses from 0.75 to 1.00 mm as shown in Table 8.3. These materials combine excellent resistance to harsh weather, chemical exposure and oil exposure but have a limited track record.

Table 8.4 Minimum property values for CoolGuard™ from Cooley Group (Cooley, 2008)

Grade	Thickness (mm)	Tensile strength (N/m) (ASTM D-751 [2])	Tear strength (N) ASTM D-5884, Method B)	Puncture resistance (N) (FTMS 101C, Method 2065)	Hydrostatic resistance kN/m^2 (ASTM D-751)
FTL30	0.91	2700; 2700	130	1450	6897
FTL40	1.17	4400; 4400	220	2450	6897
HPK80	2.03	1300; 1100	220	890	2413
HRL36	0.91	2800; 2700	530	1670	5517
MPK36	0.91	1100; 900	180	890	2413
MPK60	1.52	1100; 900	180	890	2413

Note that CoolGuard™ MPK36 is supported on a *knitted* scrim while the more common CoolGuard™ HR uses a *woven* scrim. Both products comprise two layers of PVC plasticized with a high molecular weight plasticizer laminated to a central scrim. The ply adhesion is known to be different for products containing woven versus knitted scrim.

The XPROgeo™ EIA-R membranes are reinforced with a 120 g/m^2 polyester scrim to give tear strength (ASTM D-5884) of up to 530 N and dimensional stability (ASTM D-1204) of 2.5% (maximum) in each direction. In the wicking test (ASTM D-751) the maximum extent of wicking recorded for these EIA-R geomembranes is 0.3 cm. This ensures little risk of water penetration into the scrim layer when the geomembrane is exposed to water or aqueous solutions. Note some EIA-R products come with an encapsulated edge to prevent wicking.

EIA-R geomembranes exhibit excellent weathering resistance as demonstrated by the fact that after 8000 h of exposure to ASTM G-153 (Carbon Arc) they exhibit no appreciable changes or stiffening or cracking of the coating (LG Chem, 2008).

Note that reinforced membranes such as EIA-R have different property values in different directions and so often two values are given for properties that are direction-dependent such as tensile breaking strength or tear strength.

EIA-R geomembranes are also manufactured by the Cooley Group under the CoolGuard™ brand. The perforance properties for CoolGuard™ EIA-R geomembranes are listed in Table 8.4. Note Cooley offer RCG36 as a potable water version of their MPK36 product.

A comparison of the properties of two commercial EIA-R geomembranes is shown in Table 8.5.

Sioen (Belgium) also make EIA-R geomembranes such as their E1510.

8.4.1 UV RESISTANCE

The excellent UV resistance of EIA-R geomembranes can be enhanced further by the use of newly developed additives. Tinuvin™ XT833 (Ciba, Basel, Switzerland) is an effective non-acid reactive, hindered amine stabilizer (NOR HALS) stabilizer for flexible PVC and EIA geomembranes. This additive can extend the life expectancy of flexible PVC and PVC–KEE (EIA) alloys beyond what can be achieved with conventional UV absorbers and pigments. For instance, Tinuvin™ XT833 can achieve a doubling of the expected lifetimes of PVC–KEE alloys. Figure 8.4 shows the effect of 1 phr of Tinuvin™ XT833

Table 8.5 Comparison of commercial EIA-R geomembranes (from manufacturer's published data)

Properties	Test method	XR-5® 8130 Seaman Corporation	XPROgeo™ 281CE LG Chem
Thickness	ASTM D-751	0.76 mm	0.75
Polyester base fabric weight	ASTM D-751	220 g/m²	120 g/m²
Weight	ASTM D-751	1017 g/m²	–
Tear strength	ASTM D-4533 (Trap Tear)	175–245 N	130–150 N
Breaking yield strength	ASTM D-751 (Grab Tensile)	2447 N	1100–1400 N
Dimensional stability	ASTM D-1204 100 °C for 1 h	0.5% maximum in each direction	2.5% maximum in each direction
Low-temperature brittleness	ASTM D-1790	−35 °C	−32 °C
Bursting strength	ASTM D-751 (Ball Tip)	3330 N (minimum)	2000 N (minimum) 2500 N (typical)
Puncture resistance	ASTM D-4833	1200 N (minimum)	360 N (minimum)
Bonded seam strength	ASTM D-751 (A)	2450 N (minimum)	1400 N (minimum)
Wicking resistance	ASTM D-751	0.3 cm (maximum)	0.3 cm (maximum)

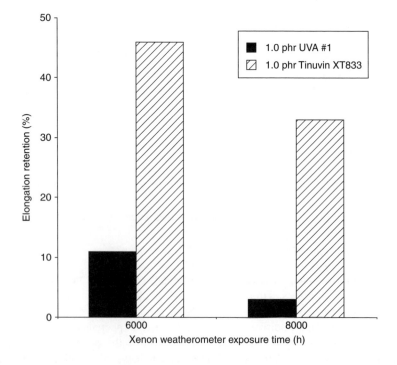

Figure 8.4 Elongation retention for PVC–KEE geomembranes after exposure to 6000 and 8000 h of UV (data from Ciba, Switzerland)

Table 8.6 Comparison of the properties of EIA-R (XR-5®) and CSPE-R (Hypalon™) geomembranes ([http://www.xr-5.com/comparisons/xr-vs-hypalon.html], 2008)

Property	0.75 mm XR-5®	0.9 mm Hypalon™	1.12 mm Hypalon™
Tensile strength (ASTM D-751, Grab Method)	2446 N	890 N	1110 N
Puncture resistance (ASTM D-751, Ball Tip)	2891 N	840 N	890 N
UV resistance	Excellent	Excellent	Excellent
Warranty given	Weathering and chemical compatibility	Weathering only	Weathering only
Blocking resistance	Excellent	Fair	Fair
Water absorption (100 °C, 7 days)	+9.9%	+85.8%	+85.8%
Hydrostatic resistance (ASTM D-751)	5512 kN/m^2	1724–2410 kN/m^2	1724–2410 kN/m^2
Resistant to wicking	Yes	No	No
Sheet edge lock required (to prevent wicking)	No	Yes	Yes
Thermal expansion (mm/30 m with 65 °C temperature change)	20 mm	75 mm	75 mm
Repairable with thermal welding	Yes	No	No

on a PVC–KEE alloy geomembrane where the % elongation retention after 6000 h and 8000 h of xenon lamp UV exposure is 46% and 33%, respectively, compared to only 11% and 3% retained elongation, respectively for a PVC–KEE geomembrane containing 1 phr of a conventional UV absorber.

8.5   COMPARISON OF EIA-R (XR-5®) VERSUS CSPE-R (HYPALON™) GEOMEMBRANES

The main competitor of EIA-R geomembranes in terms of long-term resistance to UV degradation and broad chemical resistance is CSPE-R. A comparison of the performance properties of EIA-R (XR-5®) and CSPE-R (Hypalon™) geomembranes is shown in Table 8.6.

8.6   APPLICATION AREAS FOR EIA-R GEOMEMBRANES

Typical application areas for EIA-R geomembranes include:

- Tank liners (this is a major application area)
- Secondary Containment
- Hazardous Waste Impoundments
- Crude Oil Holding Pits
- Brine Ponds

- Solar Ponds
- Emergency Containments
- Industrial Lagoons
- Wastewater Lagoons
- Chemical Holding Basins
- Tailing Ponds
- Leach Pads

EIA-R geomembranes can even be used to contain acidic mining chemicals such as those used in copper heap leaching ponds. The heap leaching process is used to recover copper at many mines worldwide. The recovery process employs a solvent extraction system using sulfuric acid and kerosene to extract minerals from ore. The ore is piled on an impermeable surface and sprayed and showered with leaching solution which leaches out the mineral content. The metal laden solution, termed the 'pregnant' solution, is then transferred to a mixing-settling tank where the kerosene removes the copper. The copper is then removed from the kerosene by sulfuric acid to produce an electrolytic solution from which the copper is plated in an electrowinning process. Finally, the 'barren' solution is re-acidified and sprayed back over another load of ore to complete the cycle.

Such heap leach pads can easily be over 80 acres in size. The chemical action of sulfuric acid, combined with temperatures reaching 71 °C on the black surface, can seriously soften most geomembrane liner materials, especially HDPE and LLDPE which are swollen by the kerosene. Moreover dumping hundreds of tonnes of ore on the liner necessitates a strong membrane that is resistant to abrasion and punctures. Finally the steep, angular design of the collecting ponds requires a strong, flexible product. Also the leaching ponds are concrete reinforced structures that require the liner to be sealed around the edges at odd angles and thus a flexible and compliant liner is required. EIA-R geomembranes are increasingly used for leach pads due to their excellent resistance toward kerosene and other hydrocarbons. EIA-R liners also exhibit greater strength than equivalent thickness HDPE liners. In fact, a 0.75 mm EIA-R liner is stronger than a liner made from polyethylene which is three times thicker.

REFERENCES

Cooley [www.cooleygroup.com] (2008).
[http://www.xr-5.com/comparisons/xr-vs-hypalon.html] (2008).
LG Chem [www.lgchem.com.cn/upload/chemProdFile/20080410/lg_xprogeo_281ce.pdf] (2008).
Qlining [www.qlining.com/seaman.html] (2008).
Seaman Corporation [www.xr-5.com/downloads/library/geomembranes-tech-manual.pdf] (2008).
XR-5 [http://www.xr-5.com/tech-library/technical-library.html] (2008).

9

EPDM Geomembranes

9.1 INTRODUCTION

EPDM is a vulcanized (i.e. crosslinked) rubber sheet that is classified as a polyolefin. The long rubber molecules in EPDM are crosslinked to give an elastic and chemically stable membrane. As a result, EPDM rubber geomembranes can withstand deformation better than many other geomembrane materials. Their strength and elasticity are not seriously affected by high or low temperatures and physical properties are practically unchanged over decades of service without becoming brittle, cracking or shrinking.

EPDM is a durable geomembrane with a plus 20 year proven performance history in exposed applications. The anticipated lifetime of the exposed liner is 25–40 years. EPDM, due to its rubbery nature, exhibits excellent elongation characteristics and does not require a soil cover. It possesses high tensile strength and excellent resistance to punctures, UV radiation and microbial attack. Since EPDM is a highly flexible material with a low coefficient of thermal expansion and contraction it will lay flat in a wide range of temperatures and substrates thereby conforming well to irregularities in the subgrade.

Field seaming of EPDM generally involves thermal fusion (e.g. with a hot air gun) or the use of in-seam tape or cover strip tape. Rolling with a hand held roller helps to ensure good contact and bonding. EPDM is used as liners, caps and covers in a wide range of environmental containment applications.

9.2 ATTRIBUTES OF EPDM GEOMEMBRANES

The attributes of EPDM geomembranes include:

- Excellent elongation characteristics and does not require a soil cover.
- High degree of flexibility and conformability.
- Excellent resistance to weathering and UV exposure. This is important around the waterline where the membrane can become periodically exposed with rising and falling

A Guide to Polymeric Geomembranes: A Practical Approach   J. Scheirs
© 2009 John Wiley & Sons, Ltd

water levels, and in situations where other materials suffer from antioxidant/stabilizer extraction.

- Ability to expand and contract, while maintaining its integrity as water levels and temperatures change.
- High tensile strength and good resistance to punctures (working strain >300%).
- Easy to install – lays flat in a wide range of temperatures and terrains (little or no wrinkles).
- Field weldable by conventional thermal fusion methods or by tape seaming methods.
- Low maintenance – remains stable in soil or water.

The positive performance attributes of EPDM geomembranes and liners are summarized in Table 9.1.

Some advantage and limitations of EPDM Liners are listed in Table 9.2.

Table 9.1 Performance attributes of EPDM

Attributes	Rating
UV resistance	✔✔✔
Ozone resistance	✔✔✔
Cold temperature resistance	✔✔✔
Lay flat behaviour and conformability	✔✔✔
Oxidation resistance	✔✔✔
20 year warranty	✔✔✔

Table 9.2 Advantages and limitations of EPDM liners

Advantages	Limitations
Good flexibility, toughness and ageing resistance	Relatively high cost
Resistant to most chemicals and abrasion	Requires protection from mechanical damage and vandalism if exposed
Can withstand extreme changes in temperature	
Excellent UV resistance	Service life is dependent on careful and consistent fabrication which may be difficult for large projects
Low coefficient of thermal expansion and so good conformance to substrate	EPDM can undergo swelling and plasticization by animal fats and oils
Low degree of wrinkling which can otherwise interfere with drainage and run-off of water	Low tear strength unless scrim reinforced
A relatively high mass per unit area which provides resistance against wind uplift.	EPDM is also susceptible to swelling and attack by petroleum, grease, oil, solvents, vegetable or mineral oil. EPDM is also sensitive to aromatic solvents and chlorinated hydrocarbons

9.3 STRUCTURE AND CHEMISTRY

EPDM rubber (ethylene propylene diene M-class rubber) is an elastomer where the E refers to ethylene, P to propylene, D to diene and M refers to its classification in ASTM standard D-1418 (see structure in Figure 9.1). The 'M' class includes rubbers having a saturated chain of the polyethylene type. The dienes currently used in the manufacture of EPDM rubbers are DCPD (dicyclopentadiene), ENB (ethylidene norbornene) and VNB (vinyl norbornene).

EPDM is comprised mainly of saturated polymeric chains of ethylene and propylene molecules. The polymer structure has good resistance to ozone and ageing. The presence of a third monomer – a diene (i.e. a chemical with reactive double bonds) typically ethylidene norbornene (ENB) provides chemically active cure sites for vulcanization (i.e. crosslinking). Relatively high levels of carbon black are added to the formulation to increase the UV resistance and build properties such as tear resistance. Certain lubricating oil extenders and processing aids are also added to the formulation prior to the vulcanization process.

The ethylene content is around 45 to 75%. The higher the ethylene content, the higher the loading possibilities of the polymer, better mixing and extrusion. During peroxide curing these polymers attain a higher crosslink density compared with their amorphous counterpart. The diene groups act as crosslinking sites and the diene content can vary between 2.5 wt% up to 12 wt%.

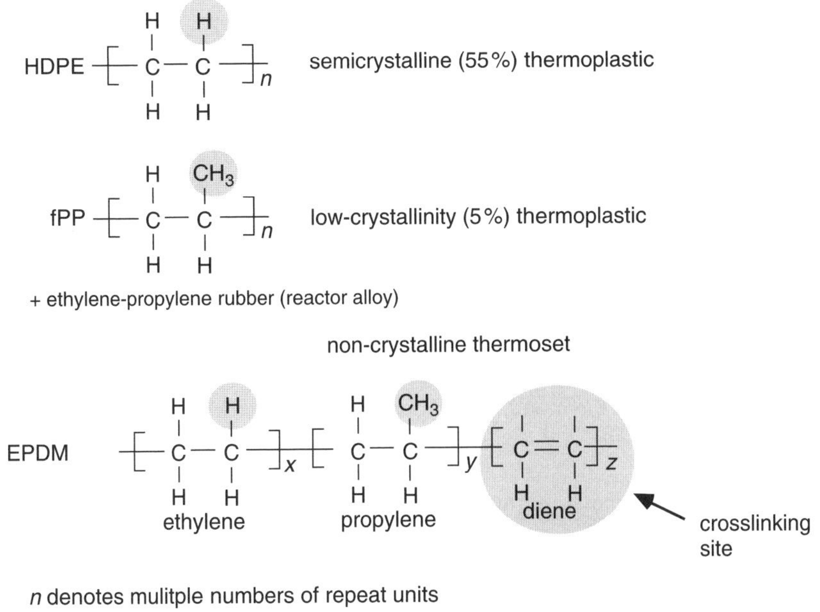

Figure 9.1 Chemical structure of EPDM geomembranes when compared with HDPE and fPP

Table 9.3 General composition of EPDM geomembranes

Ingredient	Amount (%)	Function
EPDM polymer	30	Base rubber precursor
Carbon black	30–40	For reinforcement and UV stability
Mineral oils	25	For processability and flexibility
Vulcanizing Agent	4	To crosslink the EPDM
Zinc oxide	1	Vulcanizing aid

EPDM has a crosslinked polymer structure which imparts to it a unique elasticity and unsurpassed ageing resistance without the need for problematic stabilizers.

The general composition of EPDM Geomembranes is shown in Table 9.3.

9.4 MECHANICAL PROPERTIES

EPDM rubber membranes have no defined yield points under strain, whereas thermoplastic geomembranes such as HDPE yield, become thinner at a relatively low tensile force. In contrast, EPDM rubber geomembranes will elongate in a linear fashion to a maximum of about 500% and can be elongated equally in all directions at the same time (this is known as the 'multiaxial response').

EPDM rubber membranes are not susceptible to stress cracking unlike semicrystalline HDPE which can stress crack even at a low stress, at locations where the surface is scratched or contains notch defects. In some grades of HDPE such stress cracking can occur at elongations of 20–30% of the yield point which translates to elongations of just 2–4%.

EPDM geomembranes exhibit viscoelastic properties which enables them to withstand an very high pressure load. At low loads, the elastic properties dominate whereas at high loads the viscous properties dominate. Owing to their crosslinked elastomeric structure these geomembranes can be deformed to extreme limits and still return to their original size and shape. The opposite is true for the semicrystalline thermoplastic such as HDPE where a concentration of load or force can result in a permanent deformation or thickness reduction (Trelleborg, 2008).

There are two main types of EPDM geomembranes – nonreinforced geomembranes and scrim-reinforced geomembranes. Both types are normally available in two thicknesses, 1.14 mm and 1.52 mm. In the manufacturing and curing of EPDM no rework is permitted. Once the EPDM membrane is cured in the autoclave it is impossible to add other cured product.

Unreinforced EPDM geomembranes are produced by Carlisle Syntec Inc. Trelleborg and Firestone and their typical properties are shown in Table 9.4.

Note the tear strength of the unreinforced EPDM membrane is rather low (i.e. 40–53 N).

Reinforced EPDM geomembranes are produced for example by Firestone Specialty Products and their typical properties are shown in Table 9.5.

Table 9.4 Typical performance properties of unreinforced EPDM geomembranes (Carlisle Syntec Inc., 2008)

Properties	1.14 mm EPDM	1.5 mm EPDM
Thickness (ASTM D-5199)	1.14 mm	1.5 mm
Break strength (ASTM D-882)	9.6 kN/m	11.4 kN/m
Break elongation (ASTM D-882)	500%	500%
Tear resistance (ASTM D-1004)	40 N	53 N
Puncture resistance (ASTM D-4833)	125 N	155 N
Multiaxial elongation (ASTM D-5617)	100%	100%
Brittleness temperature (ASTM D-746)	−45 °C	−45 °C
Density (ASTM D-792)	1.1 g/cc	1.1 g/cc
Dimensional stability (ASTM D-1204)	1%	1%

Table 9.5 Typical performance properties of reinforced EPDM geomembranes (Firestone Specialty Products, 2008)

Properties	1.15 mm EPDM–R	1.5 mm EPDM–R
Thickness (ASTM D-1593 or D-5199)	1.15 mm	1.5 mm
Reinforcement type	PET	PET
Density (ASTM D-792)	1.15 g/cc	1.15 g/cc
Ply adhesion (ASTM D-413)	1.2 kN/m	1.2 kN/m
Dimensional stability (ASTM D-1204)	1%	1%
Tensile strength (ASTM D-751)	850 N	970 N
Tear resistance (ASTM D-5884, Method B)	580 N	750 N
Puncture resistance (FTMS 101C, Method 2031)	60 N	80 N

The effect of the inclusion of polyester scrim is apparent with a major increase in the tear strength of EPDM geomembranes from 40 N to 580 N (an increase of 14 times).

A standard specification for 'Test Methods, Properties, Frequency and Recommended Warranty for Ethylene Propylene Diene Terpolymer (EPDM) Nonreinforced and Scrim Reinforced Geomembranes' is published as GRI GM-21 (Geosynthetic Research Institute).

EPDM can be factory fabricated into panels up to 15.2 m wide and 60 m long. Panels are seamed in the field, using 150 cm wide seam tape.

A comparison of the performance properties of seven geomembranes is shown in Table 9.6.

9.5 ADVANTAGES OF EPDM OVER HDPE GEOMEMBRANES (PROPERTY COMPARISON)

EPDM geomembranes and liners exhibit a number of performance property advantages over HDPE geomembranes. This section discusses the various issues and limitations of HDPE geomembranes and contrasts those with the attributes of EPDM geomembranes.

Table 9.6 Comparison of geomembranes (Trelleborg, 2008)[a]

Property	HDPE	LLDPE	EPDM	R-CSPE	PVC	R-fPP	fPP
Water tightness	A	A	A	A	A	A	A
UV resistance	A	D	A	A	D	A	A
Service life	A	C	A	A	D	A	A
Cold temperature impact	C	B	A	B	D	B	B
High temperature resistance	B	D	A	B	D	A	A
Flexibility	D	B	A	C	A	C	B
Elasticity	D	D	A	C	D	D	C
Tensile strength	A	B	C	A	C	A	B
Chemical resistance	A	B	B	B	C	B	B
Resistance to hydrocarbons	B	C	D	D	C	C	C
Stress crack resistance	D	B	A	A	B	A	A
Yield point	D	C	A	B	C	B	B
Resistance to plasticizer extraction	A	A	A	A	D	A	A
Root resistance	A	A	A	B	B	A	A
Resistance to microbiological attack	A	A	A	A	C	A	A
Puncture resistance	C	B	B	B	B	B	A
Surface friction	D unless textured	D unless textured	A	B	B	B	B
Slope stability	C unless textured	B unless textured	A	B	A	B	B
Thermal stability	C	B	A	A	C	A	B
Dimensional stability	D	D	A	A	B	A	B
Multiaxial strain	D	C	A	C	B	C	B
Resistance to settlements	C	B	A	C	A	C	B
Seamability	C	B	A	B	B	A	B
Seamability at cold temperatures	D	D	A	B	D	B	A
Seam strength	A	A	A	A	B	A	A
Seam testing	A	A	A	B	A	B	A
Ease of installation	C	C	A	B	A	B	B
Permeability	A	B	B	B	C	B	B
Environmental properties	A	A	A	B	D	A	A
Repairability	C	B	B	D	C	B	B
Details, design and installation	D	C	B	B	B	C	B
Conformance to substrate	D	C	A	B	B	C	B

[a] A, excellent; B, good; C, fair; D, poor; NS, not stated; NA, not applicable.

9.5.1 STRESS CRACKING

Stress cracking is a brittle fracture phenomenon associated with some HDPE geomembranes. It is a fundamental property of crystalline HDPE and can occur at stresses that are only 20–30% of the yield stress. Stress cracking occurs primarily at the welds of HDPE, at stress concentrations and thickness irregularities. Thickness irregularities can be caused by thickness variations in membrane, by scratches, abrasions or grinding and at overlaps or cap strips.

EPDM on the other hand is a stable, elastic product, which can be stressed and elongated either uniaxially or multiaxially during its lifetime without stress cracking occurring.

9.5.2 THERMAL EXPANSION/CONTRACTION

HDPE has a high coefficient of thermal expansion and expands significantly during moderate to high temperature differentials, such as from day to night. Increased ambient and surface temperatures will cause waves in HDPE liners due to expansion and stress due to contraction. Such continual flexing causes problems that can lead to stress cracking or cause the liner to be pulled out of anchoring trenches. It can also cause welding problems on site and could lead to site weld quality issues.

In contrast EPDM has very low thermal expansion characteristics, is extremely flexible and exhibits excellent lay flat properties. Waves and 'fishmouths' are not an issue with EPDM geomembranes.

9.5.3 UV RESISTANCE

HDPE has a restricted natural UV resistance. By adding 2–3% carbon black and HALS stabilizers, UV resistance is improved significantly.

EPDM Geomembranes have by nature a high content (approximately 1/3 by weight) of carbon black which imparts outstanding UV resistance.

9.5.4 SEAM STRENGTH

Welding of HDPE is a sensitive, precise operation, and there is a narrow heat window available for welding. This heat window varies with sheet thickness, with outdoor temperature and with variations of equipment. Therefore control of welding parameters, CQA systems and skill of welders are critical factors for membrane performance.

Certain grades of EPDM geomembranes can be welded with thermal welding methods giving reliable results that are largely independent of the ambient temperature or material variations. The seams are as elastic as the membrane and can easily be tested for continuity and mechanical strength by non-destructive and destructive methods.

9.5.5 RIGIDITY

HDPE sheets exhibit a high degree of rigidity in both cold and warm temperatures. This makes installation difficult, especially corners, pipe boots, flashings, overflows and penetration details. The attempt to use a rigid material in a flexible application results in basic design, engineering and installation problems.

EPDM on the other hand is an extremely flexible geomembrane material with excellent conformance properties. It is pliable and easy to work with in detailed applications.

9.5.6 YIELD POINT

HDPE polymers exhibit a significant yield point, where the material will draw and thin out at its weakest point when under stress. The yield point occurs at elongations of only 10–12%, despite the fact that the membrane can have elongation at break as high as 700–800%.

EPDM on the other hand has no yield point and will continue to elongate up to its break elongation of approximately 400–500%.

9.5.7 INSTALLATION

HDPE is produced in large-sized rolls, with typical dimensions of 7 m × 100 m , for example. Heavy equipment can be required for positioning and unrolling rolls of HDPE geomembranes.

EPDM can be prefabricated to panels of 800–1500 m², with each panel measured to fit exactly into the installation site. The flexible membrane can easily be transported and positioned on site, by small work crews.

9.5.8 SITE SEAMING

HDPE geomembranes have a coefficient of expansion of approximately 0.017%/°C. Since the membrane is quite stiff, the expansion creates wrinkles when deployed before welding. For example, a 100 m long HDPE membrane is loose laid in the morning with a temperature of +10 °C and welded midday when the membrane has reached a temperature of +80 °C. In this case the HDPE membranes would have increased in length by about 1.2 m and the entire length difference manifests as wrinkles in the edges. This high degree of liner expansion places limitations on when seaming can be done.

EPDM minimizes wrinkles since the membrane has a low coefficient of expansion and is an extremely flexible product.

9.5.9 FRICTION ANGLES

Increased landfill capacity can be realized by taking advantage of the higher friction angles provided by EPDM. When the volumetric difference is multiplied by an average compacting factor and then multiplied by the anticipated dumping charge per cubic metre, the increased revenue can be substantial.

HDPE geomembranes (untextured) have a typical friction angle of approximately 16–18 degrees. EPDM has, due to its structured surface, a typical friction angle of approximately 24–27 degrees. This gives the designer the possibility to design a construction with rather high slope surroundings to create a containment facility, that can contain a higher volume on the same land surface.

9.5.10 WELDING PERFORMANCE AND SPEED

The welding of HDPE geomembranes is very sensitive to ambient conditions and environmental factors. This creates problems in terms of optimizing the optimum welding speed and welding temperature. The installer needs to adjust these welding parameters on-site and the weld integrity needs to be carefully checked.

EPDM is vulcanized by tape seams or thermal welded. The welds have the same elasticity and flexibility as the membrane itself. Thickness variations are of less importance when using these welding methods with EPDM geomembranes (Trelleborg, 2008). Tape seams are slower to produce than thermal welding and quite dependent on site conditions (e.g. dust, moisture, humidity).

9.5.11 SUBSTRATES

HDPE geomembranes generally demand substrates with high compaction and a smooth and flat surface due to its rigidity, thermal expansion and the risk of stress cracking.

In contrast, EPDM is a fully elastic membrane with exceptionally good properties for irregular surfaces. The membrane conforms to any irregularity and can tolerate movement and settlement in the ground without major problems.

Table 9.7 compares the roll dimensions and available thicknesses for HDPE versus EPDM geomembranes.

A comparison of the performance properties of HDPE and EDPM geomembranes is presented in Table 9.8.

A comparison of the general chemical resistance of HDPE and EPDM geomembranes is presented in Table 9.9.

Table 9.7 Comparison of roll dimensions and available thicknesses for HDPE versus EPDM geomembranes (Trelleborg, 2008)

Roll measurement	HDPE	EPDM
Most popular thickness	1.50 mm	1.00 mm
0.80 mm thickness	Not used	Available
1.00 mm thickness	Available	Available
1.20 mm thickness	Available	Available
1.50 mm thickness	Available	Available
2.00 mm thickness	Available	Available but not used
2.50 mm thickness	Available	Not used
Roll width	5–10 m	1.7 m (1.3–3.0 m)
Roll length	50–200 m	20–125 m
Prefabricated to specified size	No	Yes, maximum 1500 m^2

Table 9.8 Comparison of physical properties of HDPE and EPDM geomembranes (Trelleborg, 2008)

Property	HDPE	EPDM
Density	0.94 g/cm^3 (floats on water)	1.15 g/cm^3
Tensile strength at break	30.0 MPa	10.0 MPa
Elongation at break	700%	400%
Tensile strength at yield	15 MPa	No yield point
Elongation at yield	12%	No yield point
Thermal expansion	Significant	insignificant
Low temperature brittleness (ASTM D-746)	$-70\,°C$	$-50\,°C$
Carbon black content	2–3%	30–40%
Environmental stress crack resistance (ASTM D-5397)	300 h (value of time to failure under load of 30% yield stress of membrane)	Not prone to stress cracking

Table 9.9 Comparison of chemical resistances of HDPE and EPDM geomembranes (Trelleborg, 2008)[a]

Chemical	HDPE	EPDM
Hydrocarbons	2	3
Inorganic salts	1	1
Animal oils	1	1
Bases	1	1
Organic salts	1	1
Vegetable oils	1	1
Weak inorganic acids	1	1
Alcohols	1	1
Aldehydes	1	1
Amines	1	1
Esters	1	1
Ketones	1	1
Organic acids	1	1
Ethers	3	2
Phenols	2	2
Mineral oils	1	3

[a] 1, resistant; 2, moderately resistant; 3, non-resistant.

A comparison of the installation and welding factors of HDPE and EPDM geomembranes is presented in Table 9.10.

A comparison of the relative costs of HDPE and EPDM geomembranes are shown in Table 9.11.

9.6 COMPARISON BETWEEN EPDM AND PVC GEOMEMBRANES

9.6.1 DURABILITY AND UV RESISTANCE

EPDM is a synthetic rubber material that is a stable elastic product designed for decades of outdoor exposure to the elements and inert to the effects of buried environments containing microorganisms. The high carbon black content (30–40%) guarantees UV weatherability beyond the normal life of many containments requiring exposed membranes.

PVC on the other hand contains plasticizers, normally phthalates such as DEHP and dibutyl phthalates, to impart flexibility. These plasticizers can migrate and evaporate from the membrane over time, depending on the compound quality of the PVC and the service environment. When the plasticizer is lost the membrane can revert, during a short time interval, to a brittle, hard product with poor mechanical properties and with a total loss of elongation. PVC is also quite sensitive to heat, sunlight and microorganisms and hence only buried applications can be recommended.

Organic compounds in contact with PVC can result in microbiological degradation of the membrane and there is a risk of migration when in contact with other plastic materials and concrete.

Due to the loss of plasticizers, PVC is also susceptible to weight loss and shrinkage over time.

Table 9.10 Comparison of installation and welding of HDPE and EPDM geomembranes (Trelleborg, 2008)

Installation characteristics	HDPE	EPDM
Welding type	Fusion or extrusion	Fusion or tape seams
Weldability	Require specialized operators and is very sensitive to external factors such as weather and temperature	Uncomplicated operation, sensitive only to wet weather
Membrane characteristics	Rigid, relatively inflexible	Flexible, conforms to irregular shapes
Effective permanent elongation installed on site	3–8%	50–75%
Effective puncture resistance	Restricted (crystalline material, can only absorb stress in one direction)	Excellent (absorbs substantial irregularities in substrate)
Repairability	Generally by extrusion welding	Easily repaired by patching with custom tape
Effect of high temperatures (exposed black surfaces reach 80–100 °C under the sun in hot climates)	Severe loss of physical properties if not adequately stabilized	No significant change of physical properties

Table 9.11 Comparison of relative costs of HDPE and EPDM geomembranes[a] (Trelleborg, 2008)

Relative pricing	HDPE	EPDM
Price indication – membrane (1.5 mm HDPE, 1.0 mm EPDM)	100%	150–200%
Price indication – installed lining (1.5 mm HDPE, 1.0 mm EPDM)	100%	125–130%

[a]Note: dependent on many factors such as oil pricing.

9.6.2 SUBSTRATES

EPDM is a fully elastic membrane with exceptional elongation properties for irregular surfaces. The membrane conforms to any irregularity and movements or settlements in the subgrade even after years of service.

On the other hand, unreinforced PVC in thicknesses in the range 0.50–1.00 mm are quite susceptible to tearing from rocks, sharp stones or rough substrates. PVC membranes must be protected with geotextiles or protective soil layers for guaranteed long-term performance.

9.6.3 INTERFACE FRICTION ANGLE

EPDM membranes with a textured surface have a typical friction angle of 24–27 degrees with most soil types. This gives the designer the possibility to design a construction with steeper slopes or to create a reservoir or capping that can contain a higher volume for a given land area. Increased water storage or landfill capacity can be realized. Increased slopes provide increased capacity that translates to higher revenue.

However, the risk of slippage of cover soils and slope stability must be carefully considered by designers.

9.6.4 LOW-TEMPERATURE ENVIRONMENTS

EPDM is not generally affected by low temperature extremes, even if exposed. It remains flexible and can be installed and seamed at below freezing temperatures. Low temperature resistance is down to $-50\,°$C, and thus applications in extreme Northern hemisphere climates is acceptable.

PVC has a limited resistance to low temperatures. The membrane will become stiff and brittle, with increased susceptibility to tear and puncture. The best quality PVC can have a brittle point of $-25\,°$C, but in most cases the effective low temperature resistance is restricted to $-17\,°$C.

9.6.5 COST EFFECTIVENESS

EPDM provides a cost effective product and system. Panels are prefabricated to between 800 and 1500 m^2, with custom sizes to fit exactly into the excavation or installation site. The flexible membrane can easily be transported and positioned on site, by small working crews. Installation of single panel projects, up to 2000 m^2 in size, can be performed by general building/excavating contractors.

PVC is prefabricated to large panels with similar methods, but 0.75–1.00 mm PVC is not as durable and resistant to installation related stress, and will not take as much abuse from rough installation surfaces. In addition, protection from environmental degradation will generally require a minimum 50 cm of soil cover. EPDM can be installed exposed to the atmosphere and will not be damaged by placement of landscaping rock directly on the membrane.

9.6.6 CHEMICAL RESISTANCE

EPDM resists chemical leachates, acids, alkalis, nitrates and phosphates in solution. Additionally it is highly resistant to microbiological attack and root penetration. The service life and performance of EPDM geomembranes and liners is quite exceptional as their strength and elasticity remains virtually unchanged even after decades of service (Trelleborg, 2008).

Notable characteristics of EPDM geomembranes are shown in Table 9.12.

9.7 SEAMING EPDM GEOMEMBRANES

Specialized EPDM membranes such as Trelleborg Elastoseal™ EPDM utilizing the Thermobond™ seaming technique can be easily welded to produce high quality thermally welded seams. Onsite seaming can be performed using a thermal hot wedge. During production of the EPDM, a thin layer of a thermoplastic rubber (TPE) is laminated to one side of the membrane resulting in a vulcanized EPDM membrane which can be

Table 9.12 Notable characteristics of EPDM geomembranes

Attributes	Comments
Elasticity and strength	Irrespective of temperature the membrane will perform well under maximum working load
Multidimensional strain characteristics	Superior resistance to earth settlements and movements
No defined yield point and no yielding behaviour	Puncture resistant with full flexibility up to the maximum tensile strength
Lay flat characteristics	Adheres to and provides close and intimate contact to any substrate
Excellent low temp	Unaffected by low temperatures and fully thermally seamable even in freezing temperatures
Optimal surface friction characteristics	Soft textured rubber surface provides high interface friction
Excellent UV and ozone resistance	Superior service life for exposed installations
Large panels can be factory fabricated to specified sizes according to site drawings.	Reduced field seaming and short installation times
Broad chemical resistance	The geomembrane has good chemical resistance needed in landfills, wastewater reservoirs and ponds
Can be hot wedge welded (e.g. the Thermobond™ seaming technique)	For maximum installation control and quality control (can be air channel tested)
Easily Repairable	Damage is easily repaired, even after long service life in exposed installations
Longest service history of all geomembranes	EPDM membranes have been used for over 40 years in lining applications

heat seamed. This technique combines the advantages of both thermoset elastomers and thermoplastics. Quality control of seams is performed by means of air lance testing. Seams with air pressure test channels can be produced with dual hot wedges at seaming speeds of 3 m/min. The dual wedge technique provides for easy air pressure testing of seams, both immediately after installation and after many years of service.

9.8 APPLICATIONS

EPDM is a durable geomembrane material with a proven performance history in exposed applications such as ponds, lagoons, lakes, canals, etc. It is also used as caps and covers in a wide range of environmental containment applications (Frobel, 1999).

9.9 SERVICE LIFE

The estimated lifetime of an exposed EPDM liner is 25–40 years.

The service life of EPDM geomembranes is highly dependent upon the quality of the seams. Factory seams are often vulcanized and therefore the welds have high integrity in terms of strength and quality. Field seams and patches on the other hand are thermally fused or glued with a rubber cement or seam tape, and subsequent strength may be lower than that of the parent material.

REFERENCES

Carlisle Syntec Inc. [www.carlisle-syntec.com] (2008).

Firestone Specialty Products [www.firestonebpe.com] [www.firestonebpe.com/lining/syst_comp/epdm_geomembrane/_en/index.shtm] (2008).

Frobel, R. K. *et al*. 'EPDM Geomembrane Caps, Commercial Landfill', Geotechnical Fabrics Report, pp. 36–41 (October/November, 1999).

Geosynthetic Research Institute, 'Standard Specification for Test Methods, Properties, Frequency and Recommended Warranty for Ethylene Propylene Diene Terpolymer (EPDM) Nonreinforced and Scrim Reinforced Geomembranes' published as GRI GM21.

Trelleborg [www.trelleborg.com] (2008).

10

Bituminous Geomembranes (BGMs)

10.1 INTRODUCTION

Bitumen is a tarry residue obtained from the distillation of crude oil and well known for paving roads and waterproofing. It is a thick, highly viscous hydrocarbon. Modern bitumen products have been developed that are less viscous and more flexible than traditional hard bitumen and these can be used as flexible membranes. There are two types of flexible bitumen membrane liners – those that are prefabricated and the spray-in-place bitumen liners. The bitumen used in BGM is a special grade (e.g. from Shell) that offers a broader range of chemical resistance and pliability than traditional bitumen.

10.2 PREFABRICATED BITUMEN MEMBRANES

Prefabricated bitumen membrane linings were developed mainly for small-scale lining installations that did not warrant the use of the minimum truck size loads and necessary skilled labour to supply hot bitumen directly to the site. These linings are supplied in similar thickness to the range recommended for the sprayed-in-place lining. Prefabricated bitumen-based liners combine modified bitumen with a polyester and/or glass fibre rein-forcement to produce a composite liner. The liner generally uses high-quality bitumen, different to that of road construction combined with a styrene–butadiene–styrene poly-mer modifier and a polyester reinforcement to provide puncture and tear resistance. The membrane is prefabricated into rolls about 4–5 m wide which reduces quality control issues in the field with spray applied membranes.

Bituminous geomembranes (BGMs) are comprised of a thick non-woven geotextile that is completely impregnated with bitumen to give an impervious (yet flexible) sheet. The manufacturing process completely impregnates the geotextile support in a bath of bitumen. The geotextiles range in weight from $200 \, \text{g/m}^2$ to $400 \, \text{g/m}^2$, thus increasing the thickness and amount of bitumen on the roll. While the bitumen is still hot, sand is placed in the hot bitumen. The sanded surface allows for contractor safety on site (i.e. the contractor is able to walk with snow or water present and not slip) and in addition provides increased friction angles of 32–34 degrees.

Specifically BGMs are engineered geomembranes consisting of a fibreglass fleece and a non-woven geotextile impregnated with blown oxidized- or rubber-modified bitumen. They can tolerate rough subgrades and have very low thermal expansion coefficients. They are particularly well suited to sub-freezing climates and can be installed and welded at temperatures as cold as −25 °C.

BGMs exhibit outstanding oxidative and UV resistance without the need for antioxidants and stabilizers such as those used in polymeric geomembranes such as HDPE, LLDPE, fPP, PVC and so on. BGMs' stability to weathering and oxidation is an inherent property of the bitumen which is intrinsically stable due to its structure being composed of highly condensed polycyclic aromatic hydrocarbons.

Bituminous geomembranes have been used in cold climates for many years in such diverse applications as lining of dam faces in high altitude regions such as the Alps, lining mine process liquor dams and robust applications like waste rock containment pads for mines. Their flexibility makes them particularly well suited for liner and caps for landfills.

📌 *Bituminous geomembranes consist of a synthetic fabric saturated by a modified bituminous – or an oxidized bituminous blend.*

10.3 COMPOSITION AND CONSTRUCTION

A prefabricated bituminous geomembrane (PFBGM) is defined as a material fabricated in a production facility and consisting principally of a synthetic fabric, an oxidized or an elastomeric modified bitumen blend incorporating a filler.

These BGMs are multilayer composite liners (see Figure 10.1) that comprise:

- a bitumen-impregnated fibreglass reinforcing layer;
- a bitumen-impregnated nonwoven polyester geotextile core;
- an SBS[1] elastomer-modified bitumen layer;
- a top roughened surface layer for enhanced friction.

Bituminous geomembranes are very thick (3–5 mm) compared to conventional geomembranes and are able to withstand installation-related mechanical damage. Their coefficient of thermal expansion is essentially zero (the CTE of a BGM is approximately 1×10^{-6} cm/cm/ °C, approximately 100–200 times less than for HDPE) and so they lie flat in close contact with the subgrade. Advantageously they can also cope well with differential settlement. Their high relative mass per unit area (3.5–6 kg/m^2) means they are not prone to uplift in heavy winds and therefore require less ballasting (i.e. sandbagging). Importantly, they can be thermally welded.

Due to the viscoelastic nature of the elastomer-modified bitumen compound, bituminous geomembranes are 'self-healing' since the viscous mass can flow to seal small penetrations. The puncture resistance is important when the geomembrane is installed under onerous conditions, for example, on a machine graded subgrade with angular stones under

---

[1] Where SBS is a synthetic rubber based on styrene–butadiene–styrene.

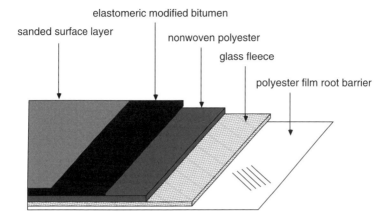

Figure 10.1 Structure of bituminous geomembranes (BGMs) which consist of a nonwoven polyester geotextile impregnated with modified bitumen. BGMs are engineered geomembranes consisting of a non-woven geotextile and a fibreglass fleece impregnated with blown oxidized- or rubber-modified bitumen. They can tolerate rough subgrades and have very low thermal expansion coefficients. They are particularly well suited to sub-freezing climates and can be installed and welded at well below 0 °C

high loads. Tests have been performed on bituminous geomembranes using 20–40 mm crushed stone on the geomembrane with applied head equivalent to 40 m of refuse, with no resulting puncturing of the membrane.

Bituminous geomembranes are able to operate at higher strain due to the viscous nature of the bitumen and the tensile properties of the long fibre non-woven geotextile which result in a significant stress relaxation after loading (a stress relaxation of 70% occurs after 3 h). Field seams in bituminous liners can be made with a propane torch, however the quality of field seams is sometimes poor making the seams the weak points.

Whilst bitumen/polyester liners are a more costly material than HDPE this is offset by its comparative ease of installation and lack of the need for specialized thermal welding equipment since seams can be joined simply by using a gas torch.

An example of a field application is where bituminous liners comprising sheets measuring 1 m × 4 m in size and 13 mm thickness have been used in South America. These were composed of 65% bitumen and 35% mineral fillers reinforced with fibreglass. The sheets were spliced together with 100–150 mm overlap and bound together with hot applied oxidized bitumen and bituminous mastic.

Typical properties of bituminous geomembranes are shown in Table 10.1.

A comparison of the mechanical properties of bituminous geomembranes with other reinforced geomembranes is presented in Table 10.2.

10.4 LONGEVITY OF BITUMINOUS GEOMEMBRANES

The tensile properties of bituminous geomembranes are primarily derived from the polyester nonwoven geotextile at its core. This polyester geotextile is well protected

Table 10.1 Typical properties of bituminous geomembranes (Coletanche)

Property	Value
Thickness (ASTM D-5199)	3.8 mm (minimum)
Mass per unit area (ASTM D-5261)	4000 g/m^2 (minimum)
Density (ASTM 1505)	1.1 g/cm^3
Tensile properties (ISO 1421, Stress at Break)	22.8 kN/m (longitudinal direction)
	18.9 kN/m (transverse direction)
Elongation at break	50% (longitudinal direction)
	50% (transverse direction)
Tear resistance (ASTM D-751)	165 N (longitudinal direction)
	203 N (transverse direction)
Static puncture resistance (FTMS 101C)	>2.0 kN
Cold bending (ASTM D-746)	$-15\,°$C
Dimensional stability (ASTM D-1204)	<+0.20% (longitudinal direction)
	<+0.06% (transverse direction)

Table 10.2 Comparison of the mechanical properties of bituminous geomembranes with other reinforced geomembranes (Geosynthetics Specifier Guide, 2008)

Geomembrane	Thickness	Tensile strength (kN) (ASTM D-751)	Elongation (%) (ASTM D-751)	Tear strength (kN) (ASTM D-5884B)	Puncture strength (kN) (FTMS 101C)
RPP	1.14 mm	1.2	~25	0.31	1.33
EIA	0.75 mm	2.46	–	0.55	1.56
BGM	4.9 mm	25	>55	0.60	0.6

RPP = reinforced flexible polypropylene and EIA = ethylene inter alloy polymers.

from ageing and degradation since it is totally impregnated and encased with bitumen and thus no significant degradation of its properties is expected.

The life expectancy of bituminous geomembranes is more than 20 years based on French field data on dams that were waterproofed with bituminous membranes that were exposed to an unprotected environment for more than 20 years. Furthermore, testing of a bituminous geomembrane exhumed from motorway ditches and the Guazza Reservoir in Corsica after 20 years and 7 years, respectively, showed no significant reduction in tensile strength (IFC, 2008).

10.5 SPRAY-IN-PLACE BITUMEN MEMBRANES

Spray-in-place bitumen linings consist of a film of bitumen mixed with a synthetic polymer spayed at high temperature (approximately 200 °C) to form an integral waterproof barrier. The bitumen for the liner is mixed at or near the job and transported by trucks to the point where it is applied.

Some problems that can occur with spray-in-place bitumen liners include cracking in cold weather, weed eruption/penetration and breaching. Often a geotextile is laid on the soil before spraying or otherwise the bitumen may not bond well to the subgrade. It is important that the geotextile substrate is anchored in place with geotextile stakes or else the sprayed section may wash away, particularly in channel lining applications (Swihart *et al.*, 1994).

10.6 DESIGN LIFE OF BITUMINOUS GEOMEMBRANES

Bituminous geomembranes such as Coletanche™ ES3 are expected to have a design life of hundreds of years. For instance, on the basis of their expected long life bituminous geomembranes have been used by the French Atomic Energy Authority to cap a low- and medium-level radioactive waste stockpile where the waste has a radiation half-life of 300 years. This expected long service life contrasts with the uncertainty over the expected life of some HDPE geomembranes which can be susceptible to failure by environmental stress cracking and oxidative stress cracking.

The degradation of the polyester geotextile and loss of strength has not been observed to be a significant problem with bituminous geomembranes, as the geotextile core of these geomembranes is completely protected by bitumen (i.e. the polyester fibres are completely encapsulated by the bitumen coating). For this reason UV, oxidative, hydrolysis, chemical and biological processes are prevented from acting on the geotextile scrim. Testing of geotextiles exhumed from permanent installations and subject to various degrading processes has shown very little loss of mechanical properties. The major contributions to loss of required properties were from installation damage. It is concluded that the expected lifetime of bituminous geomembranes is likely to be in excess of the lifetimes normally expected from geosynthetics.

REFERENCES

Coletanche, 'Bituminous Geomembranes Property Guide (Coletanche™ ES3)' [www.coletanche.com].

'Geosynthetics Specifier Guide' (2008).

IFC, Presentation entitled 'Bituminous Geomembranes (BMGs), Coltanche™ ES3' [http://www.ifc.org/IFCExt/spiwebsite1.nsf/b7a881f3733a2d0785256a550073ff0f/f95106404992f6418525716c0059694e/%24FILE/AttachmentC.AppendixA.Tech_Doc_ES3%20_2_.pdf] (2008).

Swihart, J., Comer, A. and Haynes, J. 'Deschutes Canal Lining Demonstration Project: Durability Report Year 2', US Bureau of Reclamation, R-94-06 (1994).

11

Specialty Geomembranes and Liners

This chapter covers those geomembranes and liners that are used in specialty or niche applications, on account of their unique combination of performance properties. These materials include ethylene vinyl acetate (EVA), butyl rubber (BR), nitrile rubber (NBR), chlorinated polyethylene (CPE), polyurethane (PU), acrylic, poly(vinylidene fluoride) (PVDF) and chemical vapour barriers (CVBs).

11.1   EVA LINERS/GEOMEMBRANES

Ethylene vinyl acetate has the highest flexibility of all polyolefin liners and is often used as a replacement for PVC membranes in pond linings and metallic tank linings. Its softening point ($\sim$85–95 $^\circ$C) however is lower than that of LDPE so it is not recommended for exposed use (i.e. uncovered) in warm climates. Plastika Kritis produces various types of co-extruded EVA membranes, with vinyl acetate levels between 9 and 18% (note the higher the VA level, the more elastic the membrane but the lower the softening point) (Plastika Kritis, 2008). The properties of EVA geomembranes are listed in Table 11.1 while Table 11.2 compares the performance properties of a 1 mm EVA liner with a 1 mm HDPE liner.

Note that EVA liners have shortcomings owing to their 'sticky' nature, such as blocking (i.e. the liner sticking to itself on rolls) and taste issues (i.e. imparting taste to potable water). To overcome these limitations a coextruded membrane of LLDPE/EVA/LLDPE (10:80:10) is available where the skin layers of LLDPE prevent blocking and taste problems while the main core layer of EVA gives outstanding flexibility. This also overcomes most of the issues associated with LLDPE tank liners, being 'star fractures' due to crinkling.

EVA liners are made in widths of 7.4 m which makes it easier to weld and largely eliminates cross joints.

A Guide to Polymeric Geomembranes: A Practical Approach   J. Scheirs
© 2009 John Wiley & Sons, Ltd

Table 11.1   Properties of EVA geomembranes (Kritiflex™ from Plastika Kritis) (Plastika Kritis, 2008)

Property	Test method	0.5 mm	0.75 mm	1 mm
Density of carbon black containing sheet	ASTM D-792	0.950	0.950	0.950
Melt flow index (g/10 min)	ASTM D-1238	0.40	0.40	0.40
Tensile strength at yield (N/mm^2)	ASTM D-638IV	4.5	4.5	4.5
Tensile strength at break (N/mm^2)	ASTM D-638IV	29	29	29
Elongation at yield (%)	ASTM D-638IV	24	24	24
Elongation at break (%)	ASTM D-638IV	850	850	850
Puncture resistance (N)	ASTM D-4833	140	180	220
Tear resistance (N)	ASTM D-1004C	40	60	80
Carbon black content	ASTM D-1603	2–3	2–3	2–3

Table 11.2   Comparison of properties of HDPE with EVA Geomembranes (Kritiflex™ from Plastica Kritis) (Plastika Kritis, 2008)

Property	Test method	1 mm HDPE	1 mm EVA
Density of carbon black containing sheet	ASTM D-792	0.946	0.950
Melt flow index (g/10 min)	ASTM D-1238	0.15	0.40
Tensile strength at yield (N/mm^2)	ASTM D-638IV	17	4.5
Tensile strength at break (N/mm^2)	ASTM D-638IV	32	29
Elongation at yield (%)	ASTM D-638IV	12	24
Elongation at break (%)	ASTM D-638IV	850	850
Puncture resistance (N)	ASTM D-4833	360	220
Tear resistance (N)	ASTM D-1004C	130	80
Carbon black content	ASTM D-1603	2–3	2–3

EVA liners are the most flexible type of polyolefin geomembranes. These liners are characterized by softness, ease of handling and welding and are more resistant to tear and puncture than LDPE but less resistant than HDPE. Due to their relatively low softening points, they should not be used uncovered in hot climates.

11.2 BUTYL RUBBER LINERS

Butyl rubber (BR) geomembranes were a forerunner to EPDM geomembranes. BR liners are highly flexible, highly weather resistant, highly extendable and durable materials.

Butyl rubber, an elastomer with molecular crosslinks is produced by the copolymerization of isobutylene and small amounts of isoprene. Its various properties make it an ideal material for lining water tanks and ponds. Butyl rubber is elastic, not just flexible and regardless of temperature, it reverts to its original form.

Butyl rubber actually stretches better than most other polymeric materials, because it stretches uniformly. Butyl rubber sheeting also has outstanding ageing and weather-resistant properties. Its crosslinked molecular structure gives excellent protection against ageing over a long period of time even when exposed to the atmosphere, sunlight, ultraviolet radiation and ozone.

Butyl rubber sheeting contains no plasticizers or additives that can evaporate or be leached out over time. Its strength and elasticity remain virtually unchanged over many years, without shrinking, hardening or cracking.

Butyl sheeting remains flexible and is generally unaffected by temperature over the range −30 to +120 °C.

Butyl rubber liners being highly elastic, stretch at least 300% while still retaining water-impermeable properties when fully extended. Due to butyl rubber's high degree of flexibility, damage by rough handling is minimized, and it readily adapts itself to soil subsidence and surface irregularities. It has good resistance to abrasive wear, flex-cracking and puncturing (see Table 11.3). However, it can be readily damaged by sharp tools, spades, picks, knives and the like (Tanks Australia, 2008).

It is recommended that clean rubber soled footwear is used when installing or inspecting butyl rubber liners. Since butyl rubber liners contain no plasticizers which can be leached or be washed out, it is non-toxic to animals, fish or plant life. Butyl rubber sheeting is typically available in standard thicknesses of 0.75 mm, 1.0 mm 1.5 mm in both standard and potable water approved grades.

Butyl rubber is nowadays often overlooked in favour of the cheaper and superior quality EPDM geomembranes. It still finds some application in long-life applications such as building works, channel liners and exposed pond liners where durability is an important factor. Butyl rubber geomembranes have demonstrated performance in channel linings in the United States where a study showed that there was no significant change in

Table 11.3 Butyl rubber geomembrane typical properties

Property	Value
Thickness	0.75 mm
Nominal weight per unit area	1 kg/m^2
Tensile strength	8 Mpa
Modulus at 30%	5.5 Mpa
Elongation at break	350%
Tear strength	30 N/mm
Ozone resistance (7 days @ 50 pphm and 30 °C	50% extension and no cracks
Brittle point	−30 °C
Properties after ageing for 168 h at 121 °C	
Tensile strength	6 MPa
Elongation at break	200%

mechanical properties after nine years of service in both covered and exposed applications (Irrigation Channel Lining, 2008).

The life of butyl rubber lining applications is highly dependent upon the quality of their splices and hence a high degree of care must be exercised in their fabrication. Factory seams are vulcanized and their quality and strength are generally very good. Field seams however are glued with a rubber cement, and their strength is less than the parent material. While field seams are adequate for most applications, applied peel forces on the seams can be damaging.

11.2.1 ATTRIBUTES OF BUTYL RUBBER

- excellent elongation;
- not affected by UV (hence long life outdoors);
- resists compressive loads and ground movement;
- reverts back to original form due to its highly elastic nature.

Butyl rubber geomembranes are durable liners with a combination of properties that make them particularly suitable for use in the storage, treatment, conveying and containment of water, effluent and chemical solutions. Butyl rubber has excellent weathering properties and a good general resistance to attack from most chemicals.

Butyl geomembranes have been used successfully as lining membranes for reservoirs, effluent lagoons, irrigation channels, waste disposal pits, metal tanks, ornamental pools and reed filtration beds.

Butyl liners can be fabricated to the required size using uncured butyl tape and a special heat bonding process.

Butyl rubber is often overlooked as it is more expensive than EPDM rubber, as shown in Table 11.4.

The reason why some butyl rubber is more expensive is because of the natural rubber content (approximately 28%) in the material whereas EPDM is 100% synthetic rubber (Varnamo, 2008).

Table 11.4   Relative costs of butyl rubber compared with other geomembranes (as of 2008)

Geomembrane type and thickness	Cost relative to 1.5 mm HDPE
HDPE (1.5 mm)	1.0
EPDM (1.5 mm)	2.09
Butyl rubber (1.5 mm)	3.0

11.3 NITRILE RUBBER GEOMEMBRANES

Nitrile rubber or acrylonitrile–butadiene rubber is a copolymer of butadiene and acrylonitrile. It is known for its good general resistance to oils and petroleum products. Nitrile rubber also has good mechanical properties, especially tensile strength, flexing,

Table 11.5 Typical properties of nitrile rubber liners

Property	Value
Tensile strength	10–13 MPa
% Elongation	250–350%
Tear strength	25–35 N
Working temperature	-35 to $+110\,^{\circ}C$

compression set and impermeability to gases (see Table 11.5). It has, however, only moderate resistance to cold, moderate ageing properties and relatively low tear strength.

Nitrile rubber liners have excellent resistance to general aliphatic hydrocarbons but limited resistance to aromatic hydrocarbons.

It has good resistance to inorganic chemicals except oxidizing agents such as chlorine. Since nitrile is a polar rubber means it is not recommended for use with polar liquids like ketones, ethers, amines or esters (e.g. biodiesel) which may soften and swell it.

Nitrile geomembranes are generally manufactured from nitrile rubber or alloys of nitrile with PVC. The nitrile rubber acts as a plasticizer for the PVC and provides greater chemical resistance than that of PVC alone. The nitrile rubber also gives the geomembrane greater flexibility than that of PVC.

11.3.1 TYPICAL NITRILE RUBBER FORMULATION

The carbon black content of nitrile rubbers is typically around 27% as indicated by ASTM D-3187-06 'Test Methods for Rubber – Evaluation of NBR (Acrylonitrile–Butadiene Rubber)'. See Table 11.6 for a typical NBR test recipe.

Table 11.7 gives a comparison of nitrile rubber with other rubber liners.

Nitrile rubber can vary greatly in its resistance to ozone and oxidation. Standard nitrile (or conventional nitrile) rubber without antiozonants has low or no ozone resistance (see Figure 11.1). The properties of nitrile rubber and its ozone resistance are mainly determined by the degree of unsaturation (i.e. double bonds). Hydrogenated nitrile rubber where the double bonds in the polymer backbone are removed by hydrogen saturation is far superior to standard nitrile rubber in terms of ozone resistance since the double bonds

Table 11.6 Typical ingredients in nitrile rubber liners

Ingredient	Content (phr)	Content (%)
NBR	100.0	68.4
Carbon black	40.0	27.4
ZnO	3.0	2.0
Stearic acid	1.0	0.6
TBBS	0.7	0.6
Sulfur	1.5	1.0
Total	146.2	100

ZnO = zinc oxide (activator for sulphur vulcanization) and TBBS = N-tert-butyl-benzothiazole sulfoamide (cure accelerator).

Table 11.7 Comparison of nitrile rubber with other rubber liners[a]

Property	EPDM	Butyl rubber (BR)	Nitrile rubber (NBR)
Tensile properties	3	3	3
Tear strength	2	2	2
Abrasion resistance	3	3	3
Oil/fuel resistance	1	1	3–4
Weathering/ozone resistance	5	4	1–2
Minimum temperature ($^{\circ}$C)	−35	−30	−35
Maximum temperature ($^{\circ}$C)	120	120	90

[a] 1, poor; 2, satisfactory; 3, good; 4, very good; 5, excellent.

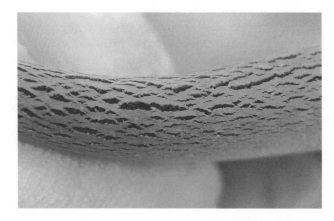

Figure 11.1 Photograph of a nitrile rubber tank liner (thickness 1.5 mm) exhibiting severe ozone cracking on folds. The material was shown to have little ozone resistance by ozone exposure testing according to ASTM D 1149-07 'Standard Test Methods for Rubber Deterioration-Cracking in an Ozone Controlled Environment'

that the ozone reacts with have been removed by hydrogen gas treatment (Nakagawa *et al.*, 1992).

The ozone resistance for regular (unprotected) nitrile rubber when tested under standard conditions (of 50 pphm ozone at 40 $^{\circ}$C and 20% strain/sample extension) is such that visible cracking is observed in less than 24 h (see Figure 11.2). In contrast, hydrogenated nitrile rubber under the same test extends this time to 168 h.

The response of nitrile rubber to ozone exposure when unprotected by antidegradants, antioxidants, antiozonants and stabilizer ingredients is a lot like natural rubber which is attacked by ozone severely. An unprotected nitrile rubber compound will not pass any of the ozone test methods, including ASTM D-1171 and the more recent ASTM D-1149. Nitrile rubber can be blended with PVC to markedly improve its ozone resistance, such as in nitrile–PVC (NBR/PVC) blends.

Nitrile rubber varies in its ozone resistance characteristics in the following order of increasing resistance to ozone cracking:

- standard unprotected nitrile (NBR);
- standard protected nitrile (NBR + antiozonants);

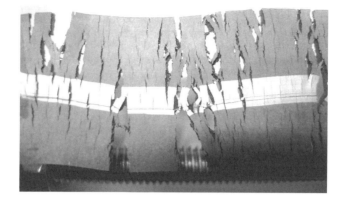

Figure 11.2   Photograph of a nitrile rubber tank liner (thickness 1.5 mm) exhibiting severe ozone cracking after exposure to an ozone concentration of 50 pphm (parts per hundred million) for 48 h under 20% strains (i.e. elongation) in accordance with standard method ASTM D-1149-07

- hydrogenated nitrile (HNBR);
- nitrile–PVC blends (NBR/PVC).

Ozone attack on a rubber begins at the surface of the rubber. Ozone exposure causes oxidative chain scission and formation of rapid cracks. As the cracks progress, 'fresh' rubber is constantly exposed to ozone and hence deeper cracks result (Keller, 1989).

Only a few parts per hundred million of ozone in air can cause rubber cracking. Ozone will attack any rubber with backbone unsaturation (i.e. double carbon bonds in the molecular chains of the rubber). Degradation results from the reaction of ozone with the rubber double bonds. Unstretched rubber reacts with ozone but does not crack. Unprotected stressed rubber will crack rapidly and extensively (as shown in Figure 11.2). As the stressed rubber chains cleave under ozone attack, a new high-stress surface is exposed. This process results in visible and progressive cracking and the cracking is always perpendicular to the applied stress (Lattimer *et al.*, 1989).

11.4   CHLORINATED POLYETHYLENE (CPE)

Chlorinated polyethylene (CPE) has a similar backbone structure to HDPE except that chlorine atoms have been randomly introduced along the sides of the HDPE backbone, replacing some hydrogen atoms. The chlorine atoms are much bulkier than the hydrogen atoms and these disrupt the formation of crystallinity. The amount of chlorine that is introduced, and the randomness of their placement determines the extent to which the resulting polymer will be non-crystalline (i.e. amorphous). As a result CPE geomembranes are more flexible materials than HDPE geomembranes.

CPE is characterized by good chemical resistance and UV resistance along with reasonable flexibility. Its chemical structure is part way between PVC and CSPE and it has been used as a transition strip between these two geomembrane types.

CPE materials with 36% chlorine are commonly used for thermoset elastomer applications. This type of CPE exhibits a good balance of chemical and solvent resistance, low-temperature properties, processability and can be peroxide-cured. CPE resins with an even lower chlorine content of 25% are available and these contain residual polyethylene crystallinity that improves the compatibility with polyethylene. CPE used in geomembranes is designed to be amorphous and contains essentially no residual crystallinity from the polyethylene structure and as such they exhibit the properties of rubbery materials.

11.5 POLYURETHANE (PU) GEOMEMBRANES

Thermoplastic polyurethane (PU) liners (e.g. CoolThane[TM] from Cooley) bridge the gap between highly flexible rubbers (e.g. EPDM or nitrile) and semi-rigid plastics (e.g. HDPE or LLDPE). PU is the reaction product of polyester polyols or polyether polyols with diisocyanates. These rubber-like liners have been used in diverse service applications from secondary containment of petroleum tanks in Alaska to portable water storage in Arizona (Cooley, 2005).

PU liners and geomembranes offer a number of advantages over other competing geomembrane materials (see Table 11.8).

PU liners and geomembranes are resistant to the following types of chemicals:

- Animal fats
- ASTM Fuel A
- Castor oils
- Cotton oils
- Crude oils
- Diesel fuels
- Gasoline
- Glycerol
- Hexane
- Kerosene
- Lubricating oils
- Mineral oils
- Naphtha
- Transformer oils
- Vegetable oils

Note that there are two different classes of PU liners. Polyester PU liners are more resistant to fuels, have a lower diffusion rate to fuels and offer better ageing (oxidation) resistance. Polyether PU liners on the other hand provide superior low-temperature properties and are also inherently stable when exposed to high humidity and are naturally more fungus resistant.

Cooley has developed a range of polyurethane (PU) geomembranes designed for primary and secondary containment of fuels and other aggressive products. Cooley has two

Table 11.8 Advantages of polyurethane liners and geomembranes over competing geomembranes

Attribute	Comments
High abrasion resistance	Higher than most other polymeric liners. PU liners can undergo 12 000 abrasion cycles (using a H18 wheel under 1000 g load in the 'Taber Abrasion test') before the surface is abraded to expose the fabric reinforcement. Actual field results are even better due to the natural rebound elasticity of PU, which is not measured in standardized laboratory tests
High scuff resistance	Higher than most other polymeric liners at equivalent hardness
Flexibility at low temperatures	PU liners have been used successfully in extremely low temperature environments around the Arctic Circle. PU remains flexible in temperatures as low as $-51\,°C$.
High strength	PUs are extremely tough compared to other thermoplastics. This strength and durability allows a thinner gauge to be used compared to competing materials
High mildew resistance	Certain PU grades are inherently resistant to mildew, bacteria and fungus growth, making them an excellent choice for marine environments
Tear resistance	PU is much more difficult to tear when it has a nick or notch in it (for example, from shipping damage or dragging over the ground) compared to other polymeric liners
Weldability	PU's unique melt characteristics allow it to be fabricated using a wide variety of methods such as radio frequency (RF), thermal and ultrasonic methods
Chemical resistance	PU is chemically resistant to ozone, petroleum products, fats, greases and a variety of solvents (see list above)

urethane products, one being ester-based and the other ether-based. Both of the products are available with several reinforcement configurations to meet specific needs.

The ester-based products consist of Cooley's exclusive extrusion laminated urethane coating on select substrates. These products are highly resistant to fuels, extremely flexible and highly puncture resistant, with excellent heat sealing characteristics. The ether-based product offers good resistance to fuels, good strength characteristics and excellent low-temperature properties. These PU products are strong and hydrolytically stable for years of service in fuel and chemical applications (Cooley, 2005).

High performance Coolthane™ liners comprise a combination of high strength fabrics, specially developed ester-based or ether-based urethane extrusion and a proprietary manufacturing process. In addition to its high fuel and chemical resistance, the liners exhibit very high physical performance (see Table 11.9).

PU-R with nylon reinforcement is produced by the Seaman Corporation under the product name 1932 PTF. It has a thickness of 0.88 mm, a very high tensile strength of 4.9 kN/m (which is some five times that of fPP-R) and seam strengths between 2.5–4.5 times those of fPP-R. It also exhibits excellent seam strength and low temperature properties (the low-temperature brittleness temperature is $-46\,°C$) (see Table 11.10).

Table 11.9 Properties of PU-R geomembranes with polyester reinforcement (CoolThane™, from Cooley) (Cooley, 2008)

Property	CoolThane™ L1612ESU	CoolThane™ L1023DEP	CoolThane™ FML 87	CoolThane™ L3284NESU
Thickness (ASTM D-5199)	0.50 mm	0.76 mm	0.89 mm	1.14 mm
Tensile strength (ASTM D-751) (MD/TD)	1.4 kN/m 1.1 kN/m	1.6 kN/m 1.3 kN/m	1.8 kN/m 1.6 kN/m	4.4 kN/m 4.4 kN/m
Tear strength (ASTMD-5884, Method B)	290 N	710 N	670 N	180 N
Hydrostatic resistance (ASTM D-751 A)	3448 kN/m^2	2758 kN/m^2	3447 kN/m^2	4138 kN/m^2
Ply adhesion (ASTM D-413)	2600 N/m	3500 N/m	3500 N/m	2600 N/m
Low-temperature brittleness (ASTM D-1790)	$-43\,°C$	$-54\,°C$	$-43\,°C$	$-46\,°C$

Table 11.10 Comparison of Properties of PU-R with fPP-R

Property	0.9 mm fPP-R	0.88 mm PU-R
Reinforcement type	Polyester	Polyamide (nylon)
Specific gravity	0.92–0.96	1.20
Tensile strength (ASTM D-751)	1.0 kN/m	4.9 kN/m
Indicative seam strength in shear (ASTM D-751)	0.78 kN/m	3.56 kN/m
Indicative seam strength in peel (ASTM D-413)	3.50 kN/m	8.77 kN/m
Low-temperature brittleness (ASTM D-1790)	$-40\,°C$	$-46\,°C$

11.6 ACRYLIC-COATED GEOMEMBRANES/GEOTEXTILE COMPOSITES

Geomembranes are impermeable but have relatively poor tear/puncture resistance while geotextiles are permeable and have excellent tear/puncture resistance. By combining these two products the benefits of both can be achieved. The main issues in creating geomembrane and geotextile composites are (i) permanently bonding the two materials which have very different melting points and (ii) the limited UV resistance of geotextiles.

Infrastructure Technologies (Australia) Pty Limited (Milsons Point, NSW) has addressed these problems by laser bonding a polyethylene or fPP liner between two layers of polyester geotextile and then spraying the final product with a flexible UV-resistant acrylic coating (similar to long-life acrylic paint). This acrylic-based geomembrane, called the ITM Liner™, can be used for canal linings and also for lining water storages. This novel geomembrane is a recent entry into the liner industry.

This composite liner can be wedge welded and extrusion welded before application of the acrylic topcoat. The peel adhesion values are 95 MPa for fusion welds, respectively, as measured by ASTM D-4437.

Attributes of the ITM Liner™ include:

- High strength and flexible impermeable membrane structure.
- Possibility for finishing coat to be a light colour for minimizing thermal impact.
- Bonding process ensures geofabric and membrane impermeability.
- Climate control testing showed good performance down to −60 °C.
- If properly managed and maintained, the ITM Liner™ has the potential to outlast other geomembranes for applications such as concrete channel linings.
- Excellent conformability to uneven subgrade.
- Its textured surface allows 'walk-in, walk-out' for humans, livestock and fauna without slippage or damage. Hence no fencing is required to make sure humans and/or animals cannot be trapped in channel structures.
- The geofabric base has friction capabilities for batter and base integration to reduce liner movement during installation and ongoing service life.

The acrylic-coated liner also has good tear strength and low water vapour transmission (see Table 11.11).

The liner installation involves the following steps:

- Rolling out the liner.
- Joining the geofabric to headwall and related structures.
- Joining of adjacent rolls *in situ* through thermal bonding (i.e. wedge welding).
- Applying a saturating coat of the acrylic coating, 0.6 l/m^2 at an application rate of 4500 m^2/man/day and allowing 24 h to cure. Coating must be applied over all surfaces.
- Roll finished membrane into anchoring furrows and compact.

Table 11.11 Typical properties of ITM acrylic-coated liners (Infrastructure Technologies)

Property	Test method	Values
Thickness	ASTM D-1777	>1.70 mm
Melting point	ASTM D-1777	260 °C
Mean breaking load	ASTM D-4632	800 N (MD)
		900 N (TD)
Elongation at break	ASTM D-4632	1.9% (MD)^a
		1.2% (MD)^a
Wide width tensile strength	ASTMD-4885	11.5 kN/m
Tongue tear strength	ASTM D-751	120 N (MD)
		150 N (TD)
Trap tear strength	ASTM D-4533	129 N (MD)
		157 N (TD)
Water vapour transmission	ASTM E-96	0.04 g/h m^2

[a]Breakage of geotextile; note however that fPP inter layer remains intact.

Maintenance and repair of the ITM liner involves cleaning off any dirt or grime, with high pressure water and then respraying the UV-resistant acrylic coating over the area to be repaired.

11.7 PVDF LINERS

PVDF has the same basic structure as HDPE except that every carbon atom in the polymer backbone has one fluorine atom attached to it (see Figure 11.3). The fluorine atoms impart outstanding chemical and UV resistance. PVDF fluoropolymer liners provides a unique combination of chemical resistance and UV resistance coupled with a broad operating temperature range and intrinsic fire resistance.

PVDF geomembranes are available from Cooley Engineered Membranes under the tradename Coolshield™.

Table 11.12 shows the properties of the PVDF geomembranes relative to other common geomembrane materials.

PVDF liners (e.g. Coolshield™ PVDF from Cooley) have a thickness of 0.64 mm and are based on a 173 g/m² knitted polyester scrim coated with PVDF to give a total weight of 1020 g/m². Typical properties are shown in Table 11.13.

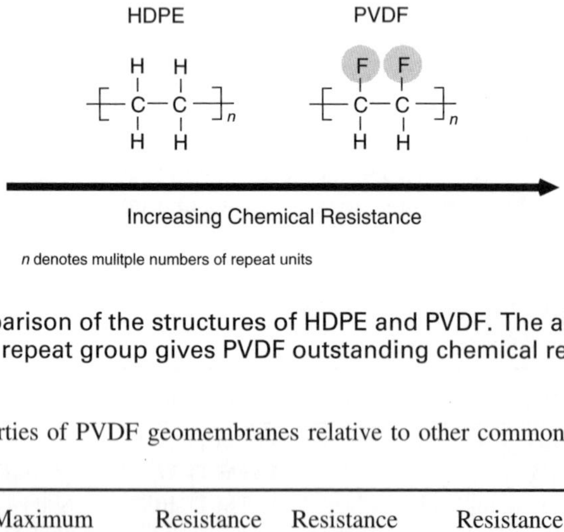

Figure 11.3 Comparison of the structures of HDPE and PVDF. The addition of two fluorine atoms per repeat group gives PVDF outstanding chemical resistance

Table 11.12 Properties of PVDF geomembranes relative to other common geomembrane materials[a]

Membrane	Maximum operating temperature (°C)	Resistance to acids	Resistance to alkalis	Resistance to strong oxidizing agents	Resistance to hydrocarbons
PVDF	150	E	E	G	E
Polypropylene	95	E	E	P	P
Polyethylene	85	E	E	P	P
PVC	80	G	G	F	F

[a] E, excellent; G, good; F, fair; P, poor.

Table 11.13 Performance properties of 0.64 mm thick Coolshield™ PVDF scrim reinforced liners (Cooley, 2005)[a]

Property	Typical value
Tensile strength (grab) (ASTM D-751A)	1560 N
Tensile strength (25 mm wide strip) (ASTM D-751B)	1070 N
Tear strength (tongue) (ASTM D-751B Mod)	712 N
Puncture (screwdriver) (ASTM D-751)	334 N
Hydrostatic resistance (ASTM D-751A)	2.8 MPa
Low-temperature limit (ASTM D-2136)	$-26\,°C$
High-temperature limit (continuous) (ASTM D-1204)	$150\,°C$
High-temperature limit (intermittent) (ASTM D-1204)	$165\,°C$
Chemical resistance	Resistant to almost all chemicals except strong alkalis

[a] A Coolshield™ PVDF membrane datasheet is available from Cooley Engineered Membranes.

The product development engineers at Bixby USA have developed a three-ply system consisting of 0.5 mm PVDF, non-woven fabric and 1.5 mm of LLDPE. This construction can be made in excess of 1800 mm in width. This three-ply system gives a combination of chemical resistance, dimensional stability and cost effectiveness. The wider the width the better to reduce the amount of field seams. The construction is as follows:

- 0.5 mm natural homopolymer PVDF;
- DuPont Sontara 8001 polyester non-woven fabric;
- 1.5 mm natural LLDPE.

The finished width is 1828 mm, which is the current maximum capability of the production equipment.

11.7.1 PVDF ANCHOR SHEET (STUDDED LINER)

Protective concrete liners based on PVDF are available in various thicknesses (from Kerton Plastics Ltd, Swansea, UK). Due to its unique anchoring system it is the long-term solution to many containment problems, particularly where aggressive chemicals are involved.

A two-layer sheet is available with a signal layer to indicate mechanical damage (i.e. if the white signal layer is visible). Strong anchoring of the studs (or anchor knobs) in the horizontal and longitudinal directions is due to the diagonal order of the studs. There is no welded joint (i.e. weak point) between sheet and stud as the studs are produced during the extrusion process, therefore, shear forces up to 1800 N/stud in the horizontal and longitudinal directions can be resisted. The extremely high pullout resistance is 42 t/m^2 due to the high number of studs (420 studs/m^2) and the specially designed profile.

11.8 CHEMICAL VAPOUR BARRIER MEMBRANES

Chemical Vapour Barrier (CVB) membranes are often used to prevent release of volatile compounds from soil contaminated with BTX or chlorinated hydrocarbons. BTX is primarily benzene, toluene, ethyl benzene and xylene. BTX and chlorinated hydrocarbon vapours can diffuse through standard geomembranes such as HDPE, LLDPE, fPP and PVC membranes and liners. A CVB geomembrane can be installed on contaminated sites to block vapour pathways.

Volatile Organic Compounds (VOCs), petroleum hydrocarbons (PHs) and chlorinated hydrocarbon (CHC) solvents are all examples of contaminants which may exist in contaminated soil and that can emit vapours that may migrate upward through unprotected building floor slabs.

Major volatile chemical contaminants associated with the need of a vapour barrier include volatile organic compounds, such as tetrachloroethylene (PCE) and trichloroethylene (TCE) which are used as industrial degreasers and dry cleaning agents. Chronic exposure to PCE or TCE has been shown to cause damage to the liver, the kidneys and the central nervous system. It may also lead to increased risk of cancer. Other contaminants such as petroleum hydrocarbons can also pose a risk.

The degree of permeation of various hydrocarbons and chlorohydrocarbons through a standard HDPE geomembrane liner is shown in Figure 11.4. It can be seen that HDPE geomembranes are a poor barrier for hydrocarbons, especially chlorohydrocarbons (such as trichloroethylene) and aromatics (such as toluene and xylene).

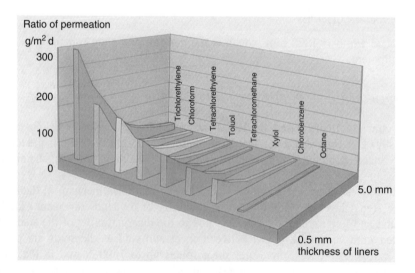

Figure 11.4 Graphical representation of permeation of various hydrocarbons and chlorohydrocarbons through a standard HDPE geomembrane liner, without an aluminium barrier. Reproduced by permission of Agru

11.8.1 INTRODUCTION TO COMPOSITE LINERS

Common vapour barrier geomembranes are comprised of various composite structures such as:

- LLDPE/aluminium layer;
- LLDPE/aluminium layer/LLDPE;
- LDPE/bonding layer/aluminium layer/bonding layer/LDPE;
- HDPE/aluminium layer/HDPE.

The aluminium vapour barrier geomembranes do however have a low coefficient of elasticity and the aluminium foil may rupture if subject to excess strains.

These membranes are generally reinforced with a polyester scrim to give protection to the encapsulated aluminium foil layers and to prevent excessive deformation.

Another option is the following construction that uses a polymeric gas barrier layer based on poly(vinylidene chloride) (PVDC). For example:

- HDPE/PVDC/HDPE
- LLDPE/PVDC/LLDPE

The LLDPE/PVDC/LLDPE systems have better multi-axial strain properties than the aluminium liner laminates. LLDPE/PVDC/LLDPE liners can tolerate higher strain; as PVDC is thermoplastic there is no danger of rupture of the barrier layer under moderate strains.

11.8.2 POLYETHYLENE ALUMINIUM COMPOSITES

Vapour barrier geomembranes often have a thin metallic sheet encapsulated between polymeric layers (generally HDPE) to provide a complete barrier to the permeation of hydrocarbons, chlorohydrocarbons and chlorofluorocarbons vapours. Composite HDPE/aluminium/
HDPE products are available in thicknesses upto 3.0 mm thick and are supplied in rolls in standard widths. These geomembranes exhibit complete impermeability to solvents such as chloroform, trichloroethylene, tetrachlorethylene, toluene, xylene, chlorobenzene, octane, etc. The bonding of the individual layers is very strong and welding of the HDPE outer layers can be achieved using conventional thermal fusing methods.

Visqueen Gas Barrier

A 'Visqueen Gas Barrier' is a split-yarn HDPE reinforced membrane with an integral aluminium foil and plies of LDPE with an impressive property profile (see Table 11.14).

The structure of the Visqueen Gas Barrier is shown in Figure 11.5.

PAG Barrier Liners

Other scrim-reinforced polyethylene and aluminium liners include 'PAG Barrier Liners' (from PAGeotechnical Ltd., Northants, UK) the properties of which are shown in Table 11.15.

Table 11.14   Properties of a Visqueen Gas Barrier

Property	Value
Overall thickness	0.30 mm
Thickness of aluminium foil	0.014 mm
Weight per unit area	350 g/m^2
Tensile strength (BS 2782: Part 3)	43.7 N/mm^2
Tear strength (BS 2782: 360B)	216 N/mm
Methane permeability	<0.001 cm^3/m^2/day
Water vapour permeability (BS 3177)	0.03 g/m^2/day
Available width	1.6 m wide

LDPE
Tough polythene outer layer provides absolute protection against moisture ingress

Visqueen Blue LDPE

Aluminium Foil
Central shield layer of aluminium provides impenetrable gas-reflecting properties

Split Yarn HDPE
Adhesive split yarn woven matrix provides 3-D tear resistance

LDPE
Protective polyethylene outer layer confers additional tensile and tear strength properties

Figure 11.5   Structure of the Visqueen Gas Barrier comprising LDPE outer skins with an aluminium foil central layer which is gas- impermeable and a split yarn woven HDPE matrix to provide tear resistance. Reproduced courtesy of Visqueen Building Products; www.visqueenbuilding.co.uk

Table 11.15   Properties of scrim-reinforced gas barrier membranes based on scrim-reinforced polyethylene and aluminium construction (from PAGeotechnical Ltd, Northants, UK)

Property	0.40 mm PAG Standard	0.46 mm PAG Super
Minimum thickness	0.40 mm	0.46 mm
Weight per unit area	370 g/m^2	415 g/m^2
Tensile strength	10.5 kN/m	9.47 kN/m
Methane permeability	<0.001 ml/m^2/day	<0.005 ml/m^2/day
Petrol vapour transmission (ASTM D-5866)	1.58 g/m^2/h	–
Moisture transmission	<0.01 g/m^2/day	<0.01 g/m^2/day
CBR burst strength	–	708 N

Monarflex™ Gas Barrier Membranes

Monarflex™ gas barrier membranes are comprised of 7–10 layers of polyethylene, adhesive bonding layers, a 1000 denier polyester scrim layer and an encapsulated aluminium foil layer. The properties of Monarflex™ gas barrier membranes are shown in Table 11.16.

The structure of a 0.9 mm Monarflex™ RAC liner is shown in Figure 11.6.

The structure of a 0.45 mm Monarflex Reflex™ Super liner (Icopal) is shown in Figure 11.7.

Table 11.16 Icopal Monarflex™ RAC/Reflex Super Gas Barrier Membranes (Icopal Monarflex™ Geomembranes Ltd, UK)

Properties	Monarflex RAC	Monarflex Reflex Super
Minimum thickness	0.9 mm	0.45 mm
Weight per unit area	900 g/m^2	500 g/m^2
Tensile strength	363 N/25 mm	270 N/25 mm
Methane permeability	<0.005 cm^3/m^2/h at 1 atmosphere differential	<0.005 cm^3/m^2/h at 1 atmosphere differential
Moisture transmission	<0.01 g/m^2/day	<0.01 g/m^2/day
Tear strength	424 N	219 N
Available width	2 m wide	2 m wide

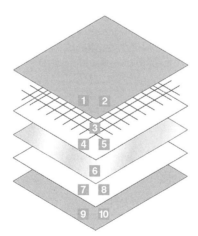

Figure 11.6 Structure of a 0.9 mm Monarflex™ RAC geomembrane (Icopal) for gas barrier applications: layer 1, 0.15 mm blown polyethylene; layer 2, 0.1 mm extruded polyethylene; layer 3, 1000 denier polyester scrim; layer 4, 0.1 mm blown polyethylene; layer 5, 0.05 mm extruded polyethylene; layer 6, 12 μm aluminium foil; layer 7, 0.05 mm extruded polyethylene; layer 8, 0.1 mm blown polyethylene; layer 9, 0.1 mm extruded polyethylene; layer 10, 0.15 mm blown polyethylene. Reproduced permission of Icopal. Note: this product has now been redeveloped as a six layer product (see www.icopal.se for new developments)

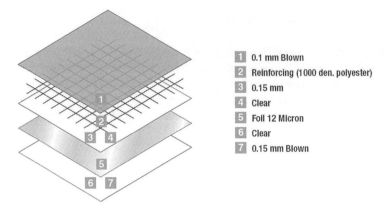

1 | 0.1 mm Blown
2 | Reinforcing (1000 den. polyester)
3 | 0.15 mm
4 | Clear
5 | Foil 12 Micron
6 | Clear
7 | 0.15 mm Blown

Figure 11.7   Structure of a 0.45 mm Monarflex Reflex Super Geomembrane (Icopal) for gas barrier applications which has aluminium foil (12 μm thick) as the gas barrier and a 1000 denier polyester scrim as the reinforcing layer. Reproduced permission of Icopal

Powerbase™ Gas Barriers

Powerbase™ gas barriers are reinforced LDPE geomembranes manufactured by cast extrusion coating LDPE onto a fabric made of strong multifilament and tape yarns. The methane barrier liner incorporates a strong aluminium foil which is encapsulated and protected with an extra layer of LDPE on both sides. This provides additional protection prevents the fabric from pin-holing the aluminium (see Table 11.17).

Table 11.17   Properties of Powerbase™ Scrim-Reinforced Gas Barrier Membranes based on scrim-reinforced polyethylene and aluminium construction (from Industrial Textiles & Plastics, York, UK)

Property	Powerbase™ MB 420	Powerbase RB 340
Minimum thickness	0.70 mm	0.65 mm
Weight per unit area	420 g/m^2	340 g/m^2
Tensile strength	500 N/50 mm	650 N/50 mm
Methane permeability	$<0.00043 \times 10^{-15}$ m^2/s/Pa	–
Radon gas permeability	–	0.00095×10^{-8} m^2/s/Pa
Temperature range	-40 to $+70\,^\circ$C	-40 to $+70\,^\circ$C

HDPE with Aluminium Barrier Layer (HDPE-Al)

HDPE geomembranes are available with a layer of aluminium sandwiched in the membrane. The aluminium barrier liner (e.g. Agrusafe™ from Agru) can be used as a liner for collection basins for containing media that may endanger ground water – particularly hydrocarbons which may diffuse through conventional HDPE. HDPE-Al liners prevent diffusion of hydrocarbons, chlorinated hydrocarbons and chlorofluorocarbons. HDPE-Al are ideal as geomembrane liners for spill containment for petrochemicals or for hazardous wastes.

11.8.3 EIA-R GEOMEMBRANES

EIA-R geomembranes also have good barrier properties to hydrocarbon vapours. For example, Aeon EIA-R geomembranes have performed as a vapour barrier to seal a contaminated building site in Brisbane. A range of hydrocarbons contaminated the site and before it could be redeveloped, a membrane barrier was needed to prevent vapours – mainly benzene – emanating from the soil. The Environmental Engineers responded to EPA requirements to select a vapour-proof membrane appropriate to the soil test profile of the site. The cover also had to be easy to fabricate and have the flexibility to fit around construction structures. The chemical resistance properties of Aeon EIA-R are excellent, giving it superior stability in contact with the compounds contaminating the site. A passive vapour extractor was installed in conjunction with the membrane, using a slotted pipe and aggregate. Aeon EIA-R was laid below concrete building slabs and fitted conveniently around foundations and support beams.

11.8.4 PVDC BARRIER MEMBRANES

PVDC geomembranes comprise polyethylene (HDPE/LDPE) material with a poly(vinylidene chloride) internal barrier layer. These liners can be constructed with a PVDC internal barrier, an HDPE scrim and an LDPE protective coating on both sides. PVDC is made by DuPont under the trademark Saran™. PVDC has extremely low permeability to all gases.

REFERENCES

Cooley, Technical Note 027-R0, 'CoolThane® Thermoplastic Urethane', available from Cooley Engineered Membranes (July 2005).

Infrastructure Technologies (Australia) Pty Limited, Milsons Point, NSW, Australia.

Irrigation Channel Lining [www.irrigation.org.au/seepage/4_2_19_9_butyl.html] (2008).

Keller, R. W. Oxidation and Ozonation of Rubber, Chapter 5 in N. P. Cheremisinoff (Ed.), *Handbook of Polymer Science and Technology*, Vol. 2, Marcel Dekker, New York, NY, USA, p. 143 (1989).

Lattimer, R. P., Layer, R. W. and Rhee, C. K., Antiozonant Protection of Rubber Compounds, Chapter 8 in N. P. Cheremisinoff (Ed.), *Handbook of Polymer Science and Technology*, Vol. 2, Marcel Dekker, New York, NY, USA, p. 243 (1989).

Nakagawa, T., Toya, T. and Oyama, M., Ozone resistance of highly saturated nitrile rubber (HNBR), *Journal of Elastomers and Plastics*, **24**(3), 240–261 (1992).

Plastika Kritis [http://www.plastikakritis.com] (2008).

Tanks Australia [www.tanksaustralia.com] (2008).

Varnamo, Trelleborg Butyl Liners Technical Data, Trelleborg, Sweden, [www.trelleborg.com].

12

Key Performance Properties of Geomembranes

This chapter provides an overview of the key performance properties of geomembranes and compares and contrasts their in-service behaviour.

12.1 SPECIFIC GRAVITY (DENSITY)

The specific gravity (SG) (i.e. density) of a geomembrane will determine the area that a given kilogram of material can cover at a given thickness. For instance, flexible polypropylene (fPP) has the lowest density of any commonly used geomembrane material. This means that a kilogram of fPP will yield 5% more square metres than HDPE and 30% more square metres than PVC (at equivalent thickness). Thus since geomembranes are used on a square metre basis the cost of various geomembranes needs to be compared on a $/m^2 basis and not on a $/kg basis. Figure 12.1 shows the density values of some common geomembrane resins.

Another advantage of low-density geomembranes is that they are easier to handle and deploy in the field. For instance, CSPE geomembranes which have one of the highest densities of commercial geomembranes ($1.47-1.50$ g/cm^3) are very heavy and cumbersome to move manually on site.

The low-density geomembranes, namely fPP, LLDPE and HDPE have densities or specific gravities less than 1.0 (i.e. the density of water) (see Table 12.1) and therefore will tend to float in water. This may be seen as a positive or negative attribute depending on the service application. The SG of EIA geomembranes for instance, is 1.3 g/cm^3 making it easier to sink in applications where this is desirable.

Note: the halogen (i.e. chlorine and fluorine)-containing geomembranes all have relatively high densities due to the higher molecular mass of the chlorine and fluorine atoms.

The density classifications for polyolefins as per ASTM D-1248 are shown in Table 12.2.

Table 12.3 shows the density ranges as those commonly used and accepted in the geomembrane community. Note the density of the carbon black raises the resin density to the geomembrane density.

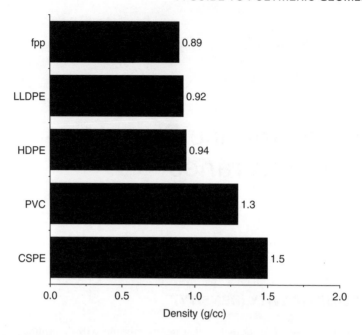

Figure 12.1   Density of geomembrane resins

Table 12.1   Densities of common geomembrane materials

Geomembrane type	Density (g/cm^3)
fPP	0.89–0.90
LLDPE	0.92–0.93
HDPE	0.94
EPDM	1.15
Bitumenous	1.25
Nitrile	1.28
EIA	1.30
PVC	1.31
CSPE	1.47–1.50
PVDF	1.77

Table 12.2   Density classification for polyolefins used in geomembranes

Resin density	ASTM D-883 description	ASTM D-1248 description
0.910–0.925	LDPE	LDPE (Type I)
0.919–0.925	LLDPE	LDPE (Type I)
0.926–0.940	MDPE	MDPE (Type II)
0.941–0.960	HDPE (copolymer)	HDPE (Type III)
>0.961	HDPE (homo polymer)	HDPE (Type IV)

Table 12.3 Density of common polyolefin resins and geomembranes

Polymer	Density of polymer resin (g/cm^3)	Density of geomembrane (i.e. polymer resin +2.5 wt% carbon black (g/cm^3))
HDPE (MDPE)	0.930–0.940	0.941–0.951
LLDPE	0.915–0.930	0.926–0.941
VLDPE	0.880–0.906	0.891–0.917

Note: The addition of 2.5% wt. carbon black raises the density of the polymer by 0.011 g/cm3.

12.2 MASS PER UNIT AREA

Geomembranes with higher mass per unit area have improved resistance to uplifting in windy weather thus making them easier to position and more secure pending covering, under windy conditions. For instance, bitumenous geomembranes (e.g. Coletanche ES 3) have a mass per unit area of about 6.4 kg/m^2, which is more than three times that of 2 mm HDPE and four times that of 1.5 mm HDPE [IFC, 2008].

Typically an HDPE geomembrane with a specific gravity of between 0.950 and 0.970 g/cm^3 (average 0.960 g/cm^3) will be used (including additives). In this case the mass per unit area will be as follows for each thickness:

$$1.5 \text{ mm} = 1.0 \times 1.0 \times 960 \times 0.0015 = 1.44 \text{ kg/m}^2$$

$$2.0 \text{ mm} = 1.92 \text{ kg/m}^2$$

$$2.5 \text{ mm} = 2.39 \text{ kg/m}^2$$

$$3.0 \text{ mm} = 2.88 \text{ kg/m}^2$$

Mass per unit area also is important in terms of the final cost of the geomembrane. Table 12.4 shows the cost of various geomembranes relative to 1.5 mm HDPE. Note these relative costs are material costs and not installed costs.

Table 12.4 Relative costs of various geomembranes (as of 2008)

Geomembrane type and thickness	Cost relative to 1.5 mm HDPE
HDPE (0.75 mm)	0.43
HDPE (1.5 mm)	1.0
LLDPE (1.5 mm)	1.17
fPP (0.75 mm)	0.72
fPP (1.0 mm)	1.12
R-fPP (1.1 mm)	1.5
EPDM (1.5 mm)	2.09
Butyl rubber (1.5 mm)	3.0

12.3 THICKNESS

Thicker geomembranes are generally preferred for a number of reasons including the ability to weld without damage to the liner, increased strain to tensile yield, greater stress crack resistance and less susceptibility to folding which can lead to stress cracking. The benefits of thickness however have to be weighed against the higher cost of the thicker material.

Thickness is an important factor in geomembrane selection from a welding consideration. Geomembranes with a thickness of under 1.5 mm (i.e. 0.5–1.0 mm) are more susceptible to welding problems. If welding conditions are not optimum or, if the welding machine is not set correctly, it is easy for holes to be burnt in these thinner geomembranes (see Figure 12.2). In general, thicknesses of 1.5–2.0 mm are preferred from a welding perspective. Especially in cold climates where welding is to take place at very cold or freezing conditions, then a geomembrane with a minimum thickness of 1.5 mm is recommended.

The thickness of a geomembrane also dictates performance criteria such as its tear resistance, puncture resistance and its resistance to installation damage. All these properties increase with increasing geomembrane thickness (see Figures 12.3 and 12.4). On this basis many geomembrane specifiers set a minimum thickness for a given geomembrane application. Some materials are not viable geomembrane products below a minimum thickness. The tear strength of HDPE geomembranes as a function of thickness is shown in Figure 12.4.

In fact the geomembrane property most involved with resistance or susceptibility to tear, puncture and impact damage, is thickness. At least a linear, and sometimes even an exponential, increase in resistance to the above properties is observed as the thickness increases. For this reason many agencies require a minimum thickness under any

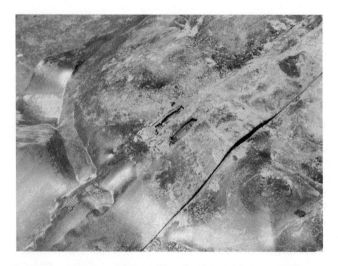

Figure 12.2 Photograph of 'burn through' or 'burn out' on weld tracks in a 1 mm HDPE geomembrane

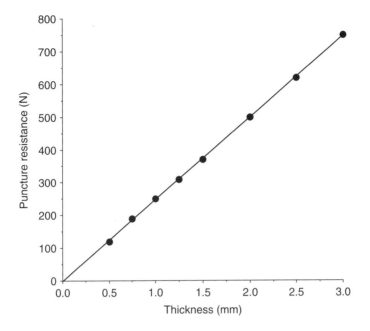

Figure 12.3 Puncture resistance for LLDPE geomembranes (ASTM D-4833) (data from Geosynthetics Specifier's Guide, IFAI publication, 2009)

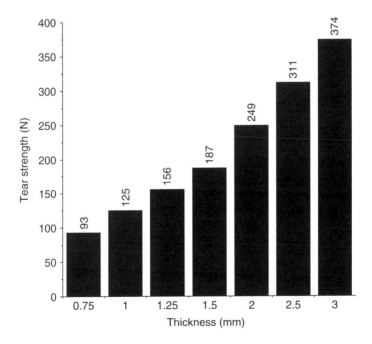

Figure 12.4 Tear strength for HDPE geomembranes (ASTM D-1004) (data from Geosynthetics Specifier's Guide, IFAI publication, 2009)

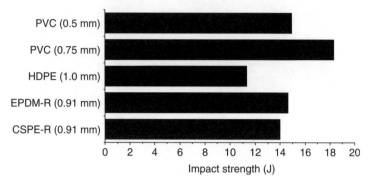

Figure 12.5 Impact tear resistance of various geomembranes (ASTM D-1424 – 45 degree cone angle). Note that scrim-reinforced geomembranes have 10 × 10 scrims

circumstances. For example, the US Environment Protection Agency (US EPA) recommends that 'The primary geomembrane be at least 0.75 mm thick where covered by a protective soil and/or geotextile layer. For an uncovered geomembrane a thickness of at least 1.14 mm is recommended' (referring to scrim-reinforced geomembranes). While in Germany, a minimum thickness of 2.5 mm is required for landfill liners.

Figure 12.5 shows the impact resistance of various geomembranes and demonstrates that both 0.5 and 0.75 mm PVC can outperform 1.0 mm HDPE in the impact test with a 45 degree cone, while Figure 12.6 shows the dramatic increase in falling pendulum impact strength that accompanies increasing sheet thickness of HDPE geomembranes. For this reason thicker HDPE geomembranes are specified to avoid installation damage such as from falling tools or gravel cover being dumped on the liner.

The geomembrane thickness can also affect the durability of the geomembrane due to the extent of surface exposure. Some modes of degradation, such as oxidation and UV degradation, are a function of the total surface area exposed, while other factors such as absorption and diffusion are inversely related to thickness. For instance the migration of a plasticizer from a plasticized PVC geomembrane is faster in the case of thinner membranes (i.e. high surface area to volume ratio).

Thicker geomembranes are better able to resist chemical attack, temperature fluctuations and gradients, stress corrosion cracking and environmental stress cracking, etc.

In summary the reasons for using a thicker geomembrane include:

- to protect the geomembrane from a harsher subgrade during the installation;
- to provide better protection of the liner from welding related damage (e.g. burn through);
- to protect the geomembrane from abrasion;
- to increase liner mechanical durability;
- to provide greater protection from environmental factors and in-service conditions.

HDPE geomembranes are commercially available in a number of thicknesses (usually from 0.75 mm to 3.0 mm) and there is continuing debate over the minimum thickness

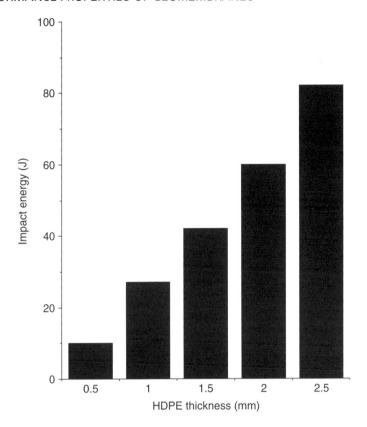

Figure 12.6 Impact strength of HDPE of varying thickness (ASTM D-3998 – falling pendulum impact)

required for different applications. Clearly there needs to be a trade-off between mechanical durability, ease of installation/welding and affordability. A thick HDPE membrane such as 3.0 mm will have excellent mechanical properties and durability but will be very stiff and difficult to install and of course more costly. On the other hand HDPE membranes with thickness of less than 1.5 mm will be more economical but are more likely to present problems with fusion welding such as 'burn through'.

Table 12.5 summarizes some of the considerations when choosing the optimum thickness of HDPE geomembranes.

The specified (regulated) allowable thicknesses of geomembranes for liners and caps for municipal waste landfills and hazardous solid waste (HSW) landfills in various countries are shown in Tables 12.6 & 12.7. Note that Germany has the most stringent regulations. In Germany, landfill liners must be at least 2.5 mm thick (BAM Requirement).

PVC geomembranes used in waste containment systems are typically 0.75 to 1.50 mm thick while HDPE geomembranes are normally 1.5 to 2 mm thick.

Table 12.5 Factors to consider when choosing the optimum thickness of HDPE geomembranes

Relative thickness	Advantages	Potential disadvantages
Thinner membranes (e.g. 1.0–1.5 mm HDPE)	Lower purchase price	Higher repair and maintenance costs in the long term
	Better lay-flat characteristics and conformance to the subgrade to aid installation	Difficult to extrusion weld
Thicker membranes (e.g. 2.5–3.0 mm HDPE)	Lower repair and maintenance costs in the long term	Higher purchase price
	Less difficulties for extrusion welding	Very stiff and rigid and difficult to conform to certain subgrades

 The thickness tests for geomembranes are shown in the following table:

Standard	Title
ASTM D-5199	Standard Test Method for Measuring the Nominal Thickness of Geosynthetics/Geotextiles
ASTM D-1593	Specification for Nonrigid Vinyl Chloride Plastic Sheeting (thickness)
ASTM D-5994	Standard Test Method for Measuring Core Thickness of Textured Geomembranes
EN 1849	Flexible Sheets for Waterproofing. Determination of Thickness and Mass per Unit Area
ISO 2286-3	Rubber- or Plastics-Coated Fabrics – Determination of Roll Characteristics – Part 3: Method for Determination of Thickness
DIN 53370	Testing of Plastic Films: Determination of Thickness by Mechanical Feeling

The instrument used to measure thickness, such as an anvil micrometer, is required to have an accuracy of 0.005 mm (i.e. 5 μm).

12.4 MELTING POINT

The melting temperature (T_m) range of the polymers used to manufacture geomembranes has important practical implications with respect to their welding behaviour (see Table 12.8). Figure 12.7 shows the melting behaviour for some different polymer geomembranes. It is important to note, for instance, that LLDPE geomembranes cannot be welded to HDPE pipe penetrations because of the different melting temperatures of these two polymers (122 °C for LLDPE and 132 °C for HDPE). Too high an application of temperature will hence result in excessive melting of the LLDPE geomembrane.

The melting temperature range can influence the ease and reliability of thermal fusion welding. For instance, fPP has a very broad melting curve and a relatively high melting

Table 12.6 The specified (regulated) allowable thicknesses of geomembranes for liners and caps for *municipal waste landfills* in selected countries. Note that Germany has the most stringent regulations (Hölzlöhner, 1995; Koerner and Koerner, 1999)

Country	Landfill base liner[a]	Landfill cap[a]
Austria	≥2.5 mm HDPE	Not designated
Canada (New Brunswick)	≥2.5 mm HDPE	Not designated
Canada (Nova Scotia)	≥1.5 mm HDPE	≥1.5 mm HDPE
Canada (Ontario)	≥1.5 mm HDPE	Not designated
Canada (Quebec)	Not designated	≥1.0 mm FML
Germany	≥2.5 mm HDPE	≥2.5 mm HDPE
Hungary	≥2.0 mm HDPE	Not designated
Israel	≥1.5 mm FML	Not designated
New Zealand	≥1.5 mm FML	Not designated
Sweden	≥1.0 mm FML	Not designated
Switzerland	≥2.5 mm HDPE	Not designated
Taiwan	≥1.5 mm HDPE	Not designated
Thailand	≥1.5 mm HDPE	≥1.5 mm FML
United Kingdom	≥2.0 mm FML	Not designated
USA	≥1.5 mm HDPE	≥0.5 mm FML

[a]Note that FML denotes a Flexible Membrane Liner (e.g. HDPE, LLDPE, fPP, PVC, etc.).

Table 12.7 The specified (regulated) allowable thicknesses of geomembranes for liners and caps for *hazardous solid waste (HSW) landfills* in selected countries. Note that Germany has the most stringent regulations (Hölzlöhner, 1995; Koerner and Koerner, 1999)

Country	Landfill base liner[a]	Landfill cap
Austria	≥2.5 mm HDPE	Not designated
Canada (Ontario)	≥2.0 mm HDPE	Not designated
Germany	≥2.5 mm HDPE	≥2.5 mm HDPE
Hungary	≥2.0 mm HDPE	Not designated
Italy	≥2.5 mm HDPE	Not designated
South Africa	1.5–2.0 mm HDPE	Not designated
Thailand	≥1.5 mm HDPE	≥1.0 mm HDPE
United Kingdom	≥2.0 mm FML	Not designated
USA	≥1.5 mm HDPE	≥1.5 mm HDPE

[a]Note that FML denotes a Flexible Membrane Liner (e.g. HDPE, LLDPE, fPP, PVC, etc.).

Table 12.8 Melting points of various polyolefin resins

Polymer	Melting point (°C)
EVA	85
mPE (metallocene)	90–100
VLDPE	115–119
LLDPE	121–123
HDPE	126–130
fPP	140–150

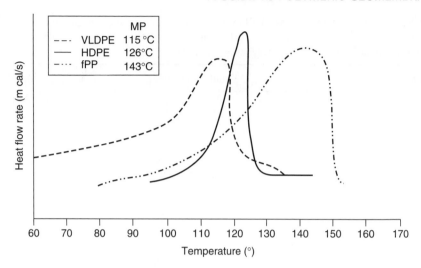

Figure 12.7 Melting curves for HDPE, fPP and VLDPE geomembrane materials as measured by differential scanning calorimetry (DSC). Note that fPP has a very wide melting transition range from an onset of 100 °C to 150 °C which gives it a broad welding window. In practice, this means there is less chance of partial or 'cold' welds due to an error with the temperature or speed calibration of the thermal welding equipment or from operator error. Reproduced by permission of Fabtech

point (of 150 °C) as shown in Figure 12.7. This therefore enables fPP geomembranes to be used in installations requiring resistance to elevated temperatures. It also makes fPP easy to weld and the broad melting range means there is less chance of error on the part of the thermal welding equipment or operator which could result from partial or 'cold' welds. The reliability of thermal fusion welds can be directly related to the melt transition range of the base polymer. The broad melting range gives fPP superior seaming behavior to HDPE; for instance, it can easily be seamed by hot air equipment at low ambient temperatures (e.g. fPP has been successfully seamed at a temperature of −9 °C in strong wind and snow).

12.5 MELT FLOW INDEX

The melt index (MI) value is defined as the mass of polymer extruded through the orifice in 10 min. The rule of thumb is, the lower the MI, the higher the molecular weight (MW). In the melt flow index (MFI) (or melt index (MI)) test, the polymer is heated in the narrow bore of a thermostatically heated steel block until it melts. Then a constant specified load is applied to the polymer melt and causes the molten polymer to extrude through a 2 mm orifice out of the base of the heating chamber. By conducting the MFI test at two different constant loads (e.g. 5 kg and 21.6 kg) the respective MI values can be converted to a melt flow ratio (MFR). The higher the value of MFR for a given polymer, the broader its molecular weight distribution. Melt flow index acquisition conditions for polyethylenes used in geomembranes and their typical values are shown in Table 12.9.

Table 12.9 Melt flow index conditions for polyethylenes

Condition in ASTM D-1238	Conditions	Typical MI values for HDPE geomembrane resins
Condition E	2.16 kg and 190 °C	0.1–0.8 g/10 min
Condition F	21.6 kg and 190 °C HLMI (high load melt index)	12–16 g/10 min (blown film) 18–24 g/10 min (cast sheet)
Condition P	5 kg and 190 °C	10–25 g/10 min

The MFI test method defined by ASTM D-1238 consists in measuring the amount of molten polymer (190 °C) that is extruded through an orifice in 10 min under a constant load (usually 2.16/5/21.6 kg). Chain scission reactions, which are one of the most important consequences of degradation, produce smaller polymer chains and this change in molecular size is reflected as an increase in the MFI value.

12.6 THERMAL EXPANSION

When exposed to sunlight or other heat sources geomembranes can expand and this can be problematic. For instance, the inherently high thermal expansion and dimensional instability of HDPE geomembranes makes them sensitive to variations in temperature during installation and service. Thus HDPE geomembranes undergo expansion and contraction (i.e. retraction) between day and night cycles. If such liners are installed with insufficient slack, they become taught and tensioned in the evening when the temperature drops, as they are constrained by anchor points present in the geomembrane installation.

In contrast, geomembranes with low coefficients of thermal expansion do not suffer significantly from the generation of stresses whilst in service as a result of temperature variations. The coefficient of thermal expansion of bituminous geomembranes, for instance, is very low and is approximately 100 times less than that of HDPE.

Due to the 'black-body' effect all black geomembranes will absorb more heat from the sun. Most geomembranes used are coloured black and black liners particularly in exposed applications, can heat up to temperatures significantly hotter than ambient. On a sunny day when the air temperature is say 25 °C a black geomembrane can over time reach temperatures of +70 °C. In fact, surface geomembrane temperatures up to 80 °C have been recorded for black HDPE despite the air temperature being only 32 °C. A temperature increase of 80 °C in an HDPE geomembrane 100 m long will cause the length to increase by upto 1.6 m. At the other end of the spectrum a bituminous geomembrane (BGM) will only increase in length by approximately 2–8 mm. The normal variation in temperature of a black geomembrane during an entire day is in the range of 40–50 °C. Such temperature cycling can affect geomembranes with a high coefficient of thermal expansion.

Table 12.10 shows that HDPE geomembranes have a coefficient of thermal expansion some 100 times that of a reinforced fPP membrane.

Figure 12.8 shows that HDPE has the worst dimensional stability of commercial geomembranes, expanding by over 1 m over a 100 m length of geomembrane when

Table 12.10 Thermal expansion coefficients for common geomembranes

Geomembrane material	Coefficient of thermal expansion	Increase in length over 10 m for a temperature rise of 50 °C
HDPE	2×10^{-4} cm/cm/°C	100 mm
FPP (unreinforced)	12×10^{-5} cm/cm/°C	60 mm
PVC	12×10^{-5} cm/cm/°C	60 mm
FPP (reinforced)	2×10^{-6} cm/cm/°C	1.0 mm
Bituminous geomembrane	2×10^{-6} cm/cm/°C	1.0 mm

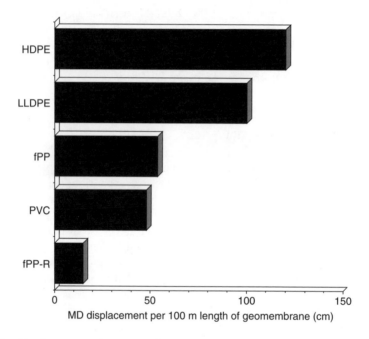

Figure 12.8 Displacement due to a 50 °C temperature change

exposed to a 50 °C temperature rise as happens from 20 °C to 70 °C when exposed to a sunny day.

The thermal expansion and contraction of a geomembrane can affect its ability to lie flat on the underlying soil (i.e. the subgrade). Expansion of the geomembrane due to temperature changes during installation can result in excessive wrinkling and the formation of 'waves'. These wrinkles and waves affect the conformity of the soil–geomembrane interface.

Thermally generated wrinkles (and resultant stresses) can thus be minimized by using low coefficient of thermal expansion liners such as fPP-R, or other reinforced geomembranes in applications where thermal stability is critical. These liners allows large panels in the field to remain dimensionally stable over a wide temperature range (e.g. from cool night temperatures to full exposure in the midday sun).

Thermal expansion and contraction process can be detrimental particularly in the case of a geomembrane installed down a slope where the membrane expands in the sun and then due to interfacial friction and load effects, it cannot fully recover/revert on cooling. This leads to the formation of stresses in the geomembrane particularly at the crest of the slope. These stresses may cause geomembranes to creep or possibly undergo necking and localized thinning. In the case of HDPE, such stresses can be particularly damaging as they can lead to brittle stress cracking.

Similarly, large wrinkles (or waves) caused by thermal expansion of some geomembranes can accumulate at the lowest fixed point on slopes. As the liner contracts during cooler temperature in the late evening and at night these wrinkles do not get pulled back up the slope. In some cases such wrinkles accumulate at the drainage channels and run-off channels causing interruption to the flow paths. This is important when geomembrane liners are placed on steep slopes where materials with poor dimensional stability could be subject to creep[1] and could fail on steep slopes.

Tautness caused from thermal contraction can affect the ability of the geomembrane to resist cracking and tearing from excessive localized stresses on the welds or at the 'toe' of slopes where bridging (or 'trampolining') occurs over the subgrade.

12.6.1   CALCULATING THE EXTENT OF THERMAL EXPANSION

The variation in liner expansion/contraction can be predicted using the following equation to calculate the maximum liner thermal movement:

$$\Delta L = L \times C \times \Delta T$$

where

$$\Delta L = \text{liner expansion/contraction (m)}$$

$$L = \text{original liner length (m)}$$

$$C = \text{linear coefficient of thermal expansion (m/m } ^\circ\text{C)}$$

$$\Delta T = \text{liner temperature variation (}^\circ\text{C)}$$

Example

To calculate the thermal expansion of a 100 m section of an HDPE liner with an expected variation in the temperature of the ambient temperature from 10 to 40 °C (black HDPE surface temperature would be expected to range from 10 to 80 °C during 24 h):

$$\Delta T = 80 - 10 = 70\,^\circ\text{C}$$

---

[1] Creep is the deformation of a material over a prolonged period of time under constant load. This phenomenon is mainly a function of the temperature, load and time and is of importance since some geomembranes are very sensitive to creep. Under sustained constant loading, the material will elongate and break. This problem can be eliminated by using a resin that is less affected by creep, and by a proper design that limits build up of high stress in the geomembrane.

$$\Delta L = L \times C \times \Delta T, \quad \text{where } C = 2 \times 10^{-4} \text{ cm/cm/}^\circ\text{C}$$

$$\Delta L = 10\,000 \text{ cm} \times 2 \times 10^{-4} \times 70 = 140 \text{ cm}$$

Wrinkles and 'fishmouths' caused by thermal expansion can make liner placement and welding difficult. Furthermore such thermally induced wrinkles do not disappear when buried, but instead can fold over (see Figure 12.9) leading to stress concentrations that act as sites for potential stress cracking to occur.

Wrinkles and waves can also interfere with the operation of the impoundment, for instance, wrinkles can create local damming and pooling of leachates and interrupt drainage channels. In another example, in the case of ponds such as those in a fish hatchery, wrinkles on the base can interfere with effective pond cleaning.

Geomembranes such as fPP and PVC possess about half the coefficient of linear thermal expansion of HDPE and LLDPE (see Figure 12.10). This is particularly advantageous during welding and installation since the geomembrane sheet will better retain its dimension and layflat characteristics under varying temperature conditions. The low thermal expansion and contraction behaviour of PVC and FPP is also advantageous during the service life of the geomembrane since there will be less stress imposed on the seams due to thermal contraction and hence less liklihood of seam failure.

Scrim reinforced geomembranes have significantly reduced thermal expansion behaviour; in fact, FPP-R exhibits only between 1 and 14% of the maximum dimensional change measured in HDPE geomembranes.

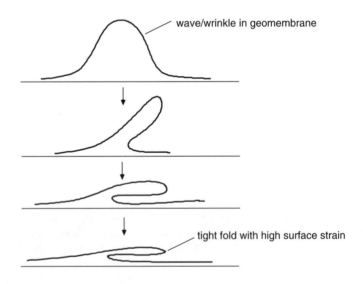

wave/wrinkle in geomembrane

tight fold with high surface strain

Wrinkles and waves (particularly those that are folded over)
induce built-in stresses in the geomembrane

Figure 12.9 Wrinkles and waves (particularly those that are folded over) induce regions of high stress on the geomembrane. This can be particularly damaging for HDPE geomembranes

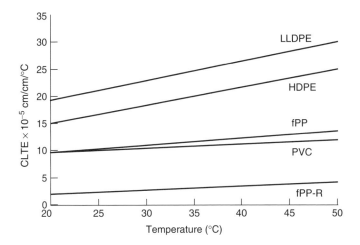

Figure 12.10 Coefficient of linear thermal expansion plots for various geomembrane types as a function of temperature

12.7 FLEXIBILITY

The flexibility of geomembranes can in some instances be a desirable attribute as it imparts the following properties:

- allows the geomembrane or liner to be fabricated in a controlled factory setting (as opposed to in the field) and to be folded into panels without damage to the sheet;
- the ability to conform to undulation in the subgrade (i.e. conformability);[2]
- reduced bridging and fewer wrinkles and waves during field installation.

Furthermore, flexible geomembranes such as fPP and PVC can also deliver significant installation time savings compared with the more rigid HDPE geomembranes. Table 12.11 compares two similar impoundments. The one lined with fPP utilizes only eight flexible fPP panels that requires 4.2 h to field weld, while the HDPE lined system would take 3.6 times longer to field weld (i.e. 15 h) (Montell, 1998).

It is now widely recognized that most leaks occur mainly at field seams therefore minimizing field seams is highly advantageous.

Table 12.11 Comparison of field welding requirements for fPP and HDPE geomembranes (Montell, 1998)

Geomembrane type	Area (m^2)	Seams (m)	Welding time (h)
FPP	15 200	756	4.2
HDPE	15 504	2268	15

---

[2] Conformibility is important as wrinkles, folds and creases in the geomembrane need to be avoided because intimate contact between the geomembrane and the subgrade is required (see Section 8).

Flexible geomembrane liners such as PVC, EIA, fPP and so on can be prefabricated into large panels in a factory environment, then folded and transported to the field site on pallets. In contrast, geomembranes made from HDPE are not permitted to be folded and must be field seamed in smaller sections instead.

12.8 CONFORMABILITY FOR INTIMATE CONTACT

Geomembranes need to be in intimate contact with the subgrade (i.e. substrate) to mini-mize the leak potential of any holes and to prevent the geomembrane being under stress. In order to achieve the desired contact between the covering geomembrane and the subgrade, the latter surface must as far as possible be planar, smooth and free of foreign matter, sharp stones or other angular protrusions. Table 12.12 shows the difference in leakage rates for an HDPE geomembrane on various substrates for different degrees of contact. For a given size hole, the leakage rate in the case of 'poor contact' on a compacted clay layer is some 5.4 times higher than in the case of 'good (intimate) contact'. For the same HDPE geomembrane on a GCL liner (e.g. Bentofix^{TM}) the leakage rate for 'poor contact' is five times greater than in the case of 'good contact'.

Table 12.12   Comparison of leakage rates (l/ha/a) of an HDPE geomembrane as a USA landfill base liner on different substrates for differing degrees of contact (NAUE)

Base on which HDPE geomembrane is laid	'Good contact' (l/ha/a)	'Poor contact' (l/ha/a)
60 cm Compacted clay layer (CCL)	96	524
Geosynthetic clay liner (GCL) (e.g. Bentofix^{TM})	18.4	92

The required flatness of the geomembranes has often been a subject of controversy. Despite some practical problems such as high thermal expansion, it is possible to achieve an evenly placed geomembrane on a flat surface if the personnel and site management are efficient and the installation is carried out conscientiously. Intimate contact between the subgrade/substrate and the geomembrane is routinely achieved by German installers.

The conformability of the geomembrane to the substrate can also increase slip resis-tance, even with textured geomembranes, improving soil retention and enabling maximum slope angles. Point loading occurs at natural undulations of a side slope and it is under these situations that the superior conformability of LLDPE and fPP provide extra shear interface points for maximum holding power on side slopes.

LLDPE and fPP demonstrate the inherent ability to elongate and conform to subgrade irregularities and protrusions under stress. This property translates to very high puncture resistance and tolerance to high working strain that enables the material to readily take the shape of a protrusion or underlying aggregate without rupture or failure.

12.9 ABILITY TO RESIST/ACCEPT STRESS AND DEFORMATION

An important aspect of geomembrane design is that the geomembrane (especially a HDPE liner) should serve only as a barrier and not serve any load-bearing or structural function.

An overriding design consideration therefore is *minimizing stresses acting* on a geomembrane.

Stresses that can be exerted on geomembranes include:

- variations and discontinuities in geomembrane thickness;
- thermally induced stresses (i.e. thermal expansion and contraction);
- damage arising from handling and installation;
- imposed and induced stress during construction (e.g. wind gusting, loading of cover material by heavy vehicles);
- inadequate welding and fixing to attachments;
- seaming induced stresses (e.g. warpage, heat affected zone).

In addition, the design of a geomembrane application needs to consider the various potential stresses imposed on the geomembrane by the in-service configuration and conditions. For instance, such stresses include the following:

- strain imposed at the anchor trench;
- strain imposed over long, steep side slopes;
- differential settlement in the subgrade or foundation soils;
- point loading by angular or rough stones.

A geomembrane liner must maintain its integrity when subject to both short-term and long-term stresses. Short-term mechanical stresses arise from thermal expansion and contraction of the geomembrane during the construction of the installation as well as equipment traffic during the construction phase. Long-term mechanical stresses include the placement of soil or stones on top of the geomembrane and also from differential or preferential settlement of the subgrade.

Ideally, geomembrane liners are designed to be installed without stress. They are intended to act only as a barrier. In practical terms a zero stress installation is impossible to achieve since amongst other considerations, some level of wrinkles and folds are virtually unavoidable.

A geomembrane must be able to accept some deformation without excessive thinning, yielding or rupture. The ductility of a geomembrane is far more important than its ultimate strength in relation to long-term performance. Strength is of prime importance during installation of a geomembrane so that it can withstand the various stresses and abuses of this phase.

Geomembranes can be subjected to uniaxial stress states from plane strain loading, such as dragdown along lined slopes and from out-of-plane loading conditions imposed by localized subsidence beneath waste containment cover and liner systems. The multiaxial tension test (ASTM D-5617) is useful in evaluating the performance of geomembranes subjected to multiaxial stress states. The multiaxial stress–strain test is designed to evaluate a geomembrane's response to out-of-plane loading as would occur during uneven or differential settlement of waste under a landfill cap (Merry and Bray, 1995).

The strain at rupture response for common geomembranes is shown in Figures 12.11 and 12.12. Note that HDPE has excellent strain capabilities under uniaxial strain (or elongation) but relatively poor multiaxial (out of plane) strain performance. In contrast, both fPP and LLDPE exhibit excellent uniaxial and multiaxial strain behaviour. Uniaxial strain

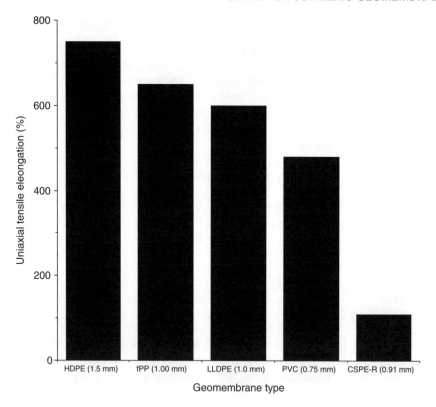

Figure 12.11   Uniaxial percentage strain at break (UTE) for different geomembranes

testing is conducted on strip tensile specimens or 'dog-bones' and multiaxial strain testing is performed on 500 mm diameter discs.

12.10   TENSILE PROPERTIES

The tensile properties are arguably the most critical mechanical properties of geomembranes since during installation and service there are a range of tensile forces that can act on the geomembrane (see Table 12.13).

Minimizing these tensile stresses on the geomembrane will enhance its longevity and durability as a long-term barrier.

12.10.1   UNIAXIAL STRAIN

HDPE exhibits a distinct yield point in its uniaxial tensile stress/strain curve at around 12% strain (i.e. elongation). Above this point further deformation can occur without additional load. This arguably represents one of the main deficiencies of HDPE geomembranes. In contrast LLDPE has a far less pronounced yield point at about 40% strain whereas PVC has no defined yield point.

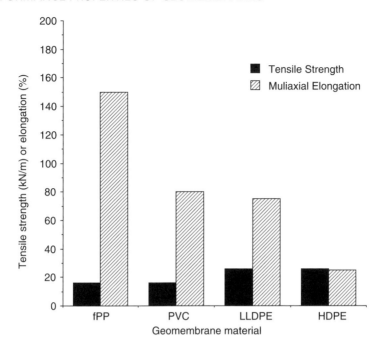

Figure 12.12 Tensile strength and elongation for 1 mm thick geomembranes

Table 12.13 Summarizes the types of tensile forces that can act on an installed geomembrane

Type of force	Examples
Uniaxial	• 'self-weight' of the geomembrane on a slope
	• 'down-drag' caused by waste settlement or tailings consolidation on side slope liners
	• contraction stress on batter due to anchor trench holding liner taut
Multiaxial	• differential settlement
	• force exerted by angular or sharp projections
	• multiaxial tensile stresses caused by consolidation of clay liners under compression
	• thermal contraction and temperature related effects

12.10.2   MULTIAXIAL STRAIN

It is misleading to assume that since a geomembrane material has a high elongation to break it follows that it will have a high survivability in service. This is because materials that perform well in uniaxial tensile tests tend to perform rather poorly in multiaxial or large scale burst tests (ASTM D-5617).

Geosynthetic engineers have thus long appreciated that uniaxial tensile tests (such as those performed on 'dog-bone' specimens in a tensile testing machine) do not represent field situations where stress states in geomembranes are biaxial (or even triaxial). To better simulate the strains found under field conditions, axi-symmetric multiaxial hydrostatic tests were developed. Here a circular geomembrane specimen is clamped at the circumference and inflated with air or water. The curve of the geomembrane deflection versus the inflation pressure is recorded and a stress–strain curve is mathematically derived assuming that the shape of the deformed geomembrane is spherical (Giroud, 2005). These multiaxial tests simulate the out-of-plane loading that is experienced by geomembranes in service where there is subsidence of the subgrade.

In applications where subsidence is likely, such as landfill caps or liners, the geomembrane's ability to elongate multiaxially and relieve stress is important.

fPP geomembranes exhibit multiaxial elongation behaviour some 2–5 times better than other geomembrane materials (see Figures 12.13 and 12.14). fPP geomembranes ultimately fail at strains greater than 160% producing a 'star-shaped rupture' pattern which is indicative of even stress distribution and balanced sheet properties. In contrast, HDPE geomembranes fail at just 20–30% elongation and typically fail by tearing in the machine

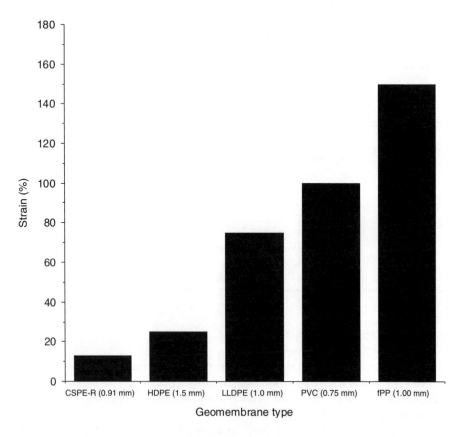

Figure 12.13 Axi-symmetric ultimate percentage strain in the multiaxial burst test

Figure 12.14 Photograph of an fPP membrane during multiaxial burst testing (as per ASTM D-5617) which simulates the out-of-plane stress and strain experienced by geomembranes in service. This test uses a large-diameter test specimen clamped to a fixture and the membrane is inflated with water or air pressure at a rate of 1 psi/min (6.9 kPa/min) until failure occurs. HDPE geomembranes fail by rupture in the machine direction at the relatively low pressures shown in the top right image corresponding to approximately 20% strain

direction indicating the sheet is highly anisotropic. Furthermore HDPE has been found to fail prematurely when scratched or scored (i.e. a cut 20% of the sheet thickness) whereas the only effect of such scratching on fPP is a moderate drop in the ultimate working strain (Montell, 1998).

12.10.3   ELONGATION AT YIELD

The elongation at yield governs the ability of the geomembrane to withstand service-induced strains and stresses.

12.10.4   ELONGATION AT BREAK

Historically, HDPE geomembranes have been regarded as being superior to other geomembranes by virtue of their 700–800% elongation at break (i.e. break strain) in a uniaxial tensile test. However it is known that such large strains practically never exist in any containment application. After the occurrence of some field failures it was recognized that failures under actual field conditions are governed instead by the *yield strain* (which is approximately 12% for HDPE). As a consequence, geosynthetic engineers have started using the yield strain to design applications for HDPE geomembranes (Giroud, 2005).

Unfortunately, the yield strain of geomembranes is often not measured properly. If the strain is derived from grip separation this can lead to an overestimate of the yield strain by as much as 50%, as was demonstrated by Giroud (2005). Thus, tensile tests must be performed with an extensometer[3] to measure the tensile strain only in the central, parallel portion of the geomembrane 'dumb-bell' specimen.

It is the yield strain, not the strain at break that should be considered in design of geomembranes.

12.10.5 TENSILE PROPERTIES OF TEXTURED MEMBRANES

While textured geomembranes show reduced tensile strength at break and elongation at break compared to their smooth counterparts a comparison of textured and smooth HDPE geomembranes under tensile testing (with an extensometer) shows that both of these liner forms produce similar stress/strain curves in the initial portion leading to the yield point. It is only beyond the yield region and towards the respective break points where differences show up between textured and smooth geomembranes (where the smooth samples usually exhibit significantly higher break elongations compared to the textured sheets). Thus as discussed above, from a design perspective both textured and smooth liners are equivalent for the more important property of yield behaviour.

12.10.6 COMPARATIVE TENSILE BEHAVIOUR OF GEOMEMBRANES

The important aspects of the stress–strain curve are summarized in Table 12.14.

Comparison of the tensile stress–strain behaviour of the different geomembrane polymer types is useful in order to predict their in-service behaviour (see Table 12.15).

Note that HDPE, EIA-R and CSPE-R geomembranes fail at relatively low strains (but high stresses), whereas fPP, LLDPE and PVC fail at significantly greater strains, and lower stresses.

Figure 12.15 shows the tensile strength at break of common geomembranes.

Table 12.14 Important aspects of the stress–strain curve

Feature	Comments
Maximum stress	At yield for HDPE, at ultimate stress for PVC and LLDPE, at scrim break for CSPE-R and other reinforced geomembranes
Maximum strain	Maximum elongation
Ultimate stress at failure	Also called ultimate tensile strength (UTS)
Ultimate strain	Also called elongation at break or ultimate tensile elongation (UTE)
Modulus	The slope or gradient of the initial straight portion of the stress–strain curve

[3] An extensometer is a clip-on attachment (hence the name 'clip-on' extensometer) that attaches to a tensile 'dog-bone' specimen before uniaxial tensile testing in order to accurately measure the strain. It has a lightweight design to eliminate errors caused by resistance to the grip separation. Non-contact laser extensometers have also been developed.

Table 12.15 Comparison of the tensile stress–strain behaviour of the different geomembrane polymer types

Geomembrane type	Characteristics (under uniaxial tensile loading)
HDPE	• characteristic stress–strain curve • pronounced yield point at around 12% strain • after yield, drawdown and extension at lower load • break elongation of approximately 700–800%
PVC	• relatively smooth stress–strain response (with no discernable yield point) • exhibits gradual increase in strength • break elongation of approximately 400%
Scrim-reinforced geomembranes (e.g. fPP-R, CSPE-R, EIA-R)	• no yield point • approximately linear response at greater strain • displays abrupt failure • force does not drop down to zero as the geomembrane polymer plies on both sides of the scrim remain intact until ultimate failure occurs

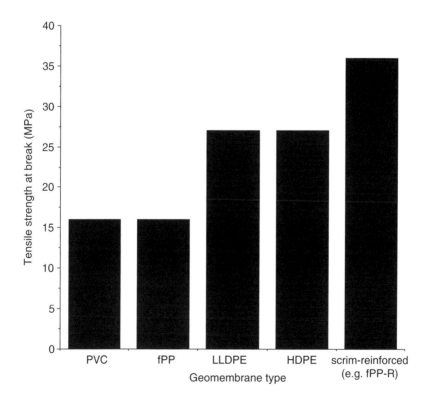

Figure 12.15 Tensile strength at break comparison of common geomembranes

12.10.7   *TENSILE PROPERTIES OF REINFORCED GEOMEMBRANES*

The tensile properties of reinforced geomembranes are derived primarily from the scrim support at the core of the product. Reinforced geomembranes show an approximately linear response at greater strain, typical of textiles – there is no yield point.

For tensile testing of geotextiles and some scrim-reinformced geomembranes a wide width sample of 200 mm is used as described in ASTM D-4885.

12.10.8   *TENSILE BEHAVIOUR OF WELDS*

For shear and peel tests on weld coupons, a 25 mm wide parallel-sided specimen is taken across the seam. The specimen is held in the grips of a tensile tester according to the testing mode. In the shear test the opposing ends of the specimens are pulled apart such that a shear stress is imparted to the seam region. In contrast for the peel test, the two adjacent strips extending from the seam (i.e. each of the two weld tracks, respectively, in the case of double wedge welded specimens) is pulled apart, exerting a seam opening force.

It is important to conduct both the shear and peel tests to ascertain weld quality and integrity. Results can vary widely depending on the type of geomembrane, the type of weld being evaluated and some techniques can even be dependent on operator skill. It should be noted that peel testing is more aggressive on the weld than the shear test.

All tensile test requirements of welds are provided in the installation specification. In addition to meeting the strength and film tear bond (FTB) requirements, a good weld will elongate more than 200% under shear. Welds that break consistently (i.e. all five shear specimens of one sample) under around 80–100% should be carefully inspected for signs of score damage if the break occurs somewhat away from the weld region, or for heat damage at the weld edge where premature breaks occur at the weld/liner interface.

For peel testing, in addition to meeting the strength, a good weld will not exhibit any peel separation. Any peel back of the weld region should be recorded (preferably also by close-up photography) and classified.

The main seam peel and shear destructive test procedure for geomembrane field welds is ASTM D-6392 (which replaced ASTM D-4437). Note although ASTM D-4437 is still widely quoted in project specifications it was actually withdrawn in 1998. ASTM D-6392 requires the measurement of shear elongation and peel separation in addition to peel and shear strength values. ASTM D-6392 also uses 'locus-of-break codes' to classify the various rupture modes of welds. Under this classification scheme adhesion failures (AD, AD1 and AD2) or break in the weld (AD-BRK) are unacceptable while a break through the fillet (denoted as AD-WLD) is acceptable only when certain minimum specification values for strength and elongation at break are met. ASTM D-6392 does not refer to FTB although it is still widely referred to in project specifications (a legacy of ASTM D-4437).

In ASTM D-6392 it is necessary to estimate the amount of seam separation in the peel test by determining the distance of separation as a percentage of the original width of the bonded seam (not the width of the test specimen). It is important to recognize that if there is an angled peel profile with no separation on one side of the 25 mm wide test specimen and complete separation on the other side, then the amount of separation is

Table 12.16 Maximum allowable strains for various geomembrane materials (Peggs, 2003)

Geomembrane type	Maximum allowable strain (%)
HDPE, smooth	6
HDPE, randomly textured	4
HDPE, structured profile	6
LLDPE, density <0.935 g/cm^3	12
LLDPE, density >0.935 g/cm^3	10
LLDPE, randomly textured	8
LLDPE, structured profile	10
fPP, unreinforced	15

100%. Even though the separation area may vary from say 10% to 50% with an angled peel profile the distance separation is 100% in each case. This is because a leak of any size (due to a 100% separation distance) is a potential 100% failure of the seam (Peggs and Allen, 2001).

12.10.9 ALLOWABLE STRAIN

The maximum allowable design strain for geomembranes is obviously far below the yield strain (e.g. 12% for HDPE) and varies widely – ranging from quite high to very conservative values. Some regulators in the USA have set a maximum of 1% strain for geomembranes. The German BAM requirements allow for a maximum global strain of 3% and a maximum local strain (e.g. at individual stone protrusions) of just 0.25% (Seeger and Müller, 1996).

It is important to draw the distinction between global strain and local strain. A maximum strain of 0.25% global strain (i.e. at any location) is sometimes mistakenly specified for HDPE geomembranes, which is very difficult to achieve in practice.

The maximum allowable strains for various geomembrane materials are shown in Table 12.16.

The measurement of strain is used as an indirect measure of the stress that exists in a geomembrane that might result in stress cracking. The objective of specifying these maximum allowable strain values is to limit the in-service stress to a sub-critical value where stress cracking will not be a problem in practice (Peggs, 2003).

While the measurement of strain is a very important consideration for HDPE geomembranes it is not as significant for other geomembranes that are not susceptible to stress cracking unless they are oxidized and embrittled.

It should also be noted that in practice, stress relaxation of the polymer will reduce the stresses to a level perhaps some 50% of the value implied by the strain.

12.11 PUNCTURE RESISTANCE

Puncture resistance of a geomembrane is obviously an important property because even small punctures can reduce the effectiveness of the installed geomembrane as a containment system. Geomembranes are often placed above or below material containing angular or sharp edges. For example, in a waste containment system, a granular drainage layer consisting of gravel may be placed above the geomembrane. As load is placed on the granular drainage layer, either by equipment or waste placement, the gravel may be pushed into the geomembrane, causing puncture. Puncture of the geomembrane may result in a breach of the containment system.

Puncture of geomembranes can also occur when containment material such as waste containing sharp projections is placed on a geomembrane without adequate protection layers. Puncture may also occur as a result of animal damage from claws and beaks. Since puncture of the geomembrane results in leakage it is important to select a geomembrane with adequate puncture resistance for a given particular application.

In practice there are two main ways that geomembranes can be punctured:

- Angular rocks or other sharp protrusions as well as hydrostatic pressure or overburden soil pressure pushing down on the geomembrane. Also, damage can be caused by the heavy equipment itself (e.g. earth movers).
- Penetration from the top where sharp or angular objects are forced into the membrane such as sharp rocks, branches or animal claws/hoofs.

Damage caused by puncture can plastically deform the material up to failure and cause leaks. There are two modes of puncture. Static puncture is due to contact of stones on the geomembrane under high static load (weight of the waste or hydrostatic loading), while dynamic puncture is due to the fall of objects mainly occurring during installation. Static puncture may be eliminated by using protective layers (cushioning geotextiles) and rounded soil particles, as well as stiff and thick geomembranes. Dynamic puncture can be eliminated by considerable care in construction (skilled workmanship is required).

Table 12.17 shows the calculated flow rates through a geomembrane with certain numbers and sizes of holes under a liquid head of 0.3 m.

Thus for many applications, geomembranes are required to exhibit excellent puncture resistance. There are two types of puncture tests – the small-scale test and large-scale test.

Puncture of geomembranes is often assessed using the truncated cone test (ASTM D-5514) which simulates field puncture (this is the large-scale test).

Table 12.17 Calculated flow rates through a geomembrane with a liquid head of 0.3 m (US EPA, 1991)

Size of holes (cm^2)	Number of holes/ha	Flow rate (l/m^2/d)
No holes	0	9.4×10^{-6}
0.1	2.5	0.31
0.1	75	9.4
1	2.5	3.1
1	75	94
10	2.5	31

12.11.1 SMALL-SCALE PUNCTURE TESTING

Index puncture tests are a useful measure of puncture resistance under standard test conditions; however the puncture resistance under the expected field conditions is of fundamental importance.

The small-scale index test, ASTM D-4833 ('Test Method for Index Puncture Resistance of Geosynthetics'), is suitable for quality control purposes but it does not approximate (nor correlate) with field performance.

The small-scale puncture test has historically been used to evaluate puncture resistance of a geomembrane but does not correlate well with out-of-plane point loading as is the situation in the field. This is because:

- the test specimen size is not large enough to evaluate the planar structure and strain response of the material;
- there are differences in the mechanical responses of stiff semicrystalline thermoplastics compared with flexible reinforced or non-reinforced thermoplastics.

The poor correlation of the small-scale puncture test with field performance can be illustrated by the fact that HDPE, when tested using ASTM D-4833, gives a comparatively high 'puncture resistance'. However, the same HDPE material tested over a large surface area and allowed to conform under load over simulated subgrade protrusions exhibits the lowest comparative puncture resistance in service (Stevens, 2008).

12.11.2 LARGE-SCALE PUNCTURE TESTING

Large scale puncture testing was developed for better simulating field stresses that geomembranes experience. A large diameter point stress test has been standardized in ASTM D-5514, entitled 'Large Scale Hydrostatic Puncture Test'. This test method can be used to evaluate the puncture resistance of a geomembrane on large aggregates or subgrades with protruding objects.

HDPE quickly yields and rapidly thins out at the top of a protrusion, leading to a puncture failure. In contrast, highly extendable materials such as LLDPE and PVC, conform under working strain and attempt to fill voids, which may instead result in a burst failure.

Comparative testing using the large-scale hydrostatic puncture test method employs standardized shape truncated cones as the puncture points. This test simulates the relative puncture resistance of a geomembrane when subjected to gradually increasing over-burden pressures. The heights of the truncated cones are varied from test to test and the cone height versus pressure at failure is plotted. From a series of plots the critical cone height (CCH) is determined and this represents the approximate maximum protrusion height before the material ruptures.

The critical cone height (CCH) is the maximum extension/elongation that a geomembrane can withstand before break.

Large-scale puncture testing shows that the maximum extension/elongation that a HDPE geomembrane can withstand before break is only 1–1.5 cm as compared to 4–4.5 cm for a reinforced fPP. Whilst the internal scrim reinforcement in FPP-R gives it high tensile strength and tear resistance, the puncture resistance of unreinforced fPP in terms of CCH is greater than 10 cm as compared to 7.5–8.0 cm for LLDPE (Stevens, 2008).

In the short-term truncated cone testing, the flexible geomembranes accommodate a greater critical cone height than the less flexible geomembranes (i.e. HPDE and CSPE-R) by about a factor of four. All geomembranes generally have a minimum failure pressure in the range of 20 to 75 kPa.

PVC and CSPE-R geomembranes are able to withstand long-term pressures better than more rigid geomembranes because the flexible material is able to deform to the truncated cone. This is important for a geomembrane because field loading is usually much slower than laboratory loading, especially in landfill and heap leach pad applications. HDPE geomembranes may not be suitable for applications with a rough subgrade in which a pressure will be applied for a long period of time (Stark *et al.*, 2007). Under point pressure loading, HDPE first yields and thins out at the top of a protruding object and then undergoes puncture.

Figure 12.16 shows the relative puncture strength of geomembranes assessed using the truncated cone penetration test according to ASTM D-5514.

The critical cone height results show that flexible geomembranes such as fPP, LLDPE and PVC have significantly better puncture resistances than rigid geomembranes such as HDPE and CSPE-R. Therefore flexible geomembranes are the liners of choice for

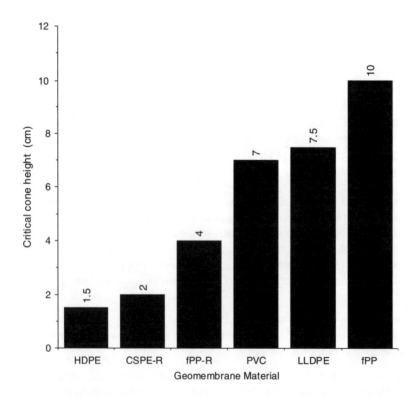

Figure 12.16   Large-scale point stress test for a truncated cone (ASTM D-5514)

subgrades that contain large and angular particles or conversely for applications where large and angular particles will be placed on top of the geomembrane.

The study by Stark has shown that there are important differences in the truncated cone test when the test is run over a short term or over a longer term. For instance, HDPE geomembranes in a long-term truncated cone test failed at pressures and cone heights that it did not fail at in the short term tests. (Stark *et al*., 2007). This is significant since field loading of geomembranes is usually much slower than laboratory loading, especially in landfill and heap leach pad applications (Stark *et al*., 2007).

12.12 TEAR RESISTANCE

Geomembranes are required to have excellent resistance to tear initiation and propagation enabling liners to resist the tensile forces exerted during service (e.g. when the geomembrane is installed on side slopes). Even in the event of a catastrophic rupture it is important that the geomembrane material resists tearing even though tearing forces are present. The most common times that geomembranes are subjected to tearing forces are during panel placement and during wind loading.

The tear strength of thin unreinforced geomembranes is relatively low. For instance, 0.50 mm LLDPE has a tear strength of only 50 N (see Figure 12.17). This highlights the

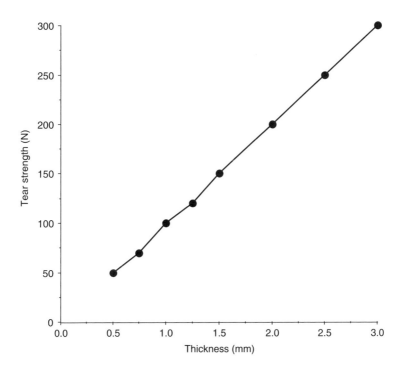

Figure 12.17 Tear strength of LLDPE geomembranes (ASTM D-1004) (data from Geosynthetics Specifier's Guide, IFAI publication, 2009)

need for careful handling and deployment of relatively thin unreinforced geomembranes in order to prevent installation damage. It is easy for such thin membranes to tear even when rolling the panels out, pulling sheets in place and securing them in anchor trenches. Fortunately the tear strength increases considerably with thickness. For instance, 1.0 mm LLDPE has a tear strength of 100 N while 2.0 mm LLDPE has a tear strength of 200 N. Thus for the thicker gauge LLDPE geomembranes, installation damage is less of an issue. Unreinforced PVC also has relatively low tear strength; for instance, 0.50 mm PVC has a tear strength of only 27 N. For this reason 0.50 mm PVC is not really regarded as a geomembrane but rather a tank or pond liner.

The inclusion of a scrim reinforcement within the geomembrane dramatically increases the tear strength. For instance, the use of a scrim in PVC lifts the tear strength from 50 to 350 N (see Figure 12.18).

When tested according to the tear propagation resistance test (DIN 53515), HDPE undergoes tear initiation at a displacement of 15 mm (and a load of 155 N/mm) whereas PVC geomembranes undergo tear initiation also at a displacement of 15 mm but under a much lower load of 40 N/mm. In contrast fPP geomembranes do not undergo tear initiation until a displacement of 25 mm is reached (at a load of 70 N/mm) (see Figure 12.19). There the onset of tear initiation for fPP occur only after almost double the deformation of HDPE. This demonstrates the large movement of the geomembrane liner that is required to propagate a tear.

The tear propagation of fPP is excellent and it will resist tearing even if the membrane is accidentally punctured and subsequently subjected to a tensile stress on a slide slope.

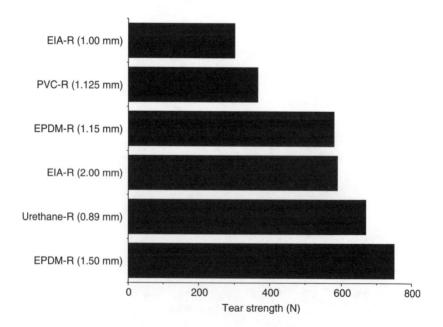

Figure 12.18 Tear strength for reinforced geomembranes (data from Geosynthetics Specifier's Guide, IFAI publication, 2009)

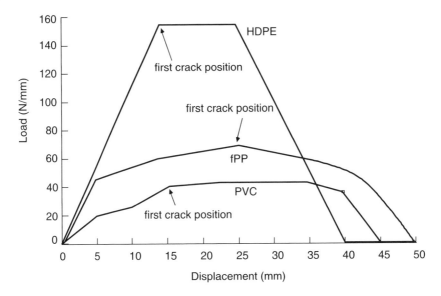

Figure 12.19 Tear properties of common geomembranes (DIN 53515)

Furthermore, the added presence of a textile scrim reinforcement in fPP significantly increases its tear strength and further reduces the tendency for tear propagation since the scrim acts as a 'rip stop'. The use of a scrim reinforcement is especially important in exposed applications.

PVC has the lowest tear strength of common thermoplastic geomembranes. Note however that nitrile rubber liners possess very low tear strength (less than 40 N). HDPE on the other hand has the best tear strength of common unreinforced geomembranes.

Figure 12.20 shows the relationship between the index tear results (ASTM D-1004) and the tensile strengths for various 1 mm thick geomembranes. Note that the index tear test does not show the deformation needed to initiate a tear (i.e. unlike test method DIN 53515) but simply reflects the tensile strength behaviour.

12.13   PLY ADHESION OF REINFORCED GEOMEMBRANES

Reinforced geomembranes are supported on a scrim reinforcement. The fabric reinforcement (i.e. scrim) is inserted into the centre of the geomembrane to provide enhanced tensile strength tear resistance and dimensional stability. Scrim consists of a woven open-mesh reinforcing fabric made from continuous filament yarn that is laminated between two geomembrane plys. The fabric scrim consists of polymer yarns in an open woven pattern sufficient to achieve the minimum specification strength and elongation values. Reinforced geomembranes are generally made by laminating two geomembrane plies with a fabric scrim interlayer. The geomembrane's performance depends on its ability to function as a single integral unit. If the layers are not adequately adhered, the performance of the reinforced geomembrane may be adversely affected.

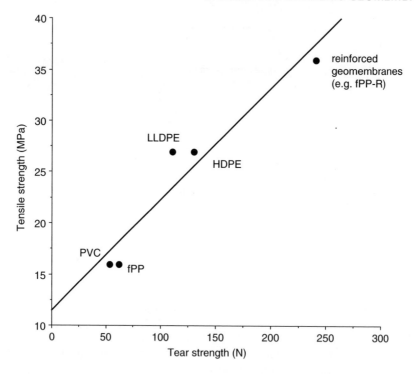

Figure 12.20 Relationship between tear strength and tensile strength for common geomembranes

Delamination describes the damage when a laminated geomembrane layer separates from the reinforcing scrim, but does not necessarily perforate the entire liner. The polymer plies must be on both sides of the fabric scrim so as to completely encapsulate it and prevent exposure of the scrim to liquid agents. The adhesion between the top plies and the scrim is termed 'Ply Adhesion' (GRI GM-15).

Ply adhesion of reinforced geomembranes can be measured in accordance with ASTM D-6636. In order to determine ply adhesion, specimen preparation must first be undertaken with the assistance of a utility knife, or some other suitable separation device which does not compromise either laminate in order to initiate separation of the laminated plies. A pair of needle-nose pliers may help in this regard. It is necessary to separate the laminated layers a minimum of 10 mm mechanically. The separated ends are held by the opposing grips of a tensile testing machine of fixed crosshead speed to determine ply adhesion [ASTM D-6636].

12.14   RESISTANCE TO STRESS CRACKING

Environmental stress cracking (ESC) of polyethylene is defined as cracking caused by an applied tensile stress which is lower than its tensile strength at yield. In the past, some HDPE geomembranes were particularly susceptible to this mode of failure. Their

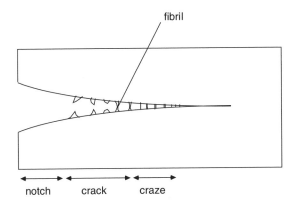

Figure 12.21 Schematic showing mechanism of cracking of a polymer from a notch defect

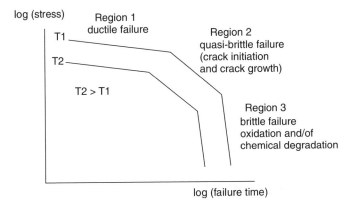

Figure 12.22 Stress rupture curves showing the three stages of failure for polymers held under a constant stress. Note that increasing temperature moves the onset of failure to earlier times. Environmental stress cracking is represented by Region 2 while Region 3 represents oxidative stress cracking

propensity to undergo stress cracking is further increased in the presence of notches and score-lines as these can transform into a crack as shown in Figure 12.21.

Stress rupture curves for polyethylene held under a constant stress are shown in Figure 12.22. Environmental stress cracking is represented by the quasi-brittle failure in Region 2. Note the accelerating effect that raising temperature plays in moving the onset of failure to shorter times. Region 3 presents rapid stress cracking associated with oxidative embrittlement (known as oxidative stress cracking).

The stress cracking resistance of HDPE is a function of the type of HDPE resin used and thus varies widely from one geomembrane manufacturer to another depending on the providence of the resin (see Figure 12.23). Most geomembrane resins now have excellent ESCR; however stress cracking can still occur due to other factors such as recrystallization,

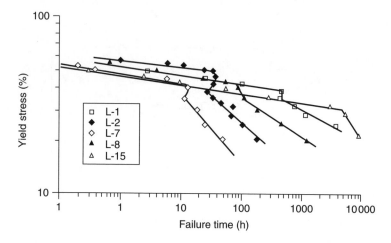

Figure 12.23 Stress rupture curves for a range of HDPE (geomembrane-grade) resins (circa 1992). The tests were conducted at various loads (all less than the yield stress) to give enough data points to plot the full curves. The 'knees' in the curves represents the transition from ductile to brittle failure. The knees thus represent the onset of stress cracking. Note that there is a 10–1000 times difference in the stress cracking resistance of these HDPE resins. Reproduced with permission from Hsuan, Y.G., Koerner, R.M. and Lord, A.E., The Notched Constant Tensile Load (NCTL) Test to Evaluate Stress Cracking Resistance, in *Proceedings of the 6th GRI Conference*, GRI, Philadelphia, PA, USA, pp. 244–256 (1992)

oxidative embrittlement and stress rupture. Figure 12.24 shows the relationship between density (and % crystallinity) and stress crack failure times for HDPE. It can be seen that polymer densities greater than 0.942 g/cm^3 produced less than acceptable stress crack resistance.

Table 12.18 below summarizes the major risk factors in the environmental stress cracking of HDPE geomembranes.

Less crystalline geomembrane materials such as LLDPE and fPP are not susceptible to ESC in their new condition. However after oxidation they too can become susceptible to stress cracking at folds and creases.

fPP geomembranes in particular exhibit outstanding resistance to ESC even in otherwise aggressive media. For instance, samples of fPP were tested using the 'Notched Constant Load Environmental Stress Crack Resistance Test', with no failures reported in an aggressive soap solution (5% soap solution skimmed from the top surface of a 52% black liquor effluent obtained from a paper-processing facility). 'Dumb-bell' shaped specimens were notched to produce a hinge thickness at 80% of the nominal thickness of the test specimen and immersed into the soap solution and maintained at 50 °C and loaded at various percentages of their room temperature yield stress at increments of 5%. The yield stresses of the material were then measured and plotted as percentage yield vs time-to-failure to determine the transition time between ductile and brittle modes of failure. No failures, surface cracking or notch propagation were observed under 200× magnification in the fPP samples. However, both the HDPE and VLDPE samples exposed to the same black liquor soap solution both exhibited premature failure due to stress cracking (Montell, 1998).

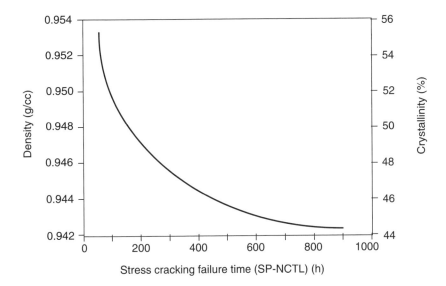

Figure 12.24 Simplistic relationship between density (and percentage crystallinity) and stress crack resistance of high-density polyethylene. Reprinted with permission from Proceedings of 6th GRI Conference, Philadelphia, The notched constant tensile load (NCTL) test to evaluate stress cracking resistance by Y. G. Hsuan, R. M. Koerner and A. E. Lord Copyright (1992) Grace Hsuan

Table 12.18   Stress cracking risk factors for HDPE geomembranes

Risk factor	Examples
Surface wetting agents (surfactants)	Detergents, soaps, black liquor, silicone oils
Stress-concentrating flaws	Notches, scratches[a], scores, nicks
Stress concentrations from irregular geometry	Variable thickness welds, abrupt thick–thin transitions, nodular inclusions
Residual stresses	Contraction stresses in weld region due to differential shrinkage and dimensional instability
Cyclic stresses	Fatigue, thermally induced cyclic stresses[b], cyclic stresses caused by wind uplift, cyclic stresses caused by swelling/desorption of organics/aromatics
Surface oxidation	Oxidation of the heat affected weld region, oxidation by oxidizing acids, UV oxidation and degradation of a geomembrane can cause it to become brittle and susceptible to stress cracking
High crystallinity	Recrystallization by slow cooling of weld region, chemi-crystallization by rearrangement of shorter oxidized polymer chains

[a]Note initiation of stress cracking is often at seams where scratches caused by grinding exist or in areas where extrusion welds are placed directly over existing wedge welds.
[b]Temperature changes of restrained geomembranes can induce tensile stresses.

🖈 *When FPP is tested in the constant load stress cracking test (ASTM D-5397) using a 5% soap solution, at 50°C solution temperature and a notch depth of 20% of the sheet thickness, they elongate but no samples fail in a brittle manner. In other words there is no 'knee' in the curve indicating that no ductile-to-brittle transition occurs. The samples will elongate over time but none of the samples show signs of cracks even when viewed under 200× magnification (Montell, 1998).*

12.14.1   SAFETY FACTORS FOR HDPE

The maximum allowable elongation that should be considered for design purposes is the elongation at yield divided by an appropriate safety factor for a given application.

For HDPE geomembranes used in critical containment applications it is common for geosynthetic engineers to use a safety factor of four or even higher.

HDPE geomembranes must be based on low design working stresses and high safety factors because of their potential for stress cracking. The US EPA recommends working stresses in HDPE be no higher than 10% of the minimum (not typical) specified yield stress for sheet material. That is a factor of safety equal to 10.

The stress cracking test for HDPE uses a stress of 30% of the yield stress.

For HDPE geomembranes the available strain (elongation) to withstand surface scratches, local stresses/deformation is approximately 5%. In contrast for geomembranes such as fPP and PVC the inherently available elongation is greater than 20%.

The usable elongation for HDPE geomembranes is thus not the yield point but an interpreted percentage of the available elongation.

12.14.2   ALLOWABLE STRAINS IN HDPE

Strains in HDPE geomembranes can originate from sharp radii of curvature of stone protrusions in the case of a basal liner, the settlement depth for a liner or capping system and local deformations.

For basal liners (i.e. base liners in a landfill for instance) the limiting local strain (i.e. permissible long-term strain limit) is 3% whereas for capping membranes the strain limit to exclude stress cracking is much higher and is usually set at 6% (Seeger and Müller, 2003). The reason for the larger strain limit for HDPE capping membranes and liners subjected to subgrade settlement is that the radius of curvature of the HDPE is always very large relative to the thickness of the geomembrane and therefore larger strain limits are permissible. The actual strain limit with respect to yielding of HDPE is 10–12%.

Indentations in the geomembrane caused by angular gravel or stones involve small radii of curvature relative to the geomembrane thickness and so can generate substantial local strains; hence the need for a conservative permissible long term strain limit (see Figure 12.25). HDPE geomembrane must therefore be protected from protrusive deformations with small radii of curvature by using cushioning materials such as geotextiles.

Geomembranes used in separation liners between old and new waste in vertical expansions of landfills may be subject to strain caused by differential settlement. In such

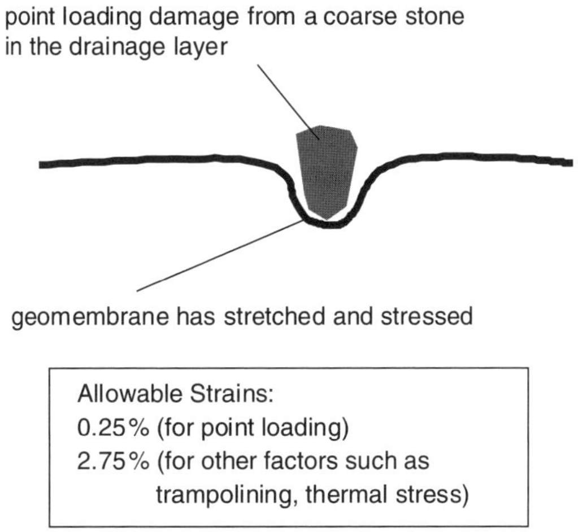

point loading damage from a coarse stone
in the drainage layer

geomembrane has stretched and stressed

Allowable Strains:
0.25% (for point loading)
2.75% (for other factors such as
          trampolining, thermal stress)

Figure 12.25   Schematic showing point loading damage to a HDPE geomembrane by a coarse stone in the drainage layer. Note that allowable strains from point loading of HDPE are 0.25% while the limit for strains from other factors is 2.75%

applications it is particularly important to define a maximum allowable strain (MAS) that any given geomembrane can tolerate without compromising its required service life. In a number of instances, some very low values have been proposed, as low as 0.25 to 1.0% for HDPE. Such low numbers are likely based on German regulations for HDPE landfill liners that require a maximum allowable global strain of 3% and a limiting local strain due, for example, to protruding drainage stones, of 0.25%. However, the low allowable strain values adopted by the German regulators were based on products and practices utilized in the 1980s and do not reflect current conditions, nor do they address geomembranes other than HDPE (Peggs *et al*., 2005).

Limiting the degree of strain (an indirect measure of stress) is very important in the case of HDPE geomembranes but it is not as significant for other geomembrane materials since these are not susceptible to stress cracking unless oxidized.

With respect to acceptable long-term strains, the threshold strain for HDPE is normally taken as 30% of the yield strain. Given that the yield strain for HDPE is around 12%, then 30% of that is an absolute strain of around 3.6%. A long-term permissible strain of up to 2.75% is acceptable for HDPE. As noted above acceptable point load strains are an order of magnitude lower at 0.25% – that for strains induced by angular rock penetration loading.

12.14.3   ESTIMATING LOCAL STRAIN

A method for estimating the local strain in a geomembrane due to the indentation of gravel particles is discussed by Tognon *et al*. (2000). The accuracy of various strain calculation methods is evaluated by a series of tests, and it is shown that the traditional arch elongation method provides only an approximate estimate of the magnitude of strain induced in the geomembrane due to indentation and does not adequately define the distribution of strain. Consideration of the combined membrane and bending strains as described was shown by Tognon and co-workers to provide a better representation of the distribution of strains and enhances the evaluation of the peak strains in the geomembrane caused by local indentations.

Large-scale tests are conducted using different protection layers, and the strains are reported based on both the arch elongation method and the combined bending and membrane theory. The results indicate that the best protection for the underlying geomembrane was provided by a sandfilled geocushion or a special rubber geomat, which limited strains induced by coarse (40–50 mm) angular gravel to 0.9% at 900 kPa and 1.2% at 600 kPa. The poorest performance was achieved using nonwoven geotextiles with a maximum strain of 8% being obtained with a 435 g/m^2 geotextile at 250 kPa and 13% with two layers of 600 g/m^2 geotextile at 900 kPa (Tognon *et al.*, 2000).

12.15 FRICTIONAL PROPERTIES

The long-term integrity of a geomembrane or liner can depend on there being adequate friction between the various components of the liner system – in particular, between the subgrade soil and the geomembrane as well as between the geomembrane and any adjacent layers such as geotextiles. In this respect the frictional characteristics can be very important performance properties.

A geomembrane of higher surface friction will increase friction angles and thus side slope stability concerns are reduced. When smooth geomembranes are pushed against other geosynthetics or soil, the degree of friction can be measured. Hard geomembranes such as smooth HDPE have little friction (hence low friction angles); however softer materials such as LLDPE, fPP, EPDM and PVC have greater friction (hence high friction angles) under the application of an applied normal stress. The friction angle for various geomembranes on sand are shown in Figure 12.26.

Adequate friction is necessary to prevent slippage or sloughing on slopes of the installation. In the case of installations with sloping sides, the geomembrane must be able to:

1. support its own weight on the side slopes;
2. withstand 'down-dragging' during and after placement of the overburden (e.g. waste in the case of a landfill liner);
3. maintain a stable state when a soil cover or a granular drainage layer is placed on top of the geomembrane;
4. maintain a stable configuration when other geosynthetic components such as geotextiles or geonets are placed on top of the geomembrane.

The critical interfaces in a lined installation may include the following:

• soil versus geomembrane;
• soil versus geotextile;
• geomembrane versus geotextile;
• geomembrane versus geonet.

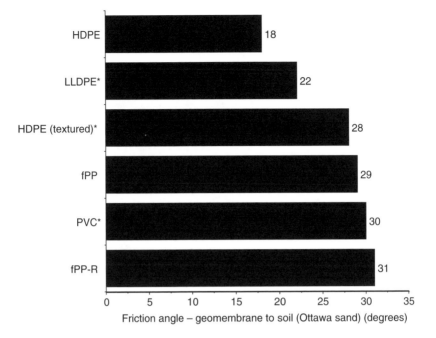

Figure 12.26   Comparison of friction angles for geomembranes on sand (* indicates tested with river sand)

Geomembrane interfacial frictional failure has been identified as the cause of numerous geosynthetic lined slope failures. As a result, the interface frictional strength of any geomembrane interface has to be determined with the utmost care. The strength of each of these interfaces therefore has to be determined experimentally using site-specific material testing.

It has been demonstrated that the higher the stiffness (or hardness) of the geomembrane (e.g. HDPE), the lower the friction angle, as compared to an inherently flexible membrane such as PVC or LLDPE (O'Rourke *et al.*, 1990). PVC geomembranes generally have excellent interface friction behavior when compared to other stiffer geomembranes due to their intrinsic flexibility. Figure 12.27 shows the relative stiffness (as measured by flexural modulus) of common geomembranes. Note the large difference in stiffness between HDPE and PVC by a factor of 20 times.

Geomembrane versus geotextile interface friction for very soft, flexible geomembrane like EPDM, medium stiffness PVC and a tough/hard geomembrane, namely HDPE, have been compared and documented (Martin *et al.*, 1984).

PVC is available in a roughened (or 'file' finish) to increase the friction angle.

Soil to geomembrane friction is a critical aspect of the design of geomembranes for side slopes in landfills, tailing dams, reservoirs and canals. A high friction angle is desirable as it allows geotechnical designers greater design freedom for such installations. The friction angle of geomembranes can be measured using a shear soil friction test which mimics the field situation.

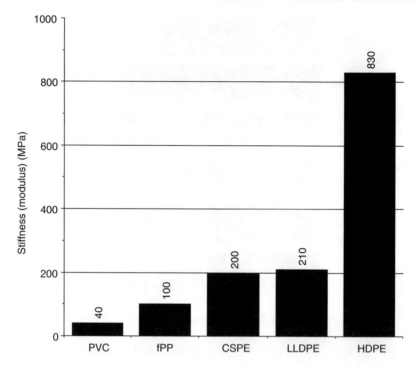

Figure 12.27 Relative stiffness (flexural modulus) of geomembrane resins

Shear properties of liners are very important for the stability of the landfill. The characteristics of the materials comprising the liners, their roughness, their stiffness, the normal load and the temperature are factors influencing interface shear strength.

Friction testing measures the frictional properties at the various interfaces. Those interfaces that are relevant to the end application need to be considered. For instance, various protective layers may be placed over the geomembrane, such as gravel layer, soil layers and a geotextile underliner. It is important to emphasise that friction angle testing should always be performed with site-specific soils and sands. That is because these natural materials have different moisture contents and particle size distributions that can significantly affect the results. The interface friction properties of geomembranes can be measured and compared using the test method ASTM D-5231 using a commonly available cohesionless Ottawa sand. It has been shown that higher friction angles are obtained with unreinforced fPP and fPP-R as compared with HDPE, LLDPE and even textured HDPE (see Figure 12.26).

Scrim-reinforced geomembranes such as fPP-R exhibit relatively high friction angles due in part to their surface texture (i.e. ribbed pattern from the underlying scrim) and additionally due to the ability of the flexible geomembrane to readily conform and 'key in' to soil surface irregularities under applied load – thus providing improved resistance to sliding.

Note that fPP achieves high friction angles without the sacrifice in elongational properties experienced with textured HDPE geomembranes (nitrogen gas blown textured

HDPE geomembranes can show reduced elongational properties due to the texturing reducing the core thickness of the liner and through the introduction of surface 'defects').

Texturing of fPP reduces its elongational properties by as much as 50%. Despite this property loss, fPP still retains both uniaxial and multiaxial elongational properties comparable to those of other common geomembrane materials.

📌 *Interface frictional failure has been identified as the cause of numerous geomembrane-lined slope failures. As a result, the interface frictional strength of any geomembrane interface has to be carefully and accurately determined. Koerner has recommended that wherever possible, the interface frictional strength for a geomembrane–soil combination be determined experimentally (Koerner et al., 1986).*

12.15.1 COMPARISON BETWEEN FRICTION ANGLES OF PVC AND HDPE

PVC geomembranes are flexible and relatively easy to handle while HDPE geomembranes are tough and inflexible (stiff). The higher the stiffness or hardness of the geomembrane, like HDPE, the lower the friction angle, as compared to a flexible membrane, like PVC (see Table 12.19). Softer geomembranes have greater friction angles than the stiffer/tougher geomembranes. Textured HDPE has the highest friction angles of the liners tested.

HDPE geomembranes often compete with PVC geomembranes as landfill covers. PVC geomembranes have exceptional interface friction behavior when compared to other geomembranes by virtue of their flexibility. PVC geomembranes are more efficient than smooth HDPE membranes in their frictional behavior (see Tables 12.19 and 12.20). PVC geomembranes are exceptional in that their adhesion to soil tends to be greater than the cohesion of the soil itself (see Table 12.20).

Figure 12.28 shows that HDPE geomembranes exhibit a sharp peak in their stress–strain curve and undergo a relatively abrupt change in properties. In contrast, PVC geomembranes undergoes a large amount of elongation before failure.

The stress–strain response of PVC is very different from that of HDPE. It has been shown that even after reaching yield stress of the interface, PVC geomembranes will not fail but maintain stability by stretching of the membrane material without loss of strength or material damage.

Table 12.19 Interface friction angle values (degrees) obtained for various interfaces (tested at 10% strain) (Bhatia, 2008)[a, b]

Geomembrane (strain level)	Fine sand	Sandy loam	Silty clay	Non-woven geotextile
0.75 mm smooth PVC (at 10% strain)	34.7	26.4	20.8	21.9
0.75 mm textured PVC (at 10% strain)	35.3	21.1	26.4	19.6
0.75 mm file-finish PVC (at 10% strain)	30.9	28.1	26.0	17.3
1.5 mm smooth HDPE (at 10% strain)	21.1	18.2	17.0	14.2
1.5 mm textured HDPE (at 10% strain)	36.6	33.8	41.8	17.4

[a]Textured PVC had roughness of 0.1–0.15 mm asperities (peaks) on its surface.
[b]File-finish PVC has a square grid etched onto one surface.

Table 12.20 Friction values (degrees) of geomembrane-to-soil interfaces for various soil types

Geomembrane	Ottawa sand	Concrete sand	Mica Schist sand
HDPE	18	18	17
CSPE-R	21	25	23
EPDM-R	20	24	24
Smooth PVC	–	25	21
Rough PVC	–	27	25

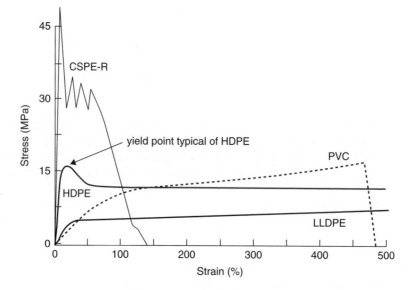

Figure 12.28 Tensile curves of common geomembranes (Koerner, 2005)

For PVC geomembranes therefore, the yield point of the interface does not represent a failure condition. This is because further shearing causes an increase in strength and not a decrease, whereas further shearing in the case of HDPE geomembranes causes reduced strength (as indicated by the drop in the curve after the yield point). Under service conditions, if the PVC geomembrane is stressed beyond the yield stress for the interface, the material stretches under the load without any loss of strength or material damage.

A further important advantage of PVC geomembranes is that due to their ability to be factory seamed they require as low as 20% of the field welds that are required by an HDPE geomembrane.

Textured HDPE was found to have a comparable interface friction angle than smooth PVC in all cases except with geotextiles, where textured HDPE gives lower friction angles. This is believed to be due to a reduction in contact area since the rough side of the geotextile and the textured surface of the geomembrane are in contact only at textured projections.

Important Facts on Friction Angles

- *As one would expect, textured geomembranes give better interface friction values than their corresponding smooth membranes.*
- *The more flexible the geomembrane, the higher the friction angle.*
- *The stiffer the geotextile, the lower the friction angle.*
- *The geotextile–geomembrane interface has the lowest friction angle in a liner system and is therefore considered the most critical.*
- *Friction angles between the geomembrane and cohesive soils vary from 15 to 26 degrees for smooth HDPE and from 25 to 36 degrees for textured HDPE.*

The friction angles between smooth HDPE and geotextiles is only 6–11 degrees; hence caution needs to be exercised to prevent slippage of these interfaces (see Table 12.21).

 It is important to check frictional properties to avoid the potential for side-slope failure that can damage the geomembrane layer.

Texturing of HDPE geomembranes to produce forced-friction HDPE geomembranes was developed for the following reasons:

- Sliding of personnel who were walking on a wet, HDPE 3:1 slope.
- Sliding of heavily loaded wet soil and waste.
- Sliding of heavily loaded polyethylene geomembrane on a wet slope.
- Sliding failures occurred at both HDPE interfaces.

Figures 12.29–12.32 show the shear interface friction angles (i.e. slope angles) of different combinations of coarse sand, clay, gravel and non-woven protective geotextiles with various structured and unstructured HDPE geomembranes. Note that 'S' denotes spiked with 6 mm high spikes on a 25×25 mm grid pattern, 'MSB' denotes a microspike bottom with 0.60 mm high spikes and 32 000 microspikes/m^2, 'MST' denotes a microspike top with 0.40 mm high spikes and 60 000 microspikes/m^2, and 'G' denotes a smooth HDPE geomembrane. Note that the spikes ('S') are for very steep slopes on clay or coarse clay.

Table 12.21 Friction values (degrees) of geomembrane-to-geotextile interfaces for various geotextile types (Martin *et al*., 1984)

Geomembrane	Nonwoven needle-punched geotextile	Nonwoven heat-bonded geotextile	Woven monofilament geotextile
HDPE	8	11	6
CSPE-R	15	21	9
EPDM-R	23	18	17
Smooth PVC	21	18	10
Rough PVC	23	20	11

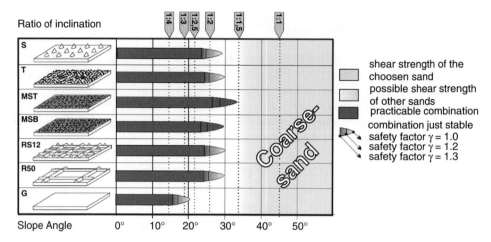

Figure 12.29 Graphic of slope stability (i.e. slope angle) of structured HDPE geomembranes on *coarse sand*. Note that 'S' denotes spiked with 6 mm high spikes on a 25 × 25 mm grid pattern, 'T' denotes textured, 'MSB' denotes the microspike bottom with 0.60 mm high spikes and 32 000 microspikes/m², 'MST' denotes a microspike top with 0.40 mm high spikes and 60 000 microspikes/m², 'R 12' denotes a spiked grid configuration, 'R 50' denotes a 50 mm × 50 mm grid pattern and 'G' denotes a smooth HDPE geomembrane. Reproduced by permission of Agru

The structured surface is formed by a calendering process in a single production step and each geomembrane is tailored with a different surface structure to support the chosen layer combination by increasing the internal shear interaction from the top to the bottom side.

12.16   LOW-TEMPERATURE PROPERTIES

Most geomembrane liners will become stiffer when cold, requiring additional care during handling, deployment and seaming. It is preferable to avoid installing geomembrane liners under cold conditions (or in cold climates); however sometimes this is inevitable. Low temperatures will usually require a modification of installation techniques. In many cases it is better to wait for warmer weather rather than to attempt an installation under cold/poor weather. Some of the complications arising during geomembrane installation in cold weather include:

- Geomembrane liners become progressively stiffer and become much more difficult to handle.
- It becomes more difficult to drape the geomembrane over complex subgrades.
- Attaching stiff liners to pipes and appurtenances is very difficult.
- Geomembrane materials that are designed to be backfilled often have low-temperature handling restrictions. For instance, a regular flexible PVC lining material may crack if folded and impacted below −5 °C. Other geomembrane materials with low temperature handling restriction are polyurethane (ester-based types) geomembranes (Layfield, 2008). The practical low-temperature handling limits are shown in Table 12.22.

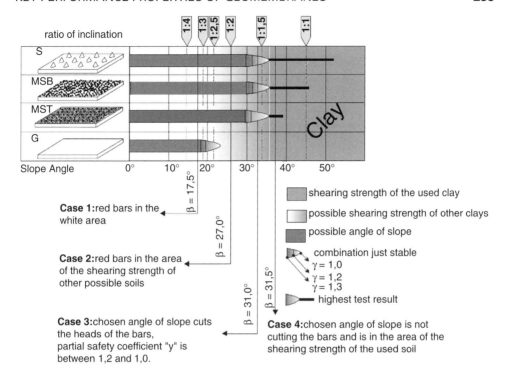

Figure 12.30 Graphic of slope stability (i.e. slope angle) of structured HDPE geomembranes on *clay*. Note that 'S' denotes spiked with 6 mm high spikes on a 25 × 25 mm grid pattern, 'MSB' denotes the microspike bottom with 0.60 mm high spikes and 32 000 microspikes/m², 'MST' denotes a microspike top with 0.40 mm high spikes and 60 000 microspikes/m² and 'G' denotes a smooth HDPE geomembrane. Reproduced by permission of Agru

Good low-temperature properties are required if geomembranes are to be deployed in cold climates. fPP membranes, for example, can be unfolded and installed at temperatures of −30 °C and the material remains ductile until −50 °C. The brittle temperature of fPP is −50 °C as tested by ASTM D-1790.

If the air temperature is very cold there are advantages in waiting for black geomembranes to heat up though solar heating. Even on a −30 °C day, dark lining materials can reach a surface temperature of over +30 °C through solar heating. Thus by waiting for a few hours or by waiting until later in the day, a liner can often be installed in otherwise prohibitive air temperatures.

The ability for a geomembrane to retain its properties at low temperatures is important for preventing damage when the product is deployed and installed under winter or arctic conditions. At temperatures below −40 °C many geomembrane materials lose their flexibility and impact resistance and become brittle. Because of the excellent low-temperature properties of fPP geomembranes they are often used in place of HDPE for cold temperature installations. HDPE geomembranes become brittle at −40 °C while fPP doesn't become brittle until below −50 °C (Montell, 1998).

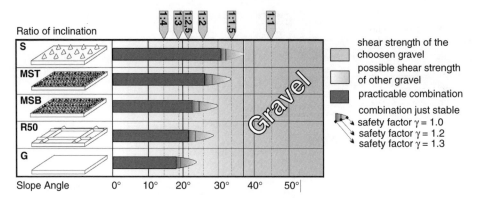

Figure 12.31 Graphic of slope stability (i.e. slope angle) of structured HDPE geomembranes on *gravel*. Note that 'S' denotes spiked with 6 mm high spikes on a 25 × 25 mm grid pattern, 'MSB' denotes a microspike bottom with 0.60 mm high spikes and 32 000 microspikes/m², 'MST' denotes a microspike top with 0.40 mm high spikes and 60 000 microspikes/m², 'R 50' denotes a 50 mm × 50 mm grid pattern and 'G' denotes a smooth HDPE geomembrane. Reproduced by permission of Agru

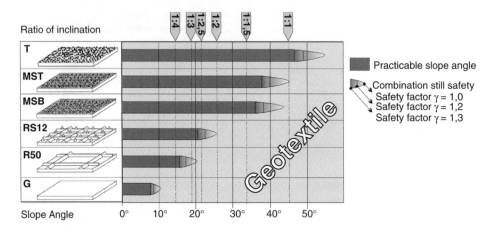

Figure 12.32 Graphic of slope stability (i.e. slope angle) of structured HDPE geomembranes on a *geotextile*. Note that 'T' denotes textured, 'MSB' denotes a microspike bottom with 0.60 mm high spikes and 32 000 microspikes/m², 'MST' denotes a microspike top with 0.40 mm high spikes and 60 000 microspikes/m², 'R 12' denotes a grid-spike pattern, 'R 50' denotes a 50 mm × 50 mm grid pattern and 'G' denotes a smooth HDPE geomembrane. Reproduced by permission of Agru

HDPE welds are susceptible to brittleness and cracking failures at low temperatures. This deficiency of HDPE welds was identified using a dynamic impact test. The dynamic impact test is preferred over other brittleness tests such as ASTM D-746 'Standard Test Method for Brittleness Temperature of Plastics and Elastomers by Impact' (which establishes the temperature at which 50% of the specimens tested fail when subjected to specific

Table 12.22 Low-temperature limits of common geomembranes (Layfield, 2008)

Geomembrane material	Practical low-temperature handling limit ($^\circ$C)
PVC	-5
EIA-R	-10
Polyurethane (PU) (ester types)	-20
HDPE	-25
fPP	-30
PE-R (reinforced polyethylene)	-40

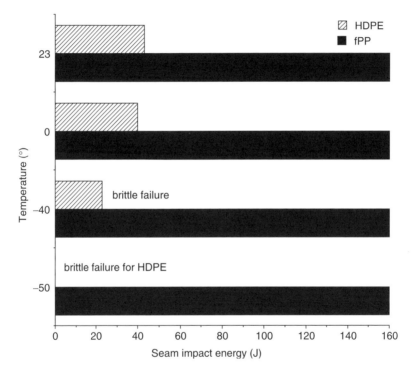

Figure 12.33 Low-temperature seam impact energy comparison for HDPE and fPP

conditions) and ASTM D-1790 'Test Method for Brittleness Temperature of Plastic Film by Impact'.

The dynamic impact test provides actual energy values required to cause failure by impact rather than simply giving pass/fail results which are less reflective of actual field behaviour.

The outstanding low-temperature toughness of fPP welds compared to HDPE welds is illustrated in Figure 12.33. HDPE welds fail in a brittle manner along the edge of the seam/weld at $-40\,^\circ$C whereas the fPP welds showed no failure down to $-50\,^\circ$C. (Montell, 1998).

12.17 HIGH-TEMPERATURE CAPABILITIES

The high-temperature capabilities of geomembranes are important for exposed applications (in full sunlight in hot climates) and also in applications where the geomembranes or liners may be in contact with hot process solutions or chemicals.

For instance, some geomembranes are used to line effluent ponds for black liquor with operating temperatures of up to 70 °C whilst others are used to line solar ponds[4] where the geomembrane liners need to withstand continuous temperatures above 80 °C. For such high-temperature applications, fPP geomembranes are well suited due to their relatively high melting points as well as EPDM and CSPE. fPP geomembranes also have the unique property of maintaining their mechanical properties across an unusually broad range of temperatures. Use of LLDPE and EVA liners on the other hand may be problematic due to their low softening points.

12.18 OXIDATIVE RESISTANCE

Oxidation (both thermal and UV induced) is generally regarded as the key degradation mechanism affecting the long-term durability of geomembranes. As oxidation proceeds, the physical and mechanical properties of the polymer start to deteriorate and eventually this process leads to failure of the geomembrane. The oxidation reactions are initiated by removal of a hydrogen atom from the polymer chains which creates a free radical that can then react with oxygen and form hydroperoxide groups. These in turn can decompose to form further free radicals (see Figure 12.34). The entire autooxidation scheme including the initiation, propagation and termination reactions is shown in Figure 12.35. To protect against oxidation during their service lifetime, antioxidants and stabilizers are added to the formulation.

The degradation of polyolefin geomembranes can be represented as a three-stage process (see Figure 12.36):

- depletion of antioxidants (and stabilizers);
- induction time to onset of polymer degradation;
- degradation of the polymer and loss of mechanical properties.

The first stage represents the period during which depletion of antioxidants and stabilizers occurs. In the second stage the antioxidant and stabilizers is effectively depleted but then follows the induction time to the onset of polymer degradation (during which time free radical yields begin to build). Finally in the third stage degradation of various properties begin to gradually develop. The point of 50% retention of mechanical properties is usually arbitrarily taken as the failure point. Note that during the majority of the degradation time frame the mechanical properties remain unchanged.

It is only after extensive oxidation has occurred that the physical and mechanical properties of the polymer start to deteriorate. For the mechanical properties, both tensile break stress and break strain decrease, whereas to a lesser extent the yield stress increases

---

[4] Solar ponds are ponds filled with a brine solution to trap heat and store solar energy. They contains layers of salt solution with increasing concentration (and therefore density) to give a density gradient which prevents heat in the lower layers from moving upwards by convection and leaving the pond.

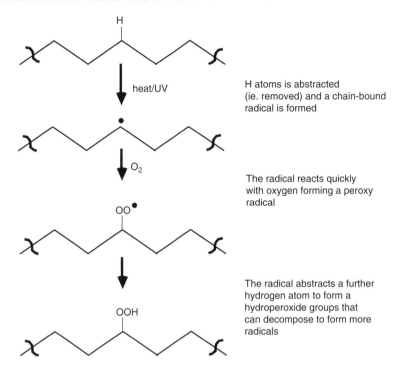

Figure 12.34 Oxidation reaction sequence showing that oxidation of polyolefins is initiated by removal of a hydrogen atom creating a chain-bound radical. This radical then reacts with oxygen forming a peroxy radical and then a hydroperoxide group which is a precursor to further radical products

and the yield strain decreases. Ultimately, the degradation becomes so severe that all tensile properties decrease and the engineering performance is jeopardized. This signifies the end of the so-called 'service life' of the geomembrane.

Although quite arbitrary, the limit of service life to polymeric materials is often selected as a 50% reduction in a specific design property. This is commonly referred to as the half-lifetime, or simply the 'half-life'. It should be noted that even at half-life, the material still exists and can function, albeit at a decreased performance level with a factor of safety lower than the initial design value.

Oxidative degradation of polymer is an auto-accelerated process. After a certain induction period during which time there are no significant changes in polymer properties, the degradation rate increases rapidly and the molecular weight of the polymer falls sharply. The material then becomes brittle and finally loses its mechanical resistance. Since the reaction rate at ambient temperatures is slow, the induction time can range up to years.

Oxidative degradation embrittles polyolefin geomembranes and can render them much more susceptible to cracking and stress cracking, thus causing regions of a geomembrane subject to relatively low tensile stresses or strains to undergo crack formation.

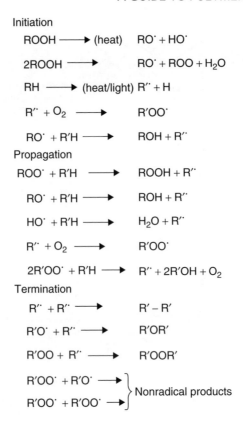

Figure 12.35 Free radical reactions responsible for the oxidative degradation of polyolefins (HDPE, LLDPE, fPP). Note that R represents the polymer, $R^{\bullet}$ is a polymer-bound free radical and ROOH is a polymer-bound hydroperoxide group which is a free radical precursor

Extraction of stabilizing additives by water, other extractive media or contact with transition metal ions such as copper, are some of the factors that greatly reduce the oxidative stability of polyolefin geomembranes (see Figure 12.37). For instance, immersion of HDPE in an alkaline solution has been found to accelerate the antioxidant depletion rate. This is because the antioxidants which are commonly used in polyolefin geomembranes (that is, the hindered phenols) have an ester bridge in their structure that is susceptible to alkaline-catalyzed hydrolysis. Greater depletion rates are recorded at higher temperatures. The temperature dependency of the depletion process follows the rule of a doubling in the depletion rate for every 10 degree increase in temperature.

The relative oxidative resistance of polyolefin geomembranes can be determined by the oxidative induction time (OIT) test. The geomembrane industry specifies that a HDPE geomembrane composition should exhibit a minimum of 100 min of standard oxidation induction time (according to ASTM D-3895).

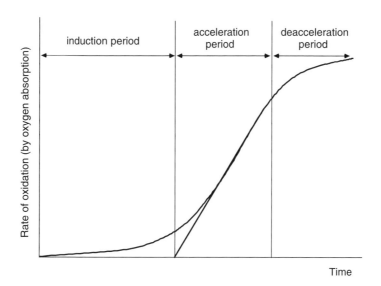

Figure 12.36   Stages of oxidation of polyolefin geomembranes. It is during the induction period that hydroperoxides (i.e. radial precursors) build up to critical levels.

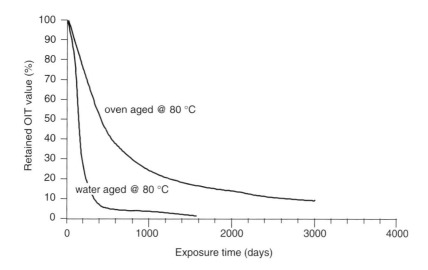

Figure 12.37   Retained OIT value (%) for HDPE geomembrane as a function of oven ageing (at 80 °C) and immersion in water (at 80 °C). Antioxidant loss is more rapid in water due to migration and extraction. Reprinted with permission from *Polymer Degradation and Stability*, Oxidative resistance of high-density polyethylene geomembranes, by W Mueller and I Jakob, **79**(1), 161. Copyright (2003) Elsevier

Table 12.23   Oven ageing requirements for HDPE and LLDPE geomembranes[b]

Polymer	Specification	Exposure conditions	Required results
HDPE	GRI GM-13	85 °C/90 days	55% S-OIT retained
			80% HP-OIT retained
LLDPE	GRI GM-17	85 °C/90 days	35% S-OIT retained
			60% HP-OIT retained

[a]Note that the performance requirement for LLDPE geomembranes is less stringent that for HDPE geomembranes, reflecting the lower oxidative stability and greater antioxidant migration potential of LLDPE geomembranes.

OIT results on geomembranes with modern stabilization packages need careful interpretation since some antioxidants/stabilizers may not give long OIT values but nevertheless may provide good service life.

The OIT (ASTM D-3895) test evaluates the relative performance of the antioxidant additive package and the resistance of a polyolefin geomembrane to oxidative degradation. The GRI GM13 standard requires that this should exceed 100 min in the case of HDPE geomembranes. The test can be performed at a lower temperature (i.e. 150 °C) in a pressurized oxygen atmosphere (ASTM D-5886) to detect the level of Hindered Amine Light Stabilizers (HALS) that are present in the additive package.

The oxidative resistance of polyolefin geomembranes can be assessed by oven ageing the geomembranes for 90 days at 85 °C. The retained stability of the antioxidant/stabilizer package is assessed using S-OIT and/or HP-OIT testing (See Table 12.23).

12.19   UV RESISTANCE

UV resistance is an important property for geomembranes that are exposed to sunlight during their service life. The outdoor degradation of geomembranes is the result of chemically irreversible oxidative reactions and physical changes that lead to the premature failure of the material. The deterioration caused by exposure to UV radiation is one of the main drawbacks of some polymeric geomembranes used in outdoor applications.

Geomembranes that are exposed to sunlight can oxidize and crack over time. This degradation is the result of the combined effects of thermal and UV radiation which can generate damaging free radicals. These free radicals quickly react with oxygen leading to unstable intermediates called hydroperoxides that can in turn cause oxidative degradation of polymers. Antioxidants, stabilizers and carbon black present in the geomembrane formulation can all scavenge or neutralize free radicals thus delaying onset of the oxidation degradation. While carbon black is a permanent filler residing in the polymer, antioxidants and stabilizers can be depleted by leaching or consumption and thus they have a limited lifetime.

The UV resistance of most geomembranes is imparted by adding carbon black (of the correct particle size) and a combination of proprietry UV stabilizers (generally hindered

Table 12.24   UV resistance requirements for HDPE and LLDPE geomembranes[a]

Polymer	Specification	Exposure conditions	Required results
HDPE	GRI GM-13	1600 h QUV exposure	50% HP-OIT retained
LLDPE	GRI GM-17	1600 h QUV exposure	35% HP-OIT retained

[a]Note that the performance requirement for LLDPE geomembranes is less stringent that for HDPE geomembranes, reflecting the lower oxidative stability and greater antioxidant migration potential of LLDPE geomembranes.

amine stabilizers). In the case of PVC geomembranes progressive embrittlement and cracking can be due to volatilization of the low molecular weight plasticizers. PVC geomembranes also employ carbon black but at much higher levels than those used in polyolefins (e.g. approximately 10 to 15%).

UV resistance of polyolefin geomembranes can be assessed by accelerated UV exposure conducted according to GRI-GM 11 using an UV-fluorescent weathering chamber. The exposure is for 1600 h, with alternating 20 h of UV at 75 °C followed by 4 h condensation at 60 °C. The retained stability of the antioxidant/stabilizer package is assessed using HP-OIT testing (See Table 12.24).

12.20   PERMEABILITY OF GEOMEMBRANES

A geomembrane is defined as a 'relatively impermeable' membrane or barrier used in a geotechnical engineering application so as to control fluid migration from a man-made project, structure or system. In the use of geomembranes as barriers to the transmission of fluids, it is important to recognize the difference between geomembranes (which are nonporous homogeneous materials) and other liner materials that are porous such as soils and concrete. The transmission of permeating species through geomembranes without holes occurs by absorption of the species in the geomembrane and diffusion through the geomembrane on a molecular basis. The driving force is the concentration gradient across the geomembrane (ASTM D-5886).

Polymeric geomembranes are therefore not absolutely impermeable (no polymeric material is totally impermeable); however they are relatively impermeable when compared to soils and clays. Geomembranes possess permeability values (as measured by water-vapour transmission tests) in the range 1×10^{-12} to 1×10^{-15} m/s which are one thousand to one million times lower than a typical clay liner. Polymeric geomembranes are thus described as being 'relatively impermeable'. Even when a liner is correctly welded and installed (that is, without installation damage or poor welds) the contained liquid may ultimately diffuse through the geomembrane liner. The rate of diffusion however is extremely low.

Permeation of gases, vapors, and liquids through polymeric geomembranes occurs on a molecular scale in a three-step process:

1. Dissolution in or absorption by the geomembrane on the upstream side.
2. Diffusion through the geomembrane.
3. Desorption on the downstream side of the barrier.

The rate of transmission of a given chemical species, whether as a single permeant or in mixtures, is driven by its concentration gradient across the geomembrane.

Factors that can affect the diffusion of an organic compound within a geomembrane include the solubility of the permeant in the geomembrane, the microstructure of the polymer, especially the percent crystallinity and the flexibility of the polymer chains, the size and shape of the diffusing molecules and the temperature at which diffusion is taking place (whether the condition at which diffusion is taking place is above or below the glass transition temperature of the polymer).

The permeability of geomembranes to liquids can be measured using ASTM E-96 where a sample of the geomembrane is affixed to the opening of a small aluminium cup containing a known volume of the test liquid (e.g. water or a solvent). The covered cup is then placed in an environmental chamber at fixed humidity (e.g. 20% or 50% RH) and temperature (21 °C). The humidity of the chamber is generally set at 20% while the humidity in the headspace of the cup is 100% and so there is a controlled concentration gradient across the geomembrane sample. Since there will be diffusion through the membrane the liquid level will drop over time. This is also known as the 'wet cup method'.

The wet cup procedure for moisture permeability uses a test cup containing water. A small air space is left below the test material resulting in a relative humidity near 100% in the test cup, which is surrounded by the 20% or 50% relative humidity environment in the chamber. In this procedure water vapour is driven from the cup into the chamber through the specimen and the cup loses mass. The permeability is calculated from the mass change of the test cup.

The rate of diffusion of moisture across a geomembrane is a function of the permeability of the polymer. HDPE has a permeability of 1×10^{-12} cm/s whereas PVC has a permeability of 1×10^{-10} cm/s.

Typical permeability values for various geotechnical materials are shown in Table 12.25.

While the 'wet cup method' test is simple and effective there exist a number of methods for determining the rate of fluid permeability through geomembranes. The ASTM D-5886 (2006) 'Standard Guide for Selection of Test Methods to Determine Rate of Fluid Permeation Through Geomembranes for Specific Applications' describes the test methods to assess the permeability of geomembranes for a proposed specific application to various permeants. The widely varying uses of geomembranes as barriers to the transport and migration of different gases, vapours and liquids under different service conditions, requires determinations of permeability by test methods that relate to and simulate the liner service. The permeating species range from a single component to highly complex

Table 12.25 Typical permeability values

Material	Permeability value
Sand drainage layer	0.01 cm/s
45 cm of clay	1×10^{-5} cm/s
Geosynthetic clay liner (GCL) soil barrier layer	1×10^{-8} cm/s
PVC geomembrane	1×10^{-10} cm/s
HDPE geomembrane	1×10^{-12} cm/s

mixtures such as those found in waste liquids and leachates. In specialized applications it may be important to measure transmission or migration of a species that would take place under specific conditions and environments including temperature, vapour pressure and concentration gradients.

Test conditions and procedures should be selected to reflect actual service requirements as closely as possible. It should be noted that field conditions may be difficult to model or maintain in the laboratory. It is important to note however that most of the leakage through a geomembrane is not due to permeation/diffusion through the liner but rather related to the quality of the welds and from installation damage (e.g. undetected holes and tears).

12.20.1 MECHANISM OF DIFFUSION AND TRANSPORT

The transport of liquids and leachates through geomembranes and liners differs fundamentally from fluid transport through soil liner materials such as compacted clay. The dominant mode of fluid transport is through holes, poor welds, seam failures and other penetrations in the geomembrane. After these physical modes of leakage, the next important mode of fluid transport is governed by molecular diffusion. This diffusion occurs in response to the presence of a concentration gradient and is termed Fickian diffision and is governed by Fick's first law.

Fick's first law relates the rate of diffusion to the concentration gradient and states that 'diffision of molecules goes from regions of high concentration to regions of low concentration, with a magnitude that is proportional to the concentration gradient'.

The transport of fluids through geomembranes by Fickian diffusion is very low compared to hydraulic flow rates for soil liners. Also the rate of Fickian diffusion of fluids through geomembranes and polymeric liners is very small compared to the flow of liquid through holes and defective wears introduced into the liner during the construction phase.

The resistance of geomembranes to chemicals is covered in detail in Chapter 14.

12.20.2 WATER VAPOUR DIFFUSION (WATER VAPOUR TRANSMISSION)

Since polymeric geomembranes are relatively impermeable to liquid water, the water vapour transmission of a geomembrane is used to estimate the amount of water that can theoretically diffuse through the membrane and escape as vapour.

Water passes through 'impermeable' geomembranes by the process of diffusion as measured by ASTM E96. Therefore the Water Vapor Transmission (WVT) is the most relevant measure of the 'permeability' of a geomembrane used for water containment.

The water vapor transmission value for a 2 mm PVC geomembrane is of the order of 0.5 g/m^2/day, while bituminous geomembranes have a water vapor transmission value of about 0.1 g/m^2/day compared to a value of <0.1 g/m^2/day for 1.5 mm HDPE. HDPE has amongst the lowest water vapour transmission rates of any geomembrane polymer material making it an ideal containment barrier for a wide range of applications (see Figure 12.38 which shows WVT rates for 0.5 mm geomembranes). Figure 12.39 shows that the water vapour permeability of HDPE increasing exponentially with increasing temperatures.

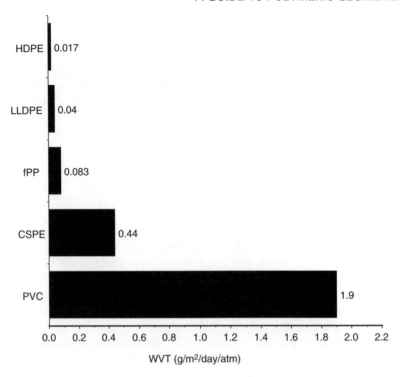

Figure 12.38 Water vapour transmission (WVT) rates for common geomembranes

Table 12.26 shows that the water vapour diffusion through a 1 mm HDPE geomembrane with a head of 300 mm is approximately 0.8 lphd.[5] This demonstrates that no single HDPE geomembrane can be considered impermeable and therefore 'leak free'.

Note the diffusion rates through LLDPE and PVC geomembranes are higher by factors of 45 and 115 times, respectively, due to their lower crystallinity and hence higher amorphous microstructure through which the water molecules can diffuse.

These leakage rates need to be viewed in perspective and are still extremely low compared with holes and punctures caused by angular stones in the subgrade or in the soil covering.

12.20.3 GAS PERMEABILITY

The permeability of polymers to gases is mainly a function of the density (hence crystallinity) and free volume. The permeability of gases in HDPE geomembranes is shown in Figure 12.40 while the permeability of methane through PVC, LLDPE and HDPE is shown in Figure 12.41 which is of particular relevance in the area of biogas covers. HDPE has a lower permeability to methane than LLDPE due to its greater crystallinity. PVC, despite having no crystallinity, has a lower permeability to methane than LLDPE due to

---

[5] lphd = litres per hectare per day.

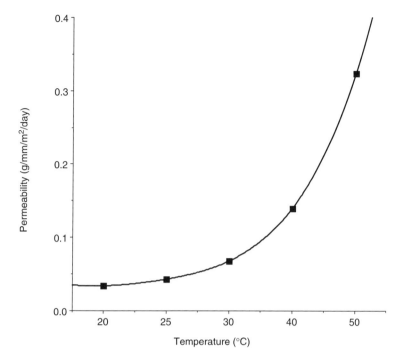

Figure 12.39 Water vapour permeability in HDPE as a function of temperature

Table 12.26 Water vapour diffusion through a 1 mm geomembrane under a head of 300 mm of water

Geomembrane material	Water vapor diffusion rate (lphd)[a]
HDPE (1.0 mm)	0.8[b]
LLDPE (1.0 mm)	36
PVC (1.0 mm)	92

[a]lphd = litres per hectare per day.
[b]Giroud and Bonaparte (1989).

the polar structure of PVC (since methane is non-polar). Figure 12.42 shows the methane gas transmission rates for PVC, LLDPE and HDPE geomembranes as a function of sheet thickness.

12.21 DURABILITY

The design life of a geomembrane is defined as the minimum expected service time where a geomembrane is intended to perform (survive) a particular containment function. The survivability of geomembranes is related to the geomembrane type, the design of the

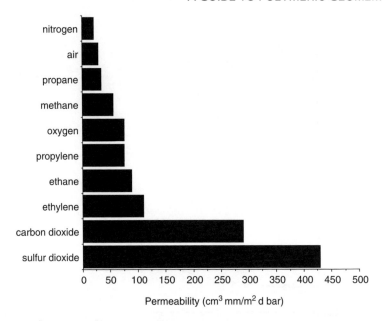

Figure 12.40 Permeability of gases in HDPE geomembranes at 25 °C (note that the data for alkanes, alkenes and SO$_2$ were obtained at 20 °C)

installation, the CQA of manufacture, the quality of installation (including welding) and the chemical nature of the products in contact with the geomembrane.

The durability of a geomembrane is related to the change of critical performance properties over time. The durability of geomembranes can also be influenced by the nature and quality of the additives and fillers, especially the type and level of carbon black used in the geomembrane sheets.

In particular the durability of geomembranes are significantly influenced by the nature and quality of the stabilizers and the UV-screening pigments. Furthermore it does not only depend on the concentration of additives present but also their dispersion. A non-uniform dispersion of additives caused by poor distributive and dispersive mixing during manufacture can lead to regions that are deficient in the protective additives and thus are more susceptible to localized degradation.

The durability of a geomembrane is a function of:

- it base polymer;
- the polymer microstructure;
- its formulation (i.e. the type of additives and fillers compounded in it, their level and their dispersion);
- characteristics of the containment product.

Sheet thickness is also an important factor governing the durability of geomembranes. For instance thermal oxidation and UV degradation are dependent on surface area to volume, while chemical uptake by diffusion and absorption are inversely related to thickness.

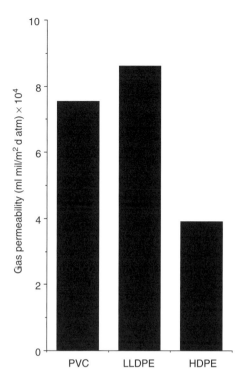

Figure 12.41   Methane permeability/migration through geomembranes

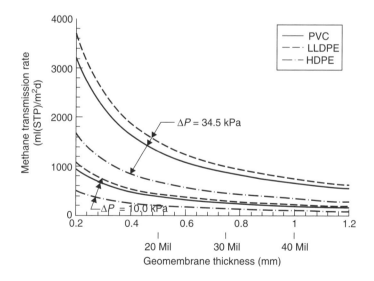

Figure 12.42   Design chart for methane gas transmission rates for HDPE, LLDPE and PVC geomembranes as a function of thickness. Reprinted with permission from *Geosynthetics International*, Methane gas migration through geomembranes, by T. D. Stark and H. Choi, **12**(2), 120. Copyright (2005) Thomas Telford Publishing

Heat ageing and weathering tests are critical factors in proper geomembrane selection and in long-term geomembrane survivability. Geomembranes are increasingly used in mining applications (e.g. for process water lagoons, tailings dams, leach pads, etc.) in arid and desert areas where consistently high temperatures are encountered.

The degradation of geomembranes leads to polymeric chain scission causing the polymer to become brittle. The following mechanical property changes are generally observed with geomembrane degradation:

- a decrease in % elongation at failure;
- an increase and then decrease in strength at failure (i.e. tensile stress at break);
- a decrease in impact strength;
- an increase in modulus of elasticity (i.e. increasing stiffness);
- increase in brittleness (i.e. general loss of ductility).

The changes in the above properties can thus be used to monitor the progressive degradation of polymeric geomembranes over time.

When examining potential degradative effects on geomembranes one has to consider synergistic factors where two or more agencies are acting simultaneously on the geomembrane (e.g. chemical degradation and stress, oxidation degradation and chemical degradation, oxidation degradation and stress).

12.21.1 CARBON BLACK DISPERSION

The carbon black dispersion in a geomembrane is determined by observing a thin slice of the geomembrane under a microscope and estimating the extent of dispersion of the carbon black particles. The observed dispersion is qualitatively compared to a reference chart containing examples of several degrees of carbon black dispersion. The majority of geomembrane specifications call for a dispersion rating of Category 1 and 2 (also known as A-1 and A-2) and one view in Category 3 (or B-1).

Poor carbon black dispersions (i.e. the presence of relatively large carbon black agglomerates) can impact negatively on the UV resistance, environmental stress crack resistance and tensile properties of the geomembrane (see Table 12.27).

Table 12.27 Effect of a poor carbon black dispersion on the performance properties of geomembranes

Effect	Comments
Reduced UV resistance	Localized regions that are deficient in carbon black, or have large interparticle distances, show increased sensitivity to UV degradation
Reduced stress crack resistance	Agglomerates of carbon black can act as stress concentrations and provide initiation sites for stress cracking
Reduced tensile properties	Agglomerates of carbon black can act as stress concentrations and provide initiation sites for fracture during tensile loading

12.21.2 WARRANTIES

HDPE and CSPE-R geomembranes come with manufacturer's warranties of up to 20 years (sometimes 30 years in the case of CSPE-R). LLDPE, fPP and PVC geomembranes on the other hand cannot tolerate long-term UV exposure to the same degree and hence generally have lower exposed lifetimes. It is good practice to cover these geomembranes with soil or with a sacrificial material such as a replaceable geotextile material.

12.22 APPLICATION SPECIFIC PERFORMANCE PROPERTIES

12.22.1 PROPERTIES REQUIRED FOR BACKFILLED GEOMEMBRANES

There are three key performance properties required for a backfilled geomembrane namely:

- Site-appropriate interface friction angles (IFAs). A high friction angle enables steeper slope construction which can optimize pond or cell containment volumes. In the case of HDPE geomembranes the interface friction angle can vary from a low of 18 degrees for smooth HDPE to a relatively high value of 36 degrees for textured HDPE.
- An optimum combination of strength and puncture resistance to increase liner protection during backfilling operations. The geomembrane is required to have a high critical cone height in puncture tests (see published values of truncated cone puncture tests (Hullings and Koerner, 1991).
- Resistance to differential settlement, which is a function of the elongation properties of a geomembrane. It should be noted that the usable elongation of the material is not the yield point but an interpreted percentage of the available elongation. Settlement is common in landfill caps or in applications where settlement in the sub-base occurs after construction. Additionally, there are cases where there is erosion of the subgrade due to water flows under the liner.

Table 12.28 Rating of flexible membrane liners (FMLs) in various situations (adapted from Ortego *et al.*, 1995)[a]

Exposure	LLDPE and fPP	PVC	CPE	CSPE	EPDM	Butyl rubber
Exposed liner	R[b]	NR	R	RR	R	R
Exposed side slope liner	RR[b]	NR	RR	RR	RR	RR
Buried liners	R	R	R	RR	R	R
Acid resitance (pH 2–7)	R	R	R	RR	R	R
Alkali resistance (pH > 7)	R	NR	R	RR	R	R
Domestic waste	RR	R	R	RR	R	R
Petroleum products	NR	NR	R	NR	NR	NR

[a]R, recommended; RR, recommended only with reinforcement; NR, not recommended.
[b]LLDPE should only be used as an exposed liner if heavily fortified with UV stabilizers.

Table 12.29 Geomembrane comparison table of performance attributes (adapted from Sadlier and Frobel, 1997)

Attribute	HDPE	LLDPE	fPP	PVC	EIA	CSPE-R	EPDM
General chemical exposure	Excellent	Good	Excellent	Fair	Excellent	Excellent (when cured)	Good
Hydrocarbon exposure	Fair–Good	Fair–Good	Fair–Good	Fair	Excellent	Good (when cured)	Good
Weathering (UV exposure)	Excellent	Fair	Fair–good	Poor–fair	Excellent	Excellent (when cured)	Excellent
Dimensional stability	Poor	Poor	Good – excellent when reinforced	Good	Good	Excellent	Excellent
Tensile properties	Good	Good	Good – excellent when reinforced	Good	Excellent	Excellent	Good
Uniaxial elongation properties	Excellent	Excellent	Excellent	Good	Fair	Good	Good
Multiaxial elongation properties	Poor	Excellent	Excellent	Excellent	Fair	Good	Good
Puncture performance	Fair	Excellent	Excellent	Excellent	Excellent	Good	Good
Installation damage resistance	Fair	Good	Excellent	Excellent	Good	Good	Excellent
Seaming methods	Thermal – excellent	Thermal – excellent	Thermal – excellent	Thermal or solvent bonding – good	Thermal – excellent	Thermal or solvent bonding – Good	Vulcanizing tape seams – good
Repair in service	Good	Good	Excellent	Good	Good	Poor – requires adhesives	Vulcanizing tape seams – good
Stress cracking	Fair	Good	Does not occur	Good	Does not occur	Does not occur	Does not occur
Flexibility for detail work	Fair	Excellent	Excellent	Excellent	Good	Good	Good

12.22.2 PROPERTIES REQUIRED FOR EXPOSED GEOMEMBRANES

There are four key performance properties required for exposed geomembranes, namely:

- Excellent resistance to ultraviolet light which is an essential requirement for all exposed geomembranes. The HP-OIT value for the geomembrane reflects the level of hindered amine stabilizers that are essential to UV durability. In addition, the level and dispersion of carbon black are important parameters in the case of black geomembranes.
- Low-temperature resistance is an important performance property in cold climates.
- Resistance to mechanical damage, which is a function of the tensile strength, elongational properties and puncture resistance of the geomembrane material.
- Adequate dimensional stability, especially for geomembranes that are placed on steep slopes since materials with poor dimensional stability may be subject to creep and could fail on steep slopes.

12.22.3 PROPERTIES REQUIRED FOR DETAIL WELDABILITY

Detail weldability is related to the technique of welding of pipes and attachments to the geomembrane and the equipment required to perform the weld. Geomembranes that can be solvent welded are the most versatile, followed by those that are hot-air weldable and then the extrusion-weldable materials. Geomembranes that require multiple pipe penetrations are best welded with materials that have excellent detail weldability.

12.23 COMPARISON OF PERFORMANCE PROPERTIES OF FLEXIBLE GEOMEMBRANES

A comparison of some performance properties of flexible geomembrane liners are shown in Tables 12.28 and 12.29.

REFERENCES

Bhatia, S. K., 'Comparison of PVC and HDPE Geomembranes (Interface Friction Performance)', Available from the PVC Geomembrane Institute (2008).

Giroud, J. P., Quantification of Geosynthetic Behavior, *Geosynthetics International*, **12**(1), 2 (2005).

Giroud, J. P. and Bonaparte, R., Leakage through Liners Constructed with Geomembranes – Part 1. Geomembrane Liners, in *Geotextiles and Geomembranes*, Vol. 8, Elsevier Science, Oxford, UK, pp. 27–67 (1989).

GRI GM-15, 'Determination of Ply Adhesion of Reinforced Geomembranes' [http://www.drexel.edu/gri/member/GM/GM15/GM15.html] (now superceded by ASTM D-6636).

Hölzlöhner, U., 'Landfill liners systems, a state of the art report', Penshaw Press, Cleadon, UK (1995).

Hsuan, Y. G. and Koerner, R. M., The Single Point-Notched Constant Load Test: A Quality Control Test for Assessing Stress Crack Resistance, *Geosynthetics International*, **2**(5), 831–843 (1995).

Hsuan, Y. G., Koerner, R. M., and Lord, A. E., The Notched Constant Tensile Load (NCTL) Test to Evaluate Stress Crack Resistance, in *Proceedings of the 6th GRI Conference*, GRI, Philadelphia, PA, USA, pp. 244–256 (1992).

Hullings, D. and Koerner, R., Puncture Resistance of Geomembranes Using a Truncated Cone Test, in *Proceedings of Geosynthetics '91*, IFAI, St Paul, MN, USA, pp 273–285 [http://www.geomembranes.com/index_resources.cfm?copyID= 35&ID=geo&type=tech] (1991).

IFC Presentation, entitled 'Bituminous Geomembranes (BMG), Coltanche ES 3' [http://www.ifc.org/IFCExt/spiwebsite1.nsf/b7a881f3733a2d0785256a550073ff0f/ f95106404992f6418525716c0059694e/%24FILE/AttachmentC.AppendixA.Tech_ Doc_ES3%20_2_.pdf] (2008).

Koerner J. R. and Koerner, R. M., 'A survey of solid waste landfill liner and cover regulations; Part II – worldwide status, GRI report #23, Geosynthetic Research Institute (1999).

Koerner, R. M., *'Designing with Geosynthetics'*, Prentice Hall (2005).

Koerner, R. M., Martin, J. P. and Koerner, G. R., Shear Strength Parameters Between Geomembranes and Cohesive Soils, *Geotextiles and Geomembranes*, **44**(1), 21–30 (1986).

Layfield Cold Temperature Installation Guide [www.layfieldgroup.com] (2008).

Martin, J. P., Koerner, R. M. and Whitty, J. E., Experimental Friction Evaluation of Slippage Between Geomembranes, Geotextiles and Soils, in *Proceedings of the International Conference on Geomembranes*, IFAI, Denver, CO, USA, pp. 191–196 (1984).

Merry, S. M. and Bray, J. D., Size Effects for Multiaxial Tension Testing of HDPE and PVC Geomembranes, *Geotechnical Testing Journal*, **18**(4), 441–449 (1995).

Montell, 'High Performance Materials for Geomembranes', Astryn fPP Booklet, Montell Polyolefins (1998).

O'Rourke, T. D. Druschel, S. J. and Netravali, A. N., Shear Strength Characteristics of Sand Polymer Interfaces, *Journal of Geotechnical Engineering*, ASCE, **116**(3), 451 (1990).

Ortego, J. D., Aminabhavibvi, T. M., Harlapur, S. F. and Balundgi, R. H., A review of polymeric geosynthetics used in hazardous waste facilities, *Journal of Hazardous Materials* **42**, 115–156 (1995).

Peggs, I. D., 'Geomembrane Liner Durability: Contributing Factors and the Status Quo' [http://www.geosynthetica.net/tech_docs/IDPigsUKpaper.pdf] (2003).

Peggs, I. D. and Allen, S., Geomembrane Seal Peel Separation: How and Why?, *GFR Magazine*, **19**(3), 1 (2001).

Peggs, I. D., Schmucker, B. and Carey, P., Assessment of Maximum Allowable Strains in Polyethylene and Polypropylene Geomembranes, in *Proceedings of Geo-Frontiers* (2005).

Peggs, I. D. and Thiel, R., Selecting a Geomembrane Material, in *Proceedings of the Sixth International Conference on Geosynthetics*, IFAI, St Paul, MN, USA, pp. 381–388 (1998).

Sadlier, M. and Frobel, R., Geomembrane Properties – A Comparative Perspective, presented at the *GeoEnvironment Conference*, Melbourne, Australia (November, 1997).

Seeger, S. and Müller, W., Limits of Stress and Strain: Design Criteria for Protective Layers for Geomembranes in Landfill Liner Systems, in *Proceedings of Geosynthetics: Applications, Design and Construction*, A.A. Balkema, Rotterdam, The Netherlands, pp. 153–157 (1996).

Stark, T. D., Boerman, T. R. and Connor, C. J., 'Puncture resistance of PVC Geomembranes – Technical Note' [http://www.geosynthetica.net/tech_docs/ pvcpunctureresistancepgi.pdf] (2007).

Stevens, 'Geomembranes Literature on Puncture Resistance [http://www.stevensgeomembrane.com/Corporate/Literature/4%20puncture%20resistance.pdf] (see also www.dowgeomembrane.com) (2008).

Tognon, A. R., Rowe R. K. and Moore, I. D., Geomembrane Strain Observed in Large-Scale Testing of Protection Layers, *Journal of Geotechnical and Geoenvironmental Engineering*, **126**(12), 1194–1208 (2000).

US EPA, 'Calculated Flow Rates Through a Geomembrane with a Liquid Head of 0.3m' (1991).

13

Testing of Geomembranes

13.1 MATERIAL PROPERTY TESTING

There are two main categories of mechanical tests for geomembranes:

- index tests;
- performance tests.

Index tests are used to measure physical and mechanical properties of geomembrane materials (such as thickness, tensile strength, puncture resistance etc.). The values obtained can be compared to specification sheet values and can be used to assess the variability between different batches and to compare geomembrane material from different suppliers. Care should be exercised if index test values are used in design calculations.

Performance tests are conducted to obtain engineering properties such as interface friction and protection efficiency that can be used directly in design calculations. These tests are conducted using site-specific materials and relevant boundary conditions. They often involve assessing the interaction between two or more geosynthetics (e.g. between geomembranes and geotextiles) and between geosynthetics and soils.

The distinction between index tests and performance tests is summarized in Table 13.1.

- The following illustrates the use of index test results for design purposes by applying safety factors.

$$\text{Design value} = \text{Laboratory value of index test}$$
$$\times 1.5 \text{ (safety factor for installation stress)}$$
$$\times 1.5 \text{ (safety factor for temperature)}$$

Such safety factors are based on the fact that both higher stresses and higher temperature can 'derate' the performance of the GM.

A Guide to Polymeric Geomembranes: A Practical Approach   J. Scheirs
© 2009 John Wiley & Sons, Ltd

Table 13.1 Distinction between index tests and performance tests

Index tests	Performance tests
These are an indicator of likely behaviour of the geomembrane and are surrogates for other properties	These are used to establish properties under site specific conditions using actual soil and/or materials from the installation site
These are useful for quality control purposes but should not be used in design calculations. Sometimes they are used in design calculation but appropriate reduction factors are applied	These results can be used directly in design calculations. Note that they don't rely on published data since no two installation sites are the same
They are used for comparing products, ensuring quality and meeting specifications. Generally they are relative values (e.g. low, medium or high)	These results must be obtained with materials from the actual installation site and under comparable anticipated service conditions
Examples include mass per unit area, mechanical properties such as tensile strength, durability tests, etc.	For example, these includes tensile strength during a soil pull-out test where the soil interacts with the geomembrane

13.2 MEASURING THICKNESS

The thickness of geomembranes can be measured by a number of different test methods which are summarized in Table 13.2.

The conditions for thickness measurements of geomembranes according to ASTM D-5199 are listed in Table 13.3.

Note when measuring the thickness of geomembranes, for example by ASTM D-5199, it is very important to ensure that the sample is completely parallel with the platen and that all measurements are made under consistent pressure and flatness. Failure to do this will produce unreliable thickness results.

13.2.1 MEASURING THICKNESS OF TEXTURED GEOMEMBRANES

The procedure to measure the core thickness of textured geomembranes is described by ASTM D-5994 'Standard Test Method for Measuring Core Thickness of Textured Geomembrane'. The method involves measuring the thickness at a specific location between two gauge points. The measured core thickness of geomembranes can vary significantly depending on the pressure applied to the specimen during measurement. To reduce this variation and the chances of getting unrealistically low values due to excessively high pressures, a specific gauge point geometry and applied force are prescribed by this test method (it requires a dead load of 0.56 N (about 58 g). The points are required to be tapered at an angle of 60 degrees to the horizontal with the tip rounded to a radius of 0.8 mm (see Figure 13.1). The gauge has a base point (base pointer) and a free-moving presser point whose axes are aligned to each other. When measuring the sample it is important to allow the presser point to come slowly into contact with the test specimen while moving the test specimen to locate the gauge points in the 'low spots' or 'valleys',

Table 13.2 Summary of thickness methods for geomembranes

ASTM method	Details of method	Comments
ASTM D-751	Thickness is measured by a deadweight thickness gauge with a 9.52 mm diameter presser foot. The presser foot exerts a 23.44 kPa pressure on the specimen during the thickness measurement	Intended mainly for COATED FABRICS and SCRIM-REINFORCED geomembranes
ASTM D-1593	The average thickness of a specimen approximately 100 mm by 100 mm square is calculated by dividing the weight of the specimen by its density and its area	Intended mainly for PLASTICIZED PVC liners and sheeting
ASTM D-3767	Method A uses an 25.4 mm dia. flat-footed micrometer on flat sheets where a test pressure between 10 and 21.8 kPa is exerted. Method A1 uses a micrometer having two contact members with domed surfaces of spherical radius 12.7 mm	Intended mainly for ELASTOMERS but can be used for geomembranes with thicknesses less than approximately 3 m
ASTM D-5199	The geomembrane between two parallel, planar surfaces under a pressure between 20 kPa and 200 kPa using a 6.35 mm diameter presser foot	Intended specifically for measuring the nominal thickness of SMOOTH geomembranes
ASTM D-5994	The core thickness of textured geomembranes is measured using a special micrometer with tapered presser feet	Intended specifically for measuring the nominal thickness of TEXTURED geomembranes[a]

[a]Note for spray-on texturing, the texture can also be scraped off and then the thickness of the smooth core sheet measured using a thickness method for smooth sheet.

Table 13.3 Conditions for thickness measurements of geomembranes according to ASTM D-5199

Parameter	Requirements
Presser foot diameter	6.35 mm
Specimen size	75 mm diameter disc (minimum)
Applied load	200 kPa for HDPE 20 kPa for non-HDPE geomembranes
Dwell time before reading	5 s
Number of specimens required	10

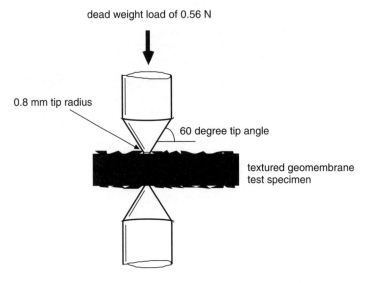

dead weight load of 0.56 N

0.8 mm tip radius

60 degree tip angle

textured geomembrane
test specimen

locate the guage points in the 'low spots' ('valleys') in between the
projections to order to obtain the local minimum thickness reading

Figure 13.1 Schematic showing the method for measuring core thickness of textured
geomembranes as per ASTM D-5994

in between the projections (or into the indentations) of the textured surfaces to obtain the
local minimum thickness reading.

13.3 DENSITY

There are two main test methods for measuring densities of geomembranes. One method
used is ASTM D-792 which is based on specific gravity. ASTM D-792 (known as the
'displacement method') is based on the classical definition of density where the density
is calculated by dividing the weight of the material in air by its loss of weight when
immersed in water. The other and less accurate method is ASTM D-1505. This technique
known as the 'density gradient column' technique is preferred for geomembrane materials
with a density less than 1 (such as polyolefins). This test is conducted by preparing a
column containing a graduated mixture of ispropanol and water). The highest density
solution is at the bottom of the column graduating to the lowest density solution at the
top. The sample for which the density is being determined is then dropped into the liquid
column and its location relative to calibrated floats is used to determine its density. Care
must be taken not to mix the various layers of solution.

The most widely used density method for geomembranes is ASTM D-792 which
requires an analytical balance with a precision of 0.1 mg or greater. The accuracy of
this method is 0.05%. Note however that when using ASTM D-792 (density by displacement), the results can be seriously compromised by air bubbles attached to the specimen,

Table 13.4 Density ranges for various polyethylene geomembrane types

Geomembrane type	Resin density range (g/cm^3)	Calculated density of compounded resin (g/cm^3)
HDPE (MDPE)	0.940–0.960	0.951–0.971
LLDPE	0.930–0.940	0.941–0.951
LDPE	0.915–0.930	0.926–0.941
VLDPE	0.900–0.906	0.911–0.917

particularly when wax or oil contamination on the polymer surface precludes full 'wetting out'.

In ASTM D-1505 a glass column usually 1 m high is filled with a solution of water and alcohol. A number of precalibrated floats (usually 5 or 6) with neutral buoyancy at a specific density position are evenly distributed along the column. For polyethylene the density gradient column typically ranges from 0.930 g/cm^3 at the top of the column to 0.950 g/cm^3 at the bottom. The position of the test specimen is recorded relative to the position of the calibration floats. From this the density of the sample is interpolated. Note that in ASTM D-1505 the results can be affected by overcrowding of the density column and by the temperature of the density gradient column. Note to maintain a constant temperature, the density gradient column should be positioned in a thermostated water bath.

The density or specific gravity depends on the base polymer of the geomembrane. The ASTM classification for HDPE, for example, requires a density at least equal to 0.941 g/cm^3.

Note the densities of HDPE geomembranes can be similar to but different to the density of the starting resin due to the incorporation of additives (e.g. carbon black) into the polymer during manufacture of the sheet. The density of most geomembrane polymers falls between 0.85 and 1.5 g/cm^3. Table 13.4 presents density ranges for various PE geomembrane resins and the associated compound density of the manufactured sheet. The values in Table 13.4 assume that the percentage carbon black in the material is 2.5% and the density of carbon black is taken as 1.8 g/cm^3.

13.4 MASS PER UNIT AREA

The mass per unit area as specified in ASTM D-1910 is determined simply by weighing a unit area of a geomembrane and then dividing the mass by the area measured.

13.5 TENSILE TESTING METHODS

Tensile testing of geomembranes is the most common way of determining key mechanical properties such as tensile strength and elongation. The tensile behaviour can be measured for purposes of quality control, for ranking candidate geomembranes or for testing retained properties after incubation or ageing to assess the residual mechanical properties.

Tensile tests measure the force required to break a specimen and the extent to which the specimen stretches or elongates to that breaking point. Tensile tests produce a stress–strain curve, which can be used to determine tensile modulus. The data is often used to specify a material, to design products to withstand application forces (i.e. care needs to be taken here to employ appropriate safety factors, as performance tests are preferred for design purposes) and as a quality control check of materials. Since the physical properties of many materials (especially thermoplastics) can vary depending on ambient temperature, it is sometimes appropriate to test materials at temperatures that simulate the intended end-use environment.

Tensile testing can be classified in two ways:

- uniaxial tensile testing (where the tensile force is exerted along one axis in opposite directions);
- multiaxial tensile testing (where the tensile force is exerted by multiple axes simultaneously).

Tensile properties of polyethylene geomembranes are particularly sensitive to the test specimen geometry and strain rate. The pre-dominant test method used by the polyethylene geomembrane industry to measure the tensile properties of smooth and textured HDPE and LLDPE geomembranes is ASTM D-638 ('Tensile Properties of Plastics') or more specifically ASTM D-6693.

A second test method is ASTM D-882 ('Tensile Properties of Thin Plastic Sheeting') which covers thin polyethylene sheeting of less than 1.0 mm (40 mil) in thickness. Both ASTM D-638 and the ASTM D-882 are index tests designed to produce tensile properties for manufacturing quality control and third party quality assurance.

A third test, ASTM D-4885 ('Determining Performance Strength of Geomembranes by the Wide Width Strip Tensile Method'), is designed to determine the performance strength of geomembranes.

Tensile tests, covered in ASTM D-638, D-882 and D-751, are commonly used to evaluate samples for quality control and quality assurance of manufactured geomembrane sheet materials.

Table 13.5 shows the various test methods for tensile testing of plastics.

13.5.1 STANDARD TENSILE TEST

In standard tensile testing, the test specimen is gripped with clamps along the top and bottom widths in a tensile testing machine. The specimen is then pulled at a constant rate (known as strain rate or crosshead speed) and the tensile strength and elongation of the specimen are recorded. The results are plotted as a stress–strain curve.

The dog-bone shape of tensile specimens ensures that failure of the specimen occurs in the narrow middle section; however despite its characteristic shape it does not represent 'plane strain' conditions on which most design work is predicated.

The tensile testing specimen has a dumb-bell or dog-bone shape as shown in Figure 13.2. It can be seen that the ends of the specimen are wide to provide easy gripping while the central section is narrower so that elongation will occur preferentially in the region (see Figure 13.3). A strain gauge extensometer can be attached to the narrow region of the specimen to give an accurate measurement of the % elongation.

Table 13.5   Various test methods for tensile testing of plastics

ASTM method	Details of method	Comments
ASTM D-638[a]	Generally used for polyethylene	Intended specifically for PLASTICS
ASTM D-6693[a]	Test to be done at 21 °C	Intended specifically for NON-REINFORCED POLYETHYLENE and NON-REINFORCED FLEXIBLE POLYPROPYLENE
ASTM D-882[a]	Generally used for PVC geomembranes	Intended specifically for THIN PLASTIC SHEETING (<1 mm thick)
ASTM D-412[a]	Generally used for vulcanized rubber geomembranes	Intended specifically for ELASTOMERIC UNREINFORCED GEOMEMBRANES
ASTM D-751	Grab tensile method for coated fabrics	Intended specifically for REINFORCED GEOMEMBRANES
ASTM D-4885[b]	Wide width tensile samples	Intended to BETTER SIMULATE FIELD PERFORMANCE
ASTM D-5617[c]	Large scale out-of-plane tensile properties; essentially a huge 'burst' test	Intended to SIMULATE MULTIAXIAL TENSILE STRESSES

[a]All these tests use a dog-bone (i.e. dumb-bell) shaped test specimen.
[b]The basic distinctions between the wide width test method and other methods measuring the tensile strength of geomembranes are the width of the specimens tested and the speed of applied force. The greater width of the specimens specified in this test method minimizes the contraction edge effect (i.e. necking) which occurs in many geosynthetics and provides a closer relationship to actual geomembrane behaviour in service. The slower speed of applied strain also provides a closer relationship to actual geomembrane behaviour in service. This test method is not intended for routine quality control testing of geomembranes.
[c]This test method covers the measurement of the out-of-plane response of geosynthetics to a force that is applied perpendicular to the initial plane of the sample. When the geomembrane deforms to a prescribed geometric shape (segmented sphere or elliptical), formulations are provided to convert the test data to biaxial tensile stress–strain values. This test method requires a large diameter pressure vessel (greater than 450 mm). Information obtained from this test method may be more appropriate for design purposes than many small-scale index tests such as Test Methods ASTM D-638 or ASTM D-751.

Crosshead speeds that can be performed on a typical hydraulic screw-driven tensile testing machine are from 0.5 mm to 500 mm/min. The resulting load–elongation curve can also be converted to a stress–strain curve. It is important to note that the crosshead displacement should not be used to determine the strain since calculating the strain from the crosshead speed rather than from an extensometer is not accurate. Due to variation in the test specimens (e.g. due to nicks and notches) and the variations in test results, a minimum of five replicates should be run under the same set of conditions so that mean and standard deviations can be reported.

According to ASTM D-638, the tensile strain for an HDPE geomembrane dumb-bell specimen when tensile tested should be measured using an extensometer to record the elongation of a defined gauge length (i.e. marked span in the central parallel portion of the dog-bone specimen) (Giroud *et al*., 1994). Figure 13.2 shows the difference between grip separation and gauge length.

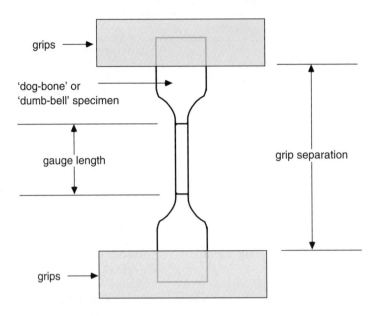

Figure 13.2 Schematic showing distinction between grip separation and gauge length on a tensile testing specimen

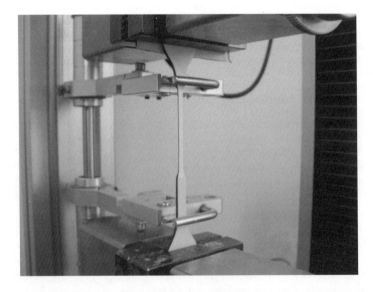

Figure 13.3 Photograph of tensile testing on a dog-bone of a black/white HDPE geomembrane. Note the necking that is occurring after the yield point

Dog-Bone Tensile Specimens

ASTM D-6693 is the industry standard test method for polyethylene geomembranes. For HDPE the test is performed using Type IV specimens (Figure 13.4) with strain rate of 50 mm/min (2 in/min). The percentage elongations are calculated based on the gauge lengths of 33 mm for yield elongation and 50 mm for break elongation. Note the tensile die (Type IV) should have a parallel 6 mm middle section with smooth sides and no nicks or deformed areas.

For HDPE, the stress versus strain curve shows a pronounced yield point, then the curve trends slightly downward before leveling out, and finally extends to some 700–1000% strain, when failure occurs.

In contrast, the curves for VLDPE and PVC geomembranes are relatively smooth with no inflection points; the stresses increase gradually until failure occurs at 700% and 450% strain, respectively.

Parallel Strip Specimens

The ASTM D-882 test method is for tensile properties of thin plastic sheeting, but is not commonly used in the polyethylene geomembrane industry. HDPE uses test specimens with a uniform width of 25 mm (1 in) and strain rate of 50 mm/min (2 in/min). The difference between this test and ASTM D-638 is the shape of the specimens (i.e. rectangular verses transitioning 'dog-bone').

ASTM D-882 uses 25×150 mm strips die cut from thin sheets or films.

Specimens are placed in the grips of a tensile testing machine and extended until failure. For ASTM D-882 the test speed and grip separation are based on the elongation

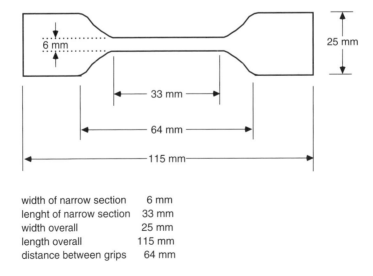

width of narrow section	6 mm
lenght of narrow section	33 mm
width overall	25 mm
length overall	115 mm
distance between grips	64 mm

Figure 13.4 Type IV tensile dog-bone specimen for testing of geomembranes as per ASTM D-638

to break of the material. Elongation and tensile modulus can be calculated using an extensometer.

13.5.2 EFFECT OF TEST PARAMETERS ON STRESS–STRAIN CURVE

It is important to note that the shape of the stress–strain curve and the magnitude of the yield stress, yield strain, break stress and break strain are all dependent on the strain rate (i.e. rate of deformation) and the test temperature.

Tensile tests performed at different strain rates on HDPE geomembranes show that both the strength and the stiffness of the geomembrane decrease with a decrease in the strain rate. The quantification of the strain rate-dependent stress–strain response of HDPE geomembranes is crucial for the design of any geomembrane structure of which the geomembranes are strained at a lower rate than the rate at which laboratory tests are performed (Wesseloo *et al.*, 2004).

The crosshead speed for HDPE testing is normally specified at 50 mm/min or 2″/min (however operations conducted in the field may increase the speed of testing to 500 mm/min in order to speed testing). The crosshead speed for PVC and fPP is normally 500 mm/min (20″/min).

Temperature is another important variable. As the temperature increases the tensile strength decreases and there is a corresponding increase in tensile elongation. The dependence of tensile properties on test temperature underscores the importance of conducting the measurements in a laboratory at 23 °C. Hence weld tensile tests conducted on-site using tensiometers will be influenced by the temperatures in the field which can have a significant bearing on the results.

Trial welds and DT cut-out coupons are tested in peel and shear modes on tensiometers (i.e. portable tensile testing machines) in a wide variety of weather conditions. These portable field tensiometers allow the contractor some confidence to proceed with installation immediately without having to wait a couple of days for specification conformance testing performed at an off-site laboratory. The field testing is generally performed at existing ambient temperatures with usually little or no conditioning of the samples prior to the test. Discrepancies between laboratory and field test results can often be related to the fact that the tensile properties of geomembranes are affected by the test temperature.

The main problem with using portable tensiometers in the field is that it is difficult to maintain an accurate test temperature. In very hot weather the tensile strength of the material decreases and minimum seam strengths may not be met on field equipment. Higher test temperatures also increase the elongation of the material. Conversely, in cold weather the material being tested may not meet a minimum tensile elongation requirement (Mills and Stang, 1997).

13.5.3 GRIPPING TEST SPECIMENS

Gripping of delicate (e.g. thin or soft) geomembranes can be a challenge in most tests. The gripping technique must prevent slipping of the geomembrane at high loads but not tear into the material and cause sites for premature specimen failure. A suitable set of grips and faces is recommended that prevent slipping of the geomembrane during tensile testing but which do not cause specimen failure.

Special adaptations may be necessary with strong geomembranes or geomembranes with extremely slick surfaces, to prevent them from slipping in the clamps or being damaged by the clamps.

13.5.4   WIDE WIDTH TENSILE TEST

To determine tensile behavior under plane strain conditions, a uniform width coupon or (strip specimen) is required. Such 'wide width' specimens are described in ASTM D-4885 which utilizes a 200 mm wide strip. The 200 mm wide strip is the best test specimen for obtaining values that will be used in design calculations. Widths of 25 mm are also sometimes used but it is important to note that the tensile strength and elongation values obtained is a function of the specimen geometry and the dog-bone, 25 mm strip and 200 mm strip all producing differing values, with the dog-bone giving the highest values and the 200 mm wide strip the lowest.

The basic distinctions between ASTM D-4885 ('Standard Test Method for Determining Performance Strength of Geomembranes by the Wide Strip Tensile Method') and other methods measuring the tensile strength of geomembranes are the width of the specimens tested and the speed of applied force. The greater width of the specimens specified in this test method minimizes the contraction edge effect (or necking) which occurs in many geosynthetics and provides a closer relationship to actual material behaviour in service. The slower rate of applied strain also provides a closer relationship to actual material behaviour in service.

The tensile properties of polyethylene geomembranes are sensitive to specimen geometry and strain rate during testing. Both ASTM D-638, D-6693 and ASTM D-882 are index test methods intended for routine acceptance testing of geomembrane by the liner manufacturers and the third party quality assurance laboratories. The shape of the specimen in the index tests significantly influences the tensile properties at break for HDPE liners. The data obtained from these tests are not intended and should not be used for design purposes.

As a performance test, the ASTM D-4885 method is not intended for routine acceptance testing of geomembranes, but to establish performance characteristics of geomembrane materials. The design of an HDPE geomembrane should be based on its tensile yield and never on tensile break properties (Poly-flex, 2008).

13.5.5   MULTIAXIAL TENSILE TESTING

Multiaxial testing (as per ASTM D-5617) is designed to simulate in-service stress conditions where there is stress on the geomembrane from multiple directions. This situation can arise for instance when a geomembrane is placed over settling waste. In this test a circular test specimen is bolted to a circular chamber and pressure exerted on the geomembrane by using compressed air or water pressure. The pressure exerted causes a change in volume of the specimen and this is plotted to produce stress–strain curves. The tensile strength values obtained by the multiaxial test are normally less than those obtained using the 200 mm strip test (ASTM D-4885). Interestingly, HDPE geomembranes under multiaxial testing, do not exhibit the well-defined yield point that they exhibit in the standard (dog-bone) tensile test.

📌 *In multiaxial testing of geomembranes a circular panel of geomembrane is inflated until failure. It is important to note that the inflated shape is not spherical.*

The multiaxial stress–strain test best simulates those out-of-plane stresses and strains that geomembranes experience in the field. The large diameter test specimen (69 cm in diameter) is inflated with air or water pressure at 1 psi/min until failure occurs. HDPE geomembranes exhibit failure at just 20% strain whereas fPP geomembranes which have low modulus and high extensibility do not failure until strains of greater than 160% are experienced. In fact fPP geomembranes exhibit the best multiaxial elongation properties of any geomembrane material.

13.6 TEAR TESTING

The tear resistance of a geomembrane determines how well the geomembrane can resist tear during installation and service. It is a measure of a geomembrane's ability to resist tear propagation. Tear testing is performed using a notched specimen (i.e. can be either a sharp notch that is an incision cut by a scalpel/scissor, or can be a more 'blunt' notch produced by a die-cut), which is tested in a tensile machine. The tear resistance value corresponds to the maximum load. For certain membranes such as thin, non-reinforced geomembranes, the tear resistance is low, from 18 to 130 N. Low tear strength values can be problematic, especially during geomembrane handling and installation since the membrane can tear after being pierced or damaged by sharp objects. However, this problem is overcome when the thickness increases. The values of tear resistance for scrim-reinforced geomembranes are significantly higher and are generally in the range of 120 to 450 N. A number of different methods exist to determine the tear resistance, for instance, ASTM D-751, D-1004, D-1424, D-1938, D-2261, D-2263 and D-5884. The trapezoidal tear test (ASTM D-4533) is often recommended for geotextiles.

13.6.1 WING TEAR

The tear strength (via wing tear) is determined in accordance with ASTM D-1004 using a rate of testing of 50 mm/min., an initial jaw separation of 25.4 mm with ten specimens each in the machine direction (MD) and traverse direction (TD), respectively. Figure 13.5 shows the tear sample cutter die and tear specimen. This tear method is normally assessed on unreinforced geomembranes. The test specimen is gripped at both ends with a notch positioned halfway between the jaws. As the tensile load is applied, the specimen begins tearing at the notch and the tear propagates across the specimen in a direction that is approximately perpendicular to the direction of the applied load. The tear resistance is reported as the maximum tensile force attained during the test.

13.6.2 TONGUE TEAR

The force required to tear a scrim-reinforced geomembrane along a reasonably defined course such that the tear propagates across the width of the specimen can be determined by the tongue tear method. This method is described in ASTM D-5884 ('Standard Test

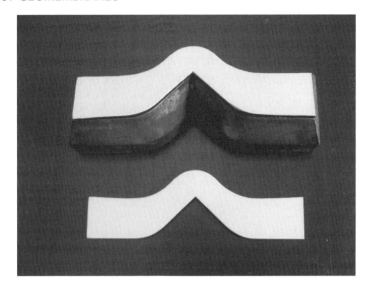

Figure 13.5 Photograph of a wing tear sample die and tear specimen as per ASTM D-1004. The tear strength is determined using a rate of testing of 50 mm/min and an initial jaw separation of 25.4 mm, with ten specimens each in the machine direction (MD) and traverse direction (TD), respectively

Method for Determining Tearing Strength of Internally Reinforced Geomembranes'). The tensile tear strength of the geomembrane needs to be measured in both the machine and the cross-machine (or transverse) direction. The tear strength is determined by measuring the maximum load (in newtons, N) when cut specimens of specific dimensions (see Figure 13.6) are tested to failure, by tearing, at a fixed crosshead displacement rate.

In this test, a 75 mm wide and 200 mm long specimen with a 25 mm initiating cut is placed in the grips of a tensile testing machine and pulled uniaxially.

Note that the tear resistance can be affected to a large degree by mechanical fibring of the membrane under stress, as well as by stress distribution, strain rate and size of specimen; the results obtained in a tear resistance test can only be regarded as a measure of the resistance under the conditions of that particular test and not necessarily as having any direct relation to in-service values. The tongue tear method is useful for estimating the relative tear resistance of different scrim-reinforced geomembranes or for assessment of tear properties in different directions for the same reinforcing textile.

 Test conditions for ASTM D-5884:

- *The speed of crosshead separation is 5 mm/s (ie. 300 mm/min).*
- *The dimensions of the gripping surfaces measure 25 by 50 mm.*
- *All edges that might cause a cutting action need to be rounded to a radius of not over 0.4 mm.*
- *The initial distance between the clamps (i.e. the grip separation) is 75 mm.*

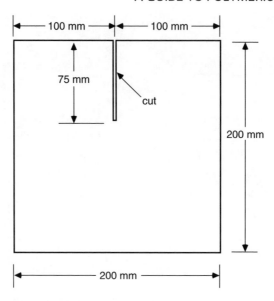

Figure 13.6   Specimen dimensions for the tongue tear testing of geomembranes as per ASTM D-5884 (Method B)

- *Each specimen comprises a square of 200 by 200 mm internally reinforced geomembrane containing a 75 mm cut at the centre of the specimen to form the 'tongues' (see Figure 13.6).*
- *Two sets of five specimens – one set cut in the machine and one set prepared in the cross-machine direction, respectively, shall be tested.*

The specimen is centred in the tensile machine with one tongue or cut strip centred in each clamp. Position the clamp parallel to the cut edge, 12.5 mm from the top of the cut edge of the geomembrane specimen. This puts the grip faces 38 mm from the converging end of the 75-mm cut.

The tearing force is the peak load of resistance registered during the separation of the tear. The tearing strength of the sample is the average of the results obtained from five specimens tested in each of the machine and cross-machine directions respectively.

Another tear test that has been used for coated fabrics (i.e. scrim-reinforced geomembranes) is the trapezoidal tear test (ASTM D-4533) which uses a specimen as shown in Figure 13.7.

13.7   PLY ADHESION

The extent of ply adhesion can be determined using ASTM D-6636 ('Standard Test Method for Determination of Ply Adhesion Strength of Reinforced Geomembranes') which is a measure of the adhesion strength (180 degrees peel) between plies of reinforced geomembranes such as internally reinforced geomembranes and coated fabrics.

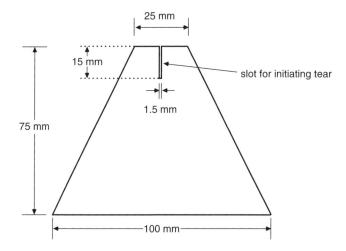

25 mm

15 mm

slot for initiating tear

1.5 mm

75 mm

100 mm

Figure 13.7 Specimen dimensions for the trapezoidal tear testing ('trap' tear) of reinforced geomembranes as per ASTM D-4533

13.8 PUNCTURE RESISTANCE

The puncture resistance is important when the geomembrane is installed under onerous conditions, for example, on a machine-graded subgrade with angular stones under high loads. Sharp stones, sticks or other debris can lead to puncture of geomembranes during installation and in service. Such punctures create points of tearing or sites for leakage.

Puncture of geomembranes can occur when they are placed on subgrades containing jagged protrusions or when containment material such as waste containing sharp projections is loaded on a liner without adequate protection layers. Puncture can also occur due to an angular gravel cover being pushed into the geomembrane by construction equipment or by top loads exerted by placement of overburden. Puncture of the geomembrane may result in a breach of the containment system. The resistance of a geomembrane to puncture is therefore very important.

There are two main tests for assessing the puncture resistance of geomembranes as shown in Table 13.6.

Both the ASTM D-4833 and FTMS 101C, Method 2065 test methods have been used to characterize geomembrane puncture strength. However it is important to note that they provide quite different results.

The FTMS 101C, Method 2065 procedure uses a tapered metal probe to determine resistance to puncture (or rupture). The test employs a pair of clamping plates with

Table 13.6 Main small-scale tests for assessing the puncture resistance of geomembranes

Method	Comments
FTMS 101C, Method 2065	Often used for REINFORCED GEOMEMBRANES
ASTM D-4833	Often used for UNREINFORCED GEOMEMBRANES

25 mm diameter holes through the centre which are affixed to a universal testing machine. The square specimen is clamped firmly between the plates. The puncture probe is a 13 mm in diameter and 125 mm long steel rod with one end tapered to a 3.2 mm radius on the puncture end. The length of the taper is 50 mm. During the test, the puncture probe is attached to the moving crosshead (which is set to travel at 500 mm/min) and pushes through the geomembrane specimen. The maximum load is recorded and the test is conducted on five replicate specimens.

ASTM D-4833 ('Test Method for Index Puncture Resistance of Geotextiles, Geomembranes and Related Products'), has effectively replaced the FTMS procedure and has been adopted into most industry geomembrane specifications such as the GRI GM13 specification for smooth and textured high-density polyethylene geomembranes and GRI GM17 for smooth and textured linear low-density polyethylene geomembranes.

The ASTM D-4833 test uses 100 mm diameter clamping plates with an open internal diameter of 45 mm and which are capable of clamping the test specimen without slippage. The puncture probe is a solid steel rod with a diameter of 8 mm and a flat end with a 45 degrees chamfered edge (see Figure 13.8). In this test the puncture probe is attached to the moving crosshead and forced into the clamped specimen until rupture occurs. The

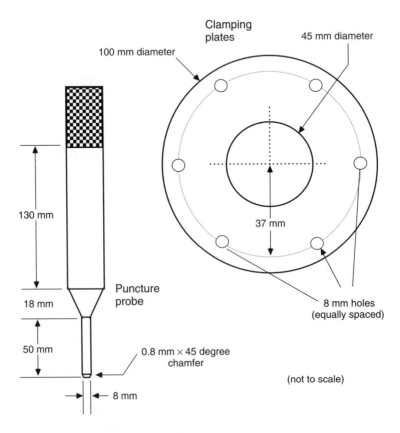

Figure 13.8 Schematic showing puncture probe and clamping plates for determining the index puncture of geomembranes as per ASTM D-4833

Table 13.7 Differences between two main small-scale puncture resistance test methods for geomembranes

Test method	Specimen dimensions	Noumber of specimens	Aperture size for clamping plates	Puncture probe dimensions	Test speed (crosshead rate)
FTMS 101C, Method 2065	50 mm × 50 mm square	5	25 mm	13 mm × 125 mm long with one end tapered to a 3.2 mm radius on the end (length of taper is 50 mm)	500 mm/min
ASTM D-4833	100 mm diameter (minimum)	15	45 mm	Diameter of 8 mm and a flat end with a 45 degrees chamfered edge	150 mm/min or 300 mm/min

puncture resistance result is taken as the maximum load recorded. The mean puncture resistance (maximum load) is calculated from the 15 specimens.

Note that the index puncture test (ASTM D-4833) can be significantly influenced by the alignment of the probe relative to the holder and also by slippage problems.

The differences between the FTMS 101C, Method 2065 puncture test and the ASTM D-4833 puncture test are summarized in Table 13.7.

An important difference between the two methods is that the flat end of the D-4833 probe creates more specimen contact area during the test, resulting in a higher value than what would be observed during the 101C, Method 2065 test (Allen, 2000). As a general rule the ASTM D-4833 puncture test gives a puncture resistance value that is some 15–35% higher than for the FTMS 101C, Method 2065.

Note if the probes are used routinely and made of mild steel, they have a tendency to become rounded. This is especially the case for the probe used in the ASTM D-4833 method. Excessive rounding tends to decrease the puncture resistance results.

When testing scrim-reinforced geomembranes the difference between the puncture probe geometry of the two tests can affect the results. For instance, there is a tendency for the FTMS 101C rounded probe to penetrate between the scrim which can underestimate puncture values. Even the ASTM D-4833 probe can prove problematic for geomembranes with reinforcing scrims spaced apart more than 6 × 6 yarns/25 mm. In this case the appropriate test method is ASTM D-6241 ('Standard Test Method for the Static Puncture Strength of Geotextiles and Geotextile-Related Products Using a 50 mm Probe) (Allen, 2000).

A criticism of the above puncture tests is that they only use a small sample size and are essentially index tests[1] and the results are difficult relate to actual field performance.

---

[1] Index tests provide a relative measure of the puncture resistance values and serve as a basis for comparison of different geomembrane types as well as geomembranes from different manufacturers.

Although index puncture tests (ASTM D-4833) are a useful measure of puncture resistance under standard test conditions, the puncture resistance under the expected field conditions is of fundamental importance.

13.8.1   INDEX PUNCTURE RESISTANCE TEST DETAILS

In the Index Puncture Resistance (IPR) test, ASTM D-4833 ('Standard Test Method for Index Puncture Resistance of Geotextiles, Geomembranes and Related Products'), a test specimen is clamped without tension between circular plates of a ring clamp attachment secured in a tensile testing machine. A force is exerted by a metal puncture rod attached to the load indicator against the centre of the unsupported portion of the test specimen until rupture of the specimen occurs. The maximum force recorded is the value of puncture resistance of the specimen. This test is used to establish an index value by providing standard criteria and a basis for uniform reporting.

The index puncture test uses a tensile/compression testing machine fitted with a ring clamp attachment consisting of concentric plates with an open internal diameter of 45 mm capable of clamping the test specimen without the sample slipping. The external diameter is suggested to be 100 mm although this is not mandated. There are six holes used for securing the ring clamp assembly and these have a diameter of 8 mm and equally spaced at a radius of 37 mm. The surfaces of these plates can consist of grooves with O-rings or coarse sandpaper bonded onto opposing surfaces. The puncture rod is a solid steel rod with a diameter of 8 mm and having a flat end with a $45° = 0.8$ mm chamfered edge contacting the test specimen's surface (see Figure 13.8).

The minimum specimen diameter is recommended to be 100 mm to facilitate clamping and it is important that the test specimen extends to or beyond the outer edges of the clamping plates. A testing speed of 150 mm/min or 300 mm/min should be used.

For the testing of composite geomembrane materials, there may be a double peak. If this occurs, the initial value should be reported even if the second peak is higher than the first one.

13.8.2   HYDROSTATIC PUNCTURE (LARGE SCALE PUNCTURE) TEST

For design applications a puncture test was developed that more closely simulates field conditions. The hydrostatic (or large scale) puncture test (ASTM D-5514) uses truncated cones (as shown in Figure 13.9) positioned in a triangular configuration to simulate a severe subgrade puncture scenario (see Figure 13.10 and Figure 13.11). Alternatively instead of the truncated cones, a sample of the actual subgrade material can be used. This so-called site specific puncture testing generally uses site specific subgrades (e.g. sand/stones/gravel) which is placed beneath the geomembrane test specimen (Stark et al., 2008). The hydrostatic puncture (large scale puncture) test apparatus is shown in Figure 13.12.

It was determined using the truncated cone puncture method (and pressures greater than 100 kPa) that 0.5 mm PVC and 1.00 mm VLDPE can withstand truncated cone heights greater than 7 cm whereas 1.5 mm HDPE and 0.9 mm CSPE can only withstand truncated cone heights less than 2 cm (Hullings and Koerner, 1991).

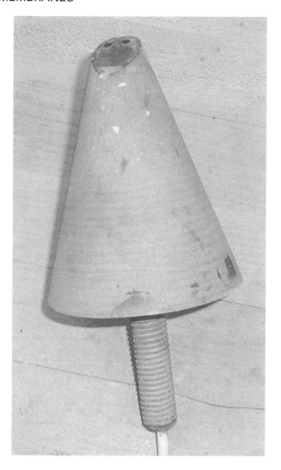

Figure 13.9 Close-up of a truncated cone used in the large scale point stress test (ASTM D-5514) for simulating a severe subgrade puncture scenario. Note that the cones have a 45 degree taper and a round 0.25 mm radius on the tip. Reprinted with permission from *Geosynthetics International*, Puncture resistance of PVC geomembranes using the truncated cone test, by T. D. Stark, T. R. Boerman and C. J. Connor, **15**(6), 480. Copyright (2008) Thomas Telford Publishing

The strains due to indentations in HDPE geomembranes caused by the overlying materials can be measured and quantified by this method also or by the so-called 'Cylinder Test' (see Figure 13.13). This test enables the selection of suitable protective materials (e.g. cushioning geotextiles) to limit strains to levels deemed to be acceptable to ensure a reasonable design life.

HDPE geomembranes are used extensively as basal and capping layers for landfills and contaminated land containment sites. After installation of the HDPE liner, a protective geotextile blanket is often placed on top before a stone drainage layer (to drain leachates) is placed on the surface. The stone drainage layer needs to be free draining and have adequate hydraulic conductivity to drain leachates over a large relatively flat area. A secondary

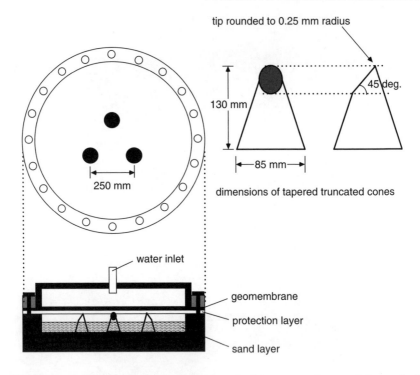

Figure 13.10 Experimental set-up for the truncated cone puncture resistance testing of geomembranes as per ASTM D-5514. The test involves placing the geomembrane specimen over a set of three truncated cones in a pressure vessel and then applying pressure over the top of the geomembrane. This test simulates the relative puncture resistance of a geomembrane when subjected to gradually increasing over-burden pressures. The heights of the truncated cones can be varied from test to test and plotted as cone height versus pressure at failure. From a series of plots the critical cone height (CCH) can be determined which represents the approximate maximum protrusion height. It is noteworthy that HDPE geomembranes undergo rapid yielding and quickly thin out at the top of the truncated cones, resulting in a puncture failure. In contrast, highly extensible materials such as fPP, LLDPE and PVC conform under working strain and reach impressive critical cone heights

function of the stone drainage layer is to provide protection of the underlying HDPE geomembrane against the placement of the first layer of waste.

Due to point pressures exerted by the coarse drainage stones on the smooth HDPE geomembrane, the relatively stiff HDPE geomembrane is susceptible to stress cracking if strained over a long period of time. The function of the protective geotextile is therefore to present a relatively smooth surface and uniformly distributing the load applied to the upper surface of the HDPE geomembrane. The cylinder test simulates as closely as possible the conditions expected on the basal layer in a landfill.

The maximum acceptable strain for HDPE as determined by the cylinder test is 0.25%. This strain level has been interpreted to equate to a 50-year design life (at 40 °C) before failure as a result of environmental stress cracking (Seeger and Muller, 1996). In other words, HDPE geomembrane lining systems as currently designed may be considered to

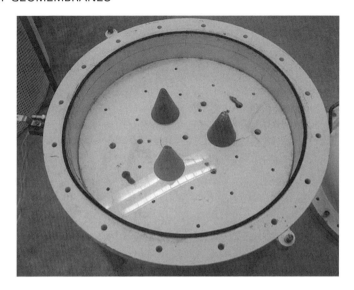

Figure 13.11   Photograph of the apparatus and cone positioning for the truncated cone (large scale point stress) test. This hydrostatic puncture test (ASTM D-5514) uses truncated cones (as shown in Figure 13.9) positioned in a triangular configuration to simulate a severe subgrade puncture scenario and more closely simulates field conditions. Reprinted with permission from *Geosynthetics International*, Puncture resistance of PVC geomembranes using the truncated cone test, by T. D. Stark, T. R. Boerman and C. J. Connor, **15**(6), 480. Copyright (2008) Thomas Telford Publishing

have a design life of 50 years. By contrast, LLDPE, fPP and PVC are able to accommodate relatively large deformations without reduction in their anticipated design life and in this critical respect they may be seen to have a significant advantage over HDPE.

Geosynthetic engineers use the cylinder test for selecting a geotextile which will adequately protect the geomembrane from the overlying stone drainage blanket. The cylinder test (or plate load test) has now been widely adopted as definitive 'proof' for the suitability of a geotextile protector for a geomembrane in the base of a landfill (Shercliff, 2008).

📌 *Note that the local strain failure criterion for avoiding long-term environmental stress cracking of HDPE is currently set at 0.25%, which includes several factors of safety.*

13.9   IMPACT RESISTANCE

The impact resistance of a geomembrane is important since the geomembrane may be damaged during installation by falling objects (e.g. tools, granular drainage from the back of a truck) and these may propagate cracks and cause leaks. Hailstones at low temperatures can also cause impact damage. Various ASTM standards used to measure impact resistance are D-746, D-1424, D-1709, D-1822, D-3029 and D-3998. Unfortunately none of these impact tests can offer direct correlation with field conditions.

Figure 13.12 Photograph of the assembled apparatus for the truncated cone (large scale point stress) test (ASTM D-5514) for simulating a severe subgrade puncture scenario. Reprinted with permission from *Geosynthetics International*, Puncture resistance of PVC geomembranes using the truncated cone test, by T. D. Stark, T. R. Boerman and C. J. Connor, **15**(6), 480. Copyright (2008) Thomas Telford Publishing

Table 13.8 lists the various ASTM test methods for assessing impact resistance of polymers.

The above tests are carried out variously by a falling weight, a free-falling dart or a pendulum strike and the type of test selected depends on the type of geomembrane tested and its thickness.

13.10 ENVIRONMENTAL STRESS CRACKING

Environmental stress cracking (ESC) can be defined as a rupture or crack in a HDPE geomembrane caused by an applied tensile stress on the geomembrane less than its tensile strength. There are several factors that may contribute to ESC:

- effects of surface wetting agents;
- effects of residual stress;
- stress cracks originating from scratches or other defects;
- stress cracks originating from irregular geometry;
- fatigue or cyclic induced cracking;
- combinations of the above.

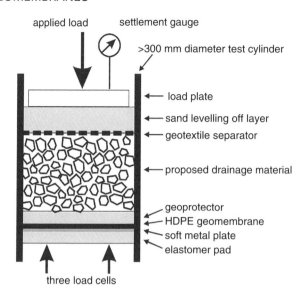

applied load   settlement gauge

>300 mm diameter test cylinder

load plate

sand levelling off layer

geotextile separator

proposed drainage material

geoprotector
HDPE geomembrane
soft metal plate
elastomer pad

three load cells

Figure 13.13 Schematic of the 'Cylinder Test' for simulating the system puncture behavior of angular gravel on geomembranes. This test is typically conducted using a 200 mm-thick drainage layer, with generally 16 to 32 mm diameter gravel, and a 2.5 mm-thick HDPE geomembrane. A 0.5 mm soft metal plate, made of 'organ pipe material' (40% lead and 60% tin), is placed beneath the geomembrane to monitor the localized geomembrane deformations that form during the test. The recommended temperature and duration load test conditions are as follows: $40\,^{\circ}$C for 1000 h at $1.50 \times$ design load; $23\,^{\circ}$C for 1000 h at $2.25 \times$ design load; $23\,^{\circ}$C for 100 h at $2.50 \times$ design load. After completion of the test, the geomembrane is inspected for damage to its upper surface (cracks or nicks), sharp-angled deformation and maximum permissible local strain using the soft metal plate. The average longitudinal strain in the geomembrane is obtained by fitting a circular segment to the indentations in the lead sheet, selecting the most crucial segment, and then calculating the 'arch elongation'. Reprinted with permission from the Proceedings of 56th Canadian Geotechnical Conference, Selection of Protective Cushions for Geomembrane Puncture Protection by Eric Blond, Martin Bouthot, Oliver Vermeersch and Jacek Mlynarek Copyright (2003) Eric Blond

The stress crack resistance of polyethylene decreases with increasing polymer density (and crystallinity). As the density increases the amount of amorphous material decreases and consequently so too does the number of tie molecules. The tie molecules bind the lamellas and thus provide the strength; hence when their concentration decreases, the polymer strength is correspondingly reduced. Increasing the comonomer content increases chain branching which in turn increases the degree of entanglement of the tie molecules and the extent of the loose loop phase. This leads to increased stress crack resistance.

Initiation of the cracking is usually at seams, for instance where scratches caused by grinding exist or where extrusion welds are placed directly over wedge welds made previously. Many reported field incidents of ESC have been in surface impoundments where geomembranes were exposed to the atmosphere and where temperature changes

Table 13.8   ASTM test methods for assessing impact resistance of polymers

ASTM test method	Type	Comments
D-746	Brittleness temperature of plastics and elastomers. The test specimen is held by a clamp as a cantilevered beam and is impacted at 2000 mm/s	Used to determine minimum impact brittleness temperature, temperature at which 50% of specimens exhibit brittle failure
D-1424	Force required to propagate a single-rip tougue-type tear (starting from a cut) by means of a falling pendulum	Also used to measure tear resistance of textiles
D-1709	Impact resistance of plastic films by a free-falling dart	Energy is expressed in terms of weight (mass) of the falling dart from a specified height to give 50% failure of specimens tested
D-1822	Tensile impact test which measures force needed to break a specimen under a high speed tensile load applied via a swinging pendulum	Tensile impact energy is recorded by the apparatus
D-3029	Impact resistance of flat rigid plastics by falling dart method	Gardner impact test

can intensify tensile stresses. Since the occurrence of ESC is also dependent on several pertinent polymeric properties (such as density, crystallinity, molecular orientation, molecular weight and distribution), blending and manufacturing processes can reduce the likelihood of ESC.

It is well known from the pipe and cable insulation industries that PE can stress crack over long periods of time. When the geomembrane industry began, this was recognized and SCR testing has been a standard specification and test ever since.

The first stress crack test method used was ASTM D-1693. This so-called 'bent strip' test was able to differentiate the first resins used to manufacture geomembranes. Small rectangular strip specimens containing a notch are bent into a U-shape, placed within a channel holder with the notches at the apex and immersed in a surface wetting agent at an elevated temperature. It is important to note that no external load is applied. Subsequent advances in resin technology have increased the SCR of PE to a performance level higher than the level that can be tested via ASTM D-1693. For this reason and the fact that the stresses at the notch relax over time, ASTM-D1693 has now been largely superceded.

🖈 *It is strongly recommended that the ASTM D-1693 bent strip test NOT be specified or performed to assess SCR of geomembranes. New generation HDPE resins have improved to the extent that the bending stress relaxes completely before stress cracking starts and therefore these specimens will not fail the bent strip test or it is of no benefit in discriminating between resin types (Hsuan et al., 1992). The appropriate test to assess SCR is the ASTM D-5397 notched constant tensile load test which as specified in GRI GM13 is typically required to exceed a single point break time of 300 h.*

Today, the most common test method used to determine SCR is ASTM D-5397, 'Standard Test Method for Evaluation of Stress Crack Resistance of Polyolefin Geomembranes Using Notched Constant Tensile Load Test'. Commonly referred to as the 'NCTL' test. In this test a centrally notched dumb-bell-shaped specimen is held under a constant load (a percentage of the sample's yield stress) in a bath containing a surfactant (i.e. surface wetting agent or detergent such as Igepal 630) at 50 °C. The entire test involves subjecting a series of test specimens to different percentages of their yield stresses in order to generate a ductile-to-brittle stress rupture curve. The comparable ISO test method is 16 700. A ductile-to-brittle behaviour is observed while tensile testing specimens at different percentages of their yield stress. The transition time varies from 10 to 5000 h depending on the material tested.

As the environmental stress crack resistance is a function of the type of resin used, the NCTL test can be performed on test samples made as plaques from the base resin or on the finished geomembrane sheet providing it is of the appropriate thickness. The procedure for the preparation of plaques is described in ASTM D-4703.

An abridged version of the stress cracking test uses only one constant load value to give a single point measurement. Accordingly this test is called the single point NCTL test (or SP-NCTL test). This test uses a notched, dumb-bell-shaped specimen (see Figure 13.14) to determine the resistance of the material to brittle fractures caused by long-term, low-level tensile stress. Testing according to the ASTM D-5397 Appendix requires that specimens are placed in a surfactant solution at 50 °C for an extended period of time, and a tensile stress is applied to the specimen equal to 30% of the material's yield stress. The material is considered acceptable if failure does not occur before 300 h of exposure.

The full test method however subjects specimens to varying constant tensile loads. Small dog-bone-shaped specimens are loaded at 20–50% of the tensile yield strength of the material. The specimens are notched 20% of their overall thickness and placed in a bath containing 10% surfactant at 50 °C. In this accelerated ageing test, the time to failure is measured. The data are reported in a plot of load versus failure time. An NCTL test provides information on both the ductile and brittle failure modes of the material. It is the transition between these two failure mechanisms that indicates the SCR of the material.

13.10.1   BENT STRIP METHOD (ASTM D-1693)

The old method for measuring the ESCR of polyethylene geomembranes (ASTM D-1693) used specimens which were notched longitudinally to a depth of approximately 20% of their thickness. These were then bent into a U-shape, held in a channel holder and immersed in a surface-active wetting agent (i.e. detergent solution) at 50 °C and thus is referred to as the 'bent strip method'. The specimens were periodically inspected for the onset of cracking. This constant strain method has been criticized on the basis that the stress relaxes and the time taken to get meaningful results is too long. This method is no longer recommended for geomembranes.

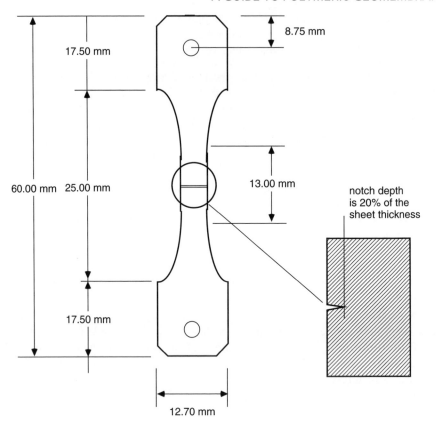

Figure 13.14   Specimen dimensions for the notched constant tensile load (NCTL) test (ASTM D-5397)

✒ *There is often confusion between the terms creep and stress relaxation. A creep test is under constant load while stress relaxation tests occur under application of a certain strain. For this reason the constant strain ESCR test (ASTM D-1693) is not an appropriate stress cracking test for geomembranes since the applied strain, although constant, allows the stresses to relax due to viscoelastic flow.*

13.10.2   *NOTCHED CONSTANT-LOAD STRESS CRACKING RESISTANCE TEST (NCTL) (ASTM D-5397)*

The NCTL test produces a stress rupture curve (log of the applied tensile stress versus log of time to failure) by immersing test specimens under a constant tensile load in a bath of 10% (by volume) surfactant (usually Igepal CO-630) and 90% water at 50 °C. Generally test specimens are loaded with tensile stresses varying from 20–65% of the nominal yield stress applied in 5% increments. The resulting stress–rupture curve then shows the transition from ductile failure (the relatively horizontal portion of the curve) to brittle failure (where the curve drops down sharply. The 'knee' on the curve is taken as the onset of brittle failure. The full NCTL test generates a curve that also shows when the sample undergoes embrittlement due to oxidative degradation. The third portion of the curve where the stress curve drops vertically is indicative of depletion of the antioxidants/stabilizers and totally brittle failure due to oxidation. At this last stage the polymer undergoes brittle failure even at very low stress levels. Thus the three stages of the stress rupture curve can be characterized as:

- REGION 1: Ductile failure (with deformation);
- REGION 2: Brittle failure (with antioxidant);
- REGION 3: Brittle failure (without antioxidant).

Note from Figure 13.15 the effects of temperature. The stress–rupture behaviour observed at higher temperatures can be extrapolated to the longer-term behaviour at lower temperatures.

The full NCTL test is very lengthy and involved and for this reason a simplified test was developed based on a single point result using a single fixed constant load (i.e. fixed at 30% of the yield stress).

The NCLT-SP (single point) test is an excellent test for screening different HDPE resins since these can vary widely in their stress crack resistance. The test has the advantages of being able to give results in a relatively short time (compared to the time taken for the full NCTL test) and the test specimens can be punched directly from the geomembrane sheets.

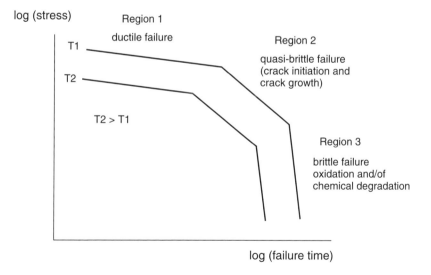

Figure 13.15 Stress–rupture curves showing the three stages of failure for polymers held under a constant stress. Note increasing temperature moves the onset of failure to earlier times. Environmental stress cracking is represented by Region 2 while Region 3 represents oxidative stress cracking

The NCTL test specimens (see Figure 13.14) are notched along their width to a depth of 20% of the sheet thickness with a razor blade. Since the geometry of the notch and the quality of the notch tip can affect the results, special care needs to be taken during notching of the specimens. The notch is a 'controlled imperfection' to offset the effects of any other small notches or scores that may be introduced during sample preparation. Given the notch extends 20% into the sheet thickness the 'ligament thickness' of the notched specimen is then 80%.

For the measurement of SCR it is necessary to melt the resin and cast it into a thin sheet (plaque) from which the test specimens are die cut. The cooling rate after casting is critically important. One standard (ASTM D-4703) suggests cooling rates of 15°C/min, 5°C/h and quenching. Quenching will not allow the development of a crystalline microstructure, thereby providing very optimistic results. The most appropriate cooling rate, most similar to that occurring during liner production, is 15°C/min (Peggs, 2008).

The stress cracking test results when tested according to ASTM D-5397 are very dependent on the following variables:

- the rate of cooling after casting the test plaque;
- the reproducibility of the specimen notching;
- the degree of temperature control of the surfactant bath;
- the concentration of surfactant (i.e. Igepal) and ensuring that the surfactant solution is changed regularly with fresh solution;
- the accurate calculation of the mass to hang on the mechanical advantage arm.

Sample Preparation

The technique used for sample preparation of tensile and stress cracking specimens can have a significant impact on the test results due to the introduction of flaws and notches. Failure often occurs on the edge of the specimen at small flaws.

Table 13.9 shows the trend of long-term tensile stress results for various sample preparation techniques.

Table 13.9 also shows that the time to failure results are very dependant on the surface quality of the specimens.

A well-defined notch should be introduced into stress cracking samples as a 'deliberate flaw'. However even the technique used to introduce the notch can affect the stress crack resistance times. It can be performed by milling, sawing or cutting by a razor blade.

Table 13.9 Failure times for various sample preparation techniques

Type of sample preparation	Failure times
Punched specimens	Rapid failure
Sawn specimens by high-speed saws	Long failure times
Milled specimens by high-speed milling	Long failure times
Specimens with edges planned off by a microtome knife	Long failure times

13.11   DIMENSIONAL STABILITY

Dimensional stability (also known as dimensional change) is a reversion phenomenon of polymeric geomembranes due to 'frozen-in' stresses and in-built orientation during manufacture. Dimensional stability of geomembranes is a measurement of the linear dimensional change resulting from exposure to temperature. With subsequent heating such as solar heating by the sun (where black membranes can achieve temperatures greater than $73\,°C$) or thermal seaming, these stresses and orientations can relax. This is accompanied by a change in dimensions of the sheet. The dimensional change can be either negative (i.e. shrinkage) or positive (i.e. expansion). Generally there will be shrinkage in the machine direction (i.e. the extrusion direction). This is especially true for geomembranes produced by the blown film process since the bubble is pulled upward by nip rollers and the speed of the haul off can induce orientation in the machine direction. There can be either shrinkage or expansion in the transverse direction (TD) (i.e. the cross-machine direction). It is important to note that the degree of dimensional change is a function of the manufacturing method. In addition, the dimensional change is not consistent along the width of a geomembrane roll.

There is an important distinction between thermal expansion and dimensional change. When geomembranes are heated, thermal expansion occurs. This phenomenon is largely reversible. However dimensional change due to relaxation of inbuilt stresses and orientation is not reversible.

Localized dimensional changes can cause distortion and waviness around thermal welds where the heat of welding causes permanent shrinkage and sets up stresses in the weld region. Such stresses may be damaging in terms of stress cracking of HDPE geomembranes.

Geomembranes made by the flat die extrusion method (i.e. flat bed extrusion) and also by calendering can possess significant orientation. If the degree of orientation is not controlled through proper processing techniques then there can be significant associated shrinkage. Shrinkage after installation and welding can lead to permanent distortion and wrinkles.

There are two main test methods that can be used to determine the coefficient of thermal contraction or expansion of a geomembrane material, namely:

- ASTM D-1042-01a 'Standard Test Method for Linear Dimensional Changes of Plastics Under Accelerated Service Conditions'.
- ASTM D-1204-02 'Standard Test Method for Linear Dimensional Changes of Non-rigid Thermoplastic Sheeting or Film at Elevated Temperature'.

These testing methods subject the test specimen to a constant temperature and then carefully measure the separation distance between two given initial locations or the change in dimension of a carefully cut sample.

The most quoted test is ASTM D-1204. Typical specimens are 25 cm × 25 cm although other sizes can be used. For HDPE the oven temperature is required to be $100\,°C$. The residence time is at least 1 h although the oven exposure time can be longer if required. Note the German DIN method DIN 53 377 requires the dimensional change to be determined by heating a test specimen of geomembrane (100 × 100 mm) in an oven for 1 h at $120\,°C$.

The dimensions of the specimen are taken at marked reference points. Talc dusted paper is inserted to form a sandwich around the specimen. The sandwiched specimens are placed in an oven for a specified amount of time. After removal from the oven, the specimens are reconditioned at room temperature for a minimum of 1 h. The distances between the reference marks are re-measured and recorded.

The percentage linear change is calculated as follows:

$$\% \text{ change} = (\text{final length} - \text{original length})/\text{original length} \times 100$$

The test gives an indication of lot-to-lot uniformity with regards to internal stress and orientation introduced during processing. The test is mostly used for geomembranes made by the flat die extrusion or calendaring processes. In particular, structured or embossed geomembranes can have relatively high levels of orientation and frozen-in stress.

There is no dimensional stability criteria listed in the GRI GM-13 specification but the BAM certification guidelines call for the dimensional changes to be within 1% (for smooth geomembranes) and <1.5% for structured geomembranes with embossed surfaces.

13.12 FRICTION ANGLES

Soil-to-membrane friction is a critical parameter because a number of well documented side slope failures have occurred. The test method for testing soil-to-geomembrane friction is defined in ASTM D-5321 and consists of a split shear box with the geomembrane/soil interface. Note that the friction angles of soil/geomembrane interfaces are always less than those for soil/soil interfaces.

☞ *A general rule of thumb is that smoother and harder geomembranes have lower friction values than softer/rougher geomembranes.*

The interfacial friction testing of geomembranes requires site-specific testing. This is because the soil or sand type will affect the interfacial friction particularly its moisture content, the particle size distribution and angularity.

13.13 MELT FLOW INDEX (MFI)

The MFI test determines the fluidity of the molten geomembrane resin by using a constant load to push the melt through a standard orifice (see Figure 13.16). The melt flow index is expressed as the mass of the molten resin that is pushed through the orifice in 10 min. The value is given in units of g/per 10 min and for HDPE geomembranes this typically ranges from 0.2 to 1.0 g/10 min (under a load of 2.16 kg). The melt flow index is directly related to the mean molecular weight of the polymer and gives an indication of the necessary processing parameters. A low melt flow index value indicates a high molecular weight polymer, and vice versa. HDPE geomembrane resins generally have MFI values of less than one (these are referred to as fractional melt index resins) and hence have a high molecular weight for good strength properties.

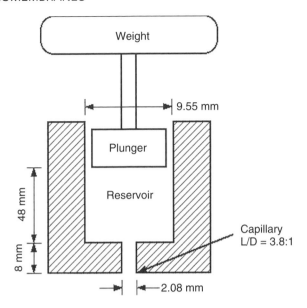

Figure 13.16 Apparatus for determining the melt flow index (MFI) of polymers as defined by ASTM D-1238. The test measures the amount of molten polymer (at 190 °C as specified for polyethylene) that is extruded through an orifice in 10 min under a constant load (usually 2.16 or 5 kg). The MFI is inversely proportional to the molecular weight of the polymer, with a higher melt flow index corresponding to a lower molecular weight

The MFI protocol defined by ASTM D-1238 consists of measuring the amount of molten polymer (at 190 °C for polyethylenes) that is extruded through an orifice in 10 min under a constant load of 2.16 kg, 5 kg or 21.6 kg. The MFI test can also be a useful indictor of polymer degradation, particularly polymer chain scission reactions in which shorter polymer molecules are formed. The relative change in molecular size of the test sample is reflected as a change in the MFI value. In the case of chain scission degradation reactions, the MFI value of the polymer will increase.

Polymer and geomembrane manufacturers use MFI to control the polymer uniformity. This test is very important for quality control and quality assurance of polyethylene resins and geomembranes. If inconsistencies exist in the molecular weight distribution of a polyolefin resin or if the polymer has undergone degradation, then these changes will be reflected in its melt index.

Note the MFI test does not give accurate results for polymers that contain plasticizers (such as PVC and EIA).

🖈 *The MFI test is useful for assessing the changes in molecular weight of the polymer due to ageing and degradation. Since MFI and molecular weight are inversely proportional to each other, a decrease in the MFI value is due to an increase in molecular weight which indicates crosslinking while an increase in the MFI value indicates chain scission. Both these reactions lead to the polymer becoming more brittle.*

13.14 DURABILITY TESTING

13.14.1 OXIDATIVE DEGRADATION (OXIDATION)

The most widely used test for assessing the resistance of polyolefin geomembranes to oxidative degradation in service is the oxidative induction time (OIT) test (ASTM D-3895) which evaluates the quality of the antioxidant additive package. A typical OIT curve is shown in Figure 13.17. The OIT is defined as the time between when oxygen gas is admitted into the sample cell to when the oxidation exotherm deviates from the baseline. The GRI standards for HDPE and LLDPE both require that the OIT should exceed 100 min. The OIT test is conducted in an apparatus called a differential scanning calorimeter (DSC) as shown in Figure 13.18.

Hindered amines stabilizers can volatilize or degrade in the standard OIT test which is conducted at 200 °C since this test temperature is above their effective temperature range (which extends to only 150 °C). Thus the test can also be performed at a lower temperature (i.e. 150 °C) in a pressurized oxygen atmosphere (ASTM D-5886) to detect the levels of HALS that are present in the additive package.

Standard OIT and HP-OIT tests are similar, except that a relatively high gas pressure and a lower temperature are used in HP-OIT tests (specimens are oxidized at 150 °C and 3500 kPa in HP-OIT, whereas 150 °C and 35 kPa are used in Standard-OIT). A correlation between the S-OIT and the HP-OIT for a HDPE geomembrane is shown in Figure 13.19.

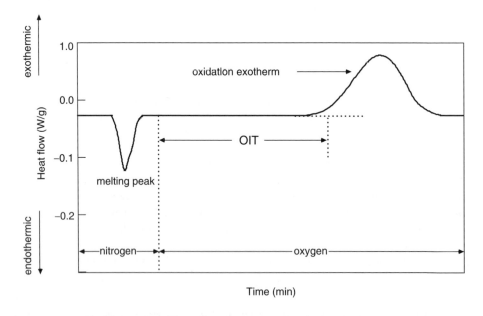

Figure 13.17 Oxidative induction time (OIT) curve showing that the OIT onset is from admission of oxygen into the sample cell. The OIT value is defined as the time when the oxidation exotherm deviates from the baseline minus the starting time when oxygen is first admitted into the sample cell (as per ASTM D-3895 and ASTM D-5885 for standard OIT and high-pressure OIT, respectively)

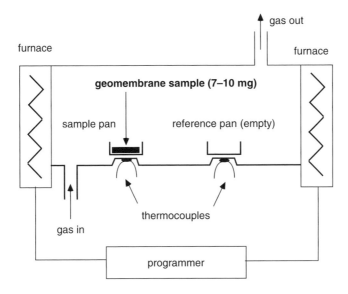

Figure 13.18 Schematic of a differential scanning calorimeter (DSC) used to determine the oxidative induction time (OIT) of polyolefin geomembrannes

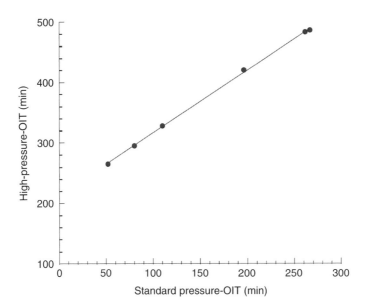

Figure 13.19 Correlation between S-OIT and HP-OIT for an HDPE geomembrane that does not contain significant levels of HALS (initial HP-OIT of 484 min). Reprinted with permission from *Geosynthetics International*, Effect of acidic mine drainage on the polymer properties of an HDPE geomembrane, by S.B. Gulec, T. B. Edil and C. H. Benson, **11**(2), 60. Copyright (2004) Thomas Telford Publishing

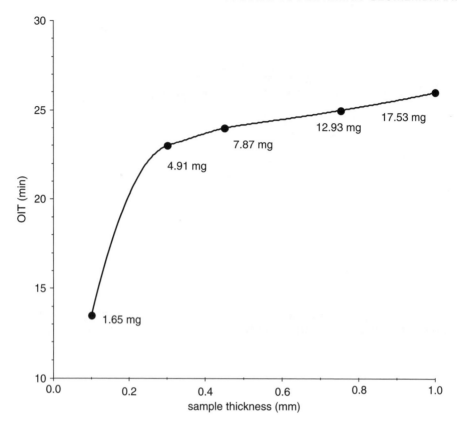

Figure 13.20   Variation in OIT with sample thickness

Note however that the slope of the correlation curve is dependent on the type of additive package used.

It is important to recognize that the OIT results are dependent on the sample thickness used with thicker specimens recording longer OIT values due to hindered oxygen diffusion into thicker samples (see Figure 13.20).

The oxidative stability of a geomembrane can be assessed by heat ageing in an air-circulating oven at 85 °C for 90 days as per ASTM D-5721 ('Standard Practice for Air-Oven Aging of Polyolefin Geomembranes'). The oven should be a controlled, forced–ventilation oven with substantial fresh air intake. Since certain metals are known to affect the thermal endurance of some polyolefins, direct contact of the specimens with metal should be minimized.

The following properties can be measured for evaluating the effects of oven exposure on geomembrane samples:

- Tensile properties (see Test Method ASTM D-638 or ASTM D-6693);
- Melt flow index (see Test Method ASTM D-1238);
- Density (see Test Method ASTM D-1792);

- Tensile impact (see Test Method ASTM D-746);
- S-OIT (see Test Method ASTM D-3895);
- HP-OIT (see Test Method ASTM D-5885).

Air-oven agring is most commonly used to evaluate and compare the performance of various heat stabilizer packages. The results are typically expressed as a percentage of the change in each physical property. The effect of the heat ageing of polyolefin geomembranes is generally determined by tensile testing for retained tensile properties and also OIT testing for retained OIT values.

For instance, the retained stability of the antioxidant/stabilizer package in HDPE and LLDPE geomembranes is assessed using S-OIT and/or HP-OIT testing following oven ageing.

The oven ageing requirements for HDPE Geomembranes after $85\,°C/90$ days as per specification GRI GM-13 are:

- 55% S-OIT retained;
- 80% HP-OIT retained.

The oven ageing requirements for LLDPE Geomembranes after $85\,°C/90$ days as per specification GRI GM-17 are:

- 35% S-OIT retained;
- 60% HP-OIT retained.

It can be seen that the performance requirement for LLDPE geomembranes is less stringent that for HDPE geomembranes reflecting the lower oxidative stability and greater antioxidant migration potential of LLDPE geomembranes.

13.14.2 ULTRAVIOLET DEGRADATION

Ultraviolet light can cause chain reactions and bond breaking of polymeric material due to the penetration of short wavelength energy. The UV spectrum wavelengths between 300 nm and 400 nm are the most damaging towards polymeric geomembranes since they cause bond rupture on a molecular structure level. Accelerated tests can be carried out in the laboratory, using ASTM D-7238, ASTM G26, ASTM G53 or ASTM D-4355 but it can be more accurate to carry out outdoor tests as described in ASTM D-4364.

UV light is composed of a range of wavelengths from the very short wavelength (180 nm) which is highly energetic and very damaging to the longer wavelength (380 nm). The wavelength ranges of the classifications are shown in Table 13.10.

Table 13.10 Range of wavelengths for UV radiation

Type of UV light	Wavelength (nanometre (nm))
UV-A	320–380
UV-B	280–320
UV-C	180–280 (screened out by the ozone layer)
Natural terrestrial sunlight	295–800

Table 13.11   Accelerated weathering equipment for indoor testing

Type of weathering chamber	Test method	Type of source	Comments
Xenon Ci-65 and Ci-5000 'Weather-O-Meters'	ASTM G155, cycle 1	Xenon arc lamp with borosilicate inner and outer filters	Best simulates the UV band of natural sunlight but is expensive to run
Carbon arc lights	Obsolete	Carbon arc lights	Unrealistic acceleration as they produce too much of the wavelengths between 325 and 425 nm
QUV	ASTM D-7238; GRI GM-11 (1600 h)	QUV fluorescent tube (UVB lamps; 280–320 nm)	Most commonly used for testing geomembranes. Note that they reduce the higher wavelengths of light

It is important that accelerated UV exposure testing is performed using light sources that best matche natural sunlight.

The main types of accelerated weathering equipment are shown in Table 13.11.

Since modern geomembranes generally have a high level of UV resistance natural weathering studies must be run for prohibitively long periods to give meaningful longevity estimates (Martin and Eng, 2005). For this reason accelerated weathering is employed to gauge the expected in-service weathering performance of geomembrane materials in a more timely fashion. However correlating an accelerated weathering exposure period to a natural weathering service life is quite difficult since UV weathering depends on many variables.

Accelerated weathering testing allows test results in shorter time periods as the exposure is continuous and irradiation more intense. Common accelerated weathering testing methods are shown in Table 13.11.

Whilst xenon arc lamps give the best match to terrestrial sunlight they are expensive to run and replace (>$US2000 each) and therefore QUV-B lamps are generally used.

QUV Testing

ASTM Standard D-7238 ('Standard Test Method for Effect of Exposure of Unreinforced Polyolefin Geomembrane Using Fluorescent UV Condensation Apparatus') covers the specific procedures and test conditions that are applicable for exposure of unreinforced polyolefin geomembranes (e.g. HDPE, LLDPE, fPP) to fluorescent UV radiation and condensation in a QUV apparatus. Test specimens are exposed to fluorescent UVA 340 lamps under controlled environmental conditions. Note while UVA 340 lamps are standard for this method other types of fluorescent UV lamps, such as UVB-313, can also be used based upon discussion between involved parties. However, if the test is run with another type of fluorescent UV lamp, such as UVB-313, this should be considered as a deviation from the standard and clearly stated in the test report. UVB-313 and UVA-340 fluorescent lamps generate different amounts of radiant power at different wavelength ranges; thus, the photochemical effects caused by these different lamps may vary. UVB-313 lamps are more damaging as they emit shorter wavelength (hence more energetic) UV radiation. The test typically runs for 1600 h during which time the sample is exposed to a 20 h UV cycle at 75 °C followed by 4 h condensation at 60 °C. After exposure, the % retained S-OIT (and/or HP-OIT) values are determined.

Xenon Arc Exposure

The xenon-arc Weather-o-Meter (also known as the Atlas Weather-o-Meter) exposes the geomembrane to UV light from a jacketed, water-cooled xenon lamp that operates at a calibrated light intensity of 350 mW/m^2 at 340 mm and over a range of wavelengths that closely match those in terrestrial sunlight. This test is formalized in ASTM G26 and ASTM D-4355 but since it is more expensive to run than the QUV panel UV exposure (ASTM D-7238/GRI GM-11) the QUV fluorescent UV condensation apparatus method is most widely used for testing geomembranes. The geomembrane samples are periodically sprayed with water to simulate rain and dew and the geomembrane temperature is kept between 60–80 °C depending on the test protocol. The xenon-arc Weather-o-Meter thus uses both UV exposure and heat exposure to simulate outdoor weathering. While it is difficult to correlate xenon-arc Weather-o-Meter testing with natural outdoor exposure correlations can be drawn as shown in Table 13.12.

The calibrated output of the xenon arc light source used in ASTM test method D-4355 is 350 mW/m^2 and the output for the fluorescent UV light source used in D-7238 is 710 mW/m^2.

The intensity of the QUV (D-7238) source is thus about twice that of the xenon arc (D-4355) source. So 500 h in the QUV is equivalent to twice as long natural exposure to 500 h in the xenon chamber (see Table 13.12).

Table 13.12 Equivalent outdoor UV irradiance exposure for 500 h of laboratory UV exposure for ASTM D-4355 (xenon arc) and ASTM D-7238 (QUV apparatus) (TenCate Geosynthetics North America)

City	Annual ean sunshine normalized to 24 h/day	Annual average UV irradiance (mW/m^2)	Equivalent time of exposure to 500 h xenon arc (years) (ASTM D-4355)	Equivalent time of exposure to 500 h QUV (years) (ASTM D-7238)
Atlanta, GA	0.34	213 701	0.8	1.7
Chicago, IL	0.28	143 607	1.2	2.5
Hartford, CT	0.28	143 607	1.2	2.5
Orlando, FL	0.37	230 797	0.8	1.5
Phoenix, AZ	0.42	317 986	0.6	1.1
San Antonio, TX	0.34	256 441	0.7	1.4
San Diego, CA	0.39	247 893	0.7	1.4
Seattle, WA	0.20	76 932	2.3	4.6
Washington, DC	0.31	157 284	1.1	2.3

For correlations between laboratory UV degradation testing and actual field UV exposure rates, we must also look at the duration of natural sunlight over a 24 h period. The natural variation in sunlight intensity due to the earth's rotation about its axis is termed 'diurnal' and refers to this 24 h cycle that is repeated daily. If the solar irradiance level is averaged over a 24 h period, the average is approximately 24% (of peak solar intensity).

Table 13.12 shows that the 500 h xenon arc UV test correlates to roughly one-half to two years of field exposure, while it takes roughly twice as long, or one to four years in the field, to achieve the same UV irradiance levels achieved in QUV testing.

EMMAQUA Testing

EMMAQUA is an acronym, which stands for 'Equatorial Mount with Mirrors for Acceleration with Water'. The EMMAQUA test is described in ASTM D-4364 ('Standard Practice for Performing Outdoor Accelerated Weathering Using Concentrated Sunlight').

EMMAQUA testing is an accelerated outdoor weathering technique which uses Fresnel-reflecting solar concentration devices to concentrate sunlight onto the test samples. It accelerates the level of exposure, yet it maintains the same wavelength distribution as sunlight. It is used widely in outdoor accelerated weathering test methods in the world today.

EMMAQUA concentrates natural sunlight via ten highly reflective, specially coated mirrors onto the specimen target area with an intensity factor of approximately 'eight suns'. The device tracks the sun and exposes specimens to the full spectrum of sunlight, making it one of the most realistic accelerated tests available.

EMMAQUA provides a spectral match to sunlight and correlates well to subtropical conditions, such as Northern Queensland, as well as an arid desert environment, such as Arizona.

The EMMAQUA test apparatus is a 'follow-the-sun' rack with mirrors positioned as tangents to simulate a parabolic trough. The target board, located at the focal line of the mirrors, lies under a wind tunnel along which a deflector directs ambient temperature cooling air across the specimens. A fan spray nozzle assembly is employed to spray the specimens with deionized water in accordance with established schedules. Night time spray cycles can be used to keep the specimens moist during the night time hours. One manufacturer (Atlas) recommend the exposure of test specimens based on an accumulated dose of ultraviolet radiation measured in MJ/m^2. However, the test can also be timed on total radiation, langleys $(Ly)^2$, or real time.

Whilst EMMAQUA has a number of desirable attributes, it is not a fully representative measure of durability of geomembranes, since the polymer samples are in the unstressed state when exposed to the UV irradiation and also the samples (although sprayed periodically with water) are not immersed in water. Thus EMMAQUA testing does not simulate two of the in-service conditions to which many geomembranes are exposed in service.

EMMAQUA test results can be correlated with real-time weathering for specific geographical regions. For instance, 3×10^6 Ly of sun exposure equates to 125 520 MJ/m^2. From annual solar radiant exposure tables it is recorded that Phoenix, Arizona receives

---

[2] A langley (Ly) is a unit of measure for radiation power distribution over area and is used to measure solar radiation. 1 Ly is 418 40.00 J/m^2 = 0.04184 MJ/m^2 while 1 kLy = 1 $kcal/cm^2$ = 41.84 MJ/m^2 and 1 kLy/year = 1.33 W/m^2.

about 8000 MJ/m^2 per year and so 3×10^6 Ly is equivalent to 15.7 years of real time exposure, while Miami, Florida receives 6500 MJ/m^2 per year of solar radiation which equates to about 19.3 years of real time exposure.

Comparison of Accelerated UV Ageing Methods

Geomembranes are usually tested by exposing them to cycles of UV radiation and condensation using a QUV Accelerated Weathering Tester, operated in accordance with ASTM G-154. The main types of light sources employed are UVB lamps. It should be noted however that UVB bulbs emit shorter wavelengths (higher energy) radiation below 300 nm (and down to 280 nm), which are not found in natural sunlight. The presence of this shorter wavelength UV radiation also makes UVB bulbs more aggressive than natural sunlight in the weathering of polymeric geomembranes. In addition UVB bulbs also emit higher levels of UV radiation, with a wavelength between 300 and 320 nm. than is found in natural sunlight. It is by virtue of this higher intensity that UVB lamps achieve their accelerated weathering.

The 300 nm wavelength radiation is considered the most damaging to polyethylene, and 320 nm is considered the most damaging to PVC.

The two typical QUV accelerated weathering cycles consists of:

- 8 h of UV light irradiance at 0.80 W/m^2/nm (measured at the peak wavelength of 313 nm) and a temperature of 60 °C, followed by a 4 h condensation cycle at 50 °C.
- 20 h of UV light irradiance at 0.80 W/m^2/nm (measured at the peak wavelength of 313 nm) and a temperature of 75 °C, followed by a 4 h condensation cycle at 60 °C.

While the UVB bulbs emit radiation with wavelengths shorter than 300 nm, and this radiation is likely to be damaging to the geomembrane samples, natural sunlight contains no significant UV radiation with wavelengths below 300 nm. For this reason any comparison between fluorescent UVB exposure and natural sunlight is obviously imperfect and underestimates the natural service life of the material [Martin, 2005].

By comparing a six year natural weathering study with a 20,000 hour accelerated weathering study it has been shown that there is a rough relationship of approximately 1000 hours of accelerated QUV weathering being equal to one year of natural exposure (Martin and Eng, 2005).

☝ *RULE OF THUMB A rough correlation used in the paint and coatings industry is 500 to 1500 h of accelerated exposure equaling approximately 1 year of real life exposure (Wagner and Ramsey, 2003). A relationship of 1000 h of accelerated weathering equating to a year of natural weathering is used as a conservative basis for the purposes of giving warranties.*

UV light intensity is measured by irradiance and is usually expressed in watts per square metre at a given wavelength. The Weather-o-Meters are run at 0.35 watts/m^2 and 340 nanometres (nm), an industry accepted protocol, which matches one of the conditions in Florida.

The temperature of exposure is also important, as the higher the temperature, the more degradation will occur. A black panel in the machine measures this temperature and is set at 63 °C. The backing on the sample may change the actual sample temperature.

In xenon testing, 2000 h in the Weather-o-Meter is approximately 140 kilolangleys (kLy) per year, which is typical of one year exposure in Miami, Florida. A kilolangley is a measure of UV exposure per square metre of surface area.

For HDPE the Weather-o-Meter on a wet and dry cycle still gives a correlation of approximately 2000 h, being roughly one year of outdoor weathering (although this is highly dependent on geographic location). Geographic location, along with changes in climate and elevation, affect actual UV performance.

It should also be noted that there is increased severity of UV radiation in higher elevation areas. In fact, for every 0.5 km increase in elevation, UV exposure increases by 3.5%. Sunshine variations remain the key variable when correlating accelerated and outdoor exposure data.

UV degradation of HDPE and PVC manifests itself by crazing and a loss of physical properties, especially the tensile elongation at break. The tensile elongation at break declines are more sensitive than tensile strength due to the the formation of oxidized species on the exposed surface of the geomembrane. This brittle surface layer form cracks that significantly affect the tensile elongational properties.

The failure point after weathering is often arbitrarily taken as a 50% reduction in the original tensile strength at break value. At the 50% point, surface crazing (a crosshatched pattern caused when the sample is flexed) become evident on the geomembrane surface.

Martin and Eng (2005) have shown, using 20 000 h accelerated exposure that thinner (0.75 mm) polyolefin geomembrane can perform equivalently to a 1.5 mm HDPE geomembrane when sufficient UV stabilizing additives are utilized (Martin and Eng, 2005).

13.15   TESTS FOR ADDITIVES

13.15.1   CARBON BLACK CONTENT

Carbon black is added to geomembranes to increase their UV resistance. To check that the right amount of carbon black has been added and that it is dispersed thoroughly within the geomembrane sheet, both carbon black content and carbon black dispersion tests are routinely performed. The amount added depends on the polymer type and varies widely, as shown in Figure 13.21. For HDPE geomembranes, typically 2 to 3% carbon black content is within acceptable limits while for PVC, approximately 8 to 15% carbon black is used.

Tests to determine the carbon black content measure the percentage of carbon black present in the geomembrane by heating the geomembrane to beyond its decomposition temperature where the polymer is pyrolysed (i.e. decomposed to gaseous degradation products) leaving only the carbon black behind – whose content is then determined gravimetrically (i.e. by weighing). The carbon black content of geomembranes is generally determined using the method ASTM D-1603 which employs a pyrolysis tube furnace with inert atmosphere. In this test a known weight of geomembrane is placed into a pre-weighed combustion boat. The sample is then placed into a 600 °C tube furnace under an oxygen-free dry nitrogen purge. After a preset time the combustion boat with the burnt residue is cooled under a nitrogen blanket and weighed. The combustion boat

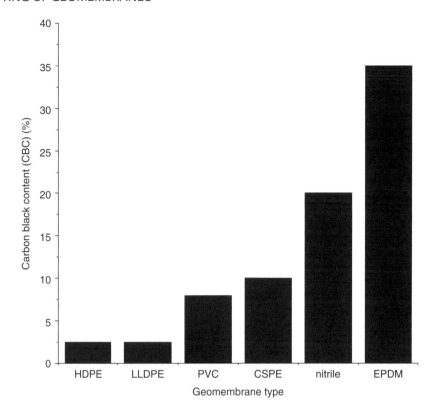

Figure 13.21 Typical carbon black content of various geomembranes

is then placed back into the tube furnace at 600 °C under oxygen to oxidize the carbon residue. When the carbon is completely oxidized the combustion boat is cooled and weighed.

Other testing methods for determining the carbon black content are ASTM D-4218 (muffle furnace method) and the microwave technique. Note these latter techniques should only be used if an appropriate correlation to the ASTM D-1603 method can be established.

These tests can be affected by moisture on the boat or crucible as well as by not adhering to the method with regard to the use of an effective desiccator. The rate of gas flow is another important variable to carefully control.

13.15.2 CARBON BLACK DISPERSION

Not only are the amount and particle size of the carbon black important to provide long-term resistance to ultraviolet-induced degradation to geomembranes, but its uniform dispersion in the geomembrane is also critical. In order to achieve this the carbon black must be dispersed and distributed uniformly throughout the as-manufactured geomembrane material.

ASTM D-5596 ('Standard Test Method for Microscopic Evaluation of the Dispersion of Carbon Black in Polyolefin Geosynthetics') covers the specimen preparation techniques and procedures for evaluating the dispersion of carbon black in polyolefin geomembranes. This test method allows for a qualitative evaluation of carbon black agglomerates and other inclusions in polyolefin geomembranes based on visual comparisons between microscopic fields of view and micrographs of the carbon dispersion reference chart (ASTM D-35). Carbon black agglomerate size rating classifications are shown in Table 13.13.

A carbon black agglomerate is defined as a cluster of physically bound and entangled aggregates of carbon black. In this test, the geomembrane is cut into thin slices using a sliding microtome which is an apparatus capable of cutting thin slices (less than 20 μm in thickness) (see Figure 13.22). The specimen's thickness range is typically from 8 to

Table 13.13 Carbon black agglomerate size rating classifications (ASTM D-5596, 2003)

Category classification	Agglomerate area (μm²)	Agglomerate diameter (μm)
Category 1	<960	<35
Category 2	>960–4390	35–75
Category 3	>4390–24 053	75–175
Category 4	>24 053–70 680	175–300
Category 5	>70 680	>300

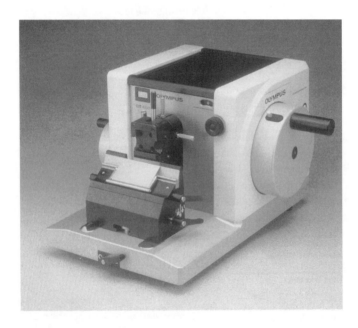

Figure 13.22 Photograph of a precision microtome for microtoming (i.e. slicing) thin sections of a geomembrane for carbon black dispersion determination as per ASTM D-5596

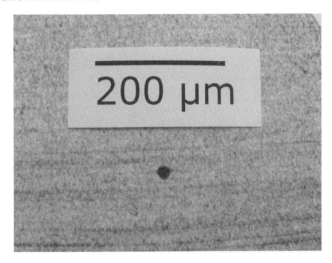

Figure 13.23 Photograph of a carbon black agglomerate in an HDPE geomembrane as viewed under magnification on a microtomed slice. The agglomerate particle has a diameter of approximately 20 µm and so falls into the Category 1 classification (i.e. diameter <35 µm; total area of <960 µm²)

20 µm. Slices 20 µm thick are usually too thick to permit adequate light transmission through the sample and therefore thin sections of 10 to 15 µm should be cut. These thin sections are then evaluated microscopically using a visual comparison between each random field of view and the carbon dispersion reference chart. Magnifications in the range 50 to 200× are required. A calibrated reticle (eyepiece micrometer) positioned in one of the eyepieces between the eyepiece lens and the objective is also required. For this test, five samples are selected randomly across the full roll width of the geomembrane.

The general specification for HDPE geomembranes for carbon black dispersions is that 90% of views need to be in Categories 1 and 2 with less than 10% of the views in Category 3. Therefore for 10 different views, 9 can be in Categories 1 or 2 and only 1 is allowed in Category 3. Note this only applies to near spherical agglomerates being accepted. A typical Category 1 classification carbon black agglomerate is shown in Figure 13.23 while Categories 2 and 5 agglomerates are shown in Figures 13.24 and 13.25, respectively.

Some non-conforming carbon black dispersion observations and the potential consequences on geomembrane properties are listed in Table 13.14.

Note also that flexible or elastomeric materials such as VLDPE or fPP generally require thin sectioning under low-temperature conditions. For instance, the clamped sample should be packed loosely in crushed dry ice for approximately 15 min or until the specimen, knife and clamp reach approximately −30 °C.

13.15.3 ANTIOXIDANT/STABILIZER LEVELS BY OIT METHODS

The oxidative induction time (OIT) technique for polyolefins is based on the fact that a linear relationship exists between the concentration of phenolic antioxidants and the

Figure 13.24 Photograph of a carbon black agglomerate in an HDPE geomembrane as viewed under magnification on a microtomed slice. The agglomerate particle has a diameter of approximately 68 μm and so falls into the Category 2 classification (i.e. diameter 35–75 μm; total area of >960–4390 μm²)

Figure 13.25 Photograph of a 'large' carbon black agglomerate in an HDPE geomembrane as viewed under magnification on a microtomed slice. The agglomerate particle has a diameter of approximately >300 μm and an approximate area of 92 000 μm² and so falls into the Category 5 classification (i.e. diameter >300 μm); area >70 680 μm²)

Table 13.14 Carbon black dispersion observations

Observation	Possible cause	Possible consequences
Large agglomerates of carbon black	A result of poor dispersive mixing (i.e. insufficient shear during extrusion)	Can lead to lower tensile properties and possibly even initiate stress cracking
Poor carbon black distribution (i.e. concentrated and dilute areas)	A result of incomplete mixing (i.e. insufficient distributive compounding during extrusion)	Can cause localized UV degradation in carbon-poor regions.
		Can also lead to separation in plane (SIP)
Uniformly large carbon black particles and significant interparticle separation	A result of an inferior carbon black with a large particle size being used	Poor UV resistance

OIT value. Polyolefins generally exhibit a linear relationship between the OIT and the concentration of non-volatile phenolic antioxidants in the concentration range which is normally employed in geomembranes (i.e. 1000–5000 ppm).

ASTM D-3895 ('Standard Test Method for Oxidative-Induction Time of Polyolefins by Differential Scanning Calorimetry') is applicable to polyolefin resins that are in a fully stabilized/compounded form. This test method is similar but not equivalent to ISO 11357–6. The *oxidative induction time* is defined as 'a relative measure of a material's resistance to oxidative decomposition; it is determined by the thermoanalytical measurement of the time interval to onset of exothermic oxidation of a material at a specified temperature in an oxygen atmosphere'.

The metal composition of the specimen holder can influence the OIT test result significantly due to associated catalytic effects. Polyolefins used in the wire and cable industries typically require copper or aluminium pans, whereas polyolefins used in geomembrane and vapour-barrier film applications exclusively use aluminium pans. Unless otherwise specified, the analysis temperature used in this test has been set arbitrarily at 200 °C.

Specimens for OIT consist of discs (6.4 mm diameter) that are cut or hole-punched from the sheet and these will typically have a weight of approximately 5 to 10 mg, depending on sample density. If the sample requires homogenization prior to analysis, a melt compounding and sheet pressing procedure is recommended. Poor sample uniformity will affect test precision adversely. Sample size also affects the OIT results as shown in Figure 13.20.

In a standard OIT test the polymer sample is heated under programmed heating and under a nitrogen gas flow of 50 ml/min from ambient temperature to 200 °C (the set point) at a rate of 20 °C/min. When the set temperature has been reached, the programmed heating is discontinued and the sample is equilibrated for 5 min at the set temperature. Once the equilibrium time has expired, the gas in the cell is changed to oxygen at a flow rate of 50 ml/min. This changeover point to oxygen flow is considered to be the zero time of the experiment. The intersection of the baseline with the exotherm signal is defined as the onset of oxidation. The time from this intersection to zero time is the OIT.

Although the OIT is a qualitative assessment of the level (or degree) of stabilization of the material tested, the OIT measurement is also an accelerated thermal-ageing test and as such can be misleading. Caution should be exercised in data interpretation since oxidation reaction kinetics are a function of temperature and the inherent properties of the additives contained in the sample.

OIT results are often used to select optimum resin formulations; however it must be emphasized that volatile antioxidants may generate poor OIT results (at the high test temperature, i.e. 200 °C) even though they may perform adequately at the intended use temperature of the finished product.

There is no accepted sampling procedure, nor have any definitive relationships been established for comparing OIT values on field samples to those on unused products; hence the use of such values for determining life expectancy is uncertain and subjective.

The OIT test is a function of a particular compound's stabilizer system and should not be used as a basis of comparison between formulations that might contain different resins, stabilizers or additive packages.

Schmutz and co-workers from Ciba Geigy have shown that the OIT at 200 °C is a valid technique for the antioxidant concentration in PP since the relationship between the OIT and the antioxidant concentration is linear. It was found however that different polymers give different OIT/antioxidant concentration relationships. Therefore the OIT values for different polymers cannot be compared (Schmutz *et al.*, 2006).

The OIT values for adequately stabilized HDPE geomembranes are >100 min at 200 °C for standard OIT and >400 min for high-pressure OIT (HP-OIT at 150 °C and 3.4 MPa) as per the GRI GM-13 specification.

The S-OIT value of 100 min has been selected as the minimum required value for polyolefin geomembranes. The basis of this was that an HDPE geomembrane which gave an S-OIT value of 80 min was predicted to have a lifetime of 200 years at 20 °C under soil burial conditions (Hsuan and Koerner, 1998). On this basis the 100 min minimum was deemed to be acceptable as a screening test. However the long-term performance of the antioxidant package should be assessed using an oven-ageing method such as ASTM D-5721 at 85 °C for 90 days followed by the measurement of the retained S-OIT value.

The OIT for unstabilized HDPE is typically a couple of minutes in the standard OIT test and 10–20 min in the HP-OIT test.

The following important rules define the boundaries for the use of the OIT technique:

- OIT is not an appropriate method to predict the long-term thermal stability of poly-olefins. This is because the solubility of antioxidants at the test temperature (generally 200 °C) can be very different to that at ambient temperatures.
- It can be particularly misleading when used as a screening test to assess the relative performance of stabilizers.
- Correlation with other accelerated tests such as oven ageing rarely exists at present.
- Extrapolation to end use temperatures (e.g. 20–70 °C) by plotting the induction time versus reciprocal temperature (i.e. the Arrhenius extrapolation) can give significant overestimations of service lives compared with oven ageing tests and real life ageing data. Such lifetime extrapolations based on OIT data are thus not valid. This is because there are changes in the oxidation mechanism when one extrapolates through a phase change, such as the melting point, due to effects of crystallization and the redistribution

of the additives in the polymer matrix. These factors lead to a discontinuity in the Arrhenius plot.

ASTM D-5885 ('Standard Test Method for Oxidative Induction Time of Polyolefin Geosynthetics by High-Pressure Differential Scanning Calorimetry') measures the oxidative induction time of a sample at a specified temperature (i.e. 150 °C) and pressure (i.e. 3.4 MPa). The oxidative induction time is defined as the elapsed time between the admission of oxygen into the sample cell and the onset of the exotherm, signalling oxidation, as measured under constant temperature (150 °C) conditions. A comparison of the S-OIT and HP-OIT methods is shown in Table 13.15.

The OIT testing temperatures must not destroy the antioxidants or stabilizers present in the geomembrane. Antioxidants have specific ranges of temperature in which they are effective (Fay and King, 1994; Lutz and Grossman, 2001). Antioxidants such as hindered phenols (primary antioxidant) have an effective temperature range of 0–300 °C, while hindered amine stabilizers are most effective below 150 °C and preferably below 135 °C (Fay and King, 1994). Hence the HP-OIT test on HDPE is normally performed at 150 °C and not 200 °C, as the hindered amine stabilizers will be destroyed at the higher temperature.

The OIT is a characteristic of a compounded polyolefin product that is dependent not only on the type and amount of additives present, but also on the type of resin. In well-defined systems, this test method can be used as a quality control measure to monitor the stabilization in geosynthetics as received from a supplier.

The OIT measurement is an accelerated thermal ageing test and, as such, interpretation of resulting data may be misleading if done by an inexperienced operator. Certain antioxidants, however, may generate poor OIT results even though they may be adequate at their intended service temperature and vice versa. The advantages and disadvantages of the Standard OIT test are shown in Table 13.16.

The use of high test temperatures, however, may have deleterious effects. The first factor to consider is the potential volatilization of the additive packages used to stabilize the test materials (Pauquet *et al*., 1993). The second concern is the potential for the influence of chemical reaction mechanisms which are not significant at end-use operation conditions. For reduction of the potential volatilization and decomposition of stabilizers, the HP-OIT test uses a reduced test temperature of 150 °C and high pressure to suppress stabilizer volatilization. The advantages and disadvantages of the HP-OIT test are shown in Table 13.17.

Table 13.15 A Comparison of the S-OIT and HP-OIT methods

Standard OIT	High-pressure OIT
ASTM D-3895 or ISO 11357	ASTM D-5885
Pressure = 35 kPa	Pressure = 3500 kPa (100 times higher)
Temperature = 200 °C	Temperature = 150 °C
This test cannot detect hindered amine stabilizers due to the relatively high test temperature	This test can be used for all types of antioxidant packages including hindered amine stabilizers and is preferred over the S-OIT test

Table 13.16 Advantages and disadvantages of the Standard OIT test

Advantages	Disadvantages
Existing ASTM test protocol	High temperatures may bias the test results for hindered phenolic antioxidants relative to hindered amine stabilizers
Short testing time (~100–150 min)	Results are in doubt for some fPP formulations
Standard test apparatus	–

Table 13.17 Advantages and disadvantages of the High-Pressure OIT test

Advantages	Disadvantages
Existing ASTM test protocol	Long testing time (>300 min)
Able to distinguish the stabilization effect of different types of stabilizers in geomembranes	Special testing cell and set-up are required
Lower test temperature relates closer to service conditions	Results are in doubt for some fPP formulations

Table 13.18 Rules to follow in regard to using OIT results

Rules	Reasons
Do not extrapolate times from the OIT test to predict lifetimes at lower temperatures	OIT data which is obtained above the melting point cannot be reliably extrapolated to temperatures below the melting point as this can give gross overestimations of lifetimes
Do not compare OIT results for different polymers	The OIT response for the same antioxidant varies depending on the polymer it is contained in
Do not compare OIT results for different antioxidant packages in the same polymer type	Certain antioxidants may generate poor OIT results even though they may be adequate at their intended use temperature and the reverse is also applicable

To avoid erroneous conclusions there are a number of rules that should be followed when using the OIT technique (see Table 13.18).

It should also be noted that in polyolefins such as HDPE and fPP the drop in S-OIT after immersion in water or service can be attributed in large part to the hydrolysis of the phosphite stabilizer (such as Irgafos 168™) (see Figure 13.26). The depletion of the phosphite occurs during the first year of water exposure of these geomembranes. For instance the steep decrease in the OIT of HDPE after exposure to hot water is largely due to the hydrolytic breakdown and leaching of the hindered phosphite stabilizers. The mid- to long-term thermal stability of HDPE is more influenced by the depletion of the phenolic antioxidant (e.g. Irganox 1010™). The GRI GM-13 standard for HDPE geomembranes

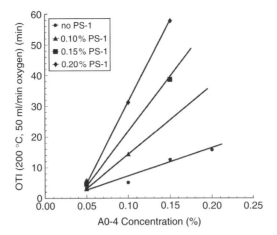

Figure 13.26   This graph demonstrates that in fPP the phosphite stabilizer (Irgafos 168™) contributes significantly to the OIT value of the polymer. Note also that a linear relationship exists between the phenolic antioxidant concentration (Irganox 1010™) and the oxidative induction time (OIT) of fPP. Reprinted with permission from Geosynthetics International, Effect of acidic mine drainage on the polymer properties of an HDPE geomembrane by S. B. Gulec, T. B. Edil and C. H. Benson, 11, 2, 60 Copyright (2004) Thomas Telford Publishing

requires that the drop in OIT value is not greater than 45% following 90 days of oven ageing at 85 °C.

13.16   CARBONYL INDEX MEASUREMENTS

The composition of polymeric materials can be readily determined by measuring their infrared spectra using a Fourier-transform infrared (FTIR) spectrometer and then comparing the results with a commercially available or specifically prepared spectral database library.

Fourier-transform infrared spectroscopy is a spectroscopic method used to detect structural changes in polymeric materials at the molecular level. A polymer specimen is subjected to infrared radiation in successively decreasing frequencies. The amount of infrared radiation absorbed at each frequency is indicated in a spectrum showing which molecules vibrate in a specific mode at a particular frequency. Polymer structure is determined by identifying the peaks using a reference library of spectra. Changes observed in the spectrum can be used as an indicator of oxidation.

The degree of oxidation of polyolefins can be measured by comparing the concentration of carbonyl groups in the exposed material relative to the unexposed material. In polyolefins the carbonyl group has an absorption band from 1710–1742 cm^{-1}.

Fourier-transform infrared spectroscopy is a technique that is used to detect the C=O carbonyl absorption which is the 'signature' of oxidation. PE, PP and EPDM normally contain no oxygen atoms; however when oxidized the oxygen atom content increases and hence a carbonyl peak forms. The carbonyl peak is therefore a 'tell-tale' by-product of the oxidation of polyethylene.

The carbonyl index is the ratio of the intensity of the carbonyl absorption band (i.e. C=O) to that of another band (such as C-H stretching which occurs at 1462 cm^{-1}) in the spectrum. An increase in a polymer's carbonyl index after exposure is a good indication that oxidation has occurred during the exposure.

The carbonyl index of EPDM can be determined by ratioing the carbonyl peak to the C-H stretch at 2919 cm^{-1}.

The carbonyl index cannot be used in the case of PVC because the plasticizer (which is an ester) absorbs at 1725 cm^{-1} and thus the plasticizer obscures any changes in the carbonyl region that may arise due to oxidation.

Polymer geomembrane samples presented for analysis are often too thick and too opaque for measurement by transmission sampling techniques. However, the relatively shallow depth of penetration of the evanescent wave (typically 0.5 to 2.0 μm) in attenuated total reflectance (ATR) generally eliminates the need to do sample preparation.

The FTIR spectra of black or opaque polymers (such as black HDPE geomembranes) can therefore be conveniently acquired using ATR. The latter is a sampling interface that enables routine analysis of solids with little to no sample preparation. The sample is sandwiched between a crystal of high refractive index and a clamp. The IR beam is directed through the crystal toward the sample at an angle that ensures total reflection at the interface between the crystal and the sample. The IR radiation penetrates into the sample a very small distance (a few microns). During this penetration, the vibrating chemical bonds in the sample absorb some of the radiation. The attenuated reflected beam is then detected by the IR detector and the resulting signal is processed to produce the reflectance spectrum of the sample. The reflectance spectra are analogous to the traditional transmittance spectra. In fact, the absorption peaks in reflectance spectra match up well with FTIR transmission spectra.

The two most popular ATR crystal materials used for polymer analysis are zinc selenide (ZnSe) and diamond. Both of these crystals have a refractive index of 2.4 and with a typical polymer refractive index of 1.5 at 45 degrees angle of incidence and at a wavenumber of 1000 cm^{-1} the ATR depth of penetration would be approximately 2 μm. The use of ATR eliminates the sample preparation generally required for FTIR analysis by transmission sampling techniques and thereby greatly speeds up the measurement. For black polymers such as geomembranes the zinc selenide crystal is not as useful as the germanium crystal, with the latter giving better resolution. A diamond ATR is a good general purpose analysis tool for geomembranes.

13.16.1   STEPS FOR DETERMINING THE CARBONYL INDEX

1. Select a geomembrane sample that is representative.
2. Place the test sample in the ATR sampling attachment and acquire the infrared spectrum.
3. Place the reference sample in the ATR sampling attachment and acquire the spectrum. The reference sample is a thin layer (20 μm) of HDPE heated in an air-circulating oven for 2 days at 85 °C in order to oxidize its surface.

Table 13.19 Peak heights and carbonyl indices for HDPE geomembranes

Sample	1462–1464 cm^{-1a}	1715–1740 cm^{-1}	Ratio (1740/1462)
Oxidized PE	0.0160	0.008	0.5
Virgin HDPE	0.0161	0.000	0.00
Geomembrane 1	0.0171	0.004	0.23
Geomembrane 2	0.0182	0.005	0.27

^{a}The peak at 1462/1464 cm^{-1} serves as a good internal reference peak.

4. Measure the carbonyl peak area (or height) at 1715 cm^{-1} (or 1735 cm^{-1} whichever is larger) relative to the internal standard peak at 1462 cm^{-1} and express the two absorbances as a ratio. This is the carbonyl index.
5. Compare the test specimen carbonyl index to that of the oxidized reference specimen. The change in carbonyl group content is a measure of the amount of oxidation (see Table 13.19).

The two exposed samples (Geomembranes 1 and 2) exhibit signs of oxidation with carbonyl indices (CI) of 0.23 and 0.27, respectively. Any CI > 0.1 indicates that oxidation has occurred. A CI > 0.2 indicate that significant oxidation has occurred.

Note that surface oxidation is possible when the HP-OIT is not zero but bulk oxidation of the sample can only occur when the HP-OIT is zero (i.e. the stabilizers are fully depleted).

In order to compare the oxidative stability of different PE geomembranes using the carbonyl index, samples of geomembrane are heated in an oven at 85 °C or 90 °C for a given time and then the samples are analysed by FTIR to monitor the evolution of the carbonyl group content. This is a rapid method for ranking the oxidative stability of different geomembrane materials or formulations (i.e. geomembranes made from different resins and with different stabilizer packages).

Peggs *et al*. (2002) found that it is necessary to heat the specimen in an oxygen rich air stream at 90 °C for at least 24 h in order to observe significant changes in the carbonyl group peak. Other tests have found that heat ageing at 85 °C in a high-pressure oxygen oven is necessary to generate sufficient oxidation in a thin sample in a reasonably short time period (e.g. 20–50 h) (Peggs *et al*., 2002).

Though not specific for geomembranes, the carbonyl index test is standardized in ASTM F-2102 - 06e1 ('Standard Guide for Evaluating the Extent of Oxidation in Polyethylene Fabricated Forms Intended for Surgical Implants'). This guide covers a method for the measurement of the relative extent of oxidation present in polymer samples where the material is analysed by infrared spectroscopy and the intensity (area) of the carbonyl absorptions (> C=O) centered near 1720 cm^{-1} is related to the amount of chemically bound oxygen present in the material.

13.17 CHEMICAL RESISTANCE TESTING

Chemical resistance is a very important property since many geomembranes are in direct long-term contact with various chemicals. To ensure adequate chemical resistance, it is recommended to assess the behaviour of a geomembrane with the leachate, or waste chemical that the geomembrane will contain. The testing conditions should be as similar as possible to the in-service exposure conditions. The results from laboratory immersion experiments as outlined in ASTM D-5322 should be plotted indicating the percentage change in the measured property from the original versus the duration of incubation. Also, ASTM D-5496 covers the 'Standard Practice for in-Field Immersion Testing of Geosynthetics'.

Ideally the incubating liquid (i.e. test immersion medium) should be site-specific. If a prepared cocktail of chemicals is used then it should consist of the highest possible concentration of the various chemicals anticipated to be in contact with the geomembrane when in service.

Incubation of geomembrane coupons is generally performed in closed containers. The container is hermatically sealed to prevent evaporation losses and the liquid contents are continuously circulated by means of a magnetic stirrer bar. The temperature and consistency are constantly monitored to ensure all coupons in the immersion medium are exposed to the same environment. This is particularly required with two-phase test solutions comprising an organic phase (e.g. kerosene) and an aqueous phase (e.g. sulfuric acid solution). Constant stirring is required to prevent phase separation. Individual test coupons are retrieved at intervals, generally at 30, 60, 90, and 120 days, according to ASTM D-5322 and ASTM D-5747. Samples removed for dimensional and weight checks can be measured and then replaced in the incubation vessels for on-going measurements since the tests are non-destructive.

The chemical resistance of a geomembrane towards hazardous waste and landfill leachate can be assessed using the US EPA Test Method 9090. This method is known as immersion or incubation testing. In this method, a geomembrane is incubated in a representative solution comprising the leachate or liquor at room temperature (23 °C) and at elevated temperatures (50 or 75 °C). In order to prevent loss of volatile components from the leachate/liquor, thus changing its composition, the geomembrane is placed in a special immersion tank made from stainless steel and fitted with a gasketed lid (see Figure 13.27). It is important that the samples are supported so that they do not touch each other or the bottom of the immersion tank. If samples touch each other the cohesive force of the liquid may cause blocking and stagnant boundary layers on the surface of the geomembrane specimens. Samples should be removed from incubation every four weeks (i.e. after 30, 60, 90 and 120 days) and the measured changes in material properties expressed as a retained percentage.

Note that chemical resistance testing does suffer from certain limitations and these are identified in Table 13.20

13.17.1 CHOOSING THE ACCELERATED TESTING INCUBATION TEMPERATURE

The highest incubation test temperature should be approximately 50 °C below the melting point of the geomembrane material. For example, the melting point of HDPE geomembranes is around 131–132 °C (as measured by a differential scanning calorimeter which is a device that measuring small changes in heat flow); therefore the highest incubation temperature that should be considered is 85 °C which is 46 °C below the melting point.

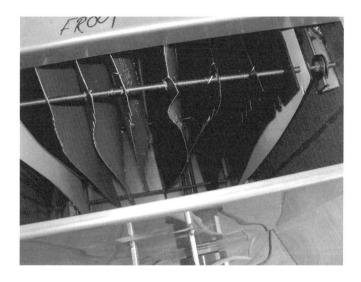

Figure 13.27   Photograph of a stainless steel incubation tank for chemical exposure testing showing the test geomembranes hanging from the stainless-steel rods. Note that it is very important that a gap is maintained between the liners during the exposure testing to ensure full contact between the liner and the incubated medium

Table 13.20   Potential limitations of chemical resistance testing by the incubation method

Potential limitation	Comments
The test time is short	The test duration is relatively short term compared to the actual service life of most geomembranes
Absence of stress	The test does not allow for the specimens to be subject to any stress during exposure
Pass/fail criteria not provided	There is no generally agreed criterion for what is acceptable property variations after exposure
Temperature to accelerate the test	The use of elevated temperatures may accelerate certain reactions that do not occur at lower service temperatures
Phase separation	Some liquids which consist of both aqueous and organic[a] components have to be kept continuously stirred to prevent phase separation
Bulk mechanical tests lack sensitivity	Bulk property measurements such as tensile strength may lack sensitivity to detect surface degradation or changes in the composition of the geomembrane sample

[a]Note that the concentrations of organics in municipal solid waste leachates that can partition, are generally low and are not aggressive towards polymeric geomembranes. However organic fractions in industrial process solutions or mining liquors can be significantly (e.g. aromatic kerosenes used in solvent extraction plants) more aggressive and can be potentially damaging towards geomembranes.

By selecting three elevated temperature for the incubation such as 55, 70 and 85 °C then the results can be used to conduct lifetime extrapolations via the Arrhenius equation. The Arrhenius extrapolation is a classical method for lifetime prediction of polymers that have been exposed to accelerated testing conditions. Arrhenius extrapolation of the higher temperature rate data predicts rate constants which can then be applied to lower temperatures to enable lifetime predications at a given temperature.

13.17.2   SITE-SPECIFIC CHEMICAL RESISTANCE TESTING

Whilst chemical resistance tables serve as a useful guide they are not appropriate surrogates for incubation liquids such as landfill leachates, heap leach pad pregnant liquors or mining process liquids. These site-specific liquids can be particularly aggressive owing to the use of strong acids, strong alkalis, hydrocarbon solvents or combinations of these.

In these cases, the actual leachate/liquor or a laboratory-prepared 'cocktail' is used as the incubation liquid. ASTM D-5322 describes the sample incubation methodology while ASTM D-5496 provides an alternative field incubation methodology.

Upon removal of the geomembrane coupons (typically 30–60 days at 50–80 °C) they are tested to ascertain the level of retained properties (that is, the percentage change from the non-incubated samples). From this an assessment of the compatibility of the geomembrane with the site-specific leachate/liquor can be made.

13.17.3   PERMEABILITY TESTING

Water Vapour Transmission

The water vapour transmission rate test measures a very critical characteristic of a geomembrane, namely its impermeability. The water vapour transmission rate of a geomembrane material can be evaluated using ASTM E-96 where a specimen is sealed over an aluminium cup with either water or a desiccant in it and then placed in a controlled relative humidity environment (see Figure 13.28). The required test time varies from 3 to 30 days. From the results of this test, the water vapour transmission, permeance and permeability are calculated.

PVC thus has a water vapour transmission some 100 times greater than HDPE (see Table 13.21).

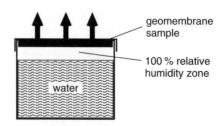

Figure 13.28   Schematic showing the testing set-up for water diffusion of geomembranes as per ASTM E-96 ('Wet Cup Diffusion Method'). Conditions: test chamber at 20% or 50% RH and 21 °C

Table 13.21 Water vapour transmission rates for various polymeric geomembranes

Geomembrane type	Water vapour transmission rate (g/m² day)
HDPE membrane (0.75 mm thickness)	0.017
LLDPE membrane (0.75 mm thickness)	0.04
fPP membrane (0.75 mm thickness)	0.083
CSPE	0.44
PVC geomembrane (0.75 mm thickness)	1.8

Table 13.22 Permeation testing methods

Class of permeant	Permeant	Examples of applications	Test methods
Gas	Oxygen, nitrogen, methane, carbon dioxide	Landfill caps, covers on aerobic ponds	ASTM D-1434
Water vapour	H_2O	Floating covers, tank liners	ASTM E-96 ASTM F-372 (H_2O vapour)
Organic vapour	Organic species such as BTX vapours	Secondary containment for organic solvents and gasoline	ASTM D-814 ASTM E-96
Organic liquid	Organic solvents	Tank liners, secondary containment	ASTM D-814

Solvent Vapour Transmission

In the presence of liquids other than water it is important to consider the solvent vapour transmission through the geomembrane. The values of solvent vapour transmission through the membrane is related to the molecular size and attraction of the liquid *vis-à-vis* the polymeric liner material. The main test method for determining the solvent vapour transmission rate is identical to the water vapour transmission test (i.e. ASTM E-96), except that the water is replaced by solvents (e.g. chloroform).

Table 13.22 lists some permeation testing methods for various chemicals including gases (e.g. methane), organic vapours (e.g. benzene vapour) and organic liquids (e.g. solvents).

REFERENCES

Allen, S., *GFR Magazine*, **18**(7), 1 (2000).

ASTM D-5596, 'Microscopic observation and comparisons with polyethylene geomembrane dispersion standards' (2003).

Blond, E., Bouthot, M. and Mlynarek, J., 'Selection of protective cushions for geomembrane puncture protection', available from CTT Group/SAGEOS, St-Hyacinthe, Québec, Canada.

Fay, J. J. and King R. E., Antioxidants for Geosynthetic Resins and Applications, in Hsuan, Y. G. and Koerner, R. M. (Eds), *Geosynthetic Resins, Formulations and Manufacturing*, GRI Conference Series, IFAI, St Paul, MN, USA, pp. 77–96 (1994).

Giroud, J. P., Monroe, M. and Charron, R., Strain Measurement in HDPE Geomembrane Tensile Tests, *Geotechnical Testing Journal*, 17(1), 65–71 (1994).

Grossman, G. W., Correlation of Laboratory to Natural Weathering, *Journal of Coating Technology*, **49**, 45–54 (1977).

Gulec, S. B., Edil, E. B. and Benson, C. H., Effect of acidic mine drainage on the polymer properties of an HDPE geomembrane, *Geosynthetics International*, 11(2), 60 (2004).

Hsuan, Y. and Koerner, R. M., Antioxidant Depletion Lifetime in High Density Polyethylene Geomembranes, *Journal of Geotechnical and Geoenvironmental Engineering*, **124**(6), 532–541 (1998).

Hsuan, Y. G., Koerner, R. M. and Lord, Jr, A. E., The Notched Constant Tensile Load (NCTL) Test to Evaluate Stress Cracking Resistance, in *Proceedings of the 6th GRI Seminar*, Philadelphia, PA, USA, pp. 244–256 (1992).

Hullings, D. and Koerner, R., Puncture Resistance of Geomembranes Using a Truncated Cone Test, in *Proceedings of Geosynthetics '91*, IFAI, St Paul, MN, USA, pp. 273–285 [http://www.geomembranes.com/index_resources.cfm?copyID=35&ID=geo&type=tech] (1991).

Lutz, J. T. and Grossman, R. F., *Polymer Modifiers and Additives*, Marcel Dekker, New York, NY, USA (2001).

Martin, D. and Eng, P., 'Advanced Thin Film Geomembrane Technology For Biocell Liners and Covers' (available from Layfield Geosynthetics and Industrial Fabrics Ltd) (2005).

Mills, J. A. and Stang, J. R., Temperature Corrected Tensile Strengths of Geomembrane Field Seams, in *Proceedings of Geosynthetics '97*, (1997).

Pauquet, J. R. Todesco, R. V. and Drake, W. O. Limitations and Applications of Oxidation Induction Time (OIT) to Quality Control of Polyolefins, in *Proceedings of the 42nd International Wire and Cable Symposium* (available from Ciba-Geigy, Ltd, Basel, Switzerland) (1993).

Peggs, I. D., 'Assuring the Quality of HDPE Geomembrane Liners: An International Perspective' [www.geosynthetica.net/tech_docs/AssuringQualityHDPEIDP.asp] (2008).

Peggs, I. D., Lawrence, C. and Thomas, R., The Oxidation and Mechanical Performance of HDPE Geomembranes: A More Practical Durability Parameter, in *Proceedings of Geosynthetics: State of the Art and Recent Developments*, AA Balkema Publishers, Lisse, The Netherlands, pp. 779–782 (2002).

Poly-flex, 'Technical Geomembrane Data' [http://www.poly-flex.com/news12.html] (2008).

Schmutz, Th., Kramer, E. and Zweifel, H., Technical paper entitled 'Oxidation Induction Time (OIT) – a Tool to Characterize the Thermo-Oxidative Degradation of a PE-MD Pipe Resin' (available from Ciba-Geigy, Ltd, Basel, Switzerland) (2006).

Seeger, S. and Muller, W., Limits of Stress and Strain: Design Criteria for Protective Layers for Geomembranes in Ponds Liner Systems, pp. 153–157 (1996).

Shercliff, D. A., 'Designing with the Cylinder Test' (available from GEOfabrics Ltd, Wellington Mills, Huddersfield Road, Liversedge, West Yorkshire WF15 7XA) [http://www.geofabrics.com/docs/] (2008).

Stark, T. D., Boerman, T. R. and Connor, C. J., 'Puncture resistance of PVC geomembranes using the truncated cone test', *Geosynthetics International*, 15(6), 480 (2008).

TenCate Geosynthetics North America, 'UV durability of TenCate Geosynthetics', Technical Note, Pendergrace, CA, USA.

Wagner, N. and Ramsey, B., 'QUV Accelerated Weathering Study: Analysis of Polyethylene Film and Sheet Samples', Technical Document by GSE Lining Technology, Inc., Houston, TX, USA (2003).

Wesseloo, J., Visserb, A. T. and Rustb, E., A mathematical model for the strain-rate dependent stress–strain response of HDPE geomembranes, *Geotextiles and Geomembranes*, **22**(4) 273–295 (2004).

14

Chemical Resistance
of Geomembranes

Geomembranes can be exposed to a wide variety of chemicals in service and their behaviour under these conditions can be predicted in part by chemical resistance testing. A number of test methods have been developed to determine the performance of geomembrane materials under specific chemical environments. Chemical immersion tests have been developed to validate long-term geomembrane material performance in a specific chemical environment.

One mechanism by which chemicals can affect polymeric geomembranes is through a softening of the material that leads to a reduction in its modulus (i.e. stiffness). This may occur with materials that are solvents or partial solvents for the geomembrane material. A related mechanism is plasticization whereby the tensile strength decreases.

Figure 14.1 shows an example of a polymeric geomembrane that has been chemically attacked leading to loss of ductility and thinning.

Simple immersion testing, where geomembrane material samples in the absence of stress are fully immersed in the chemical of interest at different temperatures, is generally used for screening purposes to ascertain the compatibility of a particular geomembrane material with a particular chemical.

Such immersion testing can provide an indication as to the effects the chemical may have on key properties. Changes in weight (weight gain or loss), volume changes (i.e. swelling) and changes in mechanical properties (e.g. tensile strength) are normally measured as a function of exposure time at various temperatures (e.g. 20 and 60 °C). The data are typically expressed in chemical resistance tables, which list the influence a particular chemical has on a general class of geomembrane materials.

Accelerated chemical immersion testing involves assessment of weight change, dimensional tests, mechanical tests (i.e. retained tensile properties and retained SCR) and chemical tests (retained antioxidant/stabilizer levels).

A Guide to Polymeric Geomembranes: A Practical Approach   J. Scheirs
© 2009 John Wiley & Sons, Ltd

Figure 14.1 Photograph of a chemically attacked geomembrane. Note the loss of ductility and thinning. The polymer became sticky and perished. Reproduced by permission of ExcelPlas Geomembrane Testing)

14.1 CHEMICAL RESISTANCE TABLES

Chemical resistance tables are available for a wide range of polymeric geomembranes. These tables show the resistance of the polymer to various chemical reagents. However an important limitation is that these values are obtained from tests made under static conditions using non-stressed specimens (i.e. in contrast to most field applications).

Unstressed immersion test chemical resistance information is applicable only when the geomembrane will not be subject to any mechanical or thermal stress. Chemicals that to do not normally affect the properties of an unstressed thermoplastic may cause completely different behaviour (such as stress cracking) when the geomembrane is under thermal or mechanical stress (such as constant internal pressure or frequent thermal/mechanical cyclic stress).

For instance, polyethylene resins are highly resistant to most chemicals and solvents in the absence of stresses. However, many polyethylenes may crack when exposed to the same chemical 'environment' under tensile or polyaxial/multiaxial stress. This effect is called environmental stress cracking (ESC) and is discussed in detail in Chapter 12.

When the geomembrane is likely to be subjected to a continuous applied mechanical or thermal stress or to combinations of chemicals, testing that replicates the expected field conditions as closely as possible should be performed on geomembrane liner samples to properly evaluate the material for use in that application.

The Rinker Company produces are producers of HDPE liners for pipes utilized in the chemical industry. The 'Rinker Chemical Resistance Chart' lists the compatibility of HDPE with various chemicals at both 24 °C and 60 °C. This chart uses the following legends:

- S – Satisfactory. The sample shall not exhibit swelling of more than 3%, a weight loss of more than 0.5% and there shall be no substantial change in the elongation at break.
- C – Conditional. Swelling of 3–8% or a weight loss of 0.5–5% and/or elongation at break reduced by up to 50%.
- U – Unsatisfactory. The sample swells more than 7%, suffers a weight loss of more than 5% and/or loses more than 50% elongation at break.

Some chemicals, such as the hydrocarbon types, have the effect of being absorbed to some extent into the geomembrane producing a softening effect.

There are three main limitations of chemical resistance charts:

- The samples are not under stress (whereas stress can greatly accelerate chemical attack).
- The chemical concentration or temperature are generally standardized and likely to be different to those experienced by the polymer in service.
- There is no indication of how the chemical exposure affects long-term properties such as stress crack resistance (SCR) and oxidative induction time (OIT).

Polymers in a chemical immersion test when tested in an unstressed condition generally perform quite well and thus may be listed in chemical resistance charts as 'resistant'. Significantly however, the same material when placed under stress may fail by cracking in a short period of time.

14.2 FACTORS AFFECTING CHEMICAL RESISTANCE

Factors such as percentage of chemical in the liquid (i.e. the concentration), combinations of chemicals (mixtures), temperature, pressure, external system applied stress, type of resin used and product quality can affect the chemical resistance of the geomembrane.

Although extensive testing has been conducted over the years to better define chemical resistance and use parameters, the possible combinations of chemical mixtures and their resultant reaction when mixed is infinite. Therefore chemical resistance data should not be regarded as definitive. Certainly, the absence of a specific chemical or solution from the chemical resistance chart does not imply that a geomembrane is suitable for use with that substance.

Many factors can affect the chemical resistance of geomembranes. These include, but are not limited to, exposure time, extremes of temperature and stress, frequency of temperature and/or stress cycling and attrition/wear due to abrasive particles. It is important to recognize that certain combinations of chemical and mechanical load can induce stress cracking in many otherwise chemically resistant polymers.

In-service characteristics are dependent upon the particular application of the geomembrane and will differ from those exposed in either laboratory testing or apparently similar field service.

Since liquors and leachates are often mixtures of various individual chemicals, it is strongly recommended that trial installations be evaluated under actual service conditions. Immersion testing in individual chemicals at a specific operating temperature does not predict the performance of geomembranes should an exothermic reaction take place when mixtures of chemicals are involved.

Mining liquors, leachates and process solutions are rarely solutions of a single chemical. Furthermore, municipal solid waste leachate is a complex cocktail of many chemicals having different concentrations.

When test data on chemical resistance of a geomembrane to a particular chemical or mixture of chemicals under specific operating conditions are not available, it is recommended that chemical immersion testing be carried out under specific conditions closely resembling those that the geomembrane will encounter in service. Chemical resistance should therefore be evaluated at the specific concentration and temperature of service to ensure compatibility.

14.2.1 POLARITY

Polarity plays a major role in determining the chemical resistance of different geomembranes to various chemical compounds. Water is a common example of a polar material. Grease is a common example of a non-polar material. These two materials do not mix. Polarity refers to the negativity and positivity of chemical compounds. Chemical compounds with similar polarity will have an affinity for each other. 'Like dissolves like' is an expression used by chemists to remember how chemical substances interact. It refers to 'polar' and 'non-polar' solvents and solutes. Since water is polar and grease is non-polar, water will not dissolve grease (hence washing clothes with grease stains in plain water achieves no cleaning). Analogously it should be noted that nitrile rubber is polar and hence not affected by oil, diesel or grease which are all non-polar. Biodiesel, however, is polar (just like nitrile rubber) and so nitrile rubber and biodiesel have a chemical affinity for each other and therefore nitrile rubber is not compatible with biodiesel from a chemically resistant liner point of view.

14.2.2 PERMEABILITY

The chemical resistance of geomembranes is related to a large degree by the permeability of the specific chemical into the geomembrane. While polymeric geomembranes are designed to be homogeneous, non-porous, impervious materials they contain 'free volume' (molecular spaces) through which small molecules can diffuse. The molecular transport of a chemical into a polymeric geomembrane is defined in terms of permeability, diffusivity and solubility coefficients where permeability is the product of diffusion and solubility.

Diffusion of chemicals such as solvents through polymeric geomembranes involves three steps:

(i)   Solubility or sorption of the liquid at the upstream surface of the geomembrane. Here, transport depends upon the solubility of the permeating species in the geomembrane and the relative chemical potential of the liquid on both sides of the interface.
(ii) Diffusion of the dissolved species through the geomembrane. Here, diffusion through a geomembrane involves factors such as size and shape of the permeating molecules in addition to structural characteristics of the polymer and the geomembrane–solvent interactions.
(iii) Evaporation or desorption of the liquid at the downstream surface of the geomembrane (Ortegoa *et al.*, 1995).

Measuring the Rate of Diffusion in Geomembranes

The rate of diffusion of chemicals into a polymeric geomembrane can be measured by exposing the polymer to solutions of the chemical under the conditions of the process for 1000 h and then microtoming the polymer so that 200 μm slices could be analysed for ions of interest (e.g. sulfate ion in the case of sulfuric acid). The results can then be fitted to the Fickian distribution curve so that the parameters C and D (i.e. the surface concentration and the diffusion coefficient, respectively) could be determined to enable predictions to be made of the time that it would take for the process solution to permeate the geomembrane. It is important to note that Fickian transport (as per Fick's Law) is driven by a concentration gradient not head pressures.

14.2.3 EFFECT OF TEMPERATURE

The equilibrium permeation rates of liquids into polymers follow the Arrhenius relationship with temperature (Aminabhavi and Naik, 1998).

The rule of thumb is that gas permeation increases by 30–50% for every increase of 5°C, whereas, water permeation increases by 10–100% for every 5°C rise in temperature.

Note that extrapolation and prediction for thicker or thinner geomembranes than those measured can be erroneous if the permeation versus thickness slope deviates from unity.

14.2.4 MOLECULAR STRUCTURE

Figure 14.2 shows that the molecular structure of the permeating gas or liquid is also an important factor as permeation depends on: (i) molecular size of the liquid, i.e. small molecules such as hexane permeate more rapidly than large molecules such as decane, (ii) molecular shape, i.e. streamlined-shaped molecules such as *p*-xylene permeate more rapidly than bulky shaped molecules such as *o*-xylene and (iii) polarity of the liquid, i.e. non-polar molecules like toluene permeate more rapidly than the polar aniline in nonpolar polymers (Ortegoa *et al.*, 1995). The latter effect is reversed in the case of polar polymers.

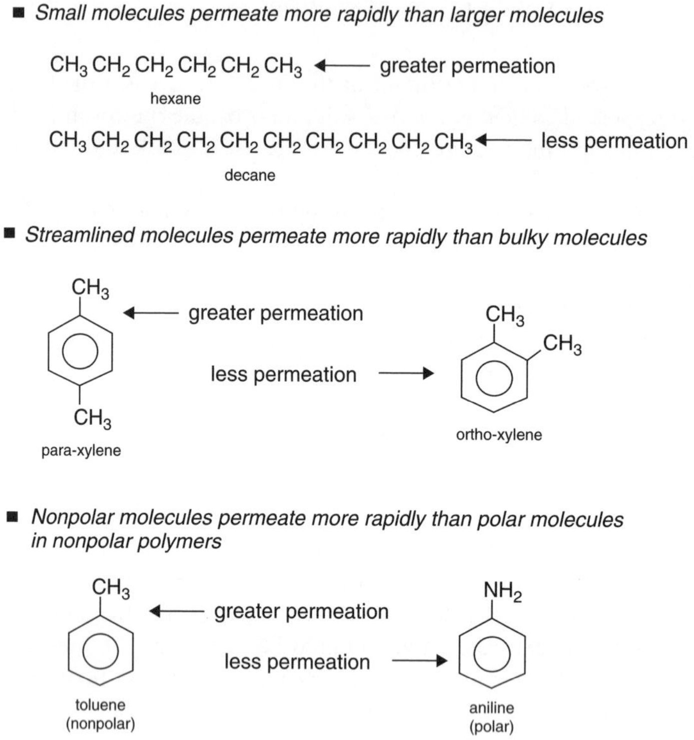
Small molecules permeate more rapidly than larger molecules
CH3 CH2 CH2 CH2 CH2 CH3 ← greater permeation
hexane
CH3 CH2 CH2 CH2 CH2 CH2 CH2 CH2 CH2 CH3 ← less permeation
decane
Streamlined molecules permeate more rapidly than bulky molecules
CH3
← greater permeation
less permeation →
CH3
para-xylene
CH3
CH3
ortho-xylene
Nonpolar molecules permeate more rapidly than polar molecules in nonpolar polymers
CH3
← greater permeation
less permeation →
toluene
(nonpolar)
NH2
aniline
(polar)

Figure 14.2 Effect of chemical structure on the permeation and diffusion of chemicals into polyethylene geomembranes

The permeability of polymers is determined by many structural and morphological properties of both the polymer (such as crystallinity, i.e. density) and the diffusing chemical (such as polarity). Additives, such as plasticizers and oil extenders, usually increase the permeation rates of polymers considerably due to their plasticizing action.

14.2.5 SOLVENT TRANSMISSION

High-density polyethylene (HDPE) is far less permeable to solvents such as xylene and chloroform than LDPE (owing to the higher crystallinity of HDPE) (see Figure 14.3) and in turn, thicker HDPE is less permeable than thinner HDPE geomembrane sheets.

HDPE is non-polar and hence is an excellent barrier for polar solvents such as acetone. As shown in Figure 14.4, the acetone transmission rate for HDPE is almost 400 times lower than that for CSPE-R. This is because the chlorine and sulphur groups increase the polarity of PE and render it more susceptible to polar solvents. In this regard PVC, which may be viewed as a chlorinated analogue of PE, is very susceptible to acetone attack and solvation. The inclusion of chlorine atoms into the basic polyethylene backbone increases its polarity and increases its susceptibility to permeation by polar solvents.

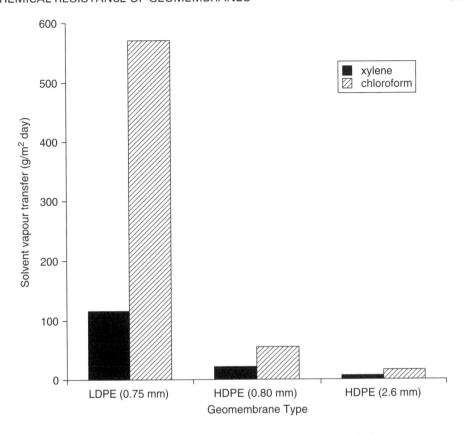

Figure 14.3   Solvent vapour transmission rates for LDPE and HDPE liners

Note the permeability of 1,3,5-trimethylbenzene in a CSPE geomembrane is some three times lower than benzene since 1,3,5-trimethylbenzene is a more bulky molecule compared with benzene. The higher polarity of chloroform means it has a permeability 65% greater than benzene in CSPE geomembranes which have some polar nature due to the presence of the chlorination. The small linear and polar molecular nature of dichloromethane means its permeability in an CSPE geomembrane is almost four times higher than that of benzene.

It is important to note that no material is 'absolutely' impermeable. Even if the geomembrane has no pinhole leaks, apparent 'leakage' may also occur through diffusion of vapour (solvent and water) through the amorphous regions of the HDPE geomembrane and can recondense on the opposite side.

The transmission rate for various solvents through a 0.8 mm HDPE geomembrane is shown in Figure 14.5.

The mass fraction of different solvents absorbed by HDPE is shown in Figure 14.6. It is clear that the order of absorption and swelling of HDPE from greatest to smallest follows

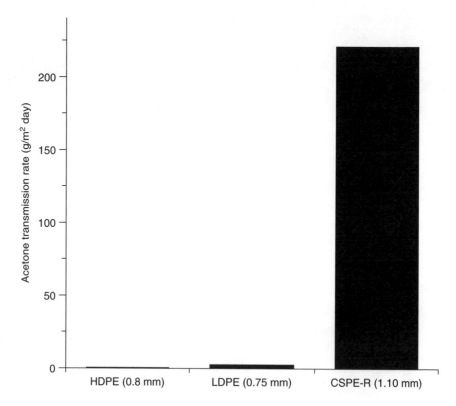

Figure 14.4   Acetone transmission rates for different geomembranes

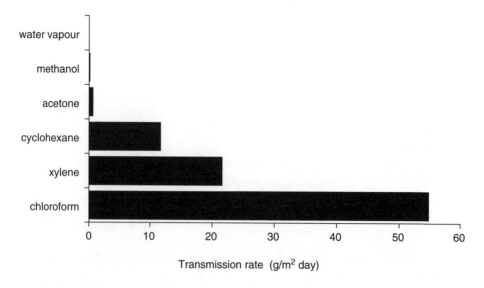

Figure 14.5   Solvent vapour permeation of HDPE geomembranes (0.8 mm)

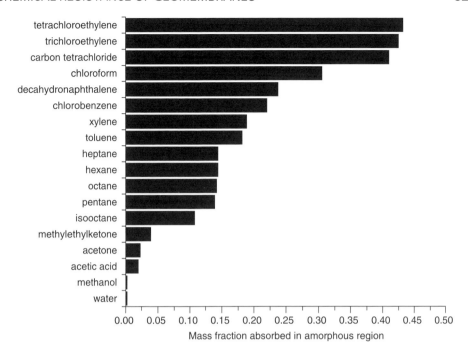

Figure 14.6   Mass fraction of liquid absorbed in the amorphous region of HDPE (crystallinity, 54–56%). Reprinted with permission from Bautechnik, Stofftransport in Deponieabdichtungssystemen. Teil 1: Diffusions- und Verteilungskoeffizienten von Schadstoffen bei der Permeation in PEHD-Dichtungsbahnen, by Müller, W.; Jakob, I.; Tatzky-Gerth, R. and August, H., 74, 3, 176–190 Copyright (1997) Wiley-VCH

the trend: chlorohydrocarbons > aromatics > alkanes > ketones > carboxylic acids > alcohols > water.

14.2.6   DESORPTION OF SOLVENTS

While non-polar liquids and solvents can be absorbed by HDPE causing swelling and softening they do not cause progressive and permanent degradation. The swelling only occurs up to a certain value (i.e. % weight gain) and when the solvent vapours volatilize out of the geomembrane the original mechanical properties are recovered. There are arguably some beneficial effects of softening/swelling of thick HDPE liners such as decreasing stresses allowing the liner to better conform to subgrade profiles and differential settlement, hence reducing the possibility of stress cracking.

There can, however, also be undesirable effects such as swell-induced waves exerting peel forces on welds and moreover, organic liquids can extract stabilizers and low molecular weight components from the polymer which can jeopardize the geomembrane's long-term stability. HDPE can undergo a significant reduction in SCR after swelling in certain liquids (e.g. creosote) and following its drying out.

In the case of PVC geomembranes, certain organic liquids and solvents can extract the plasticizer and when the PVC 'dries out' it can crack.

14.3 EFFECTS OF CHEMICALS ON GEOMEMBRANES

Mass gain and swelling of the immersed geomembrane samples are the main primary effects of chemical interaction with the geomembrane material. Mass gains (or losses) of 10% or greater can indicate that the lining will not be chemically resistant to the chemical or mixture.

One of the main criteria against which to assess the chemical resistance of polymeric geomembranes is elongation at break, which indicates the deformation behaviour. A decrease of less than 20% in elongation at break (relative decrease, not absolute decrease) indicates that no significant material change has taken place. Reduced tensile strength is another indictor of chemical interaction. Chemicals can be absorbed into a polymeric geomembrane leading to plasticization whereby the tensile strength decreases.

Note that chemical resistance tables do not generally consider secondary effects such as the changes in SCR and OIT due to immersion. These are both important parameters that should be tested in addition to simple mechanical properties. The decrease of oxidative stability and stress cracking are very relevant in assessing the chemical resistance of the material. OIT testing is of particular significance in identifying the susceptibility of the lining material to oxidation under process or service conditions.

Note: S-OIT measurements indicate the relative levels of phenolic antioxidants in the polyolefin geomembranes. Hindered phenolic antioxidants are commonly used in combination with hindered amine stabilizers and both these additives impart thermal stabilization to the material. Hindered amine stabilizers are rapidly deactivated and decomposed under acidic conditions (particularly in the presence of sulfuric acid). Under these conditions HDPE and PP liners therefore rely solely on the hindered phenolic antioxidants for oxidative stability. Therefore the S-OIT measurements are a valid way of assessing the residual oxidative stability of the HDPE samples.

14.4 CHEMICAL RESISTANCE TESTING

If geomembranes are to be in long-term contact with liquids containing strong acids, alkalis, transition metal solutions (e.g. compounds of cobalt, copper, manganese, etc.), salts or commonly used industrial chemicals, then chemical resistance charts could be used as a guide.

When the liquids are combinations of various chemicals, such as those that occur in industrial process effluents, the most aggressive of the individual chemicals to the polymeric geomembrane can be selected for the incubation (i.e. immersion) process. However, a synthesized cocktail of these chemicals duplicating the expected application situation will be preferred as the immersion medium for the candidate liner/s.

Whatever geomembrane material is proposed for a particular application, it must be thoroughly tested in the actual process solutions at the operating temperatures, with some samples additionally being tested under applied stress. It is also very important to consider the possible influence on material performance of those ambient conditions during the transition between installation and full service, such as when the geomembrane might experience extreme high and low temperatures.

If the geomembrane will be exposed to liquids or solutions that are highly heterogeneous or the full composition is unknown/not identifiable (e.g. sewage from an anaerobic treatment plant), then a precautionary approach should be adopted and the test coupons incubated in the real liquids (e.g. with the actual leachate). This allows synergistic effects to be captured in the chemical resistance testing. If these process solutions or liquors are not yet available because the project is still at the design stage, then a conservative approach is recommended and a geomembrane with excellent broad chemical resistance should be selected.

ASTM Method D-543 covers chemical degradation under the title 'Resistance of Plastics to Chemical Reagents'. The test method is based on measuring changes in weight, dimensions, surface appearance and strength after various exposure times to reagents at elevated temperatures. This standard, while not specific to geomembranes, can nevertheless be used as a basic chemical resistance screening test.

Chemical resistance testing of geomembranes using site-specific chemicals can also be performed using the following testing methods:

- ASTM D-5747: 'Standard Practice for Tests to Evaluate the Chemical Resistance of Geomembranes to Liquids'.
- ASTM D-5322: 'Standard Practice for Laboratory Immersion Procedures for Evaluating the Chemical Resistance of Geosynthetics to Liquids' which covers laboratory immersion procedures for the testing of geosynthetics for chemical resistance to liquid wastes, prepared chemical solutions and leachates derived from solid wastes.

There are two main chemical resistance tests for geomembranes: ASTM D-5747 is a test for resistance of geomembranes to chemical attack in landfills, while ASTM D-5322 is a test to determine the effects of chemicals such as the leaching of additives which can leave the unstabilized polymer vulnerable to oxidative attack.

Note that time, temperature, stress and reagent may all be factors affecting the chemical resistance of a material.

The ASTM D-5322 method requires an exposure tank, for containment of the solution and test material, made of stainless steel or glass; however, glass should not be used with strongly basic solutions. The tank or vessel needs to be sealed to prevent the loss of any volatile components.

It is generally necessary to maintain ambient atmospheric pressure in the tank using a reflux condenser open to the air, a pressure relief valve or any method allowing the movement of gas to relieve pressure while minimizing changes in the chemical composition of the test solution. This is important to prevent pressure buildup in an exposure tank from the generation of gases by chemical reactions or biological activity.

The tank requires temperature control equipment to maintain the immersion solution at the specified temperature such as by wrapping a heating coil around the tank or placing the tank/vessel on a hotplate. Placing a heating coil directly in the exposure solution is not recommended since corrosion may affect the coil and chemical reactions that may not otherwise occur may take place on the hot coil.

Stirring the immersion liquid is required unless evidence can be offered that stirring is not necessary to maintain a homogeneous solution. By keeping the solution circulating slowly, fresh solution is constantly passing over the surface of the geomembrane, thereby preventing stagnant boundary layers of reacted or used liquid from existing around the geomembrane and giving rise to stratification or settling of chemical constituents. A stirrer is thus required for mixing the solution. Magnetically moved stirring bars or mechanical stirrers entering the tank through the lid can be used to agitate the contents.

Examples of multicomponent fluids include:

- Aqueous solutions containing inorganic material such as brines, incinerator ash leachates and leach pad leachates.
- Aqueous solutions containing organic materials such as spills, hydrocarbon fuels, organic species from tanks and secondary containment.

The variables listed below need to be considered when conducting any chemical resistance testing:

- thickness of the geomembrane;
- service temperature;
- temperature gradient across the geomembrane in service;
- chemical potential across the geomembrane that includes pressure and concentration gradient;
- composition of the fluid and the mobile constituents, plus the solubility of various components of an organic liquid in the particular geomembrane that increase concentration of individual components on the upstream side of the geomembrane and can cause swelling of the geomembrane, resulting in increased permeability;
- ion concentration of the liquid;
- ability of the species to move away from the surface on the downstream side.

The immersion solution can comprise any of the following liquids:

- liquid wastes;
- leachates collected from existing installations;
- leachates made from solid wastes;
- synthetic leachates made from laboratory chemicals;
- standard chemical solutions (ASTM Test Method D-543);
- reference fuels and oils.

If it is suspected that trace amounts of chemicals may be depleted from the liquid, or that the chemical nature of the liquid changes with time, it is recommended that the immersion liquid be replaced with fresh solution after each test period in order to maintain the solution chemistry. If the liquid is to be replaced periodically, it is important that the fresh liquid be obtained from the same source as the original liquid and that the fresh liquid be as similar to the original liquid as possible.

It is necessary to position and secure in place the geomembrane materials to be immersed in the exposure container in such a manner that contact with the container and other sheets of material is limited as much as possible. This is especially important if flat sheets of geomembrane are used because unless these are held apart they will tend to adhere to one another and inhibit the flowing of the liquid over the individual surfaces.

14.4.1 INCUBATION TESTING IN THE PROCESS SOLUTIONS

Note that chemical resistance tables should not be considered as universally applicable and therefore project-specific testing should be performed for the actual geomembrane material to be installed or as a pre-qualification process for potential suppliers.

Furthermore, since the process liquid involves a mixture of various individual chemicals, it is strongly recommended that trial materials be evaluated under actual service conditions.

A related approach involves accelerated testing which simulates end-use conditions as closely as possible but with elevated temperatures used to accelerate the failures so that testing can be conducted in a reasonable time period. The test solution is also stirred so that stagnant boundary layers (depleted of oxygen) do not develop on the geomembrane surface. Actual geomembrane samples are tested due to the many variables that can influence chemical resistance such as sample geometry, presence of metal components and manufacturing or surface defects.

In practice it is necessary to identify material, formulation and manufacturing details of the geomembrane in order to provide a scientifically based means of validating material and product performance with respect to a particular chemical environment.

Unstressed immersion test chemical resistance information is applicable only when the geomembrane will not be subject to mechanical or thermal stresses that are constant or cycle frequently.

Plastics in an immersion test when tested in an unstressed condition generally perform quite well and thus are listed in chemical resistance charts as 'resistant'. However, the same material when placed under stress may fail by cracking in a short period of time.

When the geomembrane is likely to be subjected to a continuous applied mechanical or thermal stress or to combinations of chemicals, testing that duplicates the expected field conditions as closely as possible should be performed on geomembrane liner samples to properly represent the material for use in that application.

14.4.2 SERVICE SPECIFIC TESTING

The general approach with accelerated testing in the laboratory is to simulate end-use conditions as closely as possible. Elevated temperatures are used to accelerate the failures so that evaluations can be conducted in a reasonable timeframe. The test solution should be stirred so that stagnant boundary layers do not form on the geomembrane surface. Commercial geomembrane samples (representative of material delivered to site) should be tested due to the many variables that can influence chemical resistance such as the presence of metal components and manufacturing or surface defects.

Tests should also be conducted on stressed coupons because in-service geomembranes are often under stress and the failure can be driven by stress/strain dependant slow crack growth. Various extrapolation methods can then be employed to extrapolate results to end-use temperatures. Typically, testing is performed at multiple temperatures (usually three) and a multiple linear regression is conducted based on the "three coefficient rate process extrapolation" method.

The lifetime of geomembranes can be predicted using the Arrhenius relationship. The usual rule of thumb is that every 10 degrees above a certain temperature doubles the failure rate (or decreases lifetime). So if a geomembrane survives 15 years at 25 °C it will have a much reduced lifetime at 65 °C. The temperature increase is 40 °C or four decades of temperature. So the reaction rate will be 2^4 or $2 \times 2 \times 2 \times 2 = 16$ times faster and therefore the estimated lifetime will be 15/16 years or around 11 months.

In practice it is necessary to fully characterize the geomembrane material with respect to the resin type, formulation and manufacturing details in order to form a scientifically based opinion on the geomembrane performance with respect to a particular chemical environment.

Chemical resistance testing is also useful for simulating field failures and, thereby, investigating product liability claims or directing engineering resources to address material selection or formulation deficiencies.

14.4.3 CHEMICAL RESISTANCE CRITERIA

Table 14.1 shows that if the mass gain or volume change upon exposure to a chemical is greater than 10% then the geomembrane is classed as *not resistant* to the chemical/s or liquids involved. This threshold however does not apply for HDPE geomembranes. This is because HDPE geomembranes have comparatively high crystallinity and they cannot tolerate the uptake or sorption of chemicals as well as the lower crystallinity geomembranes such as PVC, fPP, EIA and so on.

Table 14.2 lists the allowable limits for changes in various properties of HDPE and LLDPE geomembranes after chemical resistance testing. Note for HDPE geomembranes that if there is swelling and mass gain greater than 1 and 2%, respectively, then they are classed as 'non-resistant'.

The German Federal Institute for Materials Research and Testing (BAM) has set the following criteria for chemical resistance of HDPE geomembranes after immersion testing at 23 °C for 90 days (or to attainment of steady-state weight), whichever occurs first on the basis of DIN ISO 175 (BAM, 1999):

- Change in weight after redrying ≤10%;
- Change in yield strength and elongation at yield (tensile tests to be conducted on redried specimens) ≤10%.

Table 14.1 Allowable limits for changes in various properties of geomembranes after chemical resistance testing (not applicable for HDPE geomembranes) (Little, 1985)

Property	Classed as 'Resistant' ✔	Classed as 'Not resistant' ✘
Change in weight (%)	<10	>10
Change in volume (%)	<10	>10
Change in tensile strength (%)	<20	>20
Change in elongation at break (%)	<30	>30
Change in 100% modulus (%)	<30	>30
Change in hardness (points)	<10	>10

Table 14.2 Allowable limits for changes in various properties of semicrystalline geomembranes after chemical resistance testing (applicable for HDPE Geomembranes) (Koerner, 2005)

Property	Classed as 'Resistant' ✔	Classed as 'Not resistant' ✘
Change in weight (%)	<2	≥ 2
Change in volume (%)	<1	≥ 1
Change in tensile strength at yield (%)	<20	≥ 20
Change in elongation at yield (%)	<30	≥ 30
Change in modulus (%)	<30	≥ 30
Change in tear strength (%)	<20	≥ 20
Change in puncture strength (%)	<30	≥ 30

14.4.4 PERMEABILITY TESTING

Because of the great number of variables, it is important to perform permeability tests of a geomembrane under conditions that simulate as closely as possible the actual environmental conditions in which the geomembrane will be in service [ASTM D-5886].

Geomembranes are nonporous homogeneous materials that are permeable in varying degrees to gases, vapours and liquids on a molecular scale in a three-step process:

1. By dissolution in or absorption by the geomembrane on the upstream side.
2. Diffusion through the geomembrane.
3. Desorption on the downstream side of the barrier.

In the transmission of a permeant through a geomembrane, Step 1 depends upon the solubility of the permeating species in the geomembrane and the relative chemical potential of the permeant on both sides of the interface. In Step 2, the diffusion through the geomembrane involves a variety of factors including size and shape of the molecules of the permeating species, and the molecular characteristics and structure of the polymeric geomembrane (ASTM D-5886).

Many of the applications of geomembranes are for barriers to the permeation of single-component permeants, that is, a single gas, vapour or liquid. With respect to water, such applications include reservoir liners, moisture vapour transmission barriers, floating covers for reservoirs, canal liners and tunnel liners. Other applications involving single-component fluids also include liners for secondary containment. Further applications might be methane barriers in tunnels, MSW landfills and buildings that are near methane and hydrocarbon sources (ASTM D-5886).

For such applications as linings for waste disposal facilities and methane barriers, the permeability to gases is important in geomembrane selection. The permeabilities of geomembranes can be assessed by measurement of the volume of the gas passing through the geomembrane under specific conditions or by measurement of the increase in pressure on the evacuated downstream side (as described in ASTM D-1434).

For applications such as reservoir covers and moisture barriers, permeability to moisture vapour can be measured by a variety of methods that reflect the service conditions. Determinations can be made by measuring the change in weight of a small cup that contains either a small amount of distilled water or a desiccant and is sealed at the mouth with a specimen of the geomembrane. In the 'Water Vapor Transmission Test' (ASTM E-96) a small cup is filled with distilled water leaving a small gap (6–18 mm) of air space between the specimen and the water. The cup is then sealed to prevent vapour loss except through the test geomembrane sample. An initial weight is taken of the apparatus and then periodically weighed over time until the results become linear. Care must be used to assure that all weight loss is due to water vapour transmission through the specimen and not due to leaks at the edges of the cup. A 100 mm × 100 mm specimen is often used since it needs to fit exactly over the liquid container. A graph of weight versus time or percentage transmission versus time is plotted using the recorded results.

Permeability of Geomembranes to Organics

The moisture vapour transmission test (ASTM E-96) can be also used to assess the permeability of various membranes to solvent vapours. In this case, the cup that is used in the moisture vapour transmission test is exposed with the solvent vapour contacting the membrane. The vapour concentration inside the cup is determined by the vapour pressure at the test temperature and the concentration outside the cup is essentially zero. Therefore, the vapour pressure gradient is the vapour pressure of the solvent at the temperature of the test if the vapour concentration is held constant (ASTM D-5886). The test method ASTM D-814 can also be used. For some materials such as fPP the membrane can be sucked right into the cup as a vacuum develops after the solvent permeates through the membrane and escapes. The fuel used for this permeation test is 'ASTM Fuel C' which is 50% isooctane and 50% toluene. This mixture is a good surrogate fluid for modelling gasoline compatibility and permeability. Regular gasoline cannot be used as the chemistry of gasoline (or any distillate fuel) as it is highly variable (private communication, Andrew Mills, Layfield, Canada).

Petroleum distillates are mixtures of as many as 400 individual chemicals usually with only an approximate formulation. Gasoline, diesel fuel and most other distillates can have significant variations in chemical composition depending on the initial crude oil source, the refining process and any blending or mixing done by the refiner. The BTEX (benzene, toluene, ethylbenzene and xylene) content in gasolines (i.e. petroleum transport fuels) can vary from under 5% to over 40%.

The ASTM Test Method F-739 is normally used to measure the resistance of protective clothing materials to the permeation of liquids or gases and is another test method that can be used for measuring permeability to organic vapours. In this method, an analytical detection system is used to measure the time to breakthrough of the permeant and the equilibrium rate of permeation (ASTM D-5886).

Direct Contact with Organic Liquids

Swelling can affect permeability and is an important factor for applications where the geomembrane will be in continuous contact with organic liquids as in the case of tank liners and liners for secondary containment. Thus it is necessary to determine the compatibility of the particular membrane with the specific organic chemical that is to be contained. The ASTM D-471 method can be used to determine whether the geomembrane under test will swell during the test and change the permeability of the test specimen during the test.

🖈 *Note: Liners for hydrocarbon storage such as tank farms are generally based on EIA-R or PU-R.*

14.5 GEOMEMBRANE CHEMICAL RESISTANCE BY CHEMICAL CLASS

14.5.1 ACIDS

HDPE and fPP are resistant to sulfuric acid (H_2SO_4) up to concentrations of 70%. Concentrations of acid greater than 80% even at room temperature can cause oxidation. While HDPE and fPP are resistant to concentrated hydrochloric acid (HCl) and hydrofluoric acid (HF), there is some diffusion of HCl (at concentrations >20%) and HF (concentrations >40%) which does not damage the geomembrane but causes secondary damage (corrosion) of the base substrates (e.g. concrete or metal tanks).

The compatibility of HDPE, LLDPE and PVC geomembranes with 98% concentrated sulfuric acid has been evaluated over a 120-day exposure programme in the laboratory. Samples of each geomembrane were exposed to 98% H_2SO_4 solutions at 50 °C and then tested each 30 days for retention of physical properties. Table 14.3 shows that both HDPE and LLDPE geomembranes and in particular their additive packages are adversely affected by the acid.

Both types of polyethylene geomembranes performed very well given the aggressive environment with tensile strength and elongation properties after 120 days generally within 10% of the original conditions (see Table 14.3). PVC, however, exhibits drastic loss of flexibility (i.e., a negative change percentage), even within the first month. The increase in tensile strength for PVC is accompanied by a reduction in elongation, which suggests that the material has lost plasticizer and becomes brittle in just 30 days (see Table 14.4). It was observed that the immersion solution turned very dark in the first 24 h, suggesting a very rapid leaching of plasticizers. On the basis of these results PVC is not suitable for use in concentrated acid pre-curing operations, even for relatively short exposure periods. While the polyethylene geomembranes retained the majority of their mechanical properties, they exhibited losses in oxidation induction time (OIT) of 73% for LLDPE and 64% for HDPE (see Table 14.3). The OIT is a key indicator for longevity and loss of OIT is a precursor to failure by ageing (Thiel and Smith, 2004). Depending on cure periods, cycle times and design life, cumulative exposures of up to 10 months are possible but 4–6 months is probably more typical. Since an OIT reduction of 64–73% was measured in 4 months the longevity of the geomembrane must be a concern for these pre-curing applications. This problem could be solved by using co-extruded HDPE with a top layer made with an additive package specifically formulated for this environment (e.g. acid-tolerant non-basic HALS stabilizers).

14.5.2 ALKALIS

HDPE and fPP do not react with strong caustic solutions even at higher temperatures and therefore may be used for such containment. Strong alkalis, can however, destroy phenolic antioxidants present in these polymers which will affect the materials long-term durability.

Table 14.3 Effect of concentrated sulfuric acid on the mechanical properties of selected geomembranes after 120 days (data are expressed in terms of the percentage change from original) (Thiel and Smith, 2004)[a]

Property	HDPE 1.5 mm	LLDPE 1.5 mm	PVC 0.75 mm
Tensile strength at yield (ASTM D-638/D-882) (%)	+2 (MD)	0 (MD)	+173 (MD)
	−4 (TD)	0 (TD)	+188 (TD)
Tensile strength at break (ASTM D-638/D-882) (%)	−4 (MD)	−7 (MD)	+54 (MD)
	−4 (TD)	−11 (TD)	+54 (TD)
Elongation at yield (ASTM D-638/D-882) (%)	−5 (MD)	−10 (MD)	NA
	−5 (TD)	+9 (TD)	
Elongation at break (ASTM D-638/D-882) (%)	+5 (MD)	−7 (MD)	−66 (MD)
	0 (TD)	−12 (TD)	−76 (TD)
Puncture resistance (ASTM D-4833) (%)	−3	+1	130
Tear resistance (ASTM D-1004) (%)	−3 (MD)	−5 (MD)	+107 (MD)
	+2 (TD)	−5 (TD)	+112 (TD)
Hardness (ASTM D-2240) (%)	0	+5	+31 (indicates loss of plasticizer)
OIT (ASTM D-3895) (%)	−64 (indicates loss of antioxidant)	−73 (indicates loss of antioxidant)	NA

[a]Notes: where two values are shown, they are for the machine direction/traverse direction (MD/TD), respectively; for PVC, tensile strength at 'yield' was taken at 100% elongation as the yield point is indeterminate; OIT is only applicable for polyethylene.

HDPE is listed as suitable for contact with 50% (w/w) sodium hydroxide solution in all published chemical resistance charts that are available. The Chemical Resistance Charts available from BP Chemicals, Rinker and Rotonics all list HDPE as being satisfactory for use with concentrated sodium hydroxide, both at 23 and 60 °C.

The German State Material Testing Institute in Darnstadt reports that a polymeric 2 mm thick Carbofol HDPE 406 geomembrane (a NAUE product) undergoes no significant property deterioration after immersion in 60% (w/w) caustic soda solution for up to 3 months (Darmstadt, 1998).

The test laboratory found that after 28 days immersion in 60% caustic soda solution the HDPE 'Carbofol HDPE 406' (thickness 2 mm) showed no weight change and only a −0.5% relative change in tensile strength at yield and only −0.3% relative change in tensile elongation at yield. These changes are considered negligible.

The test laboratory found that after 90 days immersion in 60% caustic soda solution the HDPE 'Carbofol HDPE 406' (thickness 2 mm) showed no weight change and a +4.0% relative change in tensile strength at yield and −0.8% relative change in tensile elongation at yield. These changes are considered negligible.

Table 14.4 Effect of 98% sulfuric acid on the properties of PVC geomembranes (% change from original values) (Thiel and Smith, 2004)[a]

Immersion time (days)	Tensile strength at break (ASTM D-882) (%)	Elongation at break (ASTM D-882) (%)	Puncture resistance (ASTM D-4833) (%)	Tear resistance (ASTM D-1004) (%)	Seam shear elongation (ASTM D-6392)
30	+31 (MD) +27 (TD)	−58 (MD) −74 (TD)	+129	+119 (MD) +122 (TD)	−90
60	+62 (MD) +40 (TD)	−71 (MD) −75 (TD)	+120	+122 (MD) +110 (TD)	−94
120	+54 (MD) +54 (TD)	−66 (MD) −76 (TD)	+130	+107 (MD) +112 (TD)	−86

[a]Where two values are shown, they are for the machine direction/traverse direction (MD/TD), respectively.

Based on immersion test results HDPE geomembranes should not be chemically attacked by strong caustic solutions nor are they expected to be permeated by aqueous caustic solution. Therefore no significant property deterioration is expected when 2 mm thick 'Carbofol HDPE 406' is used in contact with 60% caustic soda solution (or lower concentrations) for long-term containment at ambient temperatures.

Additive Degradation by Strong Caustic

The effect of 50% w/w sodium hydroxide (NaOH) solution on the stabilizing additives found in four commercial geomembranes has been evaluated (TRI, 2004). Coupons of the following four geomembranes were considered:

- 1.0 mm HDPE provided by Steel Dragon Enterprises, Taiwan;
- 1.0 mm HDPE provided by GSE Lining Technology, Thailand;
- 1.0 mm linear low-density polyethylene (LLDPE), manufacturer unknown;
- 1.0 mm flexible polypropylene (fPP) made by Raven Industries, USA.

These were placed in an aqueous solution of 50% w/w NaOH and placed in a forced-air oven at 85 °C for 21 days. The solution in the jar was stirred approximately every other day. After 21 days, the coupons were removed, washed thoroughly in water and cut in half. One half was saved for analysis and the other for a second exposure. The second exposure was conducted in a forced air oven at 85 °C, according to the Geosynthetic Institute's Standard Specification (GRI GM-13). The exposure time was 90 additional days. After the exposure was completed, the coupons were saved for analysis.

The purpose of the first exposure was to determine the extractability and chemical reactivity of the additives in the geomembranes. Some additives show slight to moderate solubility in water and others are ester-based and can be removed through a hydrolysis reaction when exposed to strong alkali solutions.

The purpose of the second exposure was to determine the oxidative stability of the remaining additives that were not removed during the first exposure.

Two methods were used to determine the effects of the exposures. The first one was according to ASTM D-3895, 'Standard Test Method for Oxidative-Induction Time of Polyolefins by Differential Scanning Calorimetry'. This test is commonly called the standard OIT test and involves placing a test specimen in a pure oxygen environment at 200 °C until the specimen oxidizes. The time it takes for this to happen is the induction time and the length of this time depends on the types and amounts of stabilizers present in the plastic. Since different stabilizers produce different standard OIT values and because most additive packages contain two or three antioxidants/stabilizers the standard OIT value by itself is rather meaningless. Thus there is no clear correlation between the OIT value and the resistance of the material to oxidation during its service lifetime. However, by tracking the percentage change in the standard OIT value during an oxidative exposure, one has a better chance to predict field performance. A limitation of this test method is that all types of additives are not detected with this test. In particular, hindered amine light stabilizers (HALS) do not respond to the standard OIT test.

The second test used is sensitive to HALS. It is ASTM D-5885, 'Standard Test Method for Oxidative Induction Time of Polyolefin Geosynthetics by High-Pressure Differential Scanning Calorimetry'. This one is commonly called the HP-OIT test and is different from the standard OIT test because it is performed at higher oxygen pressure and at lower temperature (150 °C). The lower temperature of the HP-OIT test allows the stability imparted by the HALS to be detected. HALS are not active at high temperatures above 150 °C.

The OIT results after 21 days immersion in 50% sodium hydroxide at 85 °C suggest that LLDPE and fPP both lost most of their antioxidants to extraction while the GSE-HDPE lost 67%. The SDE-HDPE retained 91% of the original OIT value. The HP-OIT results showed different behaviour. In this case, the LLDPE showed the greatest reduction in HP-OIT values, but the fPP retained 70% of its HP-OIT value. The GSE-HDPE lost 37% and the SDE-HDPE did not lose anything. These results are not too surprising because HALS are not easily extracted due to their high molecular weight and they are also resistant to alkalis because they do not undergo hydrolysis.

These results suggest that the SDE-HDPE, the LLDPE and the fPP all contain HALS. The GSE-HDPE used in the study did not contain HALS. The results also showed that the SDE-HDPE and the fPP are the two products most resistant to NaOH.

The HDPE liner supplied by SDE was quite resistant to the effects of NaOH and the effects of oven ageing. The GSE-HDPE lost 64% of its protective additives during the exposure to NaOH. The remaining additives performed well during oven exposure but the overall relative performance is considered poor.

The LLDPE performed poorly during both exposures. These results suggest that either HDPE or fPP could perform well during NaOH exposure if they are properly compounded with HALS additives that resist extraction and hydrolysis.

The conclusion from this work is that phenolic antioxidants are susceptible to degradation *via* base-hydrolysis in the presence of sodium hydroxide solutions. Hindered amine light stabilizers on the other hand are themselves basic and show good resistance to strong caustic solutions. For these reasons any HDPE geomembrane material considered for contact with concentrated sodium hydroxide solutions should be formulated with a combination of hindered amine stabilizers for both long-term heat and light (UV) stability.

Table 14.5 Permissible residual chlorine levels in potable
water around the world

Location	Residual chlorine level (ppm)
EU	0.6
Australia	1
USA	4
WHO recommendation	5 (e.g. for Asia)

14.5.3 CHLORINE

Polymeric geomembranes can be adversely affected by strong oxidizing agents (e.g. oxidizing acids, chlorine, etc). A common geomembrane application is for potable water containment. At the chlorine levels found in potable water throughout the world, the chlorine used to disinfect the water can promote oxidation of some geomembrane materials (see Table 14.5).

Polyolefin geomembranes can undergo stress cracking in chlorine solutions even at very low concentrations (1–5 ppm). Providing the chlorine concentration in water does not exceed 0.5 ppm then HDPE and LLDPE can be considered resistant. fPP is not resistant to low levels of chlorine unless specially stabilized.

Chlorine resistance testing of geomembranes can be used to screen a material's performance in this application. Chlorine is the most commonly used disinfectant for potable water worldwide. The most common method of addition of chlorine to potable water is the addition of chlorine gas. When chlorine gas is dissolved in water it forms hypochlorous acid (HOCl) and the hypochlorite ion (OCl$^-$) respectively (pH-dependent reaction).

Chlorine levels in potable water range from 0 to 4 ppm (or higher), with the majority of systems in the 1–2 ppm range. Testing for chlorine resistance of geomembrane materials is commonly conducted at a chlorine level of 3 ppm and a pH of 7 to provide test data under an aggressive environment.

The above illustrates the importance of understanding the precise end-use chemical environment for geomembrane applications by taking that service environment into consideration in the development of a chemical resistance testing methodology.

Geomembranes that are susceptible to chlorine-induced degradation can fail through a number of mechanisms; however, the following mechanism has been identified in many cases:

- Step 1 The chlorine consumes the antioxidants on the surface layer of the geomembrane that is exposed to the potable water.
- Step 2 The outer exposed layer begins to oxidize and degrade (becomes embrittled).
- Step 3 Microcracks are initiated on the embrittled outer exposed surface.
- Step 4 Under strain or bending stresses the cracks propagate through the thickness of the geomembrane liner *via* a slow crack growth mechanism. This may be accompanied by chlorine-induced degradation of the polymer in advance of the crack tip.
- Step 5 Ultimately brittle failure occurs.

Table 14.6 The degree of swelling of three common geomembranes when immersed in different chemicals.[a] Based on these data, fPP should not be used for secondary containment of hydrocarbons (Basell, 1998)

Chemical	HDPE (%)	fPP (%)	PVC (%)
Polar			
Methanol	1.45	0.44	−8
Water	1.60	0.5	3
Propyl alcohol	1.00	1.0	−4
Acetone	4.5	4.0	83
Hydrocarbons			
Tetralin	12.3	92	60
Iso-octane	11.5	137	1
Xylene	20	182	8
Indolene	–	198	–
Cyclohexane	25	356	4
Trichloroethylene	21	500	395

[a]Percentage swelling when immersed in 100% concentration of the specific chemical.

14.5.4   HYDROCARBONS

Polyolefins are very susceptible to swelling in aromatic and chlorinated hydrocarbons (see Table 14.6). For instance HDPE geomembranes swell readily in kerosene, even low aromatic kerosene (approximately 10% mass gain). Similarly fPP is not resistant to aromatic hydrocarbons, resulting in greater than 180% swelling in xylene just at room temperature.

14.5.5   FATS, OILS AND GREASES (FOGs)

Geomembrane materials have varying degrees of chemical resistance to fats, oils and greases (FOGs). While HDPE is quite resistant to FOGs, fPP and EPDM geomembranes can be seriously affected by interaction with FOGs leading to swelling and loss of mechanical properties.

14.5.6   SALT SOLUTIONS (INORGANIC OR IONIC SALTS)

Due to the selective nature of geomembranes, the permeation of the dissolved constituents in liquids can vary greatly, that is, components of a mixture can permeate at different rates due to differences in solubility and diffusibility in a given geomembrane. With respect to the inorganic aqueous salt solution, the geomembranes are semipermeable, that is, the water can be transmitted through the geomembranes, but the ions are not transmitted. Thus, the water that is transmitted through a hole-free geomembrane does not carry dissolved inorganics.

Solutions of metal salts comprised of metal cations and associated anions such as sodium hydroxide, calcium chloride, sodium bisulfate and so on have been found to have no measurable effect on HDPE. The very low permeability of the salt solutions is because polyethylene is a non-polar matrix and such salts cannot diffuse into the HDPE structure due to the hydrophilic nature of the charged ion.

Although inorganic salts do not permeate geomembranes, some organic species do. The rate of permeation through a geomembrane depends on the solubility of the organic species in the geomembrane and its diffusibility in the geomembrane as driven by the concentration gradient (ASTM D-5886).

14.6   GEOMEMBRANE CHEMICAL RESISTANCE BY POLYMER TYPE

The general chemical resistance of polymers towards chemicals that may swell or dissolve the polymer can be predicted from the solubility parameter (SP) which theoretically defines the ability of one material to interact with another. This concept is the basis of the well-known idiom *like dissolves like*.

The degree of interaction between materials can thus be predicted using the solubility parameter and this can be a good indication of whether a particular chemical will swell, stress crack or dissolve a polymeric geomembrane.

From a chemical resistance perspective the closer the solubility parameter of a chemical to that of the polymer, the greater their mutual compatibility – which from a chemical resistance perspective however indicates incompatibility (i.e. poor resistance).

The solubility parameters for HDPE and PVC are 16.4 and 20.5 $(MPa)^{1/2}$, respectively, while the solubility parameters for *n*-dodecane and acetone are 16 and 20.3 $(MPa)^{1/2}$, respectively. From the solubility parameters it can be predicted that *n*-dodecane will swell polyethylene while acetone will swell PVC. Indeed this occurs in practice – hence the usefulness of solubility parameter comparisons is demonstrated.

The solubility parameter of benzene is 18.8 $(MPa)^{1/2}$ and so one would predict that both PVC (SP = 16.4 $(MPa)^{1/2}$) and HDPE (SP = 20.5 $(MPa)^{1/2}$) polymers would be swelled by contact with benzene. Given that HDPE is semicrystalline and PVC is amorphous, it is anticipated that PVC would be more seriously affected. Indeed this is correct, as shown in Figure 14.7.

14.6.1   POLYOLEFINS

HDPE does not posses any functional groups in its structure which favour potential chemical attack. For polyethylene geomembranes the chemical resistance is linked with the polymer density and so for chemicals for which HDPE is already considered 'limited resistance', the use of the lower density materials (e.g. LLDPE, fPP) is not recommended. The degrees of chemical resistance of LLDPE and VLDPE are lower than those of HDPE. LLDPE has only moderate chemical resistance to hydrocarbons. However, LLDPE is frequently used for secondary containment for hydrocarbons (e.g. crude oil).

Chemical environments that pose potentially serious problems for polyethylene are strong oxidizing agents and certain hydrocarbons. Concentrated sulfuric and nitric acids are strong oxidizers, whereas hexene, benzene and xylene typify the hydrocarbons. HDPE is susceptible to attack by some chemicals which may cause stress cracking, swelling and oxidation or may permeate the polyethylene. The issue is that these interactions can reduce the physical properties of polyethylene and cause it to swell and buckle.

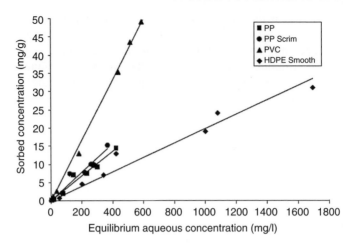

Figure 14.7 Absorption of benzene by different geomembranes. The solubility parameter for benzene is 18.8 $(MPa)^{1/2}$ whereas the solubility parameters for HDPE and PVC are 16.4 and 20.5 $(MPa)^{1/2}$, respectively. Reprinted with permission from *Geotextiles and Geomembranes*, Comparison of the equilibrium sorption of five organic compounds to HDPE, PP and PVC geomembranes, Edidia K. Nefso and Susan E. Burns, **25**(6), 360–365. Copyright (2007) Elsevier

Absorption of Liquids in HDPE

HDPE can absorb certain hydrocarbon liquids causing swelling of geomembranes. Two problems can arise due to liquids that swell HDPE:

- Absorbed liquids make it very difficult to perform good welding repairs on the liner.
- Absorption of hydrocarbon liquids causes HDPE to swell and this can place stresses on critical welds. HDPE liner swelling can thus lead to weld separation.

The rate of absorption (uptake) of chemicals (e.g. kerosene) in HDPE increases as the free volume of the polymer increases. It is widely reported in the literature that the sorption, diffusion and permeation coefficients for hydrocarbon liquids in HDPE increase as a function of temperature and follows the Arrhenius relationship (i.e. for every 10 °C increase in temperature there is a doubling in the amount of sorption). The role of temperature in this interaction is critical. Basically, the higher the temperature of the solvent in contact with HDPE, the more susceptible the material becomes to swelling. In addition, extraction of protective additives such as antioxidants and stabilizers are proportionally higher and the rate of chemical reactions (e.g. copper-catalyzed oxidation) is also expected to increase in accordance with the Arrhenius equation.

Crack-growth data generated for polyethylene by Lustiger (1983) suggests that for every 7 °C increase in temperature, the crack-growth rate is doubled. Increased temperature can therefore be regarded as another type of crack-growth accelerator. Therefore for a 10 °C temperature rise, crack growth is increased by a factor of approximately 2.8. Therefore for a 20 °C increase (i.e. from 55 to 75 °C) crack growth is increased by about 8 times.

Swell-induced waves/wrinkles/blisters can lead to problems in HDPE liners in concrete basins such as in solvent extraction tanks and mixers in the mining industry. In such applications, a cast-in HDPE liner (i.e. stud liner or anchor sheet) is used to protect the walls and floor of concrete tanks. Swelling of the HDPE liners can occur due to absorption of organic components (namely kerosene) from solvent extraction (SX) process solutions such as those used in copper and uranium ore extraction. This absorption causes large bulges to form, thereby placing peel stresses on critical welds (e.g. between the loose liner and anchored liner). This can also lead to cracking along heat-affected areas of the welds. Another consequence of the absorption and swelling of organics is that repair work on the cracked weld is very difficult (Peggs, 2007).

Case Study 1

An HDPE stud liner in a uranium mine underwent significant swelling due to diffusion of kerosene into the polyethylene. Precedents should alert the material supplier to potential compatibility problems between polyethylene and organic liquids. It should have been apparent to the materials supplier that HDPE has a tendency to soak up kerosene and expand. If the HDPE is constrained or is installed as a 'tight fit' (as in this concrete tank application), then the expansion will set up resultant stresses in the liner structure that can lead to failure, especially at the heat-affected area of welds.

An HDPE stud liner is known to absorb kerosene from mine process solutions resulting in swelling, bowing and the development of blisters in the liners on the walls and floors. This swelling leads to stressing of critical welds and separation of the studs from the back surface of the sheet. Attempts to repair swollen liners are compromised by the evolution of absorbed organic vapours as the base liner has been heated during repair welding.

A HDPE Stud liner exposed to a solvent extraction (SX) solution at an operating temperature of 45–55 °C is likely to fail by stress cracking due to:

- Swelling-induced stresses imposed on the weld region by the swelling and volumetric expansion of the HDPE sheets and "pull-off" of the studs/anchors.
- Swelling-induced stresses leading to shear and peel forces acting on the adjacent welds.

It should be emphasized that kerosene does not directly chemically attack the liner itself, but rather it provides the conditions for other agents to accelerate the degradation of the HDPE. In particular, kerosene swells the HDPE and therefore increases its permeability. The solubility parameter for kerosene is in the range 15.6–15.8 $(MPa)^{1/2}$ whereas the solubility parameter for polyethylene is 16.4 $(MPa)^{1/2}$; therefore polyethylene has a high affinity for kerosene.

The antioxidant and stabilizer additives are essential formulants in HDPE and these are susceptible to extraction by good solvents (kerosene is an excellent solvent). The loss of antioxidants by solvent extraction renders the HDPE susceptible to thermal oxidation.

Case Study 2

Interceptor ponds containing oily wastewater (30 000 mg/kg) were lined with HDPE (1.00 mm) and ultimately failed prematurely due to an "unzipping failure" failure mechanism. The oily layer of hydrocarbons floating on the water combined with solar heating of the black liner caused localized swelling, plasticization and thinning of the membrane.

The localized region of softened and thinned membrane ultimately split leading to the so-called "unzipping failure" along the perimeter of the liquid level of the ponds.

14.6.2   CHLORINATED SULPHONATED POLYETHYLENE (CSPE)

CSPE does not exhibit the same broad chemical resistance as HDPE since some hydrocarbons can readily swell and weaken it (see Figure 14.8). CSPE does, however, continue to cure (crosslink) or vulcanize over time and as a result its tensile strength and chemical resistance increases; unfortunately its ability to accept repairs also decreases with age. Special preparation and bonding agents are therefore required on older CSPE installations.

14.6.3   REINFORCED ETHYLENE INTERALLOY (EIA-R)

EIA-R geomembranes were developed to fulfil the demand for a geomembrane with better hydrocarbon and petroleum resistance than HDPE and CSPE. EIA-R geomembranes

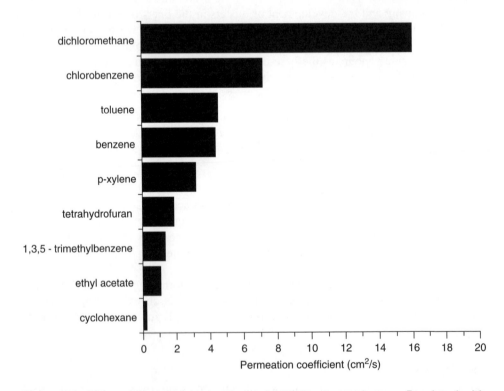

Figure 14.8 Permeability of various solvents in CSPE geomembranes. Reprinted with permission from *Journal of Hazardous Materials*, A review of polymeric geosynthetics used in hazardous waste facilities, J. D. Ortego, T. M. Aminabha, S. F. Harlapur and R. H. Balungi, **42**(2), 115–156. Copyright (1995) Elsevier

have broad chemical compatibility and have been practically tested with many different petroleum products including diesel fuel, naphtha, kerosene and crude oil.

14.6.4   POLYVINYLIDENE FLUORIDE (PVDF)

PVDF is highly resistant to swelling by hydrocarbons such as kerosene, is chemically resistant to sulfuric acid, and is not susceptible to thermal oxidation or catalysis by copper ions. Furthermore, it does not contain any additives that might be extracted by aggressive process solutions, it has better temperature resistance than HDPE, it has a higher heat distortion temperature (of $150\,^{\circ}C$) than polyethylene, and it has a higher yield stress (of 28 N/mm^2) than polyethylene.

14.6.5   ETHYLENE CHLOROTRIFLUOROETHYLENE (ECTFE)

ECTFE is highly resistant to swelling by hydrocarbons such as kerosene, is chemically resistant to sulfuric acid, and is not susceptible to thermal oxidation or catalysis by copper ions. Furthermore, it does not have any additives that might be extracted by the process solutions, has better temperature resistance, has a high heat distortion temperature (of $150\,^{\circ}C$), and it has a high yield stress (of 25 N/mm^2).

14.6.6   VINYL ESTERS

Vinyl ester resins are a class of thermoset polymers known to have excellent chemical resistance and durability. This is frequently the resin of choice for acid resistance, including hard mineral acids like sulfuric acid. Many resin tanks used for storing acid are made of vinyl ester resin and glass fibre. Vinyl ester resins can be formulated into various types of materials such as coatings, flooring, non-shrink grouts and chemical barriers for concrete.

There is a large difference between standard ester resin (fibreglass boat resin) and vinyl ester resin. In strong acid or oxidizing conditions, standard ester resins can degrade through a process called hydrolysis. Vinyl ester resins on the other hand, can be made with an epoxy backbone which are will be stable under very aggressive environments.

14.7   ENVIRONMENTAL STRESS CRACKING

Although HDPE geomembranes exhibit excellent resistance to a broad range of chemicals under conditions of applied (external or internal) stress in the presence of certain liquids, cracks may form in the material, which can cause premature failure. This phenomenon is environmental stress cracking (ESC). While not strictly chemical attack, stress cracking is accelerated by chemical exposure.

Stress cracking is the brittle fracture of the geomembrane under significantly lower stress than the material's yield strength. The factors influencing this phenomenon are:

- temperature;
- temperature gradient;

- chemical agents;
- UV oxidation;
- stress (particularly fatigue).

Stress cracking leads to small cracks and even holes in the geomembrane that allow leakage through the membrane. Stress cracking failure can be prevented by using a chemically resistant resin and by installation design to limit high stress in the liner during service.

📌 *Stress cracking of HDPE is a brittle fracture that occurs at a constant stress lower than the short-term break strength of the material. Resistance to stress cracking is the single most important performance parameter of HDPE geomembranes.*

Although the geomembrane may be able to withstand the action of either the chemical agent or the load separately (i.e. as individual factors), it is the combined forces which result in the consequence of ESC.

Typical chemicals which are considered active ESC agents include:

- polar liquids such as wetting agents (surfactants);
- soap and detergent solutions;
- oxidizing acids;
- oils and waxes;
- some organic acids;
- aromatic hydrocarbons;
- halogenated hydrocarbons;
- chlorinated solvents;
- pulp mill black liquor.

Pulp mill black liquor in particular is known to be an aggressive stress cracking agent for HDPE. Black liquor is a soapy, strongly alkaline liquid by-product from pulp and paper mills. Black liquor, also known as pulping liquor, is an alkaline spent liquor removed from the digesters in the process of chemically pulping wood.

Environmental stress cracking is thus a brittle failure mode that can be accelerated by a number of different chemical environments. The resistance of a geomembrane to this phenomenon is termed Environmental Stress Crack Resistance (ESCR). Certain agents promote stress cracking of geomembranes by promoting the slipping or untangling of the polymer chains across the advancing crack tip. ESCR testing is typically performed at elevated temperatures (50 °C) to accelerate any potential effects using the notched constant tensile load (NCTL) test.

In the NCTL test, HDPE specimens loaded at levels above 50% of the yield stress generally undergo ductile failure whilst those under approximately 30% of the yield stress can exhibit brittle fracture. The relative stress (that is, the test stress expressed as a percentage of the yield stress) is then plotted against the time to failure. The resulting stress–rupture curves for HDPE show a characteristic knee defining the transition from ductile to brittle behaviour.

📌 *Stress cracking agents (such as soaps, detergents, waxes and oils) do not weaken the polyethylene geomembrane through absorption or swelling. In fact, there is little loss of mechanical properties such as tensile strength and elongation when the polymer is immersed in the stress cracking agent.*

What makes environmental stress cracking so insidious is that certain polyethylene resins are much more susceptible to stress cracking than others, and certain liquids have a greater tendency to induce stress cracking than others.

The stress cracking potential of different chemicals toward HDPE can be expressed in terms of resistance factors. The resistance factor is a stress factor discounted for the magnitude of the constant stress required to reach the same time-to-failure value in the test media as in air (or water). Thus a resistance factor of one indicates there is no reduction in stress required for the test stress. Black liquor for instance is very aggressive and has a stress resistance factor of 0.70, while kerosene has a stress resistance factor of 0.77.

14.7.1 OXIDATIVE STRESS CRACKING

Oxidative degradation embrittles HDPE, LLDPE and fPP geomembranes, which renders them much more susceptible to stress cracking. Thus at regions of a geomembrane or liner with even low tensile stresses or strains, it is possible that stress cracks may result if oxidative embrittlement occurs under an oxidative environment.

REFERENCES

Aminabhavi, M. and Naik, H. G., Chemical compatibility study of geomembranes – sorption/desorption, diffusion and swelling phenomena, *Journal of Hazardous Materials*, **60**(2), 175 (1998).

ASTM D-5886, 'Standard Guide for Selection of Test Methods to Determine Rate of Fluid Permeation Through Geomembranes for Specific Applications' (2006).

BAM, 'Certification Guidelines for Plastic Geomembranes Used to Line Landfills and Contaminated Sites', document published by the Federal Institute for Materials Research and Testing (1999).

Basell, brochure entitled 'Astyrn FPA – high performance materials for geomembranes' (1998).

Darmstadt, 'Staaliche Materialprüfungsanstalt', Darmstadt Test report K 98 0777MPA, Darmstadt, Germany (1998).

Koerner, R. M., *Designing with Geosynthetics*, 5th Edition, Prentice Hall, NJ, USA, p. 464 (2005).

Little, A. D., 'Resistance of Flexible Membrane Liners to Chemicals and Wastes', US EPA Report PB86-119955, Cincinnati, OH, USA (1985).

Lustiger, A., The molecular mechanism of slow crack growth in polyethylene, in *Proceedings of the 8th Plastic Gas Pipe Symposium*, American Gas Association, Columbus, OH, USA, pp. 54–56 (1983).

Muller, W. W., *et al*. Stofftransport in Deponieabdichtungssystemen, Teil 1: Diffusions- and Verteilungskoeffizienten von Schadstoffen bei der Permeation in PEHD-Dichtungsbahnen. *Bautechnik* **74**, 176–190 (1997).

Ortegoa, J. D., Aminabhavibvi, T. M., Harlapur, S. F. and Balundgi, R. H., A review of polymeric geosynthetics used in hazardous waste facilities, *Journal of Hazardous Materials* **42**, 115–156 (1995).

Peggs, I. D., Caveat Venditor – When Practice is Far From Theory, *Geosynthetics*, 6 (June/July 2007).

Thiel, R. and Smith, M. E, State of the practice review of heap leach pad design issues, *Geotextiles and Geomembranes*, **22**(6), 555 (2004).

TRI, Project Report No. E2177-88-10, 'Effect of Sodium Hydroxide (NaOH) on the Additives Found in Four Commercial Geomembranes' (2004).

15

Failure Modes of Geomembranes

Failure of a geomembrane is a condition or state that prevents the liner from satisfactorily fulfilling its 'intended purpose'. Generally speaking, failure of the lining therefore refers to a leak that permits liquids to escape in significant quantities.

Since geomembranes serve an important barrier and containment function any breach of the geomembrane whether it be a brittle crack, puncture, hole, tear, breakdown or weld delamination, is considered a failure. Geomembrane liner performance issues can arise from inadequate/inexperienced design, poor material selection and poor installation. The current design philosophy is that the geomembrane serves only as a barrier and not as a load-bearing member of the system. Unfortunately geomembranes are often installed in a design or a manner where they are required to resist stresses and loads. In such instances geomembrane and installation system failure can result.

The EPA/530/SW-91/054 document states that a geomembrane should function simply as a barrier and should not be required to fulfil a load bearing function of the system.

15.1 POTENTIAL FOR GEOMEMBRANE DAMAGE

Damage to geomembrane liners can occur through the entire life cycle of the geomembrane. For instance during:

- storage and handling at the geomembrane manufacturing site;
- transportation from the manufacturing factory to the installation site;
- offloading at the installation site;
- storage at the installation site;
- deployment at the installation site;
- positioning into the final welding location;
- welding of seams (e.g. overheating, burn through);
- wind gusting;
- traffic over the membrane;
- placement of the cover material, drainage layers or soil backfill on the completed geomembrane.

A Guide to Polymeric Geomembranes: A Practical Approach   J. Scheirs
© 2009 John Wiley & Sons, Ltd

Damage from the above elements can be mitigated and largely prevented though the use of rigorous specifications, attention to detail and a comprehensive construction quality assurance (CQA) program. In addition attentive third-party CQA personnel are also required to ensure compliance to the CQA program.

Some notorious sources of geomembrane leaks are:

- soil backfilling by the earthwork contractor;
- gouges from bulldozer blades;
- surveying stakes penetrating through the underlying geomembranes;
- stone punctures.

Since landfills are considered critical containment applications they serve as a good example to illustrate the various types of failures that arise. HDPE liners are used widely in landfills and may undergo failure due to one of the following:

- inadequate welding and attachment to structures;
- imposed stresses during construction;
- mechanical damage during construction;
- stress cracking at points of stress (especially along weld edges);
- service stresses that induce weld separation (adhesion failure).

Geomembranes can fail simply by being overstressed by soil or waste-induced stresses and such failures can occur by yielding or tearing. There have been a number of instances of slope failures involving tearing and brittle fracture of HDPE geomembranes.

Today, geosynthetic design engineers recognize that installed geomembrane liners have the potential to leak. The number of leaks in a geomembrane lining system is a function of the area of the installed geomembrane and the complexity of the installation. A larger area of geomembrane typically has a lower proportion of detail work and thus less defects per unit area.

Polymeric geomembranes in general are characterized by:

- low shear strength;
- high thermal expansion coefficient;
- low-temperature rigidity;
- high-temperature flexibility;
- low coefficient of friction (for smooth liners).

There is no such thing as a perfect geomembrane; each type has its shortcomings. For this reason it is important to understand the shortcomings of specific geomembrane types and also to understand their common failure modes.

15.2 FAILURE OF INSTALLED GEOMEMBRANES

Leakage through geomembranes can originate from faulty welds, mechanical attachments such as pipe boots and batten strips, punctures from sharp objects, and from damage caused by equipment utilized during liner installation and placement of scaffolding/ladders.

15.2.1   MAIN CAUSES OF HDPE GEOMEMBRANE LINER FAILURE

The major causes of failure for HDPE geomembranes are:

- mechanical damage during construction (i.e. installation damage);
- poor welding (e.g. inadequate fusion);
- stresses imposed during construction;
- service stresses that lead to weld failure (differential settlement);
- stress cracking at high stress points.

The major causes of leaks in installed HDPE geomembranes are:

- stone puncture holes;
- gashes caused by earthmoving equipment such as bulldozers;
- installation damage;
- poor welds.

Statistically HDPE liner failures can be attributed roughly as follows:

- 60% of failures are due to inadequate designs;
- 20% to poor installation;
- 10% to inferior materials;
- 10% to miscellaneous items.

15.2.2   INSTALLATION DAMAGE

The predominant causes of failure (i.e. leaks) of geomembranes are due to installation damage (i.e. construction damage) due to the following factors:

- stone punctures;
- bulldozer damage;
- depth stake puncturing.

Electrical leak location surveys have shown (Nosko et al., 1996) that about 19% of leaks occur at seams, but over 70% of leaks occur when the liner is covered by soil or stone; therefore the covering operation is a very critical stage for a geomembrane.

Leakage through geomembranes can originate from mechanical damage caused by equipment utilized during liner installation and placement of scaffolding/ladders. Leakage can also occur where the liner/geomembrane has been accidentally damaged by the installation crew, such as by a knife blade, dropping a steel batten strip or by dropping tools on the liner, etc. Mechanical damage such as scratches and creasing of HDPE during installation can also markedly increase the susceptibility of HDPE towards stress cracking. Also it is important to avoid inducing sharp bends in the geomembrane in the anchor trench region. Slightly rounded corners are thus needed in the trench to avoid sharp bends in the geomembrane.

Straight-line scratches that penetrate the geomembrane to a depth of 0.1 mm or greater may be damaging to the durability of the geomembrane and may be regarded as a defect. Such defects should be cut out and patched. In order to avoid stress concentrations all cuts and patches should be formed with rounded edges and ends. Similarly all cuts should be ended off with a round cutout.

There is a high probability that the geomembrane may be damaged during installation, particularly in high trafficked areas. In this respect the placement and movement of the electric generator require particular attention. It is important that the generator is placed on a protective skid and that the generator is not dragged across the geomembrane. Similarly, a smooth insulating plate or geotextile should be placed under the hot welding apparatus after use to protect the geomembrane.

Field Punctures

Puncture resistance is important when a geomembrane is installed under onerous conditions, for example, on a machine-graded subgrade with angular stones under high loads.

There is a possibility of puncturing an installed geomembrane every time it is travelled on by people, vehicles and equipment, which can occur during installation, during inspections, during placement of subsequent layers (e.g. geotextiles, drainage media) and during placement of top cover soil or gravel. Relatively thin polymeric materials are more likely to be punctured by activities above the liner and suffer severe point pressures applied on its surface compared to thicker membranes. Once containment ponds are commissioned and become operational it becomes very difficult to repair the geomembrane. For this reason it is of utmost important to select a sturdy, relatively thick, polymeric material that has the least potential for puncturing.

A majority of the defects in installed geomembranes have been attributed to field punctures and faulty seaming. The ability of the geomembrane to prevent or withstand these two prevalent types of defects directly reflects on the actual impermeability of the liner.

Index puncture tests (e.g. ASTM D-4833) are a useful measure of puncture resistance under standard test conditions; however the puncture resistance under the expected field conditions is of fundamental importance. In this respect the test method ASTM D-5514 is more appropriate since it can be used to evaluate a geomembrane in combination with actual site-specific materials such as large aggregate or drainage material.

15.2.3 WELD FAILURES

Seams and welds are the weakest points of a geomembrane. Many problems encountered in landfills and lined lagoons originate at seam locations. Seams are regions of high stress concentration due to defects in seaming operations and residual stresses. Also, stress cracking and brittle fractures can deteriorate and even break seams. It is possible to reduce damage considerably at seams by using proper equipment, workmanship, quality construction and proper inspection.

In many cases wedge weld failures can be ascribed to the presence of contamination such as soil or moisture in the weld region. The welding of polyethylene geomembranes, for example, requires trained personnel and special equipment. It has been noted that many stress-cracking problems of HDPE geomembranes were immediately adjacent to and in field welds. On average, 3% of the samples (in the case of HDPE field seams) do not pass destructive weld testing.

Notches in Welds

Potentially damaging 'notches' between top and overlapped sheets can form along seam edges. This can be a potential problem in the case of HDPE geomembranes where the predominant mode of premature failure of the liner is a quasi-brittle fracture initiated at stress-concentrating surface notches.

Failure at Weld Edges

Geomembrane failures often occur at the edges of welds even though the weld itself does not fail. In many cases, the failure of the geomembrane occurs in the lower geomembrane sheet at a location directly along the edge of the weld, as shown in Figure 15.1. Failure of a geomembrane directly next to a weld can result from either (a) a loss of strength (i.e weakening) of the geomembrane material due to the welding process and/or (b) the formation of stress concentrations.

Weakening and loss of ductility of the geomembrane material can result from overheating during the welding and the formation of the so-called heat affected zone (HAZ). This overheating can cause chemical (i.e. oxidation) and structural (i.e. crystallinity) changes in the geomembrane polymer which make it more susceptible to brittle cracking. This overheating is often combined with the application of excessive pressure which also leads to a local reduction in the thickness of the geomembrane at this point.

Stress cracking can also occur in HDPE geomembranes in the heat affected zone directly adjacent to welds. This is because overheating of the HDPE during welding increases the crystallinity of the adjacent geomembrane material as well as consuming the antioxidant; both of these factors reduce the stress cracking resistance of HDPE.

Stress concentrations typically form where there are abrupt changes in geometry at the weld or in its immediate vicinity. Such changes in geometry occur where there is

Figure 15.1 Photograph of splitting along the edge of a weld track in a 1 mm HDPE geomembrane (the pen shows the location of the 'hole')

an abrupt change in thickness such as in an overlap weld. Overheating combined with excessive pressure can cause a thinning of the weld and hence a reduction in the overall thickness of the geomembrane. Extrusion fillet welds, in particular, produce areas with an abrupt transition in geometry (from thin to thick to thin again) between the weld region and the parent geomembrane.

Removal of Edge Flaps

The edge flap of the fusion welds should not be ripped off from the seam to inspect the 'squeeze out' from the weld. This pulling and ripping force places potentially damaging stresses on the weld. The edge strip should be removed by careful cutting with a hook knife in such a manner to protect the integrity of the geomembrane.

Heat Degraded Extrudates

Heat degraded extrudates (HDEs) can form in extrusion welders if they are not properly purged before beginning a weld. HDE is composed of shorter chain polymer because of its excessive heat history and is quite brittle. Extrusion welds made with HDE can crack in a brittle fashion under relatively low stresses.

Seam Area Grinding

Notches and defects can be introduced into HDPE geomembranes due to excessive seam area grinding. Overgrinding and grinding gouges parallel to the weld direction can produce stress concentrations normal to the major stresses and these can propagate into stress cracks over time and with thermal cycling.

Exposure of some geomembranes (especially HDPE) to low stress levels can lead to slow crack propagation (stress cracking), especially where the crack was initiated by notches, surface cuts, scratches, seam area grinding or sharp folds in the geomembrane. Such slow crack growth may take years to develop, thus making it difficult to predict the true service life of HDPE geomembranes. Stress cracking is defined as apparent brittle cracking that occurs at a constant stress lower than the yield or break stress of the polymer.

'T'-Seams

'T'-seams show a higher tendency for failure due to their higher heat history and the increased likelihood for contamination in the weld area. They should not be made within 100 mm of another 'T'-seam. Where 'T'-seams are within 100–300 mm of each other the entire area should be treated as a repair and patched.

Stress Cracking of Welds

By testing HDPE geomembrane sheets and welds according to ASTM D-2552 (stress rupture under constant tensile load) it was found that the sheets could resist stress cracking better than the welds with about 40% of the welds exhibiting cracking compared to only 1% of the sheet samples which cracked (Halse *et al.*, 1989). The stress cracks that formed were examined carefully and it was shown that the cracks which appeared in the welded specimens were almost always initiated near the overlapping junction of the two geomembrane sheets where the stress concentration was likely to be highest. It was also found that the cracks were preceded by crazing which formed in a direction perpendicular to the applied stress.

15.2.4 STRESS RELATED FAILURES

Cracking on Folds

Flexible geomembranes, such as PVC, fPP, fPP-R and CSPE-R, are generally factory seamed and then folded in an accordion fashion onto wooden pallets where they are strapped down. It is important that the strapping bands are applied with protective cushioning so as not to damage the folded geomembrane. Also the wooden pallet needs to extend beyond the edge of the folded panels by at least 75 mm. While this folding is not damaging to these membranes for a relatively short periods of time (e.g. a few weeks) they should not be kept stored in this fashion for long periods as surface cracking may occur. This is especially true of improperly formulated nitrile rubber membranes which are susceptible to ozone cracking on folds. Tight banding straps and improper stacking of rolls of geomembrane can also lead to buckling, rippling and sheet deformation which is undesirable. Cold temperatures can also exacerbate cracking on folds.

Abrasion of geomembrane and liners has been observed especially on thin, soft or low crystallinity materials where high points (from folds and wrinkles) occur. Damage to thin (0.25 and 0.5 mm) PVC accordion folded sheets was observed where carton abrasion and long transport times result in holes due to abrasion.

While permanent creases can form where the elastic limit of the material is exceeded on the outer radius of the fold, thin (0.5 mm) geomembranes such as fPP and PVC show little decrease in mechanical properties across permanently creased folds or star creases.

Stresses Developing at Folds

When a geomembrane is heated during the day by exposure to direct sunlight, followed by experiencing rapid and significant cooling at night, the geomembrane will be subjected to expansion and contraction leading to wrinkling. If these wrinkles or creases that arise from thermal movement become flattened by hydrostatic pressure then the creases can become knife-edge folds which can lead to premature failure due to environmental stress cracking.

A calculation can be made of the strain developed over a tight bend in a membrane of thickness 2.0 mm.

The applied strain on the tension side of the liner in a fold can be calculated using the following simple calculation:

$$\text{OUTER STRAIN} = 100T/[2(R + T)]$$

where:

$$T = \text{thickness} = 2.00 \text{ mm}$$

$$R = \text{radius of curvature} = 2 \text{ mm}$$

This gives values of percentage strains on the outer surface of tight folds of 25%.

The yield point of HDPE occurs at strains around 12–13% so that the strains developed on tight folds actually exceed the elastic limit of the material which means the sample may undergo permanent deformation.

Stress Cracking

HDPE geomembranes are generally regarded to be ductile materials that will only fail (break) at an elongation of 700–800% after yielding at an elongation of about 12%. However, low break elongation values for HDPE geomembranes suggest a sensitivity to surface flaws, and/or an increased sensitivity to brittle cracking phenomena such as stress cracking and fatigue.

The high crystallinity (i.e. high density) of HDPE geomembranes imparts excellent chemical resistance to harsh chemicals, but can be problematic with regard to its stress cracking resistance.

Stress cracking is the sudden and unexpected rupture of a plastic caused by a tensile stress lower than its short-term tensile strength. As shown in Figure 15.2, stress cracks occur after extended loading at lower stresses due to breakdown of tie molecules in the polymer. These tie molecules which are like tethers that pin together the lamellae are normally responsible for the ductile stress–strain behavior of HDPE. Stress cracking of

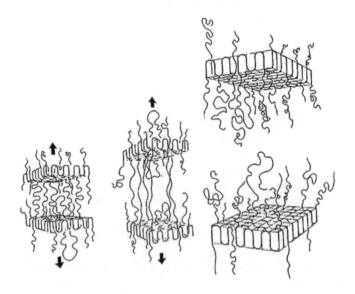

Figure 15.2 Schematic illustrating the role of tie molecules in stress cracking of HDPE geomembranes. The crystallites (i.e. lamellae) are held together by tie molecules. The region between the flat crystallites is the amorphous region. Under a low tensile stress the tie molecules are stretched and the intermediate amorphous region expands. In the case of stress cracking, the tie molecules pull out of the lamellae and brittle fracture results. Reprinted with permission from *Polymer*, Importance of tie molecules in preventing polyethylene fracture under long-term loading conditions by A. Lustiger and R. L. Markham, **24**(12), 1647. Copyright (1983) Elsevier

HDPE can be largely prevented by selecting HDPE resins that have a ductile-to-brittle transition time in excess of 300 h (Daniel and Koerner, 1993) as measured in a single-point notched constant load test (SP-NCTL) (i.e. ASTM D 5397).

While LLDPE and PP do not suffer from classic ESC (as is seen with HDPE), they do suffer from oxidative stress cracking (OSC) when their antioxidants/stabilizers are depleted and they subsequently oxidize and embrittle. (see Section 2.7.6).

Environmental stress cracking can thus lead to brittle failure of crystalline polymeric materials, especially HDPE, at stresses well below their normal yield stress, let alone failure stress. Concerns regarding the stress cracking of HDPE geomembranes have led to the adoption of rigorous precautions to protect HDPE membranes from stress in service. The main precautions have been designs limiting local strain on the geomembrane and the use of cushioning geotextile materials. Geotextiles are used as cushioning materials above geomembranes to prevent damage from drainage layers and stone overburden. Note however that since geotextiles show a propensity for biological clogging they should not be used in areas inundated by leachates such as sumps and around leachate collection pipes.

LLDPE, VLDPE and fPP are also susceptible to stress cracking but only when their antioxidants are depleted and after they have oxidized. This is a special case and is termed 'Oxidative Stress Cracking'.

How to Recognize Stress Cracking in HDPE Geomembranes

1. The breaks or cracks in the geomembrane need to be closely examined. In particular the fracture faces should be examined under moderate magnification using a microscope.
2. Smooth, featureless surfaces are typical of stress cracking which is a brittle form of cracking.
3. The presence of signs of deformation such as yield thinning, stretched ligaments and fibrils (small whip-like features) are associated with ductile failure (i.e. ductile damage mode) and suggests the geomembrane or liner have been simply overloaded by a stress which exceeded the break stress of the material.

Ductile damage to the geomembrane is often the result of simple overload tearing.

Thermal Expansion Related Stress Failures

Geomembranes with a relatively high coefficient of thermal expansion will form 'waves' (particularly in sun) leading to stresses especially in the seam areas. The amount of expansion or contraction that occurs when a membrane is heated or cooled is expressed as its coefficient of linear thermal expansion. It is determined by accurately measuring the dimensions of a test section at two different temperatures and calculating the percentage change/$°C$.

HDPE has a very high coefficient of thermal expansion of 1.7×10^{-4} cm/cm/$°C^{-1}$ ($0.017\%/°C$) which is approximately 20 times higher than steel. This means that a liner on a 100 m length of geomembrane will expand by about 1.2 m on a day when the temperature starts at $5\,°C$ and the black liner heats up to $75\,°C$ in the early afternoon sun.

To accommodate this much expansion, the liner develops large undulations or waves on the cover and exposed side slopes.

Scrim-reinforced geomembranes (e.g. RPP or CSPE-R) have a coefficient of linear thermal expansion of 0.006%/°C (1/3 that of HDPE). The greater flexibility of the RPP or CSPE-R liner combined with the lower degree of thermal expansion means reinforced liners can easily accommodate this much expansion without causing problems. A high coefficient of linear thermal expansion can add tension to the seam areas, potentially causing failed seams.

When HDPE is coextruded with a light reflective, UV-stabilized, white upper surface (approximately 0.125 mm thick) it reduces heat buildup on the liner by reflecting solar energy. Reducing liner surface temperatures leads to less thermal expansion which in turn leads to fewer wrinkles and less subgrade desiccation. This reflection of radiant heat energy reduces wrinkles in white on black geomembranes by 50% or more. In this way the thermal movement of the geomembranes then approaches that of a scrim-reinforced material.

📌 *Black HDPE sheet achieves temperatures of 75°C as measured by an infrared thermometer whereas white HDPE only reaches 35–37°C. White HDPE is a coextruded material making the white surface an integral part of the membrane and so the white surface cannot delaminate from the black base sheet.*

Ideally, the geomembrane should be immediately covered only while it is in 100% intimate contact with the subgrade. The geomembrane should not be under tension and must be fully supported by the subgrade when it is covered.

It is important that the installer allows for sufficient HDPE geomembrane to compensate for contraction of the material during lower temperatures, while at the same time preventing excessive expansion and wrinkling at higher covering temperatures. In practice, wrinkling of the geomembranes is somewhat inevitable; however the wrinkles should be regularly spaced and low in profile, rather than irregular and prominent.

The waves that form in HDPE geomembranes are the greatest during the hottest part of the day and they can be exacerbated by the placement or spreading of the drainage material. Therefore it is desirable to wait for the cooler part of the day so the waves in the HDPE geomembrane will be less pronounced and the spreading equipment is less likely to encounter a wave causing damage to the geomembrane (Giroud, 1995).

In Germany, geomembranes are installed substantially without wrinkles through the use of careful installation procedures. The wrinkle-free installation is accomplished by placing soil around the periphery of the deployed area to provide ballast and so there is minimal contraction during the cooler night temperatures and then during the early morning while the liner is still under slight tension, the soil is spread across the flat geomembrane (Averesch and Schicketanz, 2000). This method of wrinkle-free installation does however take longer and is thus more expensive than traditional installation methods. This method necessitates that only that amount of geomembrane that can be covered by the required thickness of cover soil be deployed in one day.

Large folds and wrinkles in the liner are undesirable because:

- They can be compressed flat into crease lines where the stresses generated by folding may promote stress cracking.

- In the case of earthmoving equipment spreading sand over a wrinkled geomembrane liner, there is increased likelihood that the machine scoop or tracks may contact the top of the waves and rip the liner.

It has been demonstrated that small strains (such as up to 0.25%) have no long-term effects on HDPE and any deformation is completely reversible.

Strains up to 10% are still below the nominal yield strain of HDPE but there can be some irreversible changes in the polymer microstructure (i.e. morphology) such as interspherulitic cracking (i.e. microcracking along spherulite boundaries). This incipient cracking can initiate stress cracking.

At strains greater than 10% yielding and necking of an HDPE geomembrane polymer will occur and this produces the maximum in the stress–strain curve or force–displacement curve. This necking leads to a reduction in cross-sectional area of the sample and as a result the amount of force required to maintain the constant rate of elongation reduces. Hence this is why immediately after the yield point the stress–strain curve 'drops down'.

As the extension continues the chains in the amorphous regions as well as those in the spherulites uncoil and become orientated and reach their strain limit; therefore the force required to maintain constant extension increases again and ultimately the specimen breaks at the break point strain.

The permissible local strain in HDPE geomembranes can be related to a radius of curvature.

Low Elongation at Break Failures

The high elongations at break exhibited by HDPE and LLDPE (i.e. 600–700% or even greater) are really not very relevant in practical terms. These geomembranes are rarely subjected to such high levels of elongation in service. Provided the elongation of HDPE is not below a particular minimum values (for instance, 100%) which is many times greater than the elongation at yield, then the geomembrane is considered to have adequate elongational properties.

Surface nicks and notches can dramatically reduce the maximum achieved elongation of polymers. For this reason textured geomembranes generally tend to exhibit significantly reduced elongation at break.

Sliding Failures

The sliding failure at Kettleman Hills Landfill in the USA is one of the most well known geomembrane failures documented. The failure at the Kettleman Hills Class 1 hazardous waste storage facility occurred on March 19, 1988. At the time of failure approximately 490 000 m^3 of waste (and other material) had been placed to a height of about 25 m above the base. The entire mass slid a horizontal distance of approximately 10 m toward the south-east and vertical slumps of up to 4 m along the side slopes of the landfill were recorded after the failure (Blight, 2007).

The mechanism of failure was determined to be slippage along multiple interfaces within the landfill liner system. The low liner interface strengths (with residual friction angles as low as 8 degrees) were determined to be the predominant cause of the failure.

The principal sliding surface was the HDPE geomembrane/clay interface of the 1.1 m thick secondary liner system which seemed to essentially behave in an undrained manner during the approximately one year of waste loading prior to failure. The failure demonstrated the importance of testing the undrained shear strength of the clay/geomembrane interface, since the shear strength was sensitive to the 'as-placed' moisture content of the clay.

◎ *Kettleman Hills failure, 'at a glance'*

- Slope failure at the Kettleman Hills Waste Landfill in California in 1988.
- 490 000 m³ of waste slipped down the slope.
- Landfill design was 'state of the art' incorporating multiple geomembranes, drainage layers, cushion layers and a compacted clay liner.
- Failure occurred at more than one interface.
- Primary failure was at the clay–geomembrane interface.

15.2.5 FAILURE MODES OF SCRIM-REINFORCED GEOMEMBRANES

Delamination of Scrim-Supported (i.e. Reinforced) Liners

Reinforced geomembranes are made by laminating together two geomembrane plys with a fabric scrim between them. The geomembrane's performance depends on its ability to function as a single unit. If the layers are not adequately adhered, the performance of the reinforced geomembrane may be adversely affected.

For example, an EIA-R geomembrane is supported on a scrim which is a woven open-mesh reinforcing fabric made from continuous filament yarn that is laminated (or encapsulated) between two PVC-Elvaloy™ plies. Delamination damage (or ply adhesion failure) can occur when a laminated geomembrane layer separates from the reinforcing scrim, but does not necessarily perforate the entire liner (see Figures 15.3 and 15.4).

The plys must be on both sides of the fabric scrim so as to completely encapsulate it. Bonding of the adjacent PVC/Elvaloy™ faces occurs through the apertures in the scrim. The fabric scrim consists of polymer yarns in an open woven/knitted pattern sufficient to achieve the minimum specification strength and elongation values.

Ply adhesion failure is a potential issue in any geomembrane/liner where there is a textile scrim support in the centre of the liner. If there is insufficient 'strike-through' to allow the plies to adhere, then delamination can be an issue (see Figure 15.5). It is less of an issue for monolithic HDPE liners even though many of these are made on multilayer (coextrusion) production equipment where two or three extruders feed the die.

Wicking into Scrim-Supported Membranes

It is evident from local and overseas experience that all polyester-scrim reinforced (supported) geomembrane materials (RPP, EIA-R, CSPE-R) are susceptible to delamination/degradation due to ingress of moisture provided by 'wicking' and capillary motion along the scrim fibres and which can eventually lead to hydrolysis of the scrim in the presence of heat and/or acids (the latter is termed *acid hydrolysis*. Ultimately this leads to loss of scrim properties and so the integrity of the liner is dependent on the relatively

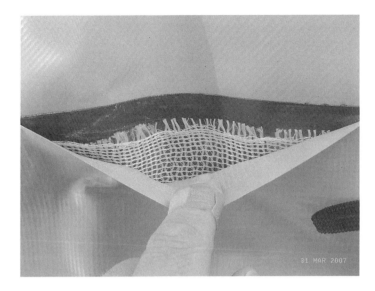

Figure 15.3 Photograph of delamination of a 0.9 mm (36 mil) EIA-R geomembrane. The top ply has fully delaminated from the polyester scrim. Reproduced by permission of ExcelPlas Geomembrane Testing

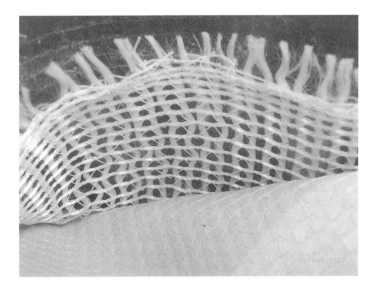

Figure 15.4 Photograph of a delamination failure of a 0.9 mm (36 mil) EIA-R geomembrane. The top ply has fully delaminated from the polyester scrim. While the scrim pattern is present on the inner surface of the top ply there has been insufficient 'strike through', i.e. bonding through the scrim. Reproduced by permission of ExcelPlas Geomembrane Testing

Figure 15.5   Digital microscopy photograph of the delaminated sites in an EIA-R geomembrane due to poor bonding ('strike-through') between adjacent PVC–Elvaloy plies through the apertures of the polyester scrim. Note that the black areas represent 'islands' of good adhesion where cohesive failure (as distinct from adhesive) occurred. Reproduced by permission of ExcelPlas Geomembrane Testing

thin top and bottom plies, each of which is typically only 0.5 mm thick and thus not load bearing. In contrast monolithic HDPE provides substantially more strength. Such scrim degradation can be a problem in floating cover applications, particularly over sewage ponds.

While scrim-reinforced geomembranes are manufactured to ensure the scrims are fully contained within two layers of polymer to eliminate wicking (fluid conduction into the liner that could delaminate the plies), there are always exposed edges, installation damage and moisture transport through the 0.5 mm plies thus leading to moisture ingress.

Bridging

If a tank liner is under load and not supported by the tank structure this is termed 'bridging'. Bridging of scrim-reinforced liners can place excessive strain on the liner and can cause delamination and fibre pull-out.

15.2.6   FAILURE MODES OF EXPOSED GEOMEMBRANES

Wind Damage

Strong winds can uplift a geomembrane pulling it out of the anchor trenches and causing tearing. It has been shown that the ability of a geomembrane to tolerate uplifting by strong winds is a function of its mass per unit area. Tensile strength and tensile stiffness (modulus multiplied by thickness) of the geomembrane are also important considerations to avoid wind uplifting and anchor trench pullout (Giroud *et al.*, 1995).

Figure 15.6 Photograph of a 'whale back' in a PVC geomembrane due to inflation by methane gas under a landfill cover. Reprinted with permission from *Geosynthetics International*, Methane gas migration through geomembranes, by T. D. Stark and H. Choi, **12**(2), 120. Copyright (2005) Thomas Telford Publishing

Whales

'Whales' or 'whale backs' are large blisters/balloons in a liner above the surface of the ground or water (see Figure 15.6). Whales are formed by the localized accumulation of generated gas (usually biogas, methane) that has not been properly vented. The tension that results when the gas lifts the liner can create damaging stresses especially at critical points such as seams and pipe penetrations. Such damage to the geomembrane liner can lead to increased gas leakage and greater whale formation.

Subsurface generated gases can push up a geomembrane liner, creating very high stresses on liners and seams. These large bulges are referred to as 'whales'.

Animal Attack

Potential 'biointrusion' from burrowing animals or kangaroos (sharp claws) is another failure mode of geomembranes. Burrowing animals are a real threat to geomembranes. There are a number of documented cases of animals breaching the liners by gnawing or cutting. These include:

- crab damage on Nauru island in a area where the geomembranes were installed in the path of migratory crabs during the breeding season;
- rodents gnawing through liners;
- pecking damage by bird beaks;
- gophers (prairie dogs) burrowing through pond liners;
- kangaroo paws scratching a liner while hopping on the liners.

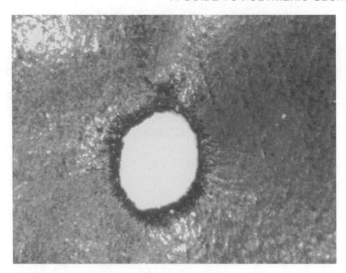

Figure 15.7 Photograph of 4 mm diameter hole crates in an fPP tank liner (0.75 mm thickness) by a New Zealand water snail with grinding jaws

If an animal with sharp and hard teeth is determined to penetrate a polymeric geomembrane there is a high likelihood that this may occur. HDPE and LLDPE are more resistant to burrowing animals than PVC due to their greater hardness. Rodents, especially rats, can eat through geomembrane liners and covers. Flexible geomembranes, such as PVC and very flexible polyethylene, are more susceptible to gnawing because they are easier to bend than an HDPE geomembrane. However it is only possible to eat the geomembrane at an edge, fold, or seam on which the animal can grip. Since PVC geomembranes may have 80% less seams than polyethylene geomembranes, there is accordingly less likelihood of having an edge, fold or weld.

The potential for penetration into the geomembrane by burrowing animals is far greater for unconsolidated, fine-grained sand/soils than for gravel and rock. A biotic barrier (BB) can therefore be installed to reduce potential intrusion by burrowing animals (for instance into landfill liners). A BB of 70 cm consisting of cobbles overlain with 30 cm of gravel can deter burrowing animals.

It has even been observed that water snails with cutting jaws could perforate fPP tank liners (see Figure 15.7).

Bird attack on geomembranes was reported by Thiel *et al*. (2003) with seagulls, in particular, pecking at the exposed cover. The two main areas that the birds pecked at are the flaps on geomembrane seams and the sandbags. If the flap left over from welding the geomembrane was kept in place, these birds felt an irresistible need to peck inside the flap, thereby damaging the geomembrane. This was solved by cutting away the flap, and the birds stopped pecking on the seams. Solving the issue for the sandbags has been more problematic. The problem became so severe that tires with covers were placed over the sandbags to protect them. The bottom side of the tires had to be cut out to fit over the sandbags and a stiff piece of polyethylene geomembrane inserted over the top of the tire to keep the birds out (Thiel *et al*., 2003).

15.2.7   ENVIROMENTAL DEGRADATION OF EXPOSED GEOMEMBRANES AND LINERS

Exposed geomembranes can be subject to ageing from a number of agencies namely:

- the ultraviolet (UV) component of sunlight (photooxidation);
- heat/elevated temperatures from the sun (thermal oxidation);
- ozone (ozonolysis);
- contained liquids (hydrolysis).

In many applications geomembranes are only exposed to UV light for a limited time before and during installation and are then covered by a protective layer of soil or sand. In some cases, however, geomembranes are exposed for the period of their service life at the tops of slopes (i.e batter) of reservoirs, ponds and impoundments.

Ageing of Geomembranes

Ageing of geomembranes is also an important problem, since environmental conditions such as temperature, UV, oxidation and chemical agents tend to accelerate deterioration of geomembranes and liners.

The modes of failure include the following:

1. Loss of physical properties due to lowering of the polymer's molecular weight by chain scission.
2. Stiffening and embrittlement due to oxidation and/or loss of plasticizers and additives.
3. Reduction of mechanical properties and increase of permeability.
4. Failure of geomembrane seams.

In the majority of cases there are combinations of these factors, which can cause damage to the liner system.

Degradation generally begins as changes in the polymer structure and eventually is manifested as changes in engineering properties (e.g. mechanical and hydraulic properties) of polymeric geomembranes. Common changes in the polymer include embrittlement, reduction or gain in molecular weight, generation of free radicals and loss of stabilizing additives.

For example, the most likely mechanisms for degradation of polyethylene geosynthetics in containment facilities are mechanical degradation, chemical degradation and oxidative degradation. Chemical degradation can occur when geosynthetics are exposed to strong chemicals (e.g. acids, bases, solvents, reactive gases) that alter the polymer by processes such as swelling, extraction and dissolution. Oxidative degradation is also a form of chemical degradation, but is considered separately owing to its significance in polymer degradation, even in environments where limited oxygen exists.

Oxidation of Polymers

Oxidation of a polymer proceeds by an auto-accelerating process, meaning that the rate is slow at first, but gradually accelerates. The period before oxidation accelerates (and measurable degradation takes place) is called the 'induction period'. When polymer degradation due to auto-oxidation chain reactions becomes severe, alteration of the physical

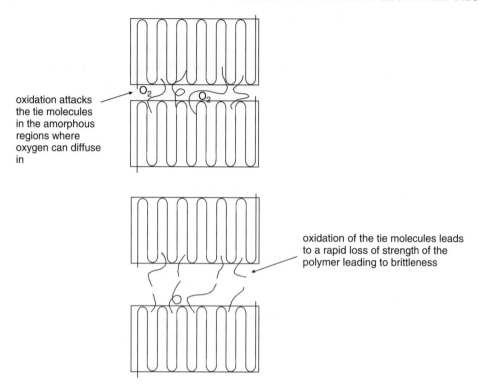

oxidation attacks the tie molecules in the amorphous regions where oxygen can diffuse in

oxidation of the tie molecules leads to a rapid loss of strength of the polymer leading to brittleness

Figure 15.8 Schematic showing the effect that oxidation has on the tie molecules which span the amorphous regions of semicrystalline polymers and which hold together the densely packed crystallites

and mechanical properties of the polymer occurs. Figure 15.8 illustrates why the oxidation of HDPE (which occurs exclusively in the amorphous region causing scission of the tie molecules) is so damaging to the mechanical properties of the polymer.

Oxidation catalysts may be present as impurities or additives in the various chemical environments that the liners are exposed to. This highlights the importance of thoroughly understanding the end use environment for the development of appropriate performance validation and chemical resistance testing methodologies. Metal ions are known to be potential oxidation catalysts for geomembrane materials. They are present in many applications and can also be introduced through metal fittings and components in the geomembrane system. The mechanism by which metal ions exert their influence is through the catalytic degradation of the hydroperoxides that are formed in the auto-oxidation cycle. The influence of metal ions on the oxidation of geomembrane materials at elevated temperatures is a further illustration of the importance of understanding the fundamental mechanisms of degradation in the design of appropriate test methodologies prior to material selection.

Polyethylenes and polypropylenes in particular are adversely affected by certain multivalent transition metal ions at elevated temperatures (especially copper, manganese and cobalt). These metal ions act as catalysts for the oxidation of polyethylene and polypropylene.

In accelerated testing the influence of metal ions on oxidation behaviour is observed at temperatures lower than 110 °C. As the test temperature is lowered from 110 °C, a greater and greater catalytic effect of metal ions is observed. This is a result of the decomposition of hydroperoxides, the step in the oxidation cycle catalyzed by metal ions. At temperatures greater than 110 °C thermal cleavage (i.e. spontaneous thermal decomposition) of the oxygen–oxygen bond in the hydroperoxides becomes more significant than the catalytic hydroperoxide decomposition caused by metal ions.

At lower temperatures, when the thermal decomposition of the hydroperoxides is not significant, metal ion catalysis of the oxidation process is observed and the presence of metal ions is seen to reduce the oxidative lifetime. Unless properly accounted for in geomembrane system design, a significant reduction in geomembrane material lifetime can be observed. For improperly stabilized materials, a 2- to 4-fold reduction in lifetime can result. Testing at multiple elevated temperatures is required in order to properly assess the influence of metal ions on the oxidative lifetime of a geomembrane material.

The combined presence of copper ions (i.e. copper in solution) and elevated temperatures can lead to accelerated thermal oxidation (Gulec *et al.*, 2004). The combined effects of heat and copper ions have caused many field failures of HDPE (particularly cable insulation applications) via a redox reaction. This reaction leads to molecular weight loss via chain scission mechanisms, surface embrittlement and oxidative stress cracking. Such copper-catalyzed degradation of HDPE is especially likely for HDPEs that have had their inventory of antioxidants exhausted or depleted.

Chemical analysis of mining process solutions and tailings in contact with geomembranes can be rich in transition metals such as:

- copper
- vanadium
- nickel
- manganese
- chromium
- iron
- cobalt

In service metal ions in mine waste liquids or mining process solutions can affect the rate of degradation of polyolefins geomembranes because oxidation of polymers is accelerated by metals or metallic compounds (Osawa and Ishizuka, 1973). As described above one of the main functions of a metallic catalyst during oxidation is the breakdown of hydroperoxides to free radicals. Osawa and Ishizuka (1973) studied the effects of various metals on oxidative degradation of polypropylene and found that the catalytic effect of acid metal salts followed the order (high activity to low activity):

$$Co > Mn > Cu > Fe > V > Ni > Zn > Al > Mg$$

Studies have shown that Fe, Cu and Mn which are commonly found elements in metallic mine waste liquids can accelerate the degradation of polyolefin geosynthetics exposed to these liquids.

Hsuan and Koerner (1995) also report that the oxidation reaction of polyethylene can be increased in the presence of transition metals, e.g. manganese, copper and iron. The source of these elements usually comes from metals present in leachates.

If the external surface of the geomembrane has been continuously subjected to attack by oxygen, moisture, sunlight and in direct contact with metal ions, the liner is very likely to have been affected. The presence of oxygen that has diffused from the outer polymer layers causes the formation of peroxides as the starting intermediates for the propagation stage of oxidation. As indicated above, different metal ions show different catalytic effects on polyolefin oxidation.

The main sources of chemical degradation of polymers are the catalytic decomposition of peroxides, direct reaction of metallic compounds with the plastic substrate, oxygen action and energy transfer during UV exposure. The largest effect is obtained when two or more of these factors are combined.

For example, in the case of a mining evaporation pond the following acceleration factors can be combined:

- heat
- UV light
- catalytic metal ions (particularly V, Ni, Cr and Fe)
- exposure time (e.g. 15 years)

Given these environmental risk factors a polyolefin liner in the evaporation pond for instance is likely to have undergone significant antioxidant/stabilizer depletion both via consumption (i.e. metal-catalyzed oxidation) and via extraction.

The OIT literature data show that the depletion rate for antioxidants from HDPE geomembranes increases with increasing exposure temperature, increased concentration of metal ions, and is higher for water immersion than for one-sided only exposure.

Extraction of Antioxidants

Extractability of antioxidants plays a part wherever geomembranes come into contact with liquids such as water or leachates. The rate of extraction is controlled by the dissolution of antioxidants from the surface and the diffusion of antioxidants from the interior structure to the surface.

A number of literature references address the premature degradation of polyolefin geomembranes. Most of these studies were performed with HDPE geomembranes. In particular the publication by Sangam and Rowe (2002) showed that rapid decay of OIT values occurred for 2 mm thick HDPE liners immersed in water at both 55 °C. and 85 °C. At 85 °C the OIT was reduced by 80% over 9 months. At 55 °C., the OIT was reduced by 60% in 2 years. Significantly, the authors concluded that antioxidants in HDPE geomembranes are depleted at rates of 1.6 to 2.4 times faster for samples in water than for air-exposed samples.

For HDPE geomembrane samples in leachates, the depletion of antioxidants is about 4 times faster than that in air and 1.6 to 3.2 times faster than that in water.

There is thus clear evidence in the literature that antioxidant depletion from polyolefin membranes/liners occurs at an accelerated rate in water as opposed to air.

In another recent study, Mueller and Jakob (2003) found that during immersion of HDPE geomembranes in water, a strong reduction in OIT occurs within the first year, after which time the curve levels off. They observed that oxidation starts when very low OIT values are reached after about 3 years. After 5 years the mechanical strength rapidly falls to values below the yield point. The available data (e.g. Mueller and Jakob (2003)) suggest that the service life of HDPE geomembranes is essentially determined by the slow loss of stabilizers due to migration. The oxidation starts only after the depletion of antioxidants and then quickly leads to embrittlement of the sample.

Sangam and Rowe (2002) considered that the antioxidants were depleted by extraction (rather than consumption) for the water and synthetic leachate exposure conditions. They also proposed that the high rate of depletion with leachate was due to the effects of a surfactant in the synthetic leachate causing an increase in the 'wettability' of the geomembrane. This enabled quick dissolution and extraction of antioxidants between the core of the sample and the surface, leading to an increase in the diffusion flux of the antioxidants.

UV Degradation

Polymeric geomembranes such as those based on HDPE, LLDPE, fPP and PVC are susceptible to degradation upon long-term exposure to sunlight. This degradation is a result of photooxidation that occurs in the polymer as a result of exposure to the UV portion of sunlight. The UV component of sunlight contains shorter wavelengths than visible light. These shorter wavelengths are of higher energy, thus they are more damaging to polymers.

The UV light creates free radicals via the scission (i.e. breakage) of carbon–hydrogen bonds, which, in turn, break the polymer chains into shorter molecules and thus produce a more brittle polymer. This reduction in molecular weight of the polymer leads to a drop in tensile elongation and impact properties with resulting embrittlement and cracking. The UV degradation of these geomembrane materials also leads to the formation of oxidized species such as carbonyl and carboxyl groups which can be detected using infrared (IR) spectroscopy. These chemical groups are markers of oxidation so that IR spectroscopic analysis can determine the extent of UV and thermal oxidation of the polymer.

UV light is composed of a range of wavelengths from very short wavelengths (180 nm) which is highly energetic and very damaging, through to longer wavelengths (380 nm). The wavelength ranges corresponding to the laboratory UV exposure lamp classifications are shown in the Table 15.1.

Table 15.1 Range of wavelengths for UV radiation

Type of UV light	Wavelength (nanometre (nm))
UV-A	320–380
UV-B	280–320
UV-C	180–280 (note this wavelength range is normally screened out by the ozone layer)
Natural terrestrial sunlight	295–800

Table 15.2 Exposed geomembrane test results (Ivy and Narejo, 2003)

Material	Age (years exposed)	Visual assessment	Retained properties	Service life prediction
HDPE (2 mm)	10	Excellent	Elongation down 90% OIT down 30%	20–25 years
PVC-R (0.75 mm)	10	Very good	Tensile strength up 30% Tensile modulus up 140% Tensile elongation down 70%	10–15 years
CSPE (1.125 mm)	10	Fair to poor	Tear strength down 60%	10–15 years
CSPE (0.9 mm)	10	Fair	Tear strength down 60%	10–15 years
EPDM (1.125 mm)	2	Excellent	Elongation down 30% Tear strength down 50%	15–20 years
LLDPE (0.75 mm)	2	Excellent	'Tensiles' down 10% Tear strength down 10%	10–15 years

It is important that accelerated UV exposure testing is performed using light sources that best match natural sunlight.

UV degradation of geomembranes can cause embrittlement and cracking of polymer sheets. Once UV degradation has occurred it is not unusual for other factors, such as the action of stress to cause failure. Heat combined with UV radiation can cause plasticizer loss from PVC geomembranes which can render them brittle and susceptible to chalking, cracking and fracture.

PVC being an amorphous thermoplastic, and not a crystalline thermoplastic, is not susceptible to classical stress cracking. It can however be susceptible to plasticizer migration and embrittlement. Note that PVC gets stiffer with time due to plasticizer loss. On the other hand, HDPE and LLDPE lose elongation and embrittle due to oxidation reactions.

Ozone Attack

Ozone (O_3) is a natural enemy of some rubber liners particularly nitrile rubber and butyl rubber compounds (see Figure 15.9). It creates such marked changes in the properties of nitrile rubbers that the oxidation itself is considered a chemical modification of the polymer. The effects of the chemical changes are cumulative over time, and since the cracks are perpendicular to the direction of strain, ozone cracking can cause rapid deterioration of nitrile liners.

Ozone is a naturally occurring gas that can become concentrated around urban and manufacturing areas. Ozone attacks the bonds in a nitrile liner at the molecular level. Bonding at the molecular level is the key to a liner's longevity. Rubber compounds are comprised of polymer chains which form a very complex and interconnected series of 'webs' within the rubber. This network of polymer chains is what provides the rubber compound with strength and integrity. Ozone contacts the outside surface of a liner and literally scissions (i.e. 'cuts') the chains exposed at the exterior surface of the rubber liner.

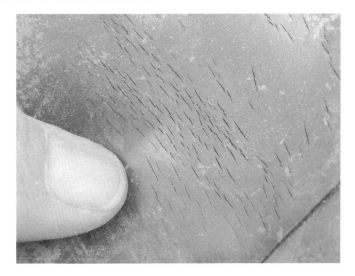

Figure 15.9 Photograph of a nitrile rubber tank liner (thickness 1.5 mm) exhibiting severe ozone cracking on the folds. Ozone cracking (also known as 'rubber ozonolysis') is characterized by the formation of small cracks or fissures on the surface of the rubber which run perpendicular to the direction of strain

This process is the initiation of the cracking. As the surface of the liner flexes under strain the cracks grow/deepen and expose more surface to the ozone attack. This mechanism of damage is progressive.

Without preventative measures, the persistent attack by ozone and surface stresses will ultimately cause the rubber liner to fail. To slow this process down, rubber manufacturers add antiozonants to the rubber. These chemical additives continuously migrate to the surface of the rubber liner and intercept ozone before it reaches the polymer chains. Ozone is not only captured by the antiozonant, but the two react, leaving behind reaction products that can form a protective film that physically shields the rubber from further ozone attack.

Traditionally, ozone attack has been the chief failure mode for nitrile rubber liners. Hence, most nitrile rubber compounds on the market today are manufactured with the inclusion of an antiozonant which protects the polymer against the damaging effects of atmospheric ozone for many years. Therefore, any ozone cracking in rubbers of current manufacture is almost always the result of failure in the manufacturing process of the rubber – specifically the failure to include the correct type (or level) of antiozonants into the rubber compound.

Oxidative Stress Cracking

When a polymeric liner is heat aged or exposed to a strong oxidant (e.g. chlorine) under the simultaneous application of a tensile stress on its surface, it can fail due to oxidative stress cracking (OSC). OSC occurs from the synergistic interaction of oxidation and applied stress.

It has been demonstrated that when HDPE is thermally aged under stress it fails rapidly when the antioxidant (AO) is consumed (Hessel, 1990). At the stresses close to the yield stress the material fails in a ductile manner before any oxidation occurs. However at lower stresses when the AO is fully consumed and oxidation has begun, brittle cracking transpires. The combined effects of oxidation and stress causes a continuously propagating crack which keeps on exposing more of the polymer to oxidation and further accelerates the oxidative embrittlement phenomenon.

LLDPE and fPP do not suffer from conventional environmental stress cracking, however they are susceptible to OSC when their antioxidants and stabilizers are depleted and they begin to oxidize.

fPP floating covers and liners particularly in potable water reservoirs have undergone oxidative stress cracking failures along the tops of folds and at other locations of elevated stress as shown by Peggs (2005). This appears to be synergism between stress and liquid environment, stress and thermal oxidation, and stress and UV radiation accelerating the loss of additives. However, the same material that has cracked in floating cover applications has performed very well in an exposed cap. fPP is not susceptible to stress cracking in its as-manufactured condition so therefore this stress cracking is an ageing phenomenon.

Stress corrosion cracking (SCC), is a related failure mechanism to oxidative stress cracking. SCC is well identified in the literature.

For instance, Wright (2001) states as follows:

"ESC and SCC are distinctly different phenomena, yet the distinction is often ignored even in modern textbooks. It is true that both involve crack initiation and growth under the simultaneous influence of fluid contact and stress, but ESC proceeds without irreversible material modification or degradation. SCC involves chemical attack, whereby the rate of attack is significantly increased by stress".

15.2.8 FAILURE MODES BY GEOMEMBRANE TYPE

Failure of HDPE Geomembrane Liners

Failure of HDPE sheets and geomembranes can generally be ascribed to one or more of the following:

- inadequate design of the HDPE geomembrane application;
- inadequate fabrication or installation of the HDPE geomembrane;
- inappropriate material grade/type/antioxidant formulation for the process solutions and temperatures;
- actual operating conditions (in particular process solutions and temperatures) different from those detailed in the original design specification (or bid request).

Critical factors which can contribute to HDPE geomembrane failure are variations in:

- minimum thickness;
- resin changes which lead to changes to SCR;
- insufficient HALS or AO;
- insufficient carbon black or pigment (e.g. titanium dioxide);
- poorly dispersed carbon black or pigment.

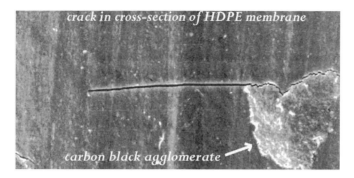

Figure 15.10   Photograph of a brittle crack in a black HDPE geomembrane initiated by a carbon black agglomerate. Reprinted with permission from Session 5D, 57th Canadian Geotechnical Conference, GeoQuebec Copyright (2004)

Figure 15.10 shows a brittle crack in HDPE initiated by a carbon black agglomerate. This highlights the importance of achieving a good carbon black dispersion.

Failure of RPP Geomembranes

There have been instances of stress cracking of reinforced polypropylene (RPP) and fPP geomembranes in some floating cover and lining applications. Effluent that is rich in fats, oils and greases (FOGs) can plasticize fPP and also can extract additives from such liners. HDPE is the most resistant to the extractive effects of FOGs-containing effluent. Fat, oils and greases can permeate flexible PP and cause a reduction in mechanical properties. As fPP is amorphous (i.e. low crystallinity) small molecules can permeate its structure.

The polyester reinforcement in fPP is susceptible to acid-catalyzed hydrolysis. Sewage is known to be acidic because of sulfur compounds. Water vapour can permeate fPP and initiate hydrolysis and breakdown of the polyester scrim reinforcement. fPP due to its amorphous nature can also have some of its additives modified and deactivated by acidic effluent. In particular the hindered amine stabilizers are generally basic in nature and these can react with acidic effluent leading to a loss of stabilizer activity.

For installations which require maximum resistance to chemicals (especially hydrocarbons and FOGs) and installations requiring higher tensile strengths, HDPE is the preferred material for geomembrane liners. fPP has certain advantages for specific applications such as flexibility and resistance to multiaxial strains but its lower density and chemical resistance generally give it a reduced lifetime compared to HDPE.

PVC Liner Failure Mechanisms

PVC liners have cracked from loss of plasticizer at elevated temperatures and under ultraviolet radiation exposure. PVC liners have also cracked due to organic liquids or solvents extracting the plasticizer from the PVC. In these cases the cracking only manifests itself after the solvent/liquid has volatilized off and the PVC is rendered brittle.

15.2.9 CHECKLIST TO PREVENT GEOMEMBRANE FAILURE

The following checklist addresses the essential considerations to prevent geomembrane failure.

Select a quality, properly formulated geomembrane that meets the functional criteria for the intended application	✔
Conduct pre-delivery compliance testing on the geomembrane	✔
Prepare a comprehensive construction quality assurance (CQA) plan	✔
As far as practical eliminate sources of stress[a]	✔
Conduct independent review of all material specifications, project specifications and project drawings, by an experienced geotechnical engineer	✔
Ensure fabricators/installers have welding training and have a level of experience which reflects the magnitude of the project	✔
Employ experienced quality assurance monitors on site when the liner is being installed to ensure compliance with the project specifications and to check the workmanship	✔

[a]In a field environment it is not possible to eliminate all sources of stress. For instance, a zero stress installation is practically impossible to achieve since wrinkles are inevitable especially in thinner/flexible liners. Given this, it is critical to select a geomembrane that has sufficient resistance to stress cracking so that it is able to tolerate the damage induced during deployment (such as scratches, nicks and notches), overheating during welding, installation stresses and service stresses that unavoidably occur.

REFERENCES

Averesch, U. B. and Schicketanz, R. T., 'Recommendations for new installation procedures of geomembranes in landfill composite sealing systems by the 'Riegelbauweise', Presented at *EUROGEO* 2000, Bologna: available from Düsseldorfer Consult GmbH, Düsseldorf, Germany and Ingenieurbüro Schicketanz, Aachen, Germany (2000).

Blight, G. E., Failures During Construction of a Landfill Lining: A Case Analysis, *Waste Management and Research*, **25**(4), 327–333 (2007).

Daniel, D. and Koerner, R., 'Quality Assurance and Quality Control for Waste Containment Facilities', US EPA, Report No. EPA/600/R-93/182, Cincinnati, OH, USA (1993).

Giroud, J. P., 'Wrinkle Management for Polyethylene Geomembranes', Geotechnical Fabrics Report, Vol. 13, No. 3, pp. 14–17 (1995).

Giroud, J. P., Pelte, T. and Bathurst, R. J., Uplift of Geomembranes by Wind, *Geosynthetics International*, **2**(6), 897 (1995).

Gulec, S. B., Edil, T. B. and Benson, C. H., Effect of Acidic Mine Drainage on the Polymer Properties of an HDPE Geomembrane, *Geosynthetics International*, **11**(2), 60–72 (2004).

Hessel, J., 'Evaluation of the Requisite Long-term Strength of Welds in PE-HD Lining Sheets', Montreal, Canada (1990).

Hsuan, Y. G. and Koerner, R., 'Long Term Durability of HDPE Geomembrane, Part 1, Depletion of Antioxidant', GRI Report 16 (1995).

Halse, Y. H., Koerner, R. M. and Lord Jr, A. E., Laboratory Evaluation of Stress Cracking in HDPE Geomembrane Seams, in R. M. Koerner (Ed.), *Durability Ageing Geosynthetics*, Elsevier, New York, NY, USA, pp. 177–194 (1989).

Hsuan, Y. G., Koerner, R. M. and Lord, A. E., The Notched Constant Tensile Load (NCTL) Test to Evaluate Stress Cracking Resistance, in *Proceedings of the 6th GRI Conference*, GRI, Philadelphia, PA, USA, pp. 244–256 (1992).

Ivy, N. and Narejo, D., Canal Lining with HDPE, *GFR Magazine*, **21**(5), 24 (2003).

Lustiger, A. and Markham, R. L., Importance of tie molecules in preventing polyethylene fracture under long-term loadings, *Polymer*, **24**(12), 1647 (1983).

Mueller, W. W. and Jakob, I., Oxidative Resistance of High-Density Polyethylene Geomembranes, *Polymer Degradation and Stability*, **79**, 161–172 (2003).

Nosko, V., Andrezal, T., Gregor, T. and Ganier, P., 'Sensor Damage Detection System (DDS) – The Unique Geomembrane Testing Method, in *Geosynthetics: Applications, Design and Construction*, Balkema, Rotterdam, The Netherlands, pp. 943–748 (1996).

Osawa, Z. and Ishizuka, T., Catalytic Action of Metal Salts in Autooxidation, *Journal of Applied Polymer* Science, **17**, 2897–2907 (1973).

Peggs, I. D., 'Factors Influencing the Durability of Polypropylene Geomembranes', GRI-18 Geosynthetics Research and Development in Progress, January (2005).

Rollin, A. L., Long Term Performance of Polymeric Geomembranes, in *Proceedings of the 57th Canadian Geotechnical Conference*, GeoQuebec, Session 5D, p. 20 (2004).

Rowe, R. K. and Sangam, H. P., Durability of HDPE geomembranes – a review', *Geotextiles and Geomembranes*, **20**, 77 (2002).

Sangam, H. O. and Rowe, R. K., Effects of Exposure Conditions on the Depletion of Antioxidants from High-Density Polyethylene (HDPE) Geomembranes, *Canadian Geotechncal Journal*, **39**(6), 1221–1230 (2002).

Stark, T. D. and Choi, H., Methane gas migration through geommebranes, *Geosynthetics International*, **12**(2), 120 (2005).

Thiel, R., Purdy, S. and Yazdani, R., Case History of Exposed Geomembrane Cover for Bioreactor Landfill, in *Proceedings of Sardinia 2003 Ninth International Landfill Symposium*; available from Vector Engineering, Grass Valley, CA, USA. (October 2003).

Wright, D., 'Failure of Plastics and Rubber Products', RAPRA, Shawbury, UK, p. 13 (2001).

16

Application Areas for Geomembranes

This chapter provides an overview of some of the common application areas for polymeric geomembranes and liners and the challenges that are faced.

16.1 LANDFILL LINERS

The primary function of a landfill liner is to create a low permeability barrier, which is one of a numbers of lines of defence in protecting the groundwater. The groundwater is in potential danger of becoming contaminated from landfill leachate (which is liquid that migrates through the landfill); either from precipitation, or from that liquid already present in the waste.

The use of abandoned quarries for landfills poses interesting design issues due to the steep slopes which are sometimes approaching the vertical. In addition, the walls of old quarries are often jagged and undulating. Therefore a protection layer needs to be considered to protect the geomembranes from localized protrusions and punctures.

Landfill leachates have been found to be quite benign in terms of chemical attack on geomembranes. Based on the US EPA 9090 'Compatibility Test for Wastes and Membrane Liners' the effect of municipal solid waste (MSW) leachate on HDPE geomembranes is innocuous in practical terms. As a result the USA EPA chemical resistance test using MSW leachate for landfill liners is now not routinely performed. There are very few examples of case studies where HDPE or PVC geomembranes are known to have failed due to the degradative effect of leachates themselves (Peggs, 2003). Extensive exposure testing with MSW leachates has shown that any degradative effect of MSW leachates on HDPE can be practically ignored although the leachate can extract antioxidants from the liners especially at higher temperatures.

It is now widely appreciated that geomembrane liners will leak to a certain degree. The maximum allowable leakage rate through single composite liners comprising geomembrane and clay liners is typically 200 l/ha per day. The US EPA requires that for critical installations, such as process water ponds and hazardous waste facilities, there should be a

A Guide to Polymeric Geomembranes: A Practical Approach   J. Scheirs
© 2009 John Wiley & Sons, Ltd

double lining system (DLS) utilized. The interspace between the liners typically includes a drainage system to prevent a liquid head on the secondary (lower) liner. In this way the total lining system is rated as 'leak proof'. Double lining systems such as those used for higher risk landfills allow continuous monitoring of leakage flow rates through primary liners into the leakage collection, drainage and removal systems.

In a wet landfill or a 'reactive' landfill the leachate is recycled back into the solid waste in the landfill to accelerate decomposition of the waste and create more airspace, thereby extending the life of the landfill. Increasingly wet landfills require a double liner system consisting of geomembrane/geosynthetic clay composites (GM/GCL) as the primary liner, GM/GCL as the secondary liner and a leak detection system in between the two GM liners. Note in the case of a GCL, it is important that GCLs are first exposed to water in order to hydrate/swell the clay and achieve its low permeability state.

The German guidelines 'TA Abfall (1991)' and 'TA Siedlungsabfall (1993)' require that landfill cappings and base liners consist of a combination of a mineral layer covered by a geomembrane (i.e. a composite liner). The composite base liner consists of a minimum 75 cm mineral layer, as designed for a domestic waste landfill, or a minimum 150 cm mineral layer for a hazardous waste landfill and a geomembrane (HDPE, thickness $\geq$2.5 mm). A protective layer over the geomembrane might be composed of heavy-weight geotextiles (e.g. 3000 g/m^2) or of a combination of protective geotextile (1200 g/m^2) overlaid with sand (0/8 mm) in a thickness of 15 cm. The drainage layer usually consists of graded permeable gravel (for example, 16/32 mm) (Averesch and Schicketanz, 2000).

For a capping liner a mineral layer of 50 cm and a geomembrane as above is specified. This is supplemented by a drainage system consisting of a protective layer and a drainage layer.

There are two main types of loads on the landfill liner:

- short-term forces (from placement of drainage gravel);
- long-term forces (overburden loads due to waste and hydrostatic loads from leachate).

Double composite liners (DCLs) represent the state-of-the-art in terms of preventing leakages for critical containment applications. The theory is that there will be a high hydraulic head on the upper geomembrane liner (e.g. in a solution pond) but by having two geomembrane liners with a gravel or geofabric leak detection layer there will be a low hydraulic head on the lower liner (see Figure 16.1) and thus no driving force for leakage. While double composite liners are excellent for solution ponds (e.g. for mine liquors) they are not common for leach pads (see below) due to the potential for unstable interfaces.

For leachate collection, geocomposite drains provide high in-plane flow and chemical resistance. A geocomposite drain is placed directly on top of the secondary liner. The geocomposite is composed of a geonet with a geotextile bonded to one or both sides to keep cover soil from clogging the geonet. The bottom geotextile clings to textured geomembrane to lock the geonet in place, especially on slopes. Some landfills require a leak detection system to limit the hydraulic head on the secondary liner. This system consists of a secondary geomembrane below the primary liner with an intermediate geonet or a geocomposite net (GAST, 2008).

Geosynthetic clay liners (GCLs) are also used beneath a geomembrane to form a composite liner system in modern landfills. Generally the geomembrane is placed directly over

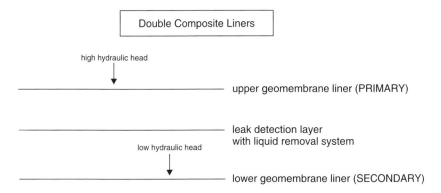

Figure 16.1 The philosophy of double liners is to have a high hydraulic head on the upper liner and a low hydraulic head on the lower liner in order to minimize leakage potential. The geomembrane/composite double liner system consists of two liners – a top primary geomembrane liner and a bottom secondary geomembrane/soil composite liner, separated by a leak-collection system

the GCL. If there are no leaks in the geomembrane the liquid is unable to pass through the geomembrane under a hydraulic gradient. However, if a hole or leak is present in the geomembrane then the GCL greatly reduces the amount of leakage.

16.1.1 DESIGN ASPECTS

The German regulations for the design and construction of landfill liners are the most stringent in the world as they reflect a very conservative, highly regulated approach to ensure long-term survivability of their engineered geotechnical installations (Averesch and Schicketanz, 2000). Table 16.1 summarizes the German design regulations for geomembranes.

Smooth surfaced liners are not optimal for installations with steep slopes. Steep side slopes in landfills means increased capacity and revenue. Steep slopes present problems for smooth liners because cover soil will not stay in place and, with overburden, the liner itself can slip down the slope and increase stresses on seams and the sheet itself. It is now well established that smoother, harder geomembranes such as those made from HDPE display lower interface friction values than softer and rougher geomembranes such as those from PVC. To overcome this problem, textured liners offer frictional characteristics to help keep the liner and cover soil in place during operations.

Increased landfill capacity can be realized by taking advantage of the higher friction angles provided by textured geomembranes or those materials which have a rubbery nature.[1] When the volumetric difference is multiplied by an average compacting factor and then multiplied by the anticipated dumping charge per cubic metre, the increased revenue can be substantial.

The importance of careful design and construction sequencing for multi-layer landfill liners consisting of a sandwich of geomembranes, geotextiles and earth materials, was

---

[1] It is very important however not to induce a shear stress in the liner.

Table 16.1 German regulations for geomembranes and the reasoning underpinning them

German requirements	Reasons
Only HDPE geomembranes are permitted	Broad chemical resistance and proven long-term track record
The minimum thickness is 2.00 mm	Thinner gauges are most susceptible to damage, especially installation damage
The use of a highly engineered compacted clay liner beneath the geomembrane	To ensure a well prepared subgrade with minimal subsidence and irregularities
A protection layer above the geomembrane	To ensure no more than 0.25% strain (point loading) is imposed on the underlying geomembrane
A drainage stone layer above the geomembrane is mandatory and it must be rounded stone (16/32 mm diameter)	To ensure no angular stones are going to puncture or stress the liner
Direct and full contact must exist between the geomembrane and the underlying compacted clay by using the installation beam method	To ensure intimate contact between the geomembrane and the subgrade
The welding of the geomembrane is only done within a stringent operating window	To ensure very low/high ambient temperatures, moisture, dirt and other factors which may adversely affect weld strength are avoided

dramatically illustrated by the failure of the Kettleman Hills landfill in 1988. This involved the failure of a multi-layer landfill lining constructed on a relatively long, steep slope (55 m long at 1:2.2 or 24°). The designers envisaged that the liner protection system would be constructed by working from the bottom of the slope towards the top. The contractor, however, chose to work from the top down, and the designers allowed him to do so. The change of sequence so altered the nature and availability of the resisting forces required for stability that two tension failures of the lining protection system occurred (Seed *et al.*, 1990; Blight, 2007).

With proper material selection, design, construction and inspection, the safe performance life of landfill liners can be considerably increased, at significant cost–benefit ratios.

16.2 LANDFILL CAPS/COVERS

Landfill caps are required to prevent rain and rainwater runoff entering the landfill which would increase the amount of leachate that has to be treated and disposed. In addition, landfill caps are required to collect and remove landfill gas (LFG) from the landfill. Flexible geomembranes such as LLDPE, fPP and PVC are well suited to accommodate the strains and resultant stresses that arise from differential settlements of the waste.

To limit seepage of rainwater into the landfill, a flexible geomembrane liner cap is required. Care needs to be exercised in the selection of the particular geomembrane cover to be used, particularly with respect to the tensile strain that may be expected to be transfered into the geomembrane as a result of settlement. One of the main reasons for settlement is the ongoing decomposition of the waste. Large settlements (5 to 30%) often occur in MSW landfills as the waste degrades due to the heterogeneity in the waste and the process of decomposition. This is evident in the undulating surface of most final covers at MSW landfills.

Leachate generation is a direct result of rainfall. Leachate pumping and treatment represent a significant cost to a landfill operation. To minimize this cost it is prudent to seal-off the completed landfill cell with a geomembrane cover. In addition to keeping rainfall out of a landfill cell, a landfill cover also serves to contain harmful methane gas which is a product of waste decomposition. Once contained, the gas can be collected for power generation or flaring. A long-term concern when employing a landfill cover is differential settlement caused by decaying and shifting waste. The geomembrane cover must facilitate deformation caused by differential settlement without failure of the containment. To meet this requirement, very flexible geomembranes such as fPP and LLPDE are particularly well suited for landfill covers.

Geomembrane landfill caps serve multiple functions including containment of landfill gas (methane), minimizing odours generated by decomposition of the waste and preventing infiltration of rainwater increasing volumes of leachate. Generally they are covered with soil and grass to protect the geomembrane and form an attractive landscape. Unfortunately there have been some cover failures due to sliding of the cover soil over the geomembrane liner. It is important to note that a geomembrane landfill cap on a slope with a soil cover is constantly under moderate, long-term stress. So stress cracking of certain geomembrane types should also be considered. When the cover soil increases in weight due to water not draining properly from it, the soil mass may slip down the slope and if the underlying geomembrane is textured then the geomembrane will have high shear stresses imposed on it and may even slip or possibly tear.

Subsidence issues can arise for sites with poor subgrade stability such as vertical expansion landfills (i.e. 'piggyback' landfills) or landfill covers above degrading waste. For these sites, geomembranes with excellent axisymmetric tensile elongational properties are required, such as fPP. Such geomembranes are especially important for withstanding the out-of-plane deformations beneath the cover.

Table 16.2 summarizes some of the advantages and shortcomings of using exposed geomembrane caps.

LLDPE and PVC are not used for exposed caps because they lack adequate oxidation and UV radiation for 30–50 years UV exposure.

The nature of MSW landfills means they are subject to relatively large waste movements and settlements. Therefore the waterproofing capping layer over a landfill must be flexible enough to accommodate all movement and settlements without cracking or rupture regardless of temperature, mechanical loads, age and some chemical- and biological-influences.

16.2.1 DESIGN CONSIDERATIONS FOR LANDFILL COVERS

Design challenges for installing an exposed geomembrane cover for a landfill include: (a) providing adequate anchorage of the geomembrane to counteract strong wind forces,

Table 16.2   Advantages and shortcomings of using exposed geomembrane caps

Advantages	Shortcomings
Less chance of geomembrane damage due to sliding soil cover on cap	Possibility of wind uplift
There is increased air space for waste	Increased chance of UV degradation (so need to select well formulated materials)
Decreased hydraulic head pressure on the cap	Landfill gas uplift issue (methane whale backs)
Decreased construction time	Damage due to animal traffic hoofs/claws (need to erect fences or barriers)
Reduced construction costs	Need to manage large amounts of rainwater run-off
Reduced seepage of rainwater into the landfill	Unsightly (unnatural) appearance; however green coloured geomembranes are available to help the exposed GM such as an exposed cap to blend into the environment such as to appear as a vegetated cover.

(b) ensuring all critical geomembrane welds are constructed so that they would only be stressed in a shear mode and not in peel mode, (c) managing rainwater runoff from the exposed geomembrane area, (d) ballast for high-wind conditions, and (e) managing the large number of penetrations through the cover geomembrane that would cause localized stresses during wind storms (Thiel *et al.*, 2003).

16.2.2   EXPOSED GEOMEMBRANE Caps (EGCs)

Traditionally, geomembrane caps were covered with soil and vegetation; however there have been a number of well publicized slope failures where the soil cover has slipped off when it became 'moisture-laden'. In some instances this has lead to damage of the cap by tearing. More recently, exposed caps are being installed. Whilst it solves the potential problem of side-slope failures there are other factors that need to be considered such as:

- UV stability;
- elevated surface temperatures;
- wind uplift;
- LFG pressure uplift (which can lead to 'whale backs').

Exposed geomembrane caps also have to possess the performance properties shown in Table 16.3 to ensure long-term service and survivability.

The main geomembranes used for EGCs are fPP-R and EIA-R. LLDPE, despite its flexibility, is discouraged in exposed caps due to its lower resistance to UV radiation. Another geomembrane which is well suited for EGC applications is a bituminous geomembrane (BGM) – particularly in cold climates. BGMs can be installed and welded at below

Table 16.3 Important properties required for exposed geomembranes

Performance property	Measured by
Ultraviolet light resistance (UV resistance)	HP-OIT after QUV exposure in the case of polyolefins (ASTM D-5885)
Resistance to mechanical damage	Tensile strength of the polymer (ASTM D-6693)
Low-temperature resistance	Low-temperature tests (ASTM D-1790)
Dimensional stability	Oven testing (ASTM D-5721)
Puncture resistance	Truncated cone puncture tests (ASTM D-5514)

freezing temperatures and because of their high mass per unit area they resist the forces of wind uplift and LFG uplift.

16.2.3 DIFFERENTIAL SETTLEMENT

In any landfill situation, there is always potential for consolidation of the waste material. If it is non-uniform, differential settlement may result, placing any cap or cover material in distress. Therefore a major long-term concern when employing a landfill cover is differential settlement caused by decaying and shifting waste. The geomembrane cover must facilitate deformation caused by differential settlement without failure of the containment.

Differential settlement causes multiaxial strains in the geomembrane, and the most commonly used liner, HDPE, is not particularly good at withstanding this stress condition (Smith and Orman, 1994). The allowable multiaxial strain for HDPE is commonly taken as 3%. However, strains caused by differential settlement may typically exceed 20%. Therefore, the geotechnical engineer needs to either design to avoid this condition which can be economically unfeasible – or select a geomembrane that can tolerate these magnitudes of strain (Smith and Orman, 1994).

PVC, LLDPE, fPP and VLDPE are the most common choices when the design criteria includes high multiaxial strains. With multiaxial elongation to break values in the 50 to 100% range, all these materials provide a factor of safety of 2 to 5 when required to accommodate actual strains of up to 20%. Flexible PP, being a propylene/ethylene copolymer geomembrane, has favorable multiaxial elongation properties to accommodate differential settlement.

Traditionally, strength-testing methods have relied on one-dimensional tensile tests such as ASTM D-6693. An alternative test method has been developed to more accurately simulate out-of-plane deformation, which occurs in many lining projects including landfill caps where subsidence may occur. The multi-axial test (ASTM D-5617) stretches a circular test specimen in all directions simultaneously. Using a pressure vessel and a constant flow of air, the geomembrane specimen gradually deforms and elongates in the form of a bubble. The peak point of deflection, flow rate and pressure are monitored until eventual rupture of the specimen. The data collected aid in determining the stress versus strain relation of the specimen by assuming either a spherical or elliptical shape to the deflection. The specimen perimeter is fixed and therefore may prevent the result from being representative of field performance.

16.3 MINING APPLICATIONS

Geomembranes are now widely used in the design and construction of various mining facilities for process solution containment (heap leach pads, solution ponds, process and overflow ponds, and tailings impoundments).

For mining facilities subjected to very high loads, such as heap leach facilities, mine waste facilities and tailings impoundments, LLDPE and HDPE geomembranes are more commonly used because of their elongation and strength properties.

Liner thickness is generally selected based on experience, anticipated ore loads and the angularity of the material placed immediately above or below the liner. The required liner thickness for a particular application is evaluated by conducting liner load testing whereby representative soil materials (overliner and underliner) and the proposed geomembrane liner are constructed according to anticipated specifications and configuration, and are loaded to the expected maximum ore loads to evaluate survivability of the liner (Lupo and Morrison, 2005). Depending on the site conditions and design, either LLDPE or HDPE are used.

LLDPE has the following benefits:

- generally higher interface friction values;
- ease of installation due to added flexibility; higher allowable strain for projects where moderate settlement may become an issue.

Conversely, HDPE may be used where:

- there is an exposure to ultraviolet radiation;
- high tensile strength is needed to address high loading and construction stresses;
- high degree of chemical resistance is needed for long-term performance;
- lower cost in parts of the world.

16.3.1 HEAP LEACH PADS

Geomembranes have become critical components in the performance of mining facilities where they are used for containment of process solutions, particularly in heap leach pads.

Heap leaching is a mineral processing technology whereby large piles of crushed or 'run-of-mine rock' (or occasionally mill tailings) are leached with acidic chemical solutions that extract valuable minerals (see Figure 16.2). The heap leaching process is used, for example, to inexpensively recover copper and uranium from their ores. The recovery process employs a solvent extraction system using sulfuric acid and other chemicals to extract minerals from ore. The ore is first irrigated on an impermeable surface where the leaching solution draws out the mineral content. The 'pregnant' solution is then transferred to a mixing-settling tank where kerosene removes the copper. The copper is then removed from the kerosene by sulfuric acid to produce an electrolytic solution from which the

Figure 16.2 Photograph of a Heap Leach pad under construction showing all layers of material used. On the left is the 0.75 mm PVC secondary liner. This is being backfilled with a select fill material for drainage. Note the drainage pipes laid on the liner surface. To the right in the foreground is a separation textile which is being backfilled with a silt layer that was used as a low permeability layer. On top of that in the far distance you can see the 1.00 mm PVC primary liner being placed. This design then called for a layer of select crushed ore to be placed directly on top of the uppermost PVC layer. The project is the Viceroy Resources gold mine and is located near Dawson City in Canada's Yukon Territory. The liner system consists of two layers of PVC liner (0.75 mm) on a silt layer as a secondary liner and a PVC liner (1.00 mm) on a silt layer as the primary liner. Reproduced by permission of Layfield, Canada

copper is plated in an electro-winning process. Finally, the 'barren' solution is re-acidified and sprayed back over another load of ore to complete the cycle (Figure 16.3).

Heap leaching employs a geomembrane lining material to catch a chemical solution that dissolve minerals in ore, and allows the solution to be collected and refined. The main liquid used for the heap leach of copper is sulfuric acid which is compatible with HDPE (at all concentrations) and with PVC (at low concentrations).

Heap leaching presents a combination of high contact pressures and high moisture/acidity conditions on the geomembrane not present in any other containment application. These applications 'push the envelope' of known geomembrane performance, including 150–180 m high heaps, equipment loading of up to 53 tons per wheel, coarse rock overliner, concentrated acid exposure, hydraulic heads of up to 60 m, liquefaction and harsh arid climates with daily temperature extremes (Thiel and Smith, 2004).

The heap leach application thus puts polymeric geomembranes to the ultimate test. The combined action of sulfuric acid and temperatures reaching $70\,^{\circ}C$ on the black sun exposed geomembrane surface, can seriously soften most liner materials. Furthermore the dumping of ore on the liner necessitates a strong membrane that is resistant to abrasions and punctures. Note also that most designs have sand protection layers. Finally the steep, angular design of the collecting ponds requires a strong, durable product.

Due to the harsh nature and location of mining projects, the performance of geomembranes are often pushed beyond the recommended general design limits. For lined mining

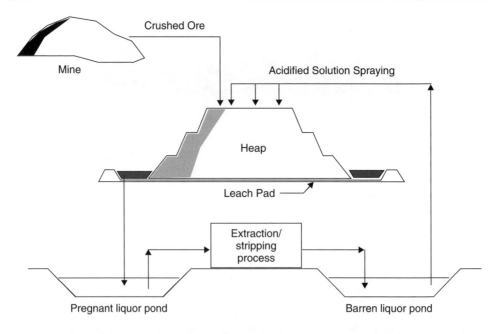

Figure 16.3   Schematic showing design of a leach pad for heap leaching of ore using acidified solutions

facilities subjected to moderate to high loads (greater than 300 kPa), geomembrane materials such as HDPE, LLDPE, and (to a lesser extent) PVC, are used because of industry experience with these materials and documented performance from constructed mine facilities. However, geomembrane-lined heap leach facilities are being designed with ore heights approaching 200 m, resulting in normal stresses in excess of 3.3 MPa (Lupo and Morrison, 2007).

The heap leach pads are generally constructed utilizing as much is possible the natural topography of the site. The pad area is cut and filled as required, and trimmed to achieve the desired slope of 0.5 to 1%. HDPE or LLDPE are normally used for the base of the pad with the liner being between 1 and 2.5 mm thick over the pad and between 2–3 mm thick in the sumps and drains. Directly over the liner drainage pipes are installed and then a granular protective soil layer about 60 cm thick is placed directly over the geomembrane–pipe system, to protect the liner during the ore stacking (Defilippis, 2005).

Geomembrane liner designs generally consist of either a single composite, or less commonly double composite systems with a leakage collection layer. The single composite liner systems generally consist of a geomembrane liner placed on top of a compacted liner bedding soil. This type of configuration is commonly used in areas that experience a low hydraulic head (typically less than 1 m).

A double composite liner system generally consists of two geomembrane liners (an upper primary geomembrane liner and a lower secondary geomembrane liner) separated by a leak collection/drainage layer. The lower secondary geomembrane is placed over a compacted liner bedding soil. A double composite liner system is typically used in mines where high hydraulic heads may occur, such as valley leach facilities (as well as some tailings impoundments and process solution ponds). Note the design of a composite liner system for heap leach pads must consider the possibility of unstable (i.e. low friction) interfaces.

Increasing heap heights, now approaching 150 m (with ore densities generally in the range of 1500–1800 kg/m^3), as well as the emerging trends of using concentrated sulfuric acid for copper ores are stretching the performance of geomembranes (Thiel and Smith, 2004). In addition, the base of heaps can reach temperatures of 40 to 45 $^\circ$C.

Types of Leach Pads

Leach pads can be divided into four categories (Thiel and Smith, 2004):

1. Conventional or 'flat' pads;
2. Valley fills;
3. On/off pads.

Conventional leach pads are relatively flat, either graded smooth or terrain contouring on gentle alluvial fans such as in the Chilean Atacama desert, Nevada and Arizona where the ore is stacked in relatively thin lifts (5–15 m typically) (Thiel and Smith, 2004).

Valley fill systems are just those – leach 'pads' designed in natural valleys using either a buttress dam at the bottom of the valley, or a leveling fill within the valley.

On/off pads (also known as dynamic heaps) are a hybrid system where a relatively flat pad is built using a robust liner and overliner system. Then a single lift of ore, from 4 to 10 m thick, is loaded and leached. At the end of the leach cycle, the spent ore is removed for disposal and the pad recharged with fresh ore. Usually loading is automated, using conveyors and stackers (Thiel and Smith, 2004).

Table 16.4 lists the characteristics of a variety of heap leach projects in South America.

Increasing Heap Heights

In the 1980s when heap leaching first became popular for gold recovery, the typical maximum ore heights were around 15 m. Then by 1990 that height had been increased by the copper industry to 50 m. Nowadays essentially all heap leach designs have ultimate target depths of at least 50 m while several are in operation or under construction with target depths of over 200–300 m (Thiel and Smith, 2004).

Materials Used

HDPE has traditionally been the material of choice for the construction of copper leach pad construction while more recently very flexible polyethylene has become more common. Although PVC and EIA-R liners are used they are not widespread.

The most common geomembrane material used for leach pad construction is 1.5–2.0 mm polyethylene (both HDPE and LLDPE). Thicker PE is used occasionally for higher heaps and 0.75–1.0 mm PVC is also occasionally used.

Geomembranes selection needs to consider the point loads from materials on top and below the geomembrane in service.

Table 16.4 Variety of heap leach projects in South America (Thiel and Smith, 2004)

Location	Type	Pad area (ha)	Base liner	Max ore height (m)
Argentina	Valley fill	150	2 mm and 2.5 mm HDPE	130
Brazil	Conventional	119	2 mm HDPE or LLDPE	75
Chile	On/off	152	2 mm HDPE	10
Chile	On/off	135	2 mm HDPE	10
Chile	On/off	100	2 mm HDPE	10
Chile	Dump	95	1.5 mm HDPE	110
Chile	Dump	125	1.5 mm HDPE	125
Chile	Conventional	200	2 mm LLDPE	145
Chile	Conventional	130	0.75 and 1.00 mm PVC	75
Peru	On/off	15	2 mm HDPE	10
Peru	Valley	55	1.5 mm HDPE	85
Peru	Valley	125	2 mm and 2.5 mm HDPE	135
Peru	Valley	75	2 mm and 2.5 mm LLDPE	230

The suitability of geomembranes in weak sulfuric acid pregnant liquor solution (PLS) has been demonstrated; however the recent trend to pre-cure the ore with concentrated (98%) sulfuric acid to increase ultimate metal recovery, significantly shorten leach cycle times, improve through-put from on/off pads and hence reduce inventory in permanent heaps can lead to polymer degradation (Thiel and Smith, 2004).

Effect of 98% Sulfuric Acid on HDPE, LLDPE and PVC

The compatibility of HDPE, LLDPE and PVC geomembranes with 98% concentrated sulfuric acid has been evaluated over a 120-day exposure programme in the laboratory. Samples of each geomembrane were exposed to 98% H_2SO_4 solution at 50 °C. and then tested each 30 days for retention of physical properties. Table 16.5 shows that both HDPE and LLDPE geomembranes and in particular their additive packages are adversely affected by the acid.

Both types of polyethylene geomembranes performed very well given the aggressive environment with tensile strength and elongation properties after 120 days generally within 10% of the original condition (see Table 16.5). PVC, however, exhibited a drastic loss of flexibility (i.e., a negative change percentage), even within the first month. The increase in tensile strength for PVC is accompanied by a reduction in elongation, which means that the material has lost plasticizer and is becoming brittle in just 30 days (see Table 16.6). It was observed that the immersion solution turned very dark in the first 24 h suggesting a very rapid leaching of plasticizers. On the basis of these results PVC is not suitable for use in concentrated acid pre-curing operations, even for relatively short exposure periods.

While the polyethylene geomembranes retained the majority of their mechanical properties they exhibited losses in oxidation induction time (OIT) of 73% for LLDPE and 64% for HDPE (see Table 16.5). OIT is a key indicator for longevity and loss of OIT is a precursor to failure by ageing (Thiel and Smith, 2004). Depending on cure periods, cycle times and design life, cumulative exposures of up to 10 months are possible but 4–6 months is probably more typical. Since an OIT reduction of 64–73% was measured in 4

Table 16.5 Effect of concentrated sulfuric acid on the mechanical properties of selected geomembranes after 120 days (% change from original) (Thiel and Smith, 2004)[a]

Property	HDPE 1.5 mm (%)	LLDPE 1.5 mm (%)	PVC 0.75 mm (%)
Tensile strength at yield (ASTM D-638/D-882)	+2 (MD) −4 (TD)	0 (MD) 0 (TD)	+173 (MD) +188% (TD)
Tensile strength at break (ASTM D-638/D-882)	−4 (MD) −4 (TD)	−7 (MD) −11 (TD)	+54 (MD) +54 (TD)
Elongation at yield (ASTM D-638/D-882)	−5 (MD) −5 (TD)	−10 (MD) +9 (TD)	NA
Elongation at break (ASTM D-638/D-882)	+5 (MD) 0 (TD)	−7 (MD) −12 (TD)	−66 (MD) −76 (TD)
Puncture resistance (ASTM D-4833)	−3	+1	+130
Tear resistance (ASTM D-1004)	−3 (MD) +2 (TD)	−5 (MD) −5 (TD)	+107 (MD) +112 (TD)
Hardness (ASTM D-2240)	0	+5	+31 (indicates loss of plasticizer)
OIT (ASTM D-3895)	−64 (indicates loss of antioxidant)	−73 (indicates loss of antioxidant)	NA

[a]Notes: where two values are shown, they are for the machine/traverse directions (MD/TD), respectively; for PVC, tensile strength at 'yield' was taken at 100% elongation as the yield point is indeterminate; OIT is only applicable for polyethylene.

months the longevity of the geomembrane must be a concern for this pre-curing applications. This problem could be solved by using coextruded HDPE with a top layer made with an additive package specifically formulated for this environment (e.g. acid-tolerant non-basic HALS stabilizers).

For the design of high load facilities such as leach pads, the foundation settlement analyses are commonly integrated into the grading plan to address potential strain development within the geomembrane liner. For example, some areas of a facility may be graded with an intentional camber (i.e. not flat) to offset potential future settlements (total or differential).

Cyanide Leaching

The Tarkwa gold mine in Ghana has a $390\,000$ m^2 heap leach pad constructed of 1.5 mm thick HDPE geomembrane (Solmax 460 smooth HDPE) which has been in operation since 2000. There are 6 pads separated by divider berms. A leaching solution of sodium cyanide is sprayed on the 8–10 m high lifts of crushed gold bearing ore to extract the gold. The function of the geomembrane liner is to capture the 'pregnant solution' at the base of the heaps and lead it via launders to a contaminant pond for processing. The lining needed to be durable enough to withstand the load of the ore, ultimately stacked up to 40 m high and withstand the chemical attack (Solmax, 2008).

Table 16.6 Effect of 98% sulfuric acid on the properties of PVC geomembranes (% change from original values) (Thiel and Smith, 2004)[a]

Immersion time (days)	Tensile strength at break (ASTM D-882) (%)	Elongation at break (ASTM D-882) (%)	Puncture resistance (ASTM D-4833) (%)	Tear resistance (ASTM D-1004) (%)	Seam shear elongation (ASTM D-6392) (%)
30	+31 (MD) +27 (TD)	−58 (MD) −74 (TD)	+129	+119 (MD) +122 (TD)	−90
60	+62 (MD) +40 (TD)	−71 (MD) −75 (TD)	+120	+122 (MD) +110 (TD)	−94
120	+54 (MD) +54 (TD)	−66 (MD) −76 (TD)	+130	+107 (MD) +112 (TD)	−86

[a]Where two values are shown, they are for the machine/traverse directions (MD/TD), respectively.

Table 16.7 Geomembrane selection general guide for leach pads (Lupo and Morrison, 2007)[a]

Foundation condition	Underliner	Overliner	Effective stress on liner	
			<1.2 MPa	>1.2 MPa
Firm	Coarse grained	Coarse grained Fine grained	2 mm LLDPE or HDPE	2.5 mm LLDPE or HDPE
Firm	Fine grained	Coarse grained Fine grained	1.5 mm LLDPE or HDPE	2.0 mm LLDPE or HDPE
Soft	Coarse grained	Coarse grained	2 mm LLDPE	2.5 mm LLDPE
Soft	Coarse grained	Fine grained	2 mm LLDPE	2.5 mm LLDPE
Soft	Fine grained	Coarse grained	2 mm LLDPE	2.5 mm LLDPE
Soft	Fine grained	Fine grained	2 mm LLDPE	2.5 mm LLDPE

[a]Notes: Foundation conditions are presented as relative stiffness and potential for settlement. Field tests are required to assess the impact of the foundation stiffness on the geomembrane; Underliner refers to the material directly beneath the geomembrane (primary geomembrane for double composite liner systems). Testing and design calculations are required to assess the impacts on the geomembrane. Overliner refers to the material directly above the geomembrane. Testing and design calculations are required to assess the impacts on the geomembrane.

Leach Pad Material Selection

Table 16.7 present a general guide for geomembrane selection for the underliners and overliners for leach pads.

For leach pads that use concrete reinforced structures and require the liner to be sealed around the edges at odd angles, EIA-R has been used owing to its chemical resistance, flexibility, strength and easy fabrication. The ability to seam the geomembrane using thermal welding and produce factory fabricated panels ensures that installation is rapid, easy and field seaming is kept to a minimum.

The following considerations should enter into the selection of geomembrane liner type and design of heap leach facility:

- High elevations (greater than 3000 m) due to higher UV exposure.

- Heating by solar radiation, exposure to UV and large temperature variations.
- Areas with high winds.
- Limitations to the maximum amount of exposed liner (prior to placement of ballast or overliner fill). Construction schedules can be adjusted to maximize liner deployment during periods of the day when winds are the lightest.
- Areas with high rainfall, liner deployment and installation must be scheduled based on the weather forecasts to maximize deployment during dry periods. This can be a difficult challenge in areas of high annual precipitation.
- For areas of high snowfall, liner deployment is also based on weather forecasts. Liner seaming can be completed in cold temperatures provided proper seaming temperatures can be maintained. Facility phasing may also be included in the design to ease construction.
- For areas with variable temperatures, care must be taken to minimize liner stresses due to expansion and contraction of the liner (Defilippis, 2005).

Important areas to pay attention to avoid failure of polymeric geomembrane heap leach pads are:

- Site specific puncture testing (e.g. hydrostatic puncture as per ASTM D5514).
- Interface Shear Testing (ASTM D-5321) to ensure stability of the liner system under the anticipated loading conditions.
- Subgrade acceptance prior to geomembrane installation (limitation of projecting rocks and limited desiccation cracking of soil liner).
- Monitoring of drainage layer placement and in particular ensuring the materials are placed according to project specifications with regard to equipment used for placement, minimum drainage layer thicknesses are maintained (function of equipment used for placement), and care taken by the contractor to limit damage to the underlying geomembrane (Defilippis, 2005).

16.3.2   TAILINGS DAMS

Tailings dams and ponds use liners to prevent the release of concentrated mine chemicals into the environment and to improve consolidation of the tailings. Solution ponds (and other specialized pond liners) retain mining chemicals and process water for retention, reuse or treatment. These ponds and dams accumulate the solids remaining from the mining operation, often in the form of a fine sand. Tailings ponds allow the sand to settle and the liquid to separate for reuse. HDPE, LLDPE and fPP are widely used to line tailings dams.

16.3.3   LINERS FOR SALT EVAPORATION PONDS

Geomembranes are used to line salt evaporation ponds in warmer climates. PVC geomembranes are the liner of choice for salt evaporation ponds even in harsh environments. This is because PVC geomembranes are durable and offer excellent chemical resistance to salts, which is important because of the long-term exposure of the geomembranes to the brine.

PVC geomembranes also exhibit much smaller wrinkles than polyethylene geomembranes when installed because of a lower thermal expansion coefficient, higher subgrade/geomembrane interface strength coupled with the flexibility of PVC geomembranes. These attributes are significant in this application because the smaller wrinkles result in substantial intimate contact between the geomembrane and subgrade and the protective salt layer. The benefit of intimate contact is a reduction in the lateral flow of the brine solution from a hole or leak in the geomembrane.

Furthermore, PVC has high elongation and tends to drape around any protrusions on the compacted layer underneath and thus a PVC liner helps minimize the occurrence of small holes and brine loss.

16.4 FLOATING COVERS

A floating cover is a polymeric membrane installed over the surface of a lagoon or reservoir that rises and falls with the changing level of the liquid. When used on water reservoirs, floating covers protect the contents from contamination and evaporation. Also used to cover chemical, industrial waste and wastewater lagoons they protect the surrounding area from the material contained and especially from its odour. Floating covers can also collect biogas (i.e. methane), which is released when anaerobic digestion takes place in concentrated animal feeding operations and food processing plants. The methane (biogas) can be converted to electricity to power the facility.

There are two main types of floating covers: (i) defined sump and (ii) mechanically tensioned floating cover systems. There are also modular floating covers for niche applications.

Most floating covers are known as the 'defined sump' type where a series of floats are arranged in parallel pairs with weights (i.e. ballast) centred between them to form defined sumps or troughs. The floats that are laid out around the cover create sufficient tension to allow workers to walk on it to direct rainwater to the troughs, then to dewatering sumps for removal. The floats are made from either rotomoulded polyethylene or encapsulated expanded polystyrene. The sumps or troughs are generally laid out in a 'double Y' configuration (see Figure 16.4).

Defined sump covers use floats and weights to create rainwater collection sumps in the cover and to accommodate changes in water level. Defined sump covers are used for all floating cover applications including potable water storage, odour control, evaporation control and contamination/dilution protection. Defined sump cover system uses carefully located floats and weights to accommodate changes in water level. The weights and floats create channels in the cover that act both as slack, and as sumps to collect rainwater. Each defined sump has two floats, one on each side of the sump, with a sump weight centred between them. As the water level rises in the pond the weight sinks and the floats move together. When the pond is full the sump floats are close together (typically about a metre apart). The sump weight pulls the slack down into the pond between the floats, making channels in the surface of the cover. Rainwater is collected in these sumps and is removed (usually by pumps) from the cover.

Companies that fabricate floating covers include Layfield Environmental Systems and MPC Containment.

While material selection for floating covers is site-specific, in all cases the membrane for the cover must have excellent resistance to UV degradation, tearing and puncture. The

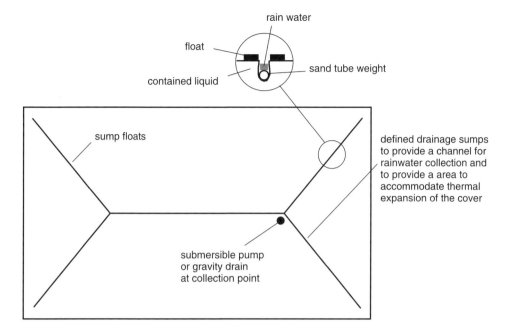

Figure 16.4 Schematic showing the basic design of a defined sump floating cover

cover material must also possess excellent seam strength (sufficient to hold up to foot traffic) and of course must be very flexible. Many floating covers are scrim-reinforced to enhance tear resistance. The type of liquid contained, its chemical composition, its temperature range, the ambient temperature range, environmental conditions and the intended function of the cover are some of the variables that must be considered. Typical materials used are fPP-R, EIA-R, CSPE-R and LLDPE-R. HDPE is also used despite its stiffness and high degree of thermal expansion, primarily for applications where good chemical resistance toward fats, oils and greases is required such as to cover sewage ponds, animal waste ponds, diary processing waste ponds, etc.

Mechanically tensioned floating cover systems consist of tension producing elements attached to the cover's interior at specific intervals all around the reservoir's perimeter. The mechanically tensioned portion of the cover is thus held in place and is protected from wind uplift and drifting. The outer cover perimeter is relaxed and forms a sump where storm water can be diverted off of the cover through a drainage system. The cover geomembrane material makes direct contact on top of the water without additional floats for buoyancy. A key performance requirement of both types of floating covers is that they function at all water levels without undue stress on the membrane.

Overall maintenance is reduced with a tensioned floating cover system. The peripheral fold is easier to clean than a fold or sump in the middle of the cover because it is not submersed in water, and can be more easily opened. Maintenance can be further reduced in the peripheral fold by the use of a protective skirt to prevent it from filling with rainwater and debris. Other floating cover designs require submersed folds (or sumps) in the working portion of the cover, creating areas which are difficult to clean and inspect for leaks, and equally difficult to repair (Layfield, 2008).

16.4.1　FLOATING RESERVOIR COVERS

Floating covers for reservoirs improve water quality, prevent water loss due to evaporation and reduce the amount of chemicals required for treatment. Floating covers provide a cost-effective solution to issues of chlorine loss, water evaporation and contamination prevention. Usually, polymeric geomembrane covers are one-eighth the cost of metal covers and feature a longer warranty. Plus, floating covers allow construction on a large-scale basis (several hectares) in a way that rigid structures are not capable of.

Floating covers for reservoirs often use patented tensioners attached around the perimeter of the cover system to prevent undue cover movement and wrinkling regardless of the reservoir's water level fluctuation. Tensioners also serve to retain slack cover material in a defined peripheral sump. Cover drains conduct storm water into a drainage system. Floats and weights are encapsulated in the geomembrane for maximum functionality. Such floating covers are typically fabricated from CSPE-R, LLDPE-R or fPP-R geomembranes. In particular, due to the excellent long-term weathering resistance of CSPE membranes and liners they are often used to manufacture floating covers. These membranes have been used as lining materials for more than 30 years now and have shown excellent durability as both a liner and a cover material.

A defined sump floating reservoir cover uses a combination of attached closed cell foam floats and sand tube ballasts. These elements are used to create drainage sumps that collect and carry storm water to either scavenging pumps or gravity drains for water removal into a drainage system.

For floating covers that are required to have traffic on the cover for maintenance and inspections, scrim-reinforced geomembranes are used. Scrim-reinforced polymeric membranes give floating covers increased strength enabling workers easy and safe access to the reservoir by walking on the cover. Floating covers can be manufactured to incorporate access hatches for reservoir divers or other workers who need to inspect or repair the facilities.

Reservoir floating covers based on polymeric membranes offer cost-effectiveness in several areas:

- low initial capital cost;
- low maintenance and operating cost;
- reduction of water loss through evaporation;
- prevention of chlorine depletion and reduction of chlorine costs.

16.4.2　INDUSTRIAL FLOATING COVERS

Industrial floating covers are often used as gas collection systems for odour control and for promoting anaerobic digestion by maintaining warm temperatures. The most popular types of geomembranes specified for this application are potable grades of either scrim-reinforced CSPE-R or fPP-R. Both are exceptionally high performance materials, CSPE-R being considerably more expensive than HDPE but it is offered with an unprecedented 30 year warranty.

Geomembrane covers for anaerobic lagoons (i.e. anaerobic digesters for sewage) must consider all aspects of sludge, scum, biogas and rainwater management. Floating covers for such lagoons can be comprised of the following typical tri-layer constructions:

- EIA-R as the top layer, for low thermal expansion and outstanding UV resistance.
- A closed-cell PE foam interlayer as a thermal insulating layer to maintain warmer temperatures in cooler climates so as to maximize biogas production (note when the lagoon temperature falls below 15 °C then biogas drops away quickly) as well as for buoyancy.
- A HDPE membrane used as the lower layer (i.e. underlayer) since its low coefficient of friction provides a slippery surface to facilitate floating scum migration to prevent scum deposition.

Alternatively:

- fPP-R (top) (low thermal expansion & outstanding UV resistance);
- PE foam (interlayer) (thermal insulator and for buoyancy);
- HDPE (underlayer) (slippery surface to prevent scum deposition).

Such covers are produced by Geomembrane Technologies Inc. (GTI), amongst others. As scum begins to accumulate in the one location it can desiccate (i.e. dry out) and become more buoyant leading to 'scum-bergs' that place an upthrust on the cover and interfere with rainwater channels. Note however that many anaerobic sewage lagoons screen their influent and so scum accumulation under the cover is less of a problem.

Industrial floating covers often have a folded design with defined sumps for rainwater management (see Figure 16.5). Because of the in-built folds such covers can rise and fall with the water level. This design however can be susceptible to wind uplift (i.e. they

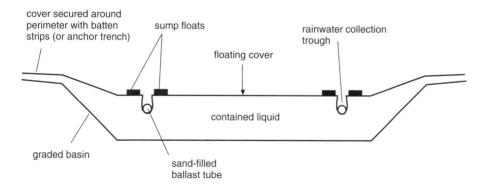

Figure 16.5 Schematic of a 'defined sump' floating cover where sand-filled ballast tubes provide defined troughs for rainwater collection

can unfold) and also buoyant scum can accumulate in the inner folds disrupting biogas collection paths.

Black geomembranes are often the preferred floating covers on sewage treatment ponds to increase the efficiency of anaerobic digestion. The black colour increases the temperature beneath the cover and this promotes further the bacterial activity and increases the volume of methane generated.

Floating covers are also used over lagoons containing food waste, dairy waste, meat processing wastes and the like. There are critical health and safety aspects with floating covers – especially those that allow human traffic on the covers to remove debris and vegetation. There have been instances of workers falling through cracks and tears in the cover into the anaerobic waste below. There have been other cases of fPP floating covers on chicken fat and on molasses cracking under foot as personnel were walking on them.

The scrim reinforced covers (i.e. EIA-R, CSPE-R and fPP-R) make the flexible membranes more dimensionally stabile against temperature variations. Some problems have however been encountered with these products due to hydrolysis of the polyester scrim at elevated temperatures caused by the presence of water vapour. These materials can wick water along the scrim fibres despite assurances from the manufacturers that anti-wicking compounds are applied.

Flexible polypropylene is permeable to fatty acids and so the surface of fPP covers can become sticky over time. There have also been problems reported of cracking of fPP-R in folds of sumps of covers on potable water storages. The polyester scrim can undergo failure (e.g. by rupture) by acid hydrolysis if subjected to high temperatures and if there are scrim-exposed edges. For this reason all exposed edges need to be sealed with extrudate. The manufacturers of these materials recommend sealing all exposed cut edges. Exposed fPP membranes are notorious for being difficult to thermally seam with age.

CSPE covers have an excellent track record in water storage covers due to outstanding UV, ozone and chlorine resistance. However CSPE covers and membranes have only a limited service record in anaerobic digestion cover applications. There are some reports of CSPE covers being susceptible to attack from fatty acids leading to softening and delamination. Other problems reported with CSPE covers and membranes are:

- autoadhesion and blocking of the sheet on rolls;
- surface crosslinking which is not readily identifiable by the welding personnel.

HDPE is not an ideal membrane for floating covers due to well-documented thermal expansion and potential stress cracking issues. Nevertheless it is gaining a good track record as a cover option for anaerobic applications owing to its good chemical resistance and the fact that it is more rigid than the reinforced membrane materials; hence there is less 'flapping' in the wind.

Table 16.8 lists some material considerations for floating cover design and longevity.

Rainwater Management Issues

Poor rainwater management places stress on the cover and blocks gas paths leading to 'whales'. Whales in turn place further stress on the cover. Folded covers with defined sumps provide a channel for rainwater collection and also provide somewhere for the cover material to go when the material thermally expands. Best practice floating covers have weighted sumps and no water pooling occurs on these.

Table 16.8 Material considerations for floating cover design and longevity

Material issues	Comments
Reinforced membranes/ covers – scrim thermal hydrolysis degradation	• Black scrim-reinforced covers can reach high temperatures that accelerate hydrolysis of exposed polyester scrims[a] • White capped scrim-reinforced geomembrane covers help to prevent high temperatures and this helps protect the scrim from thermal hydrolysis
Reinforced membranes/ covers – scrim acid hydrolysis degradation	• Hydrogen sulfide condensate against the underside of the cover can cause acid-catalyzed hydrolysis of the polyester scrim especially when the cover reaches temperatures of $80\,^{\circ}C$ in the sun

RPP degradation	• RPP when used for covers must have:
	  – the right thickness (i.e. >1.0 mm)
	  – the right additive package
	  – no folds
	 • fPP-R covers can be difficult to weld after a few years of exposure due to an oxidized surface layer. fPP patches do not bond well to aged fPP covers and patches can be peeled off even by hand pulling

CSPE degradation	• CSPE can undergo deterioration in the presence of certain fatty wastes • Sewage scum contains fatty acids that can cause 'burn marks' on CSPE • Solvent-bonded CSPE seams can cause softening and fibre pull-out that can weaken the seamed membrane • Other than the above, CSPE has outstanding weatherability[b]
EIA-R degradation	• EIA-R covers can be difficult to weld after a few years of exposure due to the oxidized surface layer • Cut samples of EIA-R immersed in $82\,^{\circ}C$ water for 14 days gave a 92% reduction in tear strength due to scrim hydrolysis. The hydrolysis of the scrim is largely due to water and heat but it is accelerated by acids and alkalis. The colour of the membrane is an important factor determining how much solar heating occurs
R-PE delamination	• Reinforced polyethylene (RPE) in which the scrim is bonded into a 'molten polyethylene bonding layer' (which is inside two plies of outer PE film) can be susceptible to blistering and delamination
LLDPE degradation	• LLDPE and VLDPE are less susceptible to brittle cracking than HDPE but can suffer from poor chemical resistance and poor UV performance
PVC degradation	• PVC can become hard and brittle when in contact with acidic sewage effluent and H_2S/H_2SO_4 (not recommended in these applications)

[a]Factory produced edges (or selvedge) encapsulate the scrim with the coating material so that only the cut edges are of concern with regard to potential wicking. Most manufacturers now use low-wicking scrim.
[b]CSPE has a 30 year track record for exposed covers and resists the elements better than any other currently known lining material. It has been the material of choice for floating covers and 'Burke Mercer Environmental' offer a 30 year guarantee.

16.4.3   FLOATING COVERS FOR GAS COLLECTION

Gas Collection Covers (GCCs) are special floating cover systems used to collect gases from wastewater treatment lagoons, sludge ponds, aeration systems, flow equalization tanks and pretreatment tanks. Such covers are manufactured by Geomembrane Technologies Inc. (GTI, New Brunswick, Canada) amongst others.

In the case of anaerobic digesters, GCCs are used to capture biogas which can be burned to generate process heat or electricity. Other benefits of a GCC system include reduction of process heat loss/gain, elimination of water evaporation and prevention of sunlight penetration. With an appropriate design nearly any water level fluctuation requirement can be accommodated and automatic rainwater removal systems can also be incorporated. Such floating covers are strong enough to safely support foot traffic, light vehicles and snow loads.

Floating covers for gas collection are being increasingly used for covering concrete pig manure holding tanks. Floating covers in pig/swine farms offer the following advantages:

- a 25% increase in tank capacity by eliminating rainwater;
- improved manure quality since the cover captures fugitive ammonia resulting in higher nitrogen content, eliminating the addition of expensive chemical nitrogen;
- virtually eliminates odours, flies and insects from the manure tank;
- potential accumulation of greenhouse gas credits due to reduced emissions;
- can cover large diameter existing concrete pig manure holding tank.

EIA-R is often the material of choice for such floating covers because of its unique combination of performance properties relating to strength, low thermal expansion/contraction, ease of installation and resistance to sunlight.

Industrial ponds have insulated covers to maintain high temperatures as anaerobic activity decreases with decreasing temperature.

Whale Backs

If gas flows in landfill caps or anaerobic lagoons are not properly managed then large gas balloons can develop (which are referred to as 'whale backs').

Case Study

One of the world's largest floating membrane cover system was installed by Geomembrane Technologies Inc. (GTI) for the Melbourne Water Corporation in Melbourne, Australia to solve a sewage and wastewater treatment odour problem (Water and Wastes Digest, 2002).

The installation covers the inlet ends of two large lagoons (25 West and 55 East) with a three-layer floating insulated cover made of EIA-reinforced geomembrane, a middle layer of polyfoam insulation and a bottom layer of high-density polyethylene. The underskin provides protection of the cover system from scum development and scum movement under the cover. The EIA-R material provides minimal expansion/contraction due to changes in ambient temperature, is UV light resistant and has high tensile strength. The three-layer nature of the cover system, with a top layer floating free and a bottom layer welded to the foam insulation, provides a lightweight but durable structural integrity to the cover system which allows buoyancy and access for maintenance and inspection.

Accumulation of rainwater and vegetation on a floating cover can be very problematic and dramatically increase the level of maintenance that is required. The cover installed for 'Melbourne Water' incorporates a network of weights, floats and self-regulating gravity drains that allow passage of rainwater into the liquid below. There are no folds in the design for collection of rainwater and there are fewer wear areas in the cover that require periodic maintenance.

The fold-less cover design allows the designer to use the entire walled perimeter of the lagoon as a transmission zone for biogas since the cover sweeps from the top of the wall, where there is a gas-tight seal, down to the liquid surface. The sweep area, along with a system of floats and weights, allows unhindered conveyance of biogas from the central areas of the cover to the perimeter where it is conveyed through wall penetrations and piping to the biogas blowers.

Attributes of the cover design include:

- provides excellent strength and stability which requires low maintenance;
- facilitates maintenance when required;
- allows for the lagoon water level to fluctuate;
- self draining for rainwater;
- long design life;
- structural integrity to withstand pressures caused by scum development;
- allows for increasing the length of the cover;
- maximizes biogas collection and minimizes infiltration;
- facilitates biogas migration/collection;
- can be used to store biogas for use during peak power demand periods.

The insulation layer is made of 12.5 mm of polyfoam insulation, providing both buoyancy and thermal properties. The underskin layer is made of 0.75 mm HDPE providing protection of the cover system from scum development and scum[2] movement under the cover. The polyfoam insulation is welded to the HDPE underskin layer, but the top membrane layer is independent from the insulation/HDPE layer. Biogas (mainly methane) is allowed to migrate between the HDPE underskin and the EIA-R cover through orifices located along the perimeter cover sweep. When biogas storage is required, the pressure beneath the cover is allowed to go positive, and biogas is forced between the cover layers. This design allows for the HDPE to remain in contact with the liquid surface, thus helping to prevent mounding of the scum layer. When the scum layer is allowed to float on the water there is a tendency for it to dry out, crust over, and solidify into 'scumbergs'.

The purpose of the floats is to keep the cover from sinking. Although floats do provide buoyancy for the cover, they also serve a secondary purpose. For covers used to protect the contents of the basin, as for potable water, they are used to define the position of the sump pump. For biogas covers, they create spaces around the edge of the cover that are used as passageways to collect the gases (Figure 16.6) where the latter shows a diagram of a modular floating cover for biogas collection) (Herman, 1999).

---

[2] Sewage scum is composed of sludge plus gas bubbles and fats which are lighter than water and hence float.

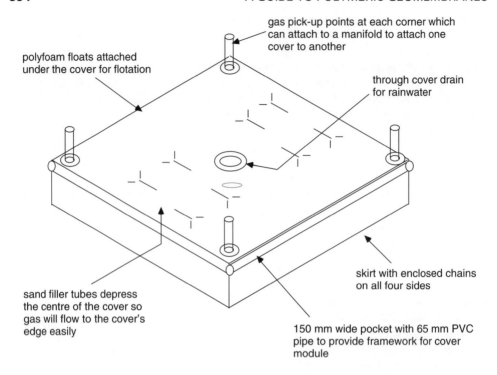

Figure 16.6 Schematic of a floating cover module for a modular floating cover for biogas collection

Ballooning (or 'whale backs') may arise if the gas collection paths become blocked or if they are installed improperly. Collection pipes, flanges or vents can clear excess gas and prevent this problem. Ballooning covers not only put stress on the material of the cover, but they can add tension to perimeter weights or cause pullout from anchor trenches and even pull over concrete ring beams (Herman, 1999).

16.5 TANK LINERS

Flexible tank liners are widely used for potable, recycled and wastewater applications by water districts, agricultural producers, greenhouses, food processors and industrial owners as well as for a variety of industrial liquids such as treated effluents, hydrocarbon liquids (e.g. diesel) and industrial chemicals (e.g. acids and alkalis).

In 'drop-in' tank liner applications, a geomembrane is used as a primary liner to ensure the system is water/fluid tight. The most common geomembrane lining materials fabricated for tank liners are LLPDE, fPP, EIA and PVC. Flexible membrane liners provide an impervious barrier between the steel structure and the aqueous environment thereby eliminating significant corrosion. In addition to using geomembrane liners in new tanks, flexible membrane liners can be used to line old tanks to eliminate any unwanted leakage, corrosion or deterioration.

Flexible tank liners are generally based on one of the following:

- fPP-R
- CSPE-R
- EIA-R
- EPDM rubber
- Nitrile rubber

fPP-R is made by laminating together two fPP plies around a reinforcing scrim layer. The resulting geomembrane is strong enough to hang vertically in the form of a baffle curtain or tank liner and durable enough to leave exposed to the elements for many years.

If wholly prefabricated liners are made then problems can occur due to placement and associated wrinkles and folds, especially in corners. If a HDPE liner is to be used in circular tanks then it is good practice to fabricate pie-shaped segments which are custom configured and welded *in situ*. The disadvantage to that is there will be an increased number of field welds.

With particular regard to the use of chlorine for water tank treatment, note sterilization according to BS-6700 of water tanks requires that 50 ppm of chlorine (completely and thoroughly dispersed throughout the tank) is held in the tank for a period of 24 h, followed by a drain down and flushing through of the system, prior to filling with clean drinking water. Such high levels of chlorine can be damaging to polyolefin liners, especially fPP.

Conventional carbon steel tanks frequently require linings to protect the metal against attack from corrosion or from corrosive solutions contained within. Polymeric and synthetic rubber linings are frequently chosen to protect these vessels. Flexible tank liners can be tailored for practically every process and storage situation and provide a service life of 20 years or more.

16.5.1   SUMMARY OF ESSENTIAL CONSIDERATIONS FOR TANKS LINER SELECTION

Selection of an appropriate tank lining for a specific application requires the identification and itemization of all available details concerning the tank and other equipment to be lined and the environment being contained. Some of the essential considerations include:

- Chemicals – the type of chemicals contained in the tank and their concentration as well as the amount of impurities/contaminants and any potential chemical reactions that might occur (e.g. oxidation).
- Temperature – maximum, minimum and operating temperatures; severity and spread of temperature change; temperature cycle time.

- Abrasion – abrasive particle type, weight and size; velocity of particle movement; proportion of solids; nature of abrasive action (e.g. sliding or impinging).
- Pressure – maximum, minimum and operating pressure or vacuum; pressure cycle time.
- Equipment – complete description of tank to be lined including physical design, dimensions and information on whether it is stationary or portable.
- Operating conditions – any other pertinent external and internal operating conditions.
- Product condition – levels of discoloration, contamination, odour or taste allowed for lining; Food and Drug Administration requirements, if any.
- Experience – past experience under similar conditions such as the lining compound used and length of previous service life.

A lining material should not be selected on the basis of just one outstanding physical property. For example, EPDM rubber liner may be known to withstand 82 °C temperatures, but it does not necessarily follow that a tank lining of this material can withstand such a high temperature over a prolonged period in the presence of a specific chemical. The specific tank lining material must be selected to be compatible with application and in the case of thermoset rubber liners the curing process may alter the properties of the cured material (Kujawski and Haines, 1983).

16.5.2 SELECTION CRITERIA

The factors that influence the choice of tank lining material include a consideration of not only the physical/mechanical properties of the liner but also its chemical resistance in its service environment (see Table 16.9). A logical starting point in selecting a suitable tank lining material is to obtain the manufacturers' literature containing chemical resistance tables for various types of lining materials. Some general guidelines on the compatibility of linings and their particular application include the following.

Chemical Environment

☞ *Natural and synthetic rubber materials as a group are not resistant to hydrocarbons such as oil products or solvents. However, exceptions do exist. Chlorinated polyethylene (CPE) and chlorosulfonated polyethylene (CSPE) synthetic rubber linings, for instance, all resist moderate amounts of oil. Nitrile rubber provides good resistance to aliphatic hydrocarbons such as kerosene; however it is expensive and can be difficult to work with as a lining material. EPDM and butyl rubber have good resistance to polar and oxygenated solvents such as methyl and ethyl ketone and acetone.*

The suitability of a tank lining for each application should be validated by laboratory testing results or a manufacturer's guarantee of suitability based on evidence of previous applications under identical conditions. Plant engineers should ask the manufacturer for additional information when the best material option is not obvious. For example, a lining that is recommended for several individual chemicals may not be suitable for a blend of those chemicals.

Table 16.9 Properties of elastomeric tank liners (Kujawski and Haines, 1983)

Lining type	Shore A durometer[a] reading	Upper temperature limit ($^\circ$C)	Thermal shock resistance	Resistance to hydrocarbons	Typical uses
Soft natural rubber	30 to 60	72	Excellent	Poor	Acid storage; transportation equipment; abrasive services; white rubber for food grades; sulfur dioxide scrubbers
'Semi-hard' natural rubber	80 to 85	82	Good	Fair	Chemical processing and plating
Hard natural rubber	90 to 100	93	Poor	Fair	Chemical processing; high temperature nickel–copper plating; steel pickling; vacuum service
Flexible hard natural rubber	90 to 100	100	Fair	Fair	Same uses as hard natural rubber; better crack and heat resistance
Graphite loaded hard rubber	95 to 100	100	Fair	Fair	Special lining for wet chlorine gas in chlorine cells and associated equipment
Three-ply (soft, hard, soft)	40 to 50	110	Excellent	Fair	Combined abrasion and corrosion services; becoming popular for steel pickling lines; phosphoric acid
Neoprene	40 to 70	110	Excellent	Very good	Chemical or abrasive services with oil present; best for strong bases; good weather resistance; fire retardant
Nitrile	60 to 90	93	Excellent	Excellent	Aliphatic hydrocarbons, kerosene, animal, vegetable, and mineral oils
Butyl	50 to 75	107	Excellent	Fair	Oxidizing acids; 70% hydrofluoric acid; 'super phosphoric acid'; best water resistance; good for alternative service
Chlorobutyl	40 to 60	93	Excellent	Fair	Much the same as butyl but easier to apply and faster curing; sulfur dioxide scrubbers
EPDM	40 to 60	82	Excellent	Poor	Hypochlorite bleach; ozone and weather resistant
CSPE	50 to 70	99	Excellent	Good	Strong acids such as chromic acid; high concentrations of nitric and sulfuric acids
CPE	50 to 75	110	Excellent	Good	Strong acids; alternative service

[a]Shore A Durometer is a measure of hardness. The higher the value, the harder the material.

Abrasive Action

When particle impingement on the lining is expected, a resilient type of rubber provides the most satisfactory service because it absorbs impact and permits the particles to bounce off without damaging the lining. A soft natural rubber with a Shore A durometer reading of 30 to 40 is generally specified in such applications. When the contained particles expose the lining to a sliding or cutting action 'in service', a tougher, harder, tear-resistant lining material with a Shore A durometer reading of 50 to 60 should be specified. Hard rubbers, such as those with a Shore A reading of 90 and above, have poor abrasion resistance and should be avoided in such applications (Kujawski, 1983).

Temperature Variations

Not all linings exhibit the same degree of resistance to thermal shock. Soft natural rubber and most synthetic linings have outstanding resistance to thermal shock. Hard rubber has less resistance to thermal shock. Hard rubber linings can be made to withstand fairly rapid temperature changes, but a thorough study of the conditions must be conducted to ensure the hard rubber is able to perform successfully (Kujawski and Haines, 1983).

Pressure or Vacuum Conditions

Normal pressure ranges are seldom a problem. When full vacuum or pulsating pressure conditions exist, hard rubber liners are used because they are less susceptible to diaphragm action. This may occur, for example, when a pinhole in the steel container allows atmospheric pressure on the underside of the lining to force the lining towards the lower pressure inside the vessel.

Special Linings for Extreme Conditions

Complex or severe service conditions, for example, those that require both chemical and abrasion resistance, may be successfully achieved with a three-ply lining. This type of lining incorporates an inner layer of semi-hard rubber for chemical and heat resistance with two outer layers of softer rubber that contribute flexibility and abrasion resistance. Three-ply linings are designed to withstand temperatures between 82 and 110 °C. When a combination of severe temperature and abrasive action is encountered, the three-ply lining is formulated with an inner layer of hard rubber and two outer layers of high temperature resistant soft rubber (Kujawski and Haines, 1983).

16.5.3   INSPECTION OF LINING INSTALLATIONS

Reputable lining installation contractors inspect and test the rubber lining extensively before permitting it to be placed in service. The newly lined tank or container is closely examined for leaks and defects such as poor adhesion. Pinhole leaks may be detected using a spark tester. However there is the potential hazard of a burn-through if the spark testing equipment is held in one place for too long. It is usually recommended for the plant engineer to supervise the conduct of this test.

Seam construction is also inspected before the lined tank is put into service. All seam edges must be bevelled. A simple way to detect air pockets behind seams is to press down at the edge of the lap seam where the under-lining sheet ends. Poor seam construction having air entrapment, can be easily confirmed as a bubble which can be felt as a soft spot or if the lining is loose and the seam edges can be lifted. For thermoset rubber liners (e.g. EPDM, CSPE, nitrile rubber) Shore A durometer readings are taken at several points; a low reading indicates that the lining has not cured completely and a high reading may indicate that the rubber is over-cured (Kujawski and Haines, 1983).

REFERENCES

Averesch, U. B. and Schicketanz, R. T., 'Recommendations for new installation procedures of geomembranes in landfill composite sealing systems by the 'Riegelbauweise', presented at *EUROGEO 2000*, Bologna; available from Düsseldorfer Consult GmbH, Düsseldorf, Germany and Ingenieurbüro Schicketanz, Aachen, Germany (2000).

Blight, G. E., Failures during construction of a landfill lining: a case analysis, *Waste Management and Research*, **25**(4), 327–333 (2007).

Defilippis, M. O., Geomembranes Used in Heap Leach SX_EW Mining: A Manufacturer's Perspective, in *Proceedings of the North American Geosynthetics Society (NAGS)/GRI19 Conference*, Las Vegas, NNV, USA; available from SL Limitada, Santiago and Antofagasta, Chile [http://www.geosynthetica.net/tech_docs/NAGS_GRI19/Ossa.pdf] (2005).

GAST [http://www.gast.co.za/landfill_applications.html] (2008).

Herman, A., Floating Cover Usage for Tanks is Growing, *GFR Magazine*, **17**(7), 1 (1999).

Kujawski, G. and Haines, F., Selecting Elastomeric Linings For Storage Tanks, *Plant Engeneering.*, **37**(21), 63–65; available from Protective Coatings, Fort Wayne, IN, USA [www.proco-fwi.com] (1983).

Layfield Floating Covers [www.layfieldgroup.com] (2008).

Lupo, J. F. and Morrison, K. F., Innovative Geosynthetic Liner Design Approaches and Construction in the Mining Industry, in *Proceedings of the Geo-Frontiers Congress Conference* [http://www.golder.com/archive/Mining%20-%20GeoFron-lup-mor-rev2.pdf] (January 24–26, 2005).

Lupo, J. F. and Morrison, K. F., Geosynthetic design and construction approaches in the mining industry, *Geotextiles and Geomembranes*, **25**, 96 (2007).

Peggs, I. D., 'Geomembrane Liner Durability: Contributing Factors and the Status Quo'; [http://www.geosynthetica.net/tech_docs/IDPigsUKpaper.pdf] (2003).

Seed, R. B., Mitchell, J. K. and Seed, H. B., Kettleman Hills Waste Landfill Slope Failure. II: Stability Analysis, *Journal of Geotechnical Engineering, ASCE*, **116**(4), 669 (1990).

Water and Wastes Digest, Scranton Gillette Communications, **42** (2) (Feb. 2002).

Smith, M. and Orman, M., 'Copper Heap Leaching – A Case for PVC Liners', *PGI Technical Bulletin* [http://www.geomembrane.com/PGI%20Tech%20Bull/97-05.htm] (1994, reprinted May 1997).

Solmax International Inc. [www.solmax.com] (2008).

Thiel, R., Purdy, S. and Yazdani, R., Case History of Exposed Geomembrane Cover for Bioreactor Landfill, in *Proceedings of Sardinia 2003 Ninth International Landfill Symposium*; available from Vector Engineering, Grass Valley, CA, USA (October, 2003).

Thiel, R. and Smith, M. E, State of the practice review of heap leach pad design issues, *Geotextiles and Geomembranes*, **22**(6), 555 (2004).

17

Welding of Geomembranes

Polymeric geomembranes can be welded or seamed by either thermal methods or chemical welding. Thermal methods rely on fusion of the surfaces to be joined using applied heat (this includes wedge welding, hot air welding, extrusion welding and less commonly, ultrasonic welding). Chemical seaming relies on the use of solvents to soften the bonded surfaces but this technique can only be used for those geomembranes that are softened by solvents.

17.1 WEDGE WELDING

Hot wedge welding is a thermal technique where the two opposing geomembrane surfaces to be welded are melted by passing a hot metal wedge or knife between them (see Figure 17.1 and Figure 17.2). Pressure is simultaneously applied to the top and/or bottom geomembranes to form an integral bond. Welds of this type can be made with dual (or 'split') bond tracks separated by a non-fused gap (known as the 'air channel') which is utilized for air pressure testing (see Figure 17.3). Such welds are referred to as dual hot wedge welds or double-track welds. Alternatively, a single hot wedge creates a solitary uniform width weld track (see Figure 17.4). Hot wedge welding is applicable for HDPE, LLDPE, fPP, PVC and EIA-R materials, but is generally not suited to EPDM and EPDM-R, although Trelleborg (Denmark) does make EPDM liners that are thermally weldable.

📌 *When hot wedge welding with dual wedges, two parallel seams are produced with an air channel in between that is used for field quality assessment via air pressure testing of the seam.*

Hot wedge welding of geomembranes utilizes a wedge-shaped, electrically heated resistance element (see Figure 17.5) that travels between the two geomembrane sheets to be joined. As the surface of the two geomembrane sheets are being melted and seamed, a shear flow field is induced across the upper and lower surfaces of the wedge. Then, as the two sheets converge at the tip of the wedge, pressure is applied via nip rollers to form the weld (see Figure 17.6).

A Guide to Polymeric Geomembranes: A Practical Approach   J. Scheirs
© 2009 John Wiley & Sons, Ltd

Figure 17.1   Photograph of seaming of HDPE geomembranes liners using a wedge welder (side view). Reproduced by permission of Sotrafa

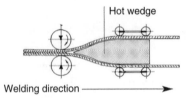

Figure 17.2   Schematic of the welding of two geomembrane sheets using a hot wedge. In hot wedge (or knife) welding a hot metal wedge is run between the two geomembrane surfaces to be seamed. Pressure is applied to the top or bottom geomembrane, or both, to form a continuous bond. Some seams of this kind are made with double weld tracks separated by a nonbonded gap. These seams are sometimes referred to as dual hot wedge seams or double-track seams. Reproduced by permission of Barry Smith, Plastic Welding Division, Plastral Pty Ltd

Modern hot wedge welding machines are largely automated and carefully control the key operating parameters of wedge temperature, travel rate and degree of pressure applied (see Figure 17.7). The non-bonded channel enables the integrity of the weld to be field assessed by pressuring the channel with compressed air and then observing whether the pressure remains constant. Any drop in air pressure can indicate there is a leak in the seam. The dual hot wedge welding methods are now recognized as the state-of-the-art in geomembrane welding.

17.1.1   THE WEDGE WELDING PROCESS

The wedge welder is a fixed overlap welder, where the welded seam is produced by the two edges of the liner going through the wedge welder (1 sheet over the wedge and one sheet under the wedge) at a fixed rate of travel (see Figure 17.8). This process melts the sheets, followed by clamping the liners together by rear pressure/drive wheels. The

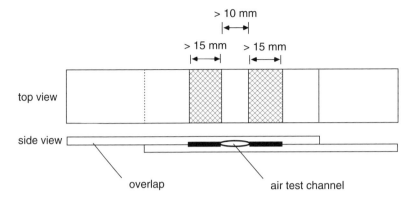

Figure 17.3 Schematic of an idealized dual wedge weld showing the optimum dimensions of the weld tracks and air channel

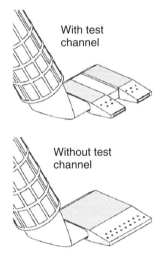

Figure 17.4 Diagrams of hot wedge designs: upper wedge produces a test channel for air testing while the lower wedge is without the test channel. Reproduced by permission of Leister

liner sections require a fixed overlap of between 8 cm to 10 cm. Trimming of the top sheet may be required to ensure that the 8–10 cm overlap is continuous along the length of the seam to be welded.

The welder travels forward with the drive wheels being at the rear of the machine (see Figure 17.9). The welder is loaded with the top liner (i.e. the sheet that will be loaded on the left-hand side when viewed from the rear of the machine) and the bottom liner (loading from the right).

The welder typically operates at a constant speed and temperature with a range between 400–430 °C once the machine is warmed up. Warmer weather will require the speed to

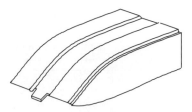

Figure 17.5   Picture of a double welding wedge for dual wedge welding of geomembranes. Note the gap in the centre which forms the air channel

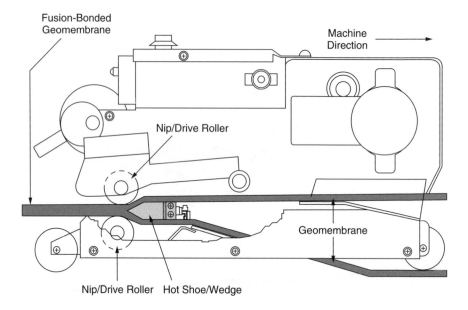

Figure 17.6   Schematic of a hot wedge welder showing the hot wedge positioned between the two geomembrane sheets and the nip/drive rollers positioned above and below the fusion-bonded sheets

be slightly adjusted upward while in colder/cooler weather it will be necessary to slow down the welding machine.

During hot wedge welding the machine runs automatically at speeds around 2.0–2.5 m/min (for HDPE) with temperatures as high as 460 °C. Considering the subgrade is often uneven, it is good practice to place one piece of a 30 cm wide HDPE strip underneath the track of the machine.

An HDPE liner of 1.00 mm or less is relatively easy to weld; however for thicknesses of 1.5–2.0 mm, welding becomes more challenging and a combination of experience, optimum welding conditions and care are required. It is important however to control the application of heat during welding of relatively thin geomembranes or else 'burn though' holes can result (see Figure 17.10).

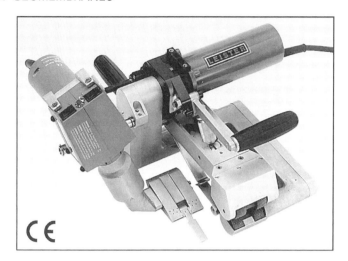

Figure 17.7   Photograph of a hot wedge welder. Reproduced by permission of Leister

Figure 17.8   Photograph of seaming of HDPE geomembranes liners using a wedge welder (front view). Reproduced by permission of Sotrafa

Unreinforced PVC and LLDPE are lightweight materials that require careful welding. Wedge welding is not recommended for unreinforced PVC and LLDPE of a thickness less than 0.3 mm.

Reinforced PVC and CSPE (Hypalon™) are arguably the easiest geomembranes to weld as they are very compliant and forgiving. For CSPE it is important to ensure that the material has not partially cured prior to welding. Curing by exposure to heat and/or UV light will alter its structure from a thermoplastic to a thermoset material. It is impossible

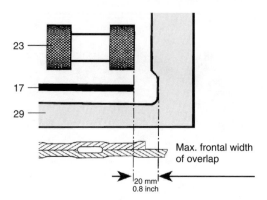

Figure 17.9 Schematic showing position of pressure roller relative to the edge of the seam. Reproduced by permission of Leister

Figure 17.10 Photograph of burn-through marks on a 1 mm HDPE geomembrane (note holes in the weld tracks)

to weld thermoset materials because of their crosslinked nature. It is important to note that the manufacturers of CSPE supply a spray or liquid solvent that reverses the curing (crosslinking) process and prepares the material for wedge welding (Novaweld, 2008).

17.1.2 WEDGE WELDING CONDITIONS

Typical wedge welding parameters for common geomembranes are summarized in Table 17.1.

While the weld parameters shown in Table 17.1 are typically employed it is important to note that the welding temperature and the welder travel speed are dependent on the

Table 17.1 Typical fusion welding parameters (Agru)

Geomembrane type	Wedge temperature (°C)	Pressure (N/mm²)	Welding speed (m/min)
HDPE	360–430	20–30	~1.6–2.8
Textured HDPE	380–420	20–30	~1.6–2.8
VLDPE	280–430	8–12	~0.8–3.5
fPP	250–480	8–12	~0.8–3.0

Table 17.2 Wedge welder temperature and speed settings (North American conditions)

Thickness/type[a]	Wedge temperature (°C)	Pressure knob (steps)	Speed (m/min)
0.75 mm HDPE	400	3	3.0
1.00 mm HDPE	400	4	2.5
1.50 mm HDPE	420	6	2.0
2.00 mm HDPE	430	8	1.3
2.50 mm HDPE	440	10	0.9
3.00 mm HDPE	450	12	0.6
1.50 mm HDPE-T	430	6	2.0
2.00 mm HDPE-T	450	8	1.3
2.50 mm HDPE-T	450	10	0.9
0.75 mm VLDPE	380	3	3.5
1.00 mm VLDPE	420	4	3.0
1.50 mm VLDPE	430	6	2.0

[a]HDPE-T, textured HDPE; VLDPE, = very low-density polyethylene.

geomembrane thickness (as shown in Table 17.2) and other factors such as the weather conditions.

Figure 17.11 shows the thermal welding temperature windows for various geomembrane types (note that fPP has a very wide seaming window).

The welding temperature and the welder travel speed are also dependent on the ambient weather conditions.

Typical wedge temperature ranges for hot wedge seaming of 0.75 mm sheet are shown in Table 17.3.

The welding temperature can be affected by ambient variations such as ambient temperature, cloud cover and wind speed. Thus it will be necessary to adjust and optimize the wedge temperature as the environmental conditions change.

Common practice when wedge welding 2.0 mm HDPE is to use a temperature of 420–430 °C and a speed of between 1.3 and 2.2 m/min depending on the ambient temperature.

17.1.3 MINIMUM REQUIREMENTS FOR A WEDGE WELDER

Wedge welding machines should have the following minimum capabilities:

- The temperature of the wedge needs to be adjustable up to 450 °C. To ensure optimal heating of the GML sheet the wedge needs to be of sufficient length. Note also that

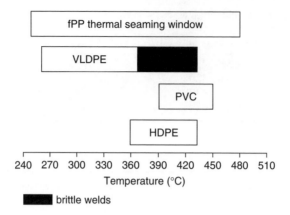

Figure 17.11 Thermal welding temperature windows for various geomembrane types

Table 17.3 Typical wedge temperature ranges for hot wedge welding of 0.75 mm polyolefin geomembranes

Liner type	Minimum warm weather temperature (°C)	Maximum cold weather temperature (°C)
HDPE	315	400
LLDPE	315	380

the gap between the wedge and the contact surface of the geomembrane liner has to be protected against direct side wind.

- The drive on the wedge welder needs to run at a constant speed of ±5 cm/min. It is necessary to ensure both nip rollers are tracking correctly or else defective welds can be created (see Figure 17.12).
- The welding pressure tolerance should be ±1 N/mm².

Figure 17.13 shows a schematic of a hot wedge welder. Figure 17.14 shows wedge welding occurring on a steep slope

The specifications and performance parameters of Leister wedge welders are listed in Tables 17.4 to 17.6.

It is advisable to turn off the welder drive motor when placing the geomembrane overlap in the wedge and roller assembly. After the overlap to be welded is properly positioned, the drive motor can be turned on and the seam locked in position.

17.1.4 WEDGE WELDING TIPS FOR SEAMING GEOMEMBRANES

The following are tips and hints to be mindful of during hot wedge welding of geomembranes.

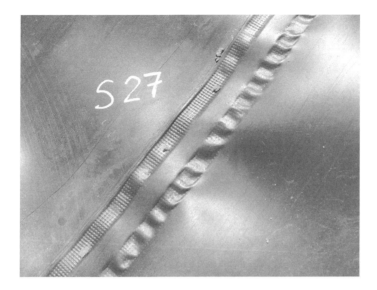

Figure 17.12 Photograph of weld tracks formed by hot wedge welding of 1 mm HDPE geomembranes. Note the poor tracking of the nip roller for the lower weld track

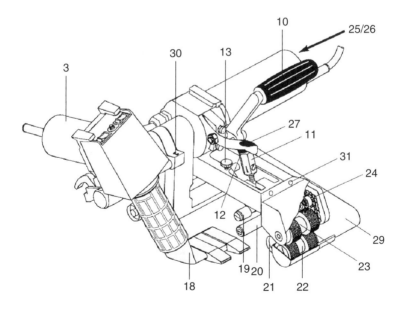

Figure 17.13 Schematic of a wedge welder (back view): hot wedge, position 18; pressure rollers, positions 22 and 23; hot air blower, position 3; lever for welding pressure, position 10; guide roller, position 20. Reproduced by permission of Leister

Figure 17.14 Photograph of a wedge welder seaming HDPE geomembranes on a steep slope. Reproduced by permission of NAUE

Table 17.4 Specifications and performance parameters of commonly used wedge welders for geomembranes (data from Leister)

Parameter	Leister Twinny T	Leister Astro	Leister Cosmo
Speed (m/min)	0.8–3.2	0.8–5.0	0.5–5.0
Temperature (°C)	20–560	20–420	20–450
Welding pressure (N) (maximum)	1000	1500	2500
Air flow (l/min)	150–190	No air flow	No air flow
Weight (kg)	7.9	23	32
Material thickness capability (mm)	1.5–3.0	1.5–3.0	1.5–3.0
Number of wedges	Single wedge	Single wedge	Double wedge
Self tracking	No	Self tracking	Self tracking

Table 17.5 Welding parameters for 1.0 mm HDPE (smooth) at 20 °C (ambient) (from Leister)

Machine	Temperature (°C)	Speed (m/min)	Pressure (N)
Leister Comet 50 mm	420	3.0	600
Leister Twinny S (short wedge)	550	3.2	500

Table 17.6 Welding parameters for 2.0 mm HDPE (smooth) at 20 °C (ambient)

Machine	Temperature (°C)	Speed (m/min)	Pressure (N)
Leister Comet 70 mm	420	1.5	1000
Leister Astro	420	2.8	1250
Leister Twinny T (long wedge)	550	1.3	1000

Note: the welding parameters shown in Tables 17.5 and 17.6 are for 'normal' weather conditions.

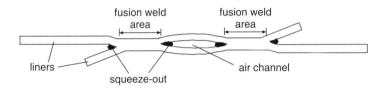

Figure 17.15 Cross-section of a dual wedge weld showing location of 'squeeze-outs' and air channel

'Squeeze-Out'

Analysis of the welds has demonstrated that there is a temperature gradient across the heated-track (i.e. the weld-track). The two edges are relatively the coolest, with the temperature increasing to a maximum at the mid-point of the track. Physically this translates into there being more molten material in the mid-region of the track. As the heated-track passes between the nip rollers which force the two molten surfaces together, a small portion of the molten material is squeezed out of the weld zone and forms what is referred to as 'squeeze-out' at each edge of the weld-track (see Figure 17.15).

In properly welded seams, the melted polymer will extrude out laterally from the seam area – referred to as 'squeeze-out'.

A small amount of 'squeeze-out' is a good sign that the proper welding temperature has been achieved. An excessive amount of extruded hot melt indicates that excessive heat and/or pressure were applied. The presence of excess 'squeeze-out' may also indicate that the seaming rate was too slow.

It should be noted that 'squeeze-out' (or 'flashing') is acceptable if it is equal on both sides and does not produce a notch effect at the edge of the weld (see Figure 17.16).

Wave Pattern

For a geomembrane with *a thickness less than or equal to* 1 mm, a long, low, sinusoidal wavelength pattern in the direction of the weld is indicative of proper welding (see Figure 17.17). If the wave peaks are too close together then the welding speed should be increased until an acceptable pattern appears. The absence of the wave pattern indicates that the welding speed is too fast and hence it should be decreased. There usually are no

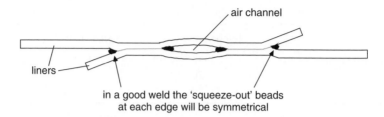

air channel

liners

in a good weld the 'squeeze-out' beads
at each edge will be symmetrical

Figure 17.16 Schematic showing the 'squeeze-out' beads that occur during wedge welding and which are considered as a sign that the correct melting temperature was achieved during welding

Figure 17.17 Photograph of a 'textbook' double wedge weld in an HDPE geomembrane (note the uniformity of the weld tracks and the slight undulation in the HDPE, both of which indicate good welding conditions were employed)

characteristic wave patterns evident for HDPE geomembranes with thicknesses greater than 1 mm because of the inherent stiffness of the material.

Nip Roller Marks

Nip/drive roller marks will always show on the welding track surface and are normal indicators of the wedge welding process. They should be noticeable and slightly embossed if welding pressures in the correct range were used.

Heat Distortion from Welding

If possible the underside of the lower liner should be inspected for heat distortion. This can be done at the end of seams, and wherever destructive weld samples are cut out of the seam. A small amount of heat distortion referred to as 'thermal puckering' is acceptable on relatively thin liners (e.g. 0.75–1.00 mm). A small degree of thermal puckering indicates

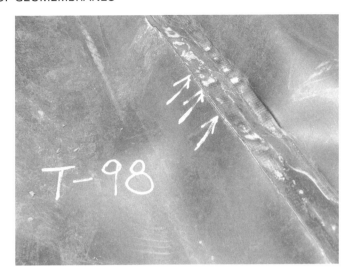

Figure 17.18   Photograph of a heat-affected weld in a 1 mm HDPE liner (note the buckling and distortion at the right side of the photograph)

that heat has penetrated through the entire sheet. On the other hand, if the lower liner is greatly distorted then too much heat has been applied and the corrective action is to either increase the welding speed or lower the welding temperature (see Figure 17.18). For geomembranes that are 2 mm or thicker, no thermal puckering should be observed.

17.2   HOT AIR FUSION WELDING

In hot air welding a high-temperature air or gas is introduced between the two geomembrane surfaces to facilitate localized surface melting (see Figure 17.19). Roller pressure is simultaneously applied to the top and/or bottom geomembrane, forcing together the two surfaces to form a continuous bond (see Figure 17.20). Hot air welding is applicable for thermoplastic geomembranes (e.g. HDPE, LLDPE, fPP, PVC and EIA-R) but cannot weld thermoset geomembranes (i.e. cured CSPE, CSPE-R, EPDM and EPDM-R).

Figure 17.21 shows a schematic of a combination hot air welder and wedge welder.

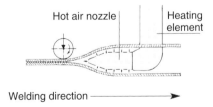

Figure 17.19   Schematic of hot air welding. Reproduced by permission of Leister

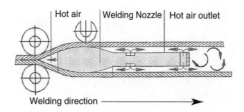

Figure 17.20 Schematic of the welding of two geomembrane sheets using hot air welding where high-temperature air is introduced between the two geomembrane surfaces to facilitate melting. Pressure is also applied to the top and bottom geomembranes, forcing together the two surfaces to form a continuous bond. Reproduced by permission of Barry Smith, Plastic Welding Division, Plastral Pty Ltd

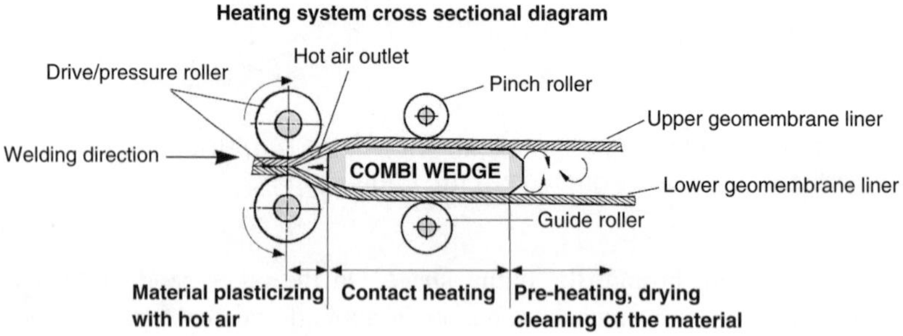

Heating system cross sectional diagram

Figure 17.21 Schematic of combination wedge welding and hot-air welding. Notice that the nip rollers squeeze the weld area together. Reproduced by permission of Leister

Table 17.7 Manual hot air welding

Welding parameter	HDPE	fPP	PVC
Air temperature ($^\circ$C)	280–320	270–500	450–500
Air flow (l/min)	ca. 230	Ca. 230	ca. 230
Pressure (joining force) (N/mm^2)	–	–	–
Welding speed (m/min)	0.1–0.2	0.1–0.2	0.2–0.4

17.2.1 HOT AIR FUSION WELDING CONDITIONS

The typical hot air fusion welding parameters for manual and automatic welding are shown in Tables 17.7 and 17.8, respectively. Note that in both tables, the results for fPP are obtained from SKZ[1] testing while the data for the other polymers are from the DVS Guidelines 225, Part 1.

Hot air fusion welders are generally used for thermally welded repair work. However these units require specialized operator training. Hand-held air fusion welders are capable

---

[1] SKZ = Sueddeutsches Kunststoffzentrum (South German Polymer Center), DVS = German Welding Society.

Table 17.8 Automatic air welding

Welding parameter	HDPE	fPP	PVC
Air temperature ($^\circ$C)	350–500	230–480	450–550
Air flow (l/min)	400–600	400–600	400–600
Pressure (joining force) (N/mm^2)	25–40	20–40	10
Welding speed (m/min)	0.5–4.5	0.5–2.5	1.0–3.0

Table 17.9 Novaweld GT-100 seam strengths for geomembrane liners (Comer, 1997)

Material	Thickness (mm)	Heat setting	Shear strength (kgf)	Peel strength (kgf)	Comments
PVC	1.0	3	33	6.3	Good
CSPE	0.9	4	21.8	2.2	Poor
HDPE (smooth)	1.5	3.5	74.5	0.45	Poor
HDPE (smooth)	1.5	5	72.2	25	Good
HDPE (textured)	1.5	4.5	69	0.9	Poor
HDPE (smooth–textured)	1.5	4	60.9	1.8	Poor
VLDPE (smooth)	1.5	4	24	1.3	Poor
VLDPE (textured)	1.5	4	38.6	11.3	Good
VLDPE (smooth–textured	1.5	4	40	6.8	Good

of producing seams with good shear strength and peel strength; however the seam quality is highly dependent on operator skill and the temperature setting. Good peel strengths are not attained until the welding temperatures and pressures are high enough. If the welding temperature is too high then the geomembrane liner material begins to 'smoke' and gives off acrid fumes.

Table 17.9 highlights the importance of temperature optimization. For 1.5 mm HDPE, extremely low peel strength values are obtained when a temperature setting on the welder (Novaweld Model GT-100) of 3.5 was used. However when the temperature setting was increased to the heat setting of 5.0, excellent peel strength values were obtained. If the temperature settings on the other types of polyolefin liners were increased then their weld strengths would have increased as in the case of the smooth HDPE. The poor performance of the CSPE is likely to be due to the material being aged and the cured surface not being removed before seaming.

17.3 EXTRUSION WELDING

The extrusion fillet welding method involves extruding a ribbon of molten resin at the edge of overlapped geomembranes to form a continuous bond (see Figure 17.22). In extrusion welding the polymer substrate upon which the molten resin is deposited must be suitably prepared by slight grinding/buffing to clean the surface. The surface must be melted and mixed with the weld bead material in order to achieve a strong bond.

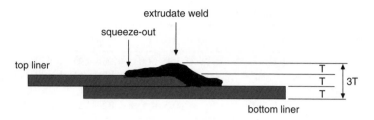

Figure 17.22   Dimensions of a typical extrusion fillet weld

🖈 *Extrusion welding involves extruding a band/ribbon of molten polymer over the edges of (or in between) the two slightly roughened geomembrane sheet surfaces to be joined. The molten extrudate causes the surfaces of the sheets to fuse and after cooling the entire weld region cools and permanently bonds together. Extrusion welding is generally performed on polyolefin geomembranes such as HDPE, LLDPE, fPP and fPP-R.*

Extrusion welding was the first welding technique developed for HDPE geomembranes. It is a thickness-dependant technique that requires a minimum material thickness to create an effective weld without distortion. Extrusion welds in sheets less than 1.0 mm thick are not recommended. Extrusion welds in 1.0 mm HDPE geomembranes can exhibit some distortion and can sometimes be very difficult to prepare around intricate pipe penetrations and mechanical attachments. However 1.5 mm HDPE can be reliably extrusion welded in most situations and is recommended in most applications. Welds in 2.0 and 2.5 mm geomembranes can be excellent and are recommended in applications that require exceptional durability.

Extrusion welding is very slow and is therefore typically used only for repairs and details. Extrusion welds are very difficult to prepare on vertical or overhead walls and require a minimum clearance of 1 m. This is especially important in sump details where a minimum clearance must be maintained underneath the lowest pipe penetration.

🖈 *Extrusion fillet seaming is effectively the only method for welding patches on polyolefin geomembrane, as well as for seams around pipes and other appurtenances, and for difficult seams in poorly accessible areas such as sump bottoms which are difficult to access with a wedge welder. Therefore, these seams should be carefully inspected to ensure they are leak free.*

The extrusion process is used primarily for detailed work around structures, pipes and other penetrations. It is also used for repair work. The hand-held extruder applies a molten layer of polyethylene material to the exposed edge of an overlapped section of liner. A large drill motor turns the extrusion screw (see Figure 17.23). Temperature is maintained by means of a temperature indicating controller which modulates the power to a heater band surrounding the barrel of the extruder. Typically, HDPE material is fed to the extruder as 4–5 mm rod. As it leaves the extruder, the molten extrudate passes through a Teflon™ and steel die (the 'welding shoe') and is deposited upon the seam in a layer about 25–40 mm wide.

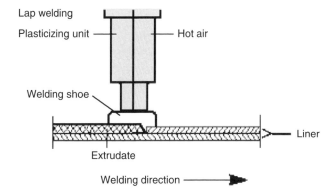

Figure 17.23 Schematic of extrusion welding. Reproduced by permission of Leister

🖈 *Extrusion fillet welding is primarily used for detailed work and patching/repairs.*

🖈 *When the extrudate is placed over the leading edge of the seam it is termed 'extrusion fillet seaming'. When the extrudate is placed between the two sheets to be joined then it is termed 'extrusion flat seaming'. Extrusion flat seaming is seldom used today.*

17.3.1 EXTRUSION WELDER REQUIREMENTS

The extrusion welder must have a calibrated temperature display (i.e. read-out) which shows the extrusion temperature.

Extrusion fillet welders employ Teflon™ dies through which the extrudate passes onto the liner. These dies must be regularly inspected for wear and notches.

17.3.2 EXTRUSION WELDING CONDITIONS

The critical parameters to be controlled in order to create an acceptable bond are temperature and welding speed. If the temperature is too high or the speed too slow then excessive melting can weaken the geomembrane or lead to excessive thinning. Localized thickness reductions can in turn lead to stress concentrations forming in those locations. Insufficient temperature on the other hand results in poor melting that causes inadequate flow of the extrudate across the weld interface and thus results in poor weld integrity and low seam strength (see Figure 17.24). Typical welding parameters for extrusion welding are shown in Table 17.10.

In extrusion welding a welding rod made of the same polymer as the liner material is fed into a hand-held extrusion gun and heated to above its melting point. The molten bead known as the 'extrudate' is then deposited onto an overlapped seam area that has been ground to remove oxidized surface material, pre-tacked and preheated by a hot air nozzle attached to the extrusion gun. A 4 or 5 mm diameter welding rod is generally used to produce an extrusion weld. The welding rod is applied as a molten bead at the edge of two overlapping liners and this produces an 'extrusion seam'.

Figure 17.24 Photograph of total delamination of an extrusion fillet weld due to too low a hot air temperature (240 °C) whereas 250–260 °C is required for a good weld. Note that the extrudate weld is the lighter coloured material

Table 17.10 Typical welding parameters for extrusion welding (Agru)

Geomembrane material	Air temperature at preheating nozzle (°C)	Extrudate temperature (°C)	Welding speed (m/min)
HDPE	240–280	220	0.4 (up to 0.55)
LLDPE	240–280	220	0.4 (up to 0.55)
VLDPE	200–240	190–200	0.4 (up to 0.55)
fPP	200–240	190–200	0.4 (up to 0.55)
PVDF	260–280	225–240	0.4

Extrusion fillet welds are generally performed for patches and around details such as pipes, sumps, fittings, appurtenances (fixtures that penetrate the liner) and 'T'- and 'Y'-shaped seams. The butt seam at a 'T' intersection for example ensures integrity of the seam at an area where a wedge welder is not able to produce a uniform seam.

The two liners to be joined must be positioned to create an overlap of at least 150 mm (Agru recommends overlapping the geomembrane panels at a minimum of 75 mm). The same general guidelines as specified for liner preparation of hot wedge welding also apply to extrusion fillet welding. A hot air welder (capable of a temperature up to 600 °C) should be used to 'tack' the liners after they are properly positioned. The hot air gun thus prepares the seam for the extrusion welder by creating a light bond between the two sheets, thereby securing their position. It is important that no heat distortion is obvious on the surface of the upper sheet.

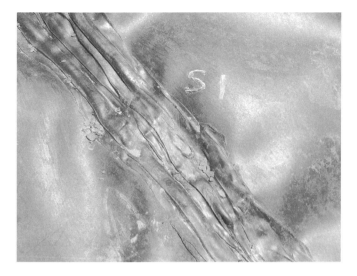

Figure 17.25   Photograph of a very poor example of an extrusion fillet weld

Extrusion fillet welding of geomembranes welding requires great operator skill and care for both the preparation step and the welding. Extrusion welding requires manual grinding of the weld area which can weaken the geomembrane due to the introduction of scores and notches. Figure 17.25 shows a 'poorly made' extrusion weld while Figure 17.26 is an example of a 'well made' extrusion weld.

📌 *When tapering the edge of the upper geomembrane, care must be taken that the lower geomembrane is not damaged by deep scoring or grooving.*

Surveys of geomembrane liners have shown that extrusion welds have been found to be a significant source of small leaks when inspected by electrical leak location. The presence of a repair patch and the rigidness of an extrusion weld introduce sites for stress to accumulate under an applied load (Darilek and Laine, 2001).

📌 *Manual extrusion welds are known to fail at a far greater rate than double wedge welds.*

Extrusion welds are mainly used to undertake geomembrane repairs. For instance, project specifications often require that double-wedge welded seams be destructively tested at regular intervals. The region of the seam where destructive test samples are removed must then be repaired using a extrusion fillet welder. In this process, typically, 1 m of a double wedge weld is replaced with about 3.5 m of a manual extrusion weld. It is well known in the geosynthetics industry that manual extrusion welds have a significantly higher failure rate than double wedge welds (Darilek and Laine, 2001).

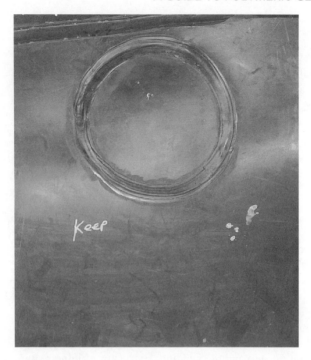

Figure 17.26   Photograph of an example of a well-made extrusion weld

It is important that the extrusion welder is purged prior to beginning a weld until the heat-degraded extrudate is removed.

Extrusion Welding Procedure

The molten extrudate should be deposited directly on top of the overlapped seam with the centre of the extrudate weld positioned directly along the edge of the upper liner. The width of the extrudate should be such that it completely covers the edge of the upper liner and the majority of the underlying grind marks. The extrudate bead should be approximately twice the specified sheet thickness as measured from the 'crown' of the extrudate to the top surface of the bottom sheet.

17.3.3 EXTRUSION WELDING TIPS

The following are tips and hints to be mindful of during extrusion welding of geomembranes.

Degraded Polymer

Since polyethylene will degrade when kept molten for extended periods of time care should be taken to ensure that the liners are not welded with degraded material which has much lower strength than virgin PE. Thus the barrel of the extrusion welder should be purged of all heat-degraded extrudate for about 30 s before starting a new weld. This should be done every time the extruder welder is restarted after being idle for more than 2 min. It is also important that the extrudate purgings are not allowed to be deposited on the surface of the liner since it may damage the liner.

Poor Weldability of HDPE

HDPE contains very low molecular weight components (known as oligomers) that can bloom (migrate to the surface) of the HDPE and produce a waxy layer that makes welding and adhesion very difficult. Before thermally welding HDPE geomembranes these surface waxes need to be removed by surface grinding in the case of extrusion welding and scrapping off by the hot wedge during thermal fusion welding. Other migratory compounds that can make welding difficult are fatty acid amide lubricants that may be present in the geomembrane formulation, such as erucamide.

Welding Rod Compatibility

Weld rods used for extrusion welding are often not manufactured from the same parent material as the geomembrane. This can lead to incompatibility issues and possible stress cracking. The two principal factors to consider when determining polymer compatibility are MFI (which indicates molecular weight) and density (which indicates the degree of chain branching). Polymers with widely different molecular weights and/or chain branching are not miscible or compatible. For instance while HDPE and PP are both polyolefins they are immiscible and cannot be melt blended since they phase separate and delaminate on cooling. This leads to 'a plane of weakness' at the interface of the two phases. It is known that different PE resins such as HDPE and MDPE or HDPEs of different density or melt flow index or carbon black concentration can lead to 'in-plane separation' of the final GM.

Case Study

A fabricator was wishing to weld Solmax HDPE geomembrane with a Huikwang WR040 welding rod made from Marlex K306. It was first necessary to determine whether the two HDPEs are compatible when welded and if they form a continuous miscible integral weld.

The two polymers to consider in this case are:

- K306 (melt index of 0.11 g/10 min and density of 0.937 g/cm^3 when unfilled with carbon black).
- Petromont S7000 (melt index of 0.12 g/10 min and density of 0.937 g/cm^3 when unfilled with carbon black).

On the basis of melt index and density results these two polymers appear substantially identical and therefore should be compatible and able to form a continuous miscible integral weld when correctly welded.

It needs to be mentioned however that MFI and density are not sufficient indicators to form the basis of a comprehensive comparison. The other two important properties are S-OIT and SCR (stress crack resistance). For instance it is known that Solmax use Petromat S7000 from Canada which has an OIT of 145 min at 200 °C. The S-OIT of K307 is known to be greater than 100 min.

In regard to SCR, the datasheets list the SP-NCTL results as >400 h and >900 h for the Petromat S7000 and K306 resins, respectively. Given that the minimum acceptable value to pass the SP-NCTL test is 400 h, both these resins pass.

The above conclusion on weld compatibility is predicated on the assumption that correct welding practices are followed, including adequate surface preparation, adequate polymer fusion and adequate melt pressure/residence time.

17.3.4 SEAM GRINDING

As mentioned above polyethylene and polyproplyene geomembranes contain low molecular weight waxes (polyoils) that can migrate and diffuse to the surface of the geomembrane and make welding difficult. These waxes and oils produce a weak boundary layer. In addition to this thin layer of wax, the outermost surface of polyolefin geomembranes become oxidized and the oxidation products which are polar compounds such as ketones and acids, can also make welding difficult.

Before extrusion fillet welding, these low molecular weight waxes and oxidation by-products need to be removed from the surface of the geomembrane by grinding/buffing and the leading edge of the upper membrane needs to be beveled or tapered to a 45 degree angle. The grinding can be done with a surface grinder but care needs to be exercised that no deep grooves or notches are introduced into the sheet as a result of the grinding process since these can act as stress concentrations which are the precursors to environmental stress cracking. Poor surface preparation can therefore leave waxy residues that can contaminate the weld and lower the integrity of the seam or track.

It is important however that grind marks are not deeper than 10% of the sheet thickness. Ideally, the grind marks should be about 5% of the liner thickness. The purpose of grinding is to remove the oxidized surface layers, additive blooms and dirt from the liner surfaces.

It should be noted that the grinding marks should not extend much beyond either side of the extrudate bead. For instance, if the final extrudate bead width is 38 mm, the width of the grinding trail should not exceed 40 mm. Furthermore the orientation of the grinding marks should be perpendicular to the seam direction rather than parallel to it (see Figure 17.27).

The welding should be performed shortly after grinding so that additive blooming and surface oxide formation does not reform. Seam grinding should be completed less than 1 h before seam welding.

Sheets of sandpaper are available with different grits (#40, #60, #80, #150, #240, #600). Sandpaper 'grit' is a reference to the number of abrasive particles per inch of sandpaper. The lower the grit number, the rougher the sandpaper and conversely, the higher the grit number, the smoother the sandpaper. The rotary grinder should use #80 grit sandpaper. Sandpaper that is coarser than #80, such as #60, is not acceptable for smooth HDPE liners as it may gouge the surface and introduce detrimental stress concentrating notches.

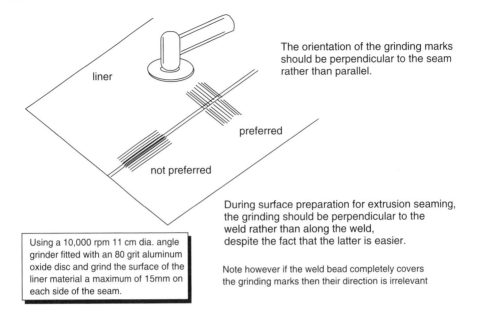

The orientation of the grinding marks should be perpendicular to the seam rather than parallel.

liner

preferred

not preferred

Using a 10,000 rpm 11 cm dia. angle grinder fitted with an 80 grit aluminum oxide disc and grind the surface of the liner material a maximum of 15mm on each side of the seam.

During surface preparation for extrusion seaming, the grinding should be perpendicular to the weld rather than along the weld, despite the fact that the latter is easier.

Note however if the weld bead completely covers the grinding marks then their direction is irrelevant

Figure 17.27   Schematic showing that grinding marks perpendicular to the extrusion weld are preferred to avoid potentially damaged score lines at right angles to tensile stresses

If the liner is 1.5 mm or thicker, the upper leading edge needs to be ground to a 45 degree bevel angle. During this procedure it is very important that the top sheet be lifted up and away from the lower sheet during the bevelling process so that no deep gouges are introduced onto the lower sheet. For this reason the bevelling procedure should be done before tack welding.

The polymer dust produced during grinding the liner must be removed from the weld area before welding.

The ends of existing welds (i.e. welds that are more than 5 min old) are ground to expose new material before restarting a weld (applies to extrusion welding only).

17.3.5   EXTRUSION FILLET WELDING CHECKLIST

Surface preparation for extrusion fillet welding requires diligent care and attention in order to obtain a satisfactory weld. This makes extrusion fillet welding labour-intensive and time consuming compared with hot wedge welding. Hot wedge welding is therefore clearly the preferred welding technique in welding situations where it is permissible. However wedge welding is not always possible in configurations such as difficult-to-access areas, around penetrations and connections in the liner, as well as for liner repair work. The quality of an extrusion fillet weld is highly dependant on the welder's skills and experience.

The following table provides a checklist of the key aspects of extrusion fillet welding.

Thin layer of surface wax and oxidation has been ground off	✔
Surface preparation has not caused deep scores, nicks or grooves	✔
Upper geomembrane has been tapered to a 45 degree bevel (if liner is 1.5 mm or thicker)	✔
All grinding marks run mainly perpendicular to the seam direction	✔
All ground material has been blown or cleaned away from weld area	✔
The extrusion welder is purged prior to beginning a weld	✔
The weld rod/extrudate is compatible with the geomembrane to be welded[a]	✔
Width of the extrusion fillet weld is sufficient to cover entire weld zone	✔
The weld is smooth and has a streak-free texture	✔
The seam has acceptable external appearance	✔
The seam has acceptable dimensions	✔
The seam has acceptable strength (both in shear and peel) (that is the seam should be ductile in shear and have no peel separation).	✔
The seam has been vacuum box[b] tested for water tightness	✔

[a]Note the weld rod should have the same or comparable melt flow rate (i.e. melt flow properties), stress crack resistance, UV stability (if used in exposed applications) and oxidative stability (i.e. oxidative induction time) as the parent geomembrane sheet. The German Guide, DVS 2207-1 'Welding of Thermoplastics – Heated Tool Welding of Pipes, Pipeline Components and Sheets made of PE-HD' (08/2007) (DVS-Verlag, 2007) specifies that HDPE resins with a melt flow rate (or melt flow index at 190 °C and 5 kg) in the range 0.3–1.7 g/10 min can be welded together.

[b]Vacuum box testing is performed in accordance with ASTM D5641 'Practice for Geomembrane Seam Evaluation by Vacuum Chamber'.

17.4   GENERAL OVERVIEW OF THERMAL WELDING METHODS

Temperature and pressure during welding should be set so that the heat and mechanical stresses on the geomembrane are minimized and long-term weld integrity is maximized.

17.4.1   RECORDING OF WELDING CONDITIONS

Welding machines need to be well maintained, regularly monitored for output consistency and all the welding parameters should be recorded. For instance in the case of hot wedge welding, variables such as welding temperature, welding speed, wedge temperature, ambient temperature, temperature of the geomembrane surface and nip roll pressure during the welding process need to be monitored to ensure that welding conditions were kept constant and also for the purpose of quality control and quality assurance. Other information to be recorded includes date, time, machine number, operator initials and installation contractor. Similar records need to be kept for fillet welding.

17.4.2   TESTING THE WELDING EQUIPMENT

Prior to embarking upon geomembrane welding, all welding equipment (both wedge and extrusion welders) must be tested in accordance with the specifications to ensure if the equipment is functioning properly. At a minimum this should be conducted at daily start-up and prior to resuming work after any break, and/or any time the machine is turned off for more than 30 min.

17.5   POTENTIAL THERMAL WELDING PROBLEMS

The quality of geomembrane welds (or seams) is significantly affected by welding temperature, welding speed, welding pressure and on-site conditions (particularly the ambient temperature). These parameters have a major influence on the long-term behaviour of the geomembrane welds. Therefore, controlling and optimizing weld process parameters are critical to achieving consistent wedge welds over a wide range of field conditions.

Theoretically, proper welding of geomembrane liners should not cause a reduction in the tensile strength across the weld as the welded sheets are expected to perform as a single geomembrane sheet. In reality, however, welding can cause a marginal reduction in tensile strength and tensile elongation relative to the adjacent parent geomembrane sheet because of the stress concentrations that arise from the weld geometry and the change in thickness across the weld. However, the converse is also observed, where the strength across the weld is higher than in the adjacent parent geomembrane (i.e. the weld performs as a local reinforcement to the liner).

Potential welding-related problems that can affect the integrity of geomembrane welds include:

- wrinkling/distortion in seam area;
- non-uniform weld;
- excess crystallinity;
- contamination in the weld such as moisture, dust, dirt, debris, wax and other foreign material within the seam;
- insufficient overlap;
- 'score' lines in the vicinity of the weld.

17.5.1   LOW AMBIENT TEMPERATURE DURING WELDING

The set point of the controller for the welding machine needs to be increased to compensate for the heat transfer effects of wind and temperature. Both welding processes are affected by heat transfer phenomena (i.e. the rate of heat transfer away from the welding process to the environment). The heat loss rates increase both with increasing temperature difference and increasing wind velocity. Increasing the temperature set point maintains the necessary temperature on the surface of the liner being welded.

Thus welding of geomembrane liners requires ongoing calibration during variable weather conditions. Failing to conduct adequate trial samples can cause incorrect calibration of the welding equipment. This can influence the welds adversely such that the sheets are inadequately welded together.

17.5.2   WELDING DURING HOT WEATHER

Black geomembranes (especially HDPE) in full sun on a hot day will expand and form wrinkles. This will make welding difficult and leads to abnormal strains when the geomembrane cools down. In extremely hot weather, geomembrane liner installation is recommended early in the morning or late in the afternoon or under special tents.

Wide variations in ambient temperature and resulting wider variations in sheet temperature coupled with the high coefficient of linear thermal expansion of HDPE can thus make maintaining good control of the welding parameters (e.g. uniform width of the joints) extremely difficult to achieve.

17.5.3   EFFECT OF MOISTURE ON WELDING

Rainfall or moisture on the geomembrane surface can interfere with the thermal fusion process by cooling down the geomembrane. Consequently the geomembrane does not reach the optimum and constant welding temperature required for a consistent quality weld. Hence it is necessary to ensure there is no free moisture in the weld area.

Surface water present on the geomembrane can vaporize during welding resulting in encapsulated bubbles and voids within the weld region that reduce weld strength and may ultimately result in leakage (see Figure 17.28).

When welding in wet or muddy conditions, it is recommended that HDPE off-cuts be used as 'drop sheets' under each join to keep mud and water from fouling the welding zone.

📌 *Moisture is a major concern in all geomembrane welding operations. Precipitation in any form, whether rain, snow, dew or fog can bring geomembrane installation to a halt and affect the integrity of welds.*

Figure 17.28   Porosity in an HDPE weld due to entrapped moisture

17.5.4   HUMIDITY DURING WELDING

High humidity can cause condensation on the welding surface which can adversely affect seam strength. At humidity levels exceeding 80%, special care needs to be taken to ensure that the difference between the air temperature and the dewpoint is a minimum of $3\,^{\circ}$C. This will prevent any substantial condensation of free water on the geomembrane being welded.

17.5.5   POROSITY IN WELDS

Porosity observed in moisture effected welds is often in the form of galleries and channels. This is consistent with water volatization in the weld interface during thermal fusion. The moisture-induced steam pressure causes elongated pathways. If elongated porosity is observed in the exposed surface of a weld that has undergone adhesive failure, then the likely cause of porosity is moisture in the weld zone during welding.

17.5.6   WINDY CONDITIONS

During windy conditions, the correct welding temperature may not be reached. To remedy this the welding temperature is typically increased by $20-30\,^{\circ}$C. If the wind is too strong, the welding area should be shielded from the wind.

17.5.7   DIRT CONTAMINATION DURING WELDING

The presence of dust/debris on the weld faces is another practical problem that needs to be controlled during the welding process. Dirt and dust in the weld region can lead to poor quality welds especially in the case of textured or structured geomembranes. Crack initiation in welds can occur at defects like grit, dust and sand trapped in thermal welds. This problem can be reduced by specifying geomembranes with smooth edges that are easily wiped clean.

Some manufacturers offer geomembranes with a removable, protective plastic edge strip to keep the edges of the geomembrane contamination-free for welding (see Figure 17.29). German guidelines call for GML to have a 15 cm protective tape strip in the smooth land area to protect against surface contamination (BAM, 1999). The plastic protective strip is applied in production. The adhesion of the strip must be sufficient to prevent separation during transport and site handling but must not leave residues on the surface of the geomembrane when removed.

17.5.8   PRESENCE OF WAXY LAYERS

HDPE and fPP geomembranes contain very low molecular weight components that migrate to the surface to form a waxy layer that can make adhesion and welding difficult. When such geomembranes are thermally welded these surface waxes need to be removed

Protection strips to keep the outside edge clean and free of dirt/dust

Figure 17.29 Photograph of a specialty HDPE geomembrane showing the white protection strips on the edge of the geomembrane to keep the edge clean and free of dirt/dust for welding. Reproduced by permission of NAUE

by scraping off by the hot wedge during thermal fusion welding or grinding prior to extrusion welding.

17.5.9 PRESENCE OF ADDITIVE BLOOM LAYERS

Additive 'blooms' can form on the liner surface due to exudation of the additives by migration/diffusion processes to the liner–air interface. These additives can interfere with the weldability of the geomembrane since they act as weak boundary layers. Before wedge welding, this bloom needs to be removed with a polar solvent such as acetone or limonene. Once a good fusion weld is formed then migration of additives over time will not interfere with the weld integrity.

The main factors affecting welding of polymeric geomembranes are summarized in Table 17.11

17.5.10 CONTRAINDICATIONS FOR WELDING

In summary, welding should *not* be performed when:

- the ambient temperature is below 5 °C unless special precautions are taken;
- the relative humidity is greater than 80%;
- the wind is causing significant dirt contamination in the weld zone;
- the wind is causing misalignment of the edges to be welded;
- the wind is causing localized cooling during the welding process;
- the temperature is below the dewpoint;
- there is significant water or/in the substrate beneath the geomembrane
- there is significant condensation/precipitation.

Table 17.11 Main factors (i.e. other than direct welding parameters) known to affect field welding and the quality/integrity of the resultant welds

Factor	Examples
Temperature	• the ambient temperature during welding • the surface temperature of the geomembrane panels being welded • the effect of clouds on the geomembrane temperature
Wind	• the effect of wind on localized cooling during the welding process • the effect of panel 'up-lift' during welding • misalignment of the edges to be welded due to wind forces
Water	• the relative humidity • moisture on/at the weld interface • condensation/precipitation • the water content of the substrate beneath the geomembrane
Contamination	• cleanliness of the weld interface with respect to windblown dust and debris • the presence of waxy or additive layers on the geomembrane • the extent/care to which surface preparation of the geomembrane to be joined is carried out
Support surface	• the nature of the subsurface on which the geomembrane is being seamed

17.6   DEFECTS THAT CAN AFFECT WELD INTEGRITY

There are a number of factors that can lead to weld defects and affect long-term weld integrity such as overheating of welds, scoring along welds, thickness reductions on or near welds, included dirt or particulate contamination, notch effects and stress concentrations in or near welds, stress cracking, welding of dissimilar geomembranes, etc. (Figure 17.30).

17.6.1   OVERHEATING OF THERMAL WELDS

During thermal welding the applied heat must be controlled carefully to avoid under- or overheating of the geomembrane sheets. Overheating of the polymer during welding leads to structural changes such as a reorganization of the polymer chains leading to an increase in the crystallinity of the polymer (in the case of HDPE). In addition, overheating causes accelerated consumption of the antioxidant reserves in the polymer. Both these factors increase the susceptibility of the geomembrane to stress cracking (see Figure 17.31).

An extruded bead (or extrusion weld) should not be applied to an already existing weld as this can lead to overheating of the adjacent geomembrane and render that region more susceptible to stress cracking. It is now well known that cracking failure is more likely to be initiated at such locations (see Figure 17.32).

The quality and integrity of thermal welds depends mainly on the experience/expertise of the operator and importantly the reliability of the welding machine's calibration and the accuracy of the temperature/pressure indicators. For instance, if the temperature calibration is inaccurate, then excess heat would have been applied to the geomembrane liner

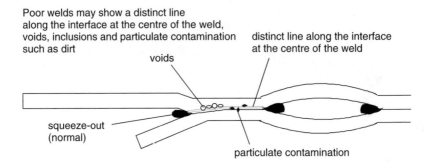

Poor welds may show a distinct line
along the interface at the centre of the weld,
voids, inclusions and particulate contamination
such as dirt

distinct line along the interface
at the centre of the weld

voids

squeeze-out
(normal)

particulate contamination

Figure 17.30 Schematic of various types of defects that can occur in wedge welds such as voids/porosity, inclusions and particulate contamination (e.g. dirt/sand)

Figure 17.31 Photograph of brittle failure along the edge of a wedge weld track in an HDPE geomembrane (see crack area indicated by white arrows)

without the operator's knowledge. Overheating of the liner can also lead to considerable distortion/buckling of the geomembrane (see Figure 17.33). Figure 17.34 shows the cross-sectional view of two welds in 1 mm HDPE and the obvious distortion has been termed a 'hot day weld' due to the failure of the operator to reduce the wedge temperature (or increase the welding speed) during hot weather conditions.

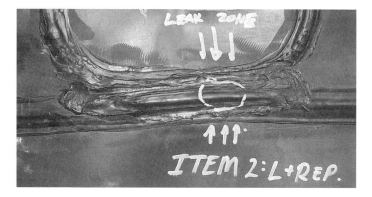

Figure 17.32 Photograph of an example of a poorly made extrusion weld which resulted in a leak zone

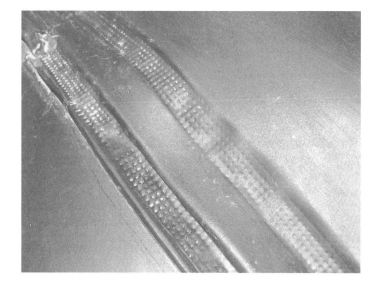

Figure 17.33 Photograph of a heat affected weld in a 1 mm HDPE liner

 Points to Check:

- *check the welder's temperature calibration is correct using an external calibrated thermocouple;*
- *check the welding machine's temperature read-out is displaying the correct temperature.*

Thermal welding PVC geomembranes may lead to localized loss of plasticizer. The localized loss of plasticizer may lead to a zone of embrittlement.

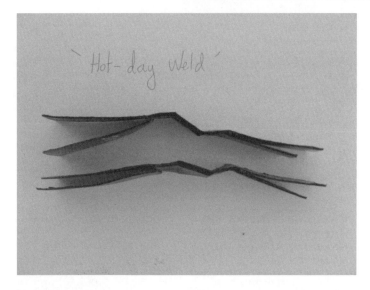

Figure 17.34 Photograph of DT weld coupons (side view) showing the distortion termed a 'hot-day weld' due to heat-affected welds

17.6.2 HEAT AFFECTED ZONE

Peggs and Carlson (1990) first identified that thermal welding of an HDPE geomembrane affects the microstructure, stress cracking resistance and durability of the adjacent geomembrane. Using constant tensile load tests for single-edge-notched specimens it was demonstrated that thermal welding of HDPE caused a reduction in the stress cracking resistance of adjacent geomembranes. This reduction in SCR was attributed to a combination of microstructural reorientation effects at the edge of the resolidified weld material and secondary crystallization of the geomembrane in the heat affected zone (HAZ) of the adjacent geomembrane (Peggs and Carlson, 1990).

17.6.3 SCORING ALONG WELDS

Welding machines can sometimes score the geomembrane in the process of welding, thereby creating a flawed region that can initiate cracking under an applied tensile force. This scoring may be detected by destructive shear testing since it will manifest itself as a sample having low elongation.

17.6.4 THICKNESS REDUCTION ON AND NEAR WELDS

Thickness reduction in geomembranes can have important implications with respect to durability of the geomembrane. Firstly a thickness reduction implies less material per cross-sectional area to resist loads, but more importantly it is the stress concentration effect that results from an abrupt change in thickness that is of concern.

Thickness reduction in geomembranes can arise from the following causes:

- Grinding down of geomembrane in preparation for extrusion fillet welding can lead to a thickness reduction immediately adjacent to the weld.
- Scoring and scratches due to improper handling and a lack of care during geomembrane installation.
- Imprinting (deformation) of a hard geonet into a softer geomembrane under the pressure of overburden loads (Giroud, 2005).

Such geomembrane thickness reductions can have a marked effect on weld behavior if they occur in the vicinity of a weld. Stress concentration can occur when there is a localized thickness reduction occurring adjacent to a weld.

In wedge welding the fact that the centre of the track has more molten material, which is subsequently squeezed out by nip pressure, is demonstrated by measuring the thickness of the weld across its width. Invariably the centre of the weld is thinnest and the thickness gradually increases towards each edge of the weld-track (see Figure 17.35).

An important parameter for determining weld quality is the 'thickness reduction' due to welding. The thickness reduction (T_r) is defined as follows:

$$T_r = (T_t + T_b) - T_w$$

where:

$$T_r = \text{thickness reduction}$$

$$T_t = \text{thickness of the top geomembrane}$$

$$T_b = \text{thickness of the bottom geomembrane}$$

$$T_w = \text{thickness of the weld}$$

The seam thickness reduction is defined as the difference between the thickness of the two original geomembranes and the welded seam thickness (as measured by an ultrasonic thickness gauge or by Vernier calipers). The reduction of thickness in the weld area (i.e.

Cross-sectional diagram of an overlap weld

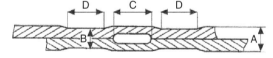

Seam thickness reduction = A – B

A : Thickness of upper and lower
    geomembrane liner
B : Thickness of welding seam
C : Width of test channel 15 +/– 2 mm
D : Width of weld ≥ 15 mm

Figure 17.35 Schematic of a cross-section of a double wedge weld. Reproduced by permission of Barry Smith, Plastic Welding Division, Plastral Pty Ltd

seam thickness reduction) can influence the long-term behaviour of a weld, particularly its strength and water tightness. The seam thickness reduction reflects an interaction of welding parameters under changing field conditions (geomembrane temperature, humidity, moisture, wind, etc.) during the welding process. The weld thickness uniformity is an indicator of how constant the welding process was maintained and the uniformity of the seam itself.

It is necessary to limit the extent of the thickness reduction to within a predetermined acceptable range (Lueders, 1998).

The allowable seam thickness reduction range must be within 0.2–0.8 mm for 2.5 mm thick HDPE geomembrane according to the German guide DVS 2225, while the allowable seam thickness reduction for 1.5–2.0 mm HDPE should be within 0.2–0.7 mm.

17.6.5   STRESS CRACKING OF WELDS

A study on comparative testing of HDPE geomembrane sheets and welds according to ASTM D-2552 (stress rupture under constant tensile load) found that the sheets could resist stress cracking better than the welds, with about 40% of the welds exhibiting cracking compared to only 1% of the sheet samples which cracked (Halse *et al.*, 1989). The stress cracks that formed were examined and it was evident that the cracks which appeared in the welded specimens were almost always initiated near the overlapping junction of the two geomembrane sheets where the stress concentration was likely to be highest. It was also observed that the cracks were preceded by crazing which formed in a direction perpendicular to the applied stress.

17.6.6   'SQUEEZE-OUT' STRESS CONCENTRATION IN WEDGE WELDS

The weld 'squeeze-out' at the edge of the extruded bead is generally not completely (i.e. 100%) bonded to the sheet leading to the propensity for notch-like stress concentrations to develop there (see Figures 17.36 and 17.37). This can cause issues in cold climates since water ingress into the notch will freeze at sub-zero temperatures causing peel forces (i.e. weld opening forces) to be exerted at the 'squeeze-out' notches.

17.6.7   NOTCH EFFECTS IN EXTRUSION WELDS

A characteristic common of extrusion welds is that the weld bead edges are not completely and intimately bonded to the geomembrane. The hottest part of the weld bead deposited on the liner is at the centre and a temperature gradient develops towards the edges due to faster cooling. As a consequence the edges of the weld bead are not as well bonded to the geomembrane as the central portion. The same is true for the start and the end of an extrusion weld track. Edges and extremities of welds have a tendency for less than 100% bonding to the parent, hence peel-back forces can develop and concentrate at these notch shaped features (see Figure 17.38).

It is important that cuts or tears in a geomembrane are cut out in a round hole and then patched using an extrusion weld as this will prevent the rip propagating (see Figure 17.39).

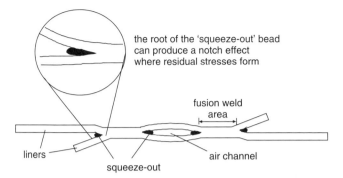

Figure 17.36 Cross-section of a dual wedge weld showing that a notch effect can occur at the root of the 'squeeze-out' bead

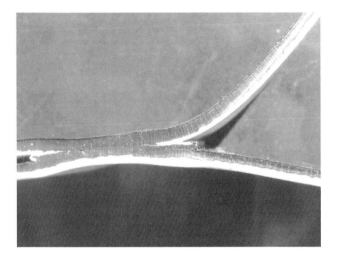

Figure 17.37 Photograph of the notch effect in a welded HDPE geomembrane

17.6.8 WELDING OF DISSIMILAR GEOMEMBRANES

Welding problems can occur when welding different geomembrane types together such as:

- LLDPE to HDPE;
- fPP to HDPE (very problematic);
- FHDPE to HDPE (note FHDPE is fluorinated HDPE);
- white HDPE to black HDPE (though not normally problematic);
- old to new GML (since old liners can have a surface layer of oxidation).

Old geomembranes can have an oxidized surface and if not removed these oxidized compounds can reside in the weld as a plane of weakness. Therefore before welding begins the oxidized surface should be removed by light grinding.

Figure 17.38   Cross-section of a fusion weld in an HDPE geomembrane showing crack
formation

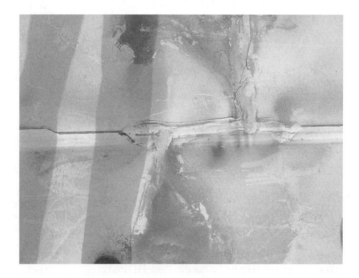

Figure 17.39   Photograph of extrusion welding used to repair a rip in a white HDPE
geomembrane. Unfortunately the rip continued to propagate (see left of image). In
order to prevent this it is necessary to cut a round hole at the end of the rip and seal it
with a patch. Reproduced by permission of NAUE

The general rule is that only geomembranes made from the same resin may be welded together. If the welding of geomembranes made from two different resins is unavoidable, then the two resins must either be in the same melt flow rate (MFR) group as defined in DIN 16776-1 or in the MFR groups 006 and 012 (BAM, 1999).

17.7 TRIAL WELDS AND FIELD WELDS

17.7.1 PANEL PREPARATION

Prior to welding, the adjoining geomembrane sheets need to be overlapped a minimum of 100 mm (and in no case less than 75 mm at any location). The overlap should not exceed 150 mm for double-wedge fusion welds. Weld overlaps on the bottom area of an impoundment or canal should be made such that the direction of flow over the lined surface is from the top sheet toward the bottom sheet to yield a 'shingle effect'.

The edges to be welded require thorough wiping and cleaning to eliminate any dirt, dust, moisture or other foreign materials. All field welds must be uniform in appearance, width and properties, with no observed warping due to overheating from welding. The peel and shear strengths of the welded welds must comply with the strength criteria shown in Table 17.12.

17.7.2 TRIAL WELDS

Trial welds (also referred to as 'qualifying' welds) are test welds on the same geomembrane as that to be installed. They are performed for the purpose of qualifying that the

Table 17.12 Recommended test method details for geomembranes and geomembrane seams in shear and in peel

Test details	HDPE	LLDPE, fPP	PVC	fPP-R, EIA-R, EPDM-R, CSPE-R
Tensile Test on Sheet				
ASTM test method	D-6693	D-6693	D-882	D-751
Specimen shape	Dumb-bell	Dumb-bell	Strip	Grab
Specimen width (mm)	6.3	6.3	25	100 (25 grab)
Specimen length (mm)	115	115	150	150
Shear Test on Seams				
ASTM test method	D-6392	D-6392	D-882	D-751
Specimen shape	Strip	Strip	Strip	Grab
Specimen width (mm)	25	25	25	100 (25 grab)
Specimen length (mm)	150 + seam	150 + seam	150 + seam	225 + seam
Peel Test on Seams				
ASTM test method	D-6392	D-6392	D-882	D-413
Specimen shape	Strip	Strip	Strip	Strip
Specimen width (mm)	25	25	25	25
Specimen length (mm)	100	100	100	100

welding contractors are using the optimum welding conditions and to identify whether the welding equipment and procedures are performing adequately. Trial seams can also be used for pre-qualifying various contractors by ranking the strength of the test welds they produce.

Trial welds seek to simulate all aspects of the on-site field welding activities intended to be performed on location using identical geomembrane material under the same climatic conditions as the actual field production welds (and on a horizontal or vertical concrete or soil substrate accordingly) in order to determine equipment and welding operator proficiency.

Trial welds are generally conducted by welding technicians prior to each seaming period as specified or as directed by the project engineer if welding problems are suspected. Unless authorized by the project engineer, once qualified by a passing trial weld, welding technicians should not change parameters (temperature, speed, wheel adjustment) without performing another trial weld.

The trial seams should be performed on strips of geomembrane from approved rolls to verify that welding conditions are adequate. Such trial welds should be made at the beginning of each welding period (start of day, midday and anytime equipment is turned off and allowed to cool down) for every welding apparatus used. A trial seam sample shall be approximately 1.0 m long by 0.3 m wide (after welding) with the seam centred lengthwise. Seam overlap is nominally 100 mm with a minimum acceptable overlap of 75 mm.

The trial welds are tested under both shear and peel conditions using a field tensiometer, and they should not fail in the seam (also needing to meet minimum specified strength criteria). Five peel and five shear specimens shall be cut from the centre of the trial seam and tested alternately in peel and shear respectively (note that the precise number of specimens to be tested in shear and peel varies and is set by the project engineer). In the event of nonconformance, an additional trial seam shall be made and tested. If this still fails to meet specifications the welding machine shall be examined and adjusted and not used for production seaming until two passing trial welds have been made.

Trial welds shall be produced each day, at the start of each workday, and thereafter at least once every 5 h of continuous operation, after each break in seaming of 30 min or longer, following a break that results in equipment replacement or shutdown. New trial welds will also be required whenever the geomembrane temperature changes by more than 25 °C.

Recommended test method details for geomembranes and geomembrane seams in shear and in peel are shown in Table 17.12.

17.7.3 FIELD WELD STRENGTH

Typical requirements for field welds include:

1. When the weld is tested in **shear** in accordance with ASTM Standard D-6392 (or ASTM D-4437), the specimen shall exhibit a film tear bond (FTB)[2] failure, and the

---

[2] FTB refers to failure only in the parent material of the geomembrane and not in the weld area. It is deemed to be an acceptable form of failure. Due to different interpretation by different people its use is discouraged today although it still appears widely in many project specifications.

Table 17.13 Typical weld strength criteria for HDPE geomembranes (kN/m)

Thickness	1.00 mm	1.5 mm	2.0 mm	2.5 mm
Peel	10.5	15.9	21.2	26.4
Shear	14.0	21.0	28.0	35.0

strength shall be equal to or greater than the values specified in Table 17.13. The tests shall be performed at a rate of displacement of 51 mm/min (2 in/min). Some specifications allow a faster rate (i.e. 510 mm/min) of travel after the maximum load is reached.

2. When the weld is tested in **peel** in accordance with ASTM Standard D-6392 (or ASTM D-4437), the specimen shall exhibit an FTB failure and/or the liner must fail before the weld, and the strength shall be equal to or greater than the values specified in Table 17.13. The peel strength criteria shall apply to both the primary and secondary welds of double-wedge fusion welds.

The following pass/fail criteria are commonly used to determine compliance of field welds with the strength criteria shown in Table 17.13:

1. All five weld specimens from a given sample or coupon tested in peel (both weld tracks in the case of double-wedge fusion welds) and all five specimens tested in shear shall exhibit the required strengths, and display film tear bond (FTB) failures (i.e. no allowance for failures based on strength and no non-FTB failures).
2. All five weld specimens from a given sample or coupon tested in peel and all five specimens tested in shear shall exhibit the required strengths and at least four out of five test specimens, for each type of test, shall exhibit a film tear bond (FTB) failure (i.e., 0% strength failures and up to 20% non-FTB failures are permitted.
3. One weld specimen from a given sample or coupon tested in peel or shear may exhibit a non-FTB failure and a strength no less than 80% of the specified strength provided that all five field weld specimens tested from the duplicate (or archive) sample coupon exhibit the required strength and no more than one test specimen out of the five additional specimens exhibits a non-FTB failure.

Despite the above pass/fail criteria it should be recommended that all welds pass. Good welders can easily achieve that these days.

If a field weld fails the criteria specified, then additional coupons may be obtained (as agreed by the various site staff), at progressively increasing distances from both sides of the failed sample, until two consecutive samples on each side of the original sample pass the field weld criteria. At that point, the extent of the original defect in both directions along the field weld will be considered isolated and the Liner Contractor may then: (i) either cap, re-weld and re-test the weld up to and including the closest of the two passing samples, and patch and weld the hole of the furthest passing sample, or (ii) cap, re-weld and re-test the entire length of sampling.

If approved by the Site Engineer, double-wedge fusion welds may be repaired by extrusion welding the flap of the top sheet to the bottom sheet if the weld non-compliance is due only to a non-FTB failure of the destructive test sample.

The pass criteria in destructive weld testing are summarized in Table 17.14.

Table 17.14 Pass/fail criteria in destructive weld testing

Pass criteria	Comments
The weld specimen must fail outside of the weld region and the entire weld must remain intact	This is denoted as a 'film tear bond' (FTB) failure. Note use of the FTB term is being phased out.
The failure force reached should be greater than the specified values for shear and peel, respectively	For the *shear* test, generally the specified value is 85 to 95% of the parent sheet value. For the *peel* test a value between 50–80% of the parent geomembrane sheet is usually specified for thermal fusion welds since the peel test is a more severe test than the shear test
The elongation at failure for the shear test should achieve the minimum threshold value	There must be clear elongation outside of the weld (e.g. at least 50% and preferably greater than 100% break elongation)
The peel test specimen should not delaminate or separate more than a specified amount at failure (e.g. not more than 10% across the weld front)	No delamination is highly preferred

17.8   CHEMICAL WELDING OF GEOMEMBRANES

Chemical welding of geomembranes uses some form of solvent or solvent cement to seam the liner. Chemical welding is very versatile allowing many complex and detailed seam configurations to be achieved. Note that chemical welds are not made with a glue but with solvents that dissolve and soften the surface of the liners.

Chemical fusion joining applies mainly for PVC, EIA-R, TPU and CSPE-R geomembranes since these polymers can be readily dissolved by appropriate solvents. A solvent (e.g. methyl ethyl ketone (MEK)) is applied between the two geomembrane sheets to be joined. The solvent causes the surfaces of the liner to soften and then roller pressure is applied to intimately bond the sheets together. The softening and solvation of the two adjacent materials allows increased mobility, interpenetration and entanglement of the polymer molecules at the surface. Note HDPE, LLDPE and fPP cannot be chemically welded as these polymers have no solvents at ambient temperature.

It is important that the amount of solvent applied is carefully controlled as too much solvent will lead to over softening, solvent entrapment and will weaken the sheet material. Sometimes the solvent is converted to a solvent cement by dissolving some of the parent polymer in the solvent to increase its viscosity and retard the evaporation rate. The solvent or solvent cement is brushed onto the geomembrane surfaces to be bonded. A list of typical solvents for chemical welding of polymeric geomembranes and liners is shown in Table 17.15.

In the case of chemical fusion welds and adhesive seams, strength testing is best done after full curing which takes several days. Alternatively a field oven may be used to accelerate the curing of the weld.

Table 17.15 List of typical solvents for chemical welding
(Layfield, 2008)

Liner material	Solvent[a]
Flexible PVC	THF
Supported PVC	THF
CSPE	Xylene
CSPE-R	Xylene
Urethane (certain grades)	Xylene or THF

[a]THF, tetrahydrofuran; xylene is not permitted in potable water applications as it is a suspected carcinogen.

Thermoset geomembranes, like EPDM and nitrile rubber, generally require bonding by chemical adhesives or vulcanization tapes, although Trelleborg modify the surface of their EDPM liners so they can be thermally welded.

PVC produces excellent solvent welds and has the versatility of being easily field solvent welded enabling excavation contractors to install a PVC liner without specialty crews. Solvent welding of seams, patches and pipe boots can all be done by the excavation contractor, thereby making PVC liners the lowest installation cost alternative.

Note there is an important distinction between a glue bond and a chemical weld. Glues or adhesives have only been found to last 7–8 years in service. A glue is actually a different polymer than the liner material and forms a distinct layer between the two liner sheets. Such glues are commonly based on epoxy or acrylic polymers. The extreme conditions that a geomembrane liner is subject to in service often limits the expected glue service life. In contrast, chemical welding is a permanent process capable of producing bonds that will last as long as the liner material.

A chemical weld is formed when the opposing faces of each liner surface are joined in a controlled fashion with a welding solvent. The welding chemical is a volatile solvent that is blended with other solvents or dissolved parent polymer to control the speed at which it dissolves the surface of the liner material and to control its rate of evaporation. The welding chemical can be brushed, poured or squirted onto the liner surface before they are pressed together. By carefully pressing the semi-dissolved liner surfaces together a bond is formed in a similar manner to a thermal bond made by thermal fusion welding. The solvent 'flashes off' or evaporates, and after about 24 h a homogeneous weld with no foreign material between the two liner surfaces is established (Layfield, 2008).

Chemical welding is also one of the most versatile welding processes. In liner installations it is easier to perform detail work with a chemical weld than with most heat welding processes. Chemical welding is suitable for small as well as large projects and has a low initial skill level; workers can be trained to make effective welds quickly and reliably.

Rather than using straight solvents, chemical welding is usually performed using a 'bodied solvent'. A bodied solvent is a chemical welding solvent that has between 5–25% of the parent liner material dissolved in solution. This additional 'body' allows the solvent to be placed on vertical surfaces, helps to fill in uneven surfaces and slows down the cure time in very hot welding conditions.

Table 17.15 shows the liner materials that can be chemically welded and the corresponding solvent. Tetrahydrofuran (THF) is a volatile solvent with an 'ether-like' odour that works very well on most PVC products but is usually limited to temperatures above $+10\,°C$. Below this temperature heat must be applied to facilitate the evaporation of the solvent. THF evaporates completely after the weld has set and the residue is undetectable. THF is the most common solvent in use on PVC liner materials.

Xylene is an aromatic solvent that is used primarily for chemical welding of CSPE liners, but has been used for some PVC work. CSPE liners crosslink (i.e. cure) over time and they become progressively more difficult to weld with age. A wipe with xylene can reactivate the surface of aged CSPE liners in most cases so that thermal welding can be accomplished. Chemical welding of 'aged' CSPE liners is dependant on the material condition (Layfield, 2008).

The two most critical aspects of chemical welding of liners such as PVC is (i) to ensure that only clean dry surfaces are bonded and (ii) that the temperature is adequate to drive the welding (i.e. to evaporate the solvent).

The bonding surfaces of the liner should be pre-cleaned using soapy water. If there is an oily residue on the liner then a commercial cleaning agent should be used. The ambient temperature must be at least $10\,°C$ to perform a solvent weld using THF. If the temperature is above $4\,°C$ but below $10\,°C$, then a ''bodied' THF/MEK blend is recommended in place of THF. Below $4\,°C$, solvent welding will not be possible without supplying heat to the process. With older, weathered (i.e. aged) geomembrane liners it is advisable to roughen the surface with a medium grit sandpaper to remove the oxidized surface layer, followed by a wipe clean with THF.

When patching using chemical welds, align the patch on the damaged area, ensuring that it conforms to the bonding surface. Then work a brush dipped in solvent between the old material and the patch material. Apply the solvent liberally to thoroughly wet both surfaces to be bonded. The solvent should flow out of the seam and onto the adjacent areas on the liner. Using a small roller, roll out the patch from the centre to the edges or press the patch down and smooth using a rag. A small bead of solvent should press out of the patch seam when pressure is applied. Excess solvent from the liner surface should be wiped off with a clean rag. It is necessary to allow 24 h curing time (at $20\,°C$) before testing the seam. The integrity of the seam can be tested by directing 0.24 MPa (35 psi) air at the edge of the seam. Any areas which 'lift' need to be resealed by applying more solvent and pressure to the area.

17.9 GENERAL WELDING INSTRUCTIONS FOR HDPE/LLDPE/fPP GEOMEMBRANES

The general instructions provided below are some recommendations on surface preparation for polyolefin geomembrane welding. It is the responsibility of the fabricator to determine the optimum method of surface preparation for consistent welding performance.

17.9.1 SURFACE INSPECTION

The welding surface of an HDPE geomembrane is the critical interface in any fabricated installation. It is estimated that nearly 24% of all geomembrane leaks occur at the field seam (weld) area. To ensure the highest possible weld integrity the welding surface of a polyolefin geomembrane needs to be carefully inspected for signs of dirt, contamination, waxy deposits (bloomed additives), moisture and so on.

A convenient method of relatively assessing the surface condition of multiple rolls is to check the surface of each roll prior to installation with a 'dyne pen' which indicates surface energy of 'wettability'. The result is an indication of any significant variation between the rolls. If variability in the 'dyne number' is detected the welder could then test weld a small portion of the differing rolls to determine the optimum weld settings prior to fabrication.

17.9.2 SURFACE CLEANING

The extent of cleaning depends on the degree of contamination at the welding surface. Commonly it is sufficient to remove adherent dirt by wiping with a dry or moist rag, also with the aid of air blowers. In these cases, water or soapy solutions, sometimes alcohol (i.e. spirit) or acetone can be employed. It is important to ensure that the weld surface is completely dry and that no residue remains after the cleaning.

In other instances, more extensive cleaning methods are required. For instance, it has been reported that outdoor exposed thermoplastic sheets could have welding problems due to factors such as industrial pollution after acid rain, surface oxidation, bloomed-out additives and oligomers (low molecular weight polymer chains).

For highly contaminated surfaces, the following procedure could be used to clean the weld surfaces:

1. Remove rough dirt using a commercial dishwashing detergent and dry surface with a rag.
2. Wipe with ethyl acetate.
3. Apply d-limonene (e.g. Citroclear, Carlisle Syntec Systems, Brussels, Belgium).
4. Rub several times using a rubbing and wiping action.
5. Immediately wipe with ethyl acetate (only), to remove the greasy residues.
6. Dry thoroughly using clean rags.

In extreme cases, light buffing with a sanding disc may be necessary: however, this should be carried out with caution since the material could be damaged. It is important that the mechanical abrasion process must not damage the liner with excessive heat, should not remove too much polymer nor damage the membrane during handling. Abrasion must be conducted only when it is called for and should be conducted just prior to the welding process.

It should be noted that sometimes it will be sufficient to clean only with ethyl acetate (e.g. in the case of blooming additives or oxidized surfaces).

The d-limonene also has excellent cleaning properties. It acts by diffusing deep into the polymer surface and extracting short molecular chain segments and blooming additives. However, it can produce a greasy film on the surface. Thus additional wiping with ethyl acetate is therefore required. The use of d-limonene requires protection against open fire, inhalation as well as eye- and skin-contact. Instructions on the safety data sheet must be followed. d-limonene is used as a substitute to xylene and is considered much less hazardous than xylene.

The extent of cleaning undertaken should be adjusted to the actual on-site situation and needs to be tested on liner fragments prior to the welding process.

17.9.3 ON-SITE CONDITIONS

On-site weather conditions can influence the seaming process and weld quality, thus requiring a high level of welding skills. The following are important considerations.

Wet surfaces compromise the formation of a secure weld. Welding or installation should not be conducted at areas of still water, or during/immediately following precipitation, unless suitable protection (e.g. a tent) is utilized.

Welding should not be performed during or after periods of high humidity since water will condense on the surface and water molecules can diffuse into the outer layers of the polymer. Heating by the welding tool can cause small water bubbles to be generated at the interface of the seam leading to unsealed spots and subsequent weld failure. Tests have revealed that an amount of 1000 ppm (0.1 wt.%) surface moisture is sufficient to hinder the formation of a good weld. The water will diffuse out of the liner over an appropriate drying time.

Welding should ideally not be conducted during strong winds since the membrane can be lifted, the welding temperature can be lowered and dust can be blown onto the weld region, unless suitable wind barriers and appropriate ballasts (e.g. sand bags) are employed.

Welding should not be attempted during prolonged and intensive sunshine which heats up the membrane and causes thermal expansion of the liner. This could result in the formation of waves, which complicate and hinder the welding operation. It also increases internal tensions around the seam region after welding. A high degree of thermal expansion will manifest itself in higher levels of residual stress at the seam area. This has contributed to the stress cracking phenomena (i.e. ESCR) exhibited by HDPE geomembranes.

Any equipment used must not damage the geomembrane due to handing, producing excessive heat, leakage of chemicals or other means. The on-site crew should not smoke in the vicinity of the installation, wear shoes that might damage the liner, or engage in activities that could otherwise damage or harm the liner.

Whenever welding is interrupted, the seam region must be protected from water, sun and blown contamination.

It is good practice for trial welds to be made on spare sections of geomembrane liners prior to the actual work on-site being started or resumed following an interruption. They should be made for each apparatus used, under the same conditions as the actual seaming operation in order to verify that the seaming equipment and conditions are adequate.

Hot air and extrusion welding technologies needs to contend with surface oxidation and radical chain scission initiated by the presence of oxygen. Using nitrogen instead of air reduces surface oxidation and degradation.

It has been noted that wedge welding requires less cleaning than hot air welding, as the direct contact of the wedge creates a melt flow which removes surface contamination (e.g. blooming products, oxide layer, fine dirt) out of the seam area into the bulge, or, mingles 'dirty' with 'clean' material. The heat transition between the wedge and the membrane is more defined and the seam is more homogeneous compared to hot air welds. However, hot air welding anneals the edge of the weld reducing residual stresses at the transition from the molten region to the unmelted region.

The above recommendations have been gathered from various industry sources and represent 'best practice' guidelines in the area of welding of HDPE geomembranes.

17.10 GENERAL WELDING INSTRUCTIONS FOR PVC GEOMEMBRANES

17.10.1 THERMAL WELDING OF PVC

The principle of thermally welding PVC geomembranes is similar to that for welding polyolefin geomembranes. Fusion is brought about by compressing the two melted surfaces together, causing an intermingling of the polymers from both sheets. The heat source itself melts the surface of the viscous polymer sheets, followed closely by the nip rollers which squeeze the two geomembranes intimately together.

Temperature controllers on the thermal welding device should be set according to PVC liner thickness, ambient temperature, rate of seaming and location of the thermocouple within the device. Ambient factors such as clouds, wind and sun require temperature and rate of travel settings to vary. Records for destructive test samples should include the temperature and travel rate settings of the thermal welder used to construct the seam.

It is necessary that the operator keep constant visual contact with the temperature controls, as well as the completed seam coming out of the machine. Occasional adjustments of temperature or speed as the result of changing the ambient conditions will be necessary to maintain a consistent seam. Constant visual and 'hands-on' inspection is also required.

A 2 m long test strip is normally fabricated and test specimens manually tested prior to constructing each seam, or at any time the seaming procedure (e.g. speed, machine temperature) has changed. A minimum of one test strip should be made each morning and afternoon prior to commencement of welding.

17.10.2 PVC ULTRASONIC WELDING

Ultrasonic welding is a thermal technique which melts the two opposing geomembrane surfaces to be welded by passing an ultrasonically vibrated metal wedge or knife between them. Pressure is simultaneously applied to the top and/or bottom geomembrane, to form a continuous bond. Some welds of this type are made with dual bond tracks separated by a non-bonded gap or channel which is used for air pressure testing (i.e. similar to double-wedge hot fusion welding). These welds are referred to as dual-track welds or double-track welds.

17.10.3 PVC SEAMING BY SOLVENTS (CHEMICAL WELDING)

A 150 mm wide overlap must be cleaned of all dust, dirt or foreign debris no more than 30 min prior to applying the chemical fusion agent. If mud has adhered to the sheet surface overlap area, it needs to be removed with clean water and allowed to dry prior to seaming. Seaming cannot be conducted in the presence of standing water. Wet surfaces must be allowed to dry. A slip sheet or seaming board may be used to lift the geomembrane above damp surfaces. If wind conditions contaminate the seaming area or displace the geomembrane sheets, temporary ballast and additional cleaning procedures will be required.

All field seams will be a minimum of 50 mm wide and a sufficient amount of chemical fusion agent should be applied such that, upon compressing the seam surfaces together, a thin excess of chemical fusion agent will be forced out. Then a high durometer rubber, nylon or steel roller can used to compress the seam surfaces together until a bond is formed. Roller action will be at a parallel direction to the seam's edge so that excessive amounts of chemical fusion agent will be purged from between the sheets. Trapped solvent or adhesive should be rolled out of the seaming area.

A continuous wet layer of chemical fusion agent is necessary to prevent a leak at the 'tie-in point' between the last chemical fusion agent application and the next. If the chemical fusion agent, which is initially shiny when applied, takes on a dull filmy appearance, this indicates that the interfaces may require a faster closing together or, the ambient temperature is too high to continue seaming.

With decreasing temperature and with increasing thickness, PVC becomes increasingly more convenient to thermally weld. For this reason, chemical seaming is not generally recommended below 10 °C.

17.10.4   REPAIRS AND PATCHES IN PVC

'Fishmouths' or wrinkles at the seam overlaps should be cut back and overlapped, then patched with an oval or round patch of the same material and thickness as the primary geomembrane. Patches are also necessary where destructive samples are removed or if material is damaged. Patches should extend 150 mm beyond the area to be repaired, be oval or rectangular with round corners and can be chemically welded a minimum of 50 mm around the perimeter.

17.10.5   PVC FIELD SEAM TESTING

Field Quality Control seam testing involves both non-destructive and destructive testing. Each seam must be checked visually for uniformity of width and surface continuity. Proper fusion chemical application visually changes the surface appearance. Usually the installer will use an air lance or blunt-end pick to check for voids or gaps under the overlapping geomembrane. For dual-track welds air channel testing is performed according to ASTM D-7177-05 'Standard Specification for Air Channel Evaluation of Polyvinyl Chloride (PVC) Dual Track Seamed Geomembranes'.

When un-bonded areas are located, they can sometimes be repaired by inserting more chemical fusion agent into the opening and applying pressure. If that is not satisfactory, a round or oval patch must be placed over them with at least 150 mm of geomembrane extending on all sides.

17.10.6   FACTORY SEAMS

Factory seams in liners such as PVC, fPP and CSPE-R are generally of a high standard since they are made in a clean and controlled environment. The large prefabricated panels are then shipped to the installation site on pallets and additional seaming is conducted in the field to produce their final configuration. In contrast field welds can be problematic and lack integrity leading to leaks due to the uncontrolled nature of the outside environment with respect to temperature/wind etc. and to the potential for contamination by dust/dirt.

REFERENCES

ASTM D-7176 'Standard Specification for Polyvinyl Chloride (PVC) Geomembranes for Buried Applications'.

ASTM D-7177, 'Standard Specification for Air Channel Evaluation of Polyvinyl Chloride (PVC) Dual Track Seamed Geomembranes'.

BAM, 'Certification Guidelines for Plastic Geomembranes Used to Line Landfills and Contaminated Sites', Federal Institute for Materials Research and Testing (BAM) (September, 1999).

Comer, A. I., Repairing Geomembranes in the Field, *Water Operation and Maintenance Bulletin*, 29 (1997).

Darilek, G. T. and Laine, D. L., Costs and Benefits of Geomembrane Liner Installation CQA, in *Proceedings of the Geosynthetics 2001 Conference*, Portland, OR, USA (February, 2001).

DVS-Verlag, The German Guide DVS 2207-1 'Welding of Thermoplastics – Heated Tool Welding of Pipes, Pipeline Components and Sheets made of PE-HD (08/2007)'; available from DVS-Verlag GmbH, Düsseldorf, Germany [www.dvs-verlag.de] (2007).

Giroud, J. P., Quantification of Geosynthetic Behavior, *Geosynthetics International*, **12**(1), 2 (2005).

Halse, Y. H., Koerner, R. M. and Lord Jr, A. E., Laboratory evaluation of stress cracking in HDPE geomembrane seams, in R. M. Koerner (Ed.), *Durability and Ageing of Geosynthetics*, Elsevier, New York, NY, USA, pp. 177–194 (1989).

Layfield, 'Chemical Welding Guidelines' (2008).

Lueders, G., 'Assessment of Seam Quality and Optimization of the Welding Process of HDPE Geomembranes', article by Bundesanstalt fuer Materialforschung und pruefung, Berlin, Germany. Published in the *Proceedings of the Sixth International Conference on Geosynthetics*, Atlanta, GA, USA (March, 1998).

Novaweld [http://www.novaweld.com/html/geomembranes.html] (2008).

Peggs, I. D. and Carlson, D. S., The effects of seaming on the durability of adjacent polyethylene geomembranes', ASTM Special Technical Report, 1081, in R. M. Koerner (Ed.), *Geosynthetics Test Waste Containment Application*, pp. 132–142 (1990).

SKZ, Testing and other polymer data from DVS Guidelines (225, Part 1); available from DVS-Verlag GmbH, Düsseldorf, Germany [www.dvs-verlag.de].

18

Geomembrane Weld Testing Methods

18.1 INTRODUCTION TO WELD TESTING

A project construction quality control (CQC) and construction quality assurance (CQA) programme for geomembrane installations generally includes observing and validating the QC program of the liner installer as well as performing destructive weld tests on samples taken from trial as well as field geomembrane welds.

Construction quality control is defined as a planned system of inspections that are used to directly monitor and control the quality of a geomembrane fabrication and installation. These actions are normally performed by the *geomembrane installer* to ensure the highest quality of the constructed system. CQC thus refers to measures taken by the installer or contractor to determine compliance with the requirements for materials and workmanship as stated in the plans and specifications for the project.

Construction quality assurance is defined as a planned system of activities that provide assurance that the geomembrane was fabricated and installed as specified in the design. CQA includes inspections, verifications, audits and evaluations of materials and workmanship necessary to determine and document the quality of the constructed facility. CQA refers to measures taken by the CQA organization to assess if the installer or contractor is in compliance with the plans and specifications for the project.

The CQC and CQA definitions and responsibilities are summarized in Table 18.1.

Geomembrane testing methods can be divided into two main categories: nondestructive testing and destructive testing. Non-destructive testing methods for geomembrane welds commonly include mechanical probing, air pressure testing on double wedge welds, air lance testing, vacuum box testing, spark testing and electrical leak location testing. Destructive testing on the other hand involves actually cutting out sections of the finished welds and testing them in shear and peel modes. Destructive testing is by its nature confined to spot checks on limited lengths of geomembrane seams. One destructive test sample per 150 m is commonly specified for double wedge welded seams although a decreased frequency is allowed if an electrical leak survey is conducted. Shorter intervals are specified for extrusion welds which have a higher manual involvement. For double wedge

A Guide to Polymeric Geomembranes: A Practical Approach   J. Scheirs
© 2009 John Wiley & Sons, Ltd

Table 18.1 CQC and CQA definitions

Type of QC/QA	Responsible party
Construction quality control (CQC)	Relates to the company installing the liner
Construction quality assurance (CQA)	Relates to a separate inspection company working on behalf of the regulatory agency but paid by and reporting to the facility's owner and/or operator

Table 18.2 Summary of weld assessment tests

Weld property	Test interval	Assessment method
Seam condition	Continuous	Visual – pocket scriber, flat screwdriver
Dimensions	Spot checks	Ultrasonic
Mechanical strength	Spot checks	Destructive tests: peel test, shear test
Water tightness	Spot checks	Vacuum box, spark test, electrical leak test, air lance
Water tightness	Continuous	Air pressure test

welds, usually a 1 m sample of a seam is taken from the liner. The current industry norm is that one third of the seam is tested by a third-party (CQA) testing laboratory, another third is utilized by the installer for QC purposes (e.g. on-site tensiometer testing) and the remaining third is stored by the facility owner. The destructive test samples are die cut into multiple specimens (typically 10 one inch (25 mm) parallel specimens) and tested under shear and peel conditions, respectively. The various weld tests are summarized in Table 18.2.

Common geomembrane weld testing methods and their features and shortcomings are shown in Table 18.3.

18.2 VISUAL INSPECTION OF WELDS

Geomembranes welds can be visually inspected to assess the quality of the workmanship and the appearance of the welded seam. In particular, for wedge welds the observer needs to observe a consistent 'squeeze-out' on the weld edge which is an indicator that the correct temperature and pressure were used during installation. Sometimes excessive 'squeeze-out' can produce a pronounced notch effect (Figure 18.1).

In the case of extrusion fillet welds, the weld appearance should be smooth, uniform and free of streaks and lumps. In addition, there should be no obvious scoring, notches or deep scratches introduced by the surface grinding process that is conducted to prepare the surfaces that are extrusion fillet welded.

18.2.1 SIGNS OF 'SQUEEZE-OUT'

A small degree of 'squeeze-out' can be a good indication that proper seaming temperatures have been applied. The molten polymer will laterally extrude out of the seam area in

Table 18.3 Common geomembrane weld testing methods and their features and shortcomings (Darilek and Laine, 2001)

Method	Area tested (percent of liner)	Speed	Test under load	Test for construction damage	Features and limitations
Air lance	Seams (0.5%)	Fast, 3–10 m/min	No	No	Economical QC test used on very flexible liners only. Requires operator skill and experience
Air pressure	Double wedge weld seams (0.5%)	Set-up time plus about 10 min/seam	No	No	Economical QC test for double wedge welds only. Welds tested at a fraction of their strength
Conductive sheet	Primarily for panels, limited test on seams (99%)	Very rapid 1–2 ha/day	No	Yes, but prior to placement of drainage layer	Rapid QC test of panels and areas that cannot be tested with a vacuum box but requires a proprietary geomembrane
Vacuum box	Primarily for extrusion welded seams	Slow and labour intensive	No	No	QC test, operator dependent. Cannot be used on wrinkles and corners. Leak may not be indicated with clay or water under liner
Destructive seam testing	0.2% of seams (0.0015%)	Very slow turnaround (days)	Yes	No	Test for maximum seam strength; about 4 m of extrusion welds are needed for every test panel taken
Electrical leak location	100% of liner (100%)	Very rapid 0.5–1.0 ha/day	Yes	Yes, after placement of drainage layer	Test is conducted after potential for construction damage has occurred

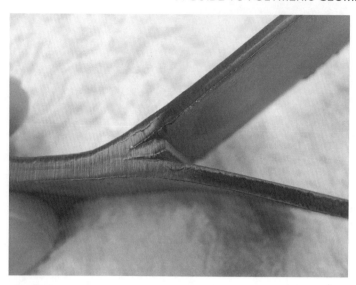

Figure 18.1   Photograph of notch-effect caused by a 'squeeze-out' bead in a wedge weld

properly welded seams. However, an excessive amount of 'squeeze-out' indicates that excessive heat and/or pressure was applied during the welding process. To rectify this situation it is necessary to reduce the temperature and/or pressure.

In the case of extrusion fillet welds excessive 'squeeze-out' (or 'flashing') is acceptable, if it is about equal on both sides of the weld and does not interfere with subsequent vacuum box testing. The presence of this 'squeeze-out' may indicate one of the following:

- that the extrusion die was not traversing directly against the liner;
- that the extrudate temperature was inappropriate for adequate flow;
- that the seaming rate was too slow (Poly-flex, 2008).

18.2.2 PUCKERING

In the case of hot wedge welds, a slight amount of heat distortion known as 'thermal puckering' on relatively thin liners (less than 1.25 mm) is acceptable and in fact desirable. This indicates that the applied heat has penetrated through the entire geomembrane sheet. If the level of distortion is too great, particularly on the underside, then it is necessary to lower the temperature or increase the rate of welding. For geomembranes of 2.00 mm or greater, no thermal 'puckering' should occur or should be observed.

Good fusion welds are characterized by a slight deformed surface known as 'thermal puckering' or 'thermal buckling'.

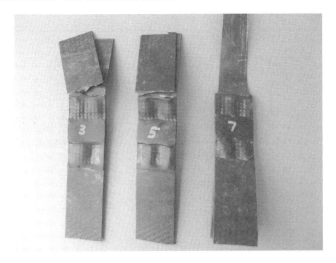

Figure 18.2 Photograph of DT weld coupons. The two coupons on the left show a lack of elongation and broke directly adjacent to the upper weld track. These two weld specimens failed in the weld boundary or just outside of the seam area in a brittle fashion without displaying any obvious elongation. Shear elongation is a very important parameter since it can determine whether the welding process has resulted in a loss of ductility of the polymer which can affect the longevity of the weld. The weld sample on the far right exhibits good elongation behaviour

18.2.3   PRESENCE OF NOTCHES OR SCORE MARKS

The presence of significant score marks in the vicinity of the weld tracks could indicate local stress concentration regions that may initiate premature failure under applied loading. If the notches are of sufficient depth then low break elongations will result under shear testing (see Figure 18.2).

18.3   NONDESTRUCTIVE SEAM TESTS

The purpose of nondestructive testing is to detect discontinuities or holes in the seam of a geomembrane. It also can indicate whether a seam is continuous and non-leaking. Nondestructive tests for geomembrane include air pressure testing for dual-track fusion welds and vacuum testing for extrusion welds. Nondestructive testing should be performed over the entire length of the seam.

   Those nondestructive weld tests that are routinely conducted are shown in Table 18.4.

18.3.1   MANUAL POINT STRESSING

Welds can be checked using a blunt tool (usually a screwdriver) to mechanically (i.e. qualitatively) test the integrity of the weld. The flat end of the screwdriver is inserted

Table 18.4 Nondestructive weld tests

Test	Description	Comments
Mechanical point stress test (MPST)	A flat head screwdriver (or other blunt tool) is levered under the top edge of the seam to detect an area of poor bonding	A rather subjective test where the results depend on the diligence and care of the person conducting the work
Air channel pressurization (ASTM D-5820 for HDPE) (ASTM D-7177 for PVC)	The air channel between the two parallel weld tracks is inflated using a hypodermic needle to a pressure of approximately 250 kPa. If no drop in pressure occurs over time then the entire seam length is properly welded	Able to test the continuity of very long seams
Air lance (ASTM D-4437)	A jet of air at a pressure of 350 kPa blown through an orifice of 5 mm diameter is directed beneath the upper edge of the over-lapped seam to detect unbonded areas of the weld. Unbonded areas can be identified by a fluttering effect and observed inflation at that point	Only really applicable for geomembranes less than 1.00 mm. Works very well on flexible geomembranes such as PVC, EIA-R and fPP. Only effective if the weld defect is open at the edge of the seam where the jet of air pressure is directed
Vacuum box	A soapy solution is applied over the seam and then the area is covered with a transparent box and a vacuum of approximately 15 kPa is applied. Any unbonded areas will show up as a cluster of bubbles. The vacuum box method is mainly used for patched areas and detailed weld areas such as sumps	The test is slow to conduct, labour intensive and it is often difficult to make a vacuum-tight seal at the bottom of the vacuum box where it passes over the weld edges. Thin, flexible geomembranes (i.e. less than 1.0 mm) can also be difficult to 'vacuum box test' on account of the tendency for upward deformations of the geomembrane into the vacuum box

along the weld edge and reasonable pressure applied to check for areas that are poorly fused or debonding. This also facilitates the visual inspection as it allows one to lift the top flap to properly observe the weld edge.

18.3.2 PRESSURIZED AIR CHANNEL TESTING

Dual-track hot wedge welders produce an air channel between the two parallel welds, which can be pressurized with air to check seam continuity.

Wedge welds are tested for continuity and integrity by air pressurization (to a pre-determined pressure) of the central void in the double wedge weld. On completion of the welded seam, the channel is sealed at one end and the other end is attached to the air nozzle of the test equipment. A hollow needle (with pressure gauge attached) is inserted into the test channel and air pressure is applied to the intermediate channel to reach a predetermined pressure (e.g. 250 kPa). The system is then allowed to stabilize and is isolated from the air pressure source. The pressure in the channel is monitored. For a satisfactory weld, the air pressure must fall initially by not more than 10% of its initial value and remain constant or with limited pressure loss for a duration preferably of least 5 min. Depending on the project specification this wait time varies from 2 to 15 min. There is a tendency to wait for shorter periods of time to speed up the test but for reliable results a period of 5 min following stabilization of the pressure is recommended.

Air channel testing is described in ASTM D-5820 'Standard Practice for Pressurized Air Channel Evaluation of Dual Seamed Geomembranes'. This method is a nondestructive evaluation of the continuity of parallel geomembrane seams separated by an air channel. All double welds with an enclosed air channel can be air pressure tested. Long lengths of seam can be evaluated in this way more quickly than by other common nondestructive tests. ASTM D-5820 supercedes ASTM D-4437 for geomembrane seams that include an air channel. The air pressure test within the channel created between dual seamed tracks can detect the presence of unbonded sections or channels, voids, nonhomogenities, discontinuities, foreign objects and the like in the weld region, which can then be identified.

Note the air pressures used are dependent on the thickness, stiffness and material type of the geomembrane. Also, the maximum allowable loss of air pressure varies depending upon thickness, stiffness and type of geomembrane material being tested.

The equipment required is as follows:

1. An air pump (manual or motor driven) capable of generating and sustaining a pressure of 450 kPa.
2. A rubber hose with fittings and connections.
3. A sharp hollow needle or other approved pressure feed device.
4. A calibrated pressure gauge capable of reading pressures up to 450 kPa, with a tolerance of less than 5 kPa.

◉ *AT A GLANCE – NONDESTRUCTIVE AIR PRESSURE TEST OF WELDS*

- *Only for double wedge welds.*
- *Seal both ends and insert air injection needle.*
- *Pressurize the air channel to test pressure (e.g. 250 kPa).*
- *Pressure should remain stable for 5 min – indicating no leaks.*
- *Cut end opposite needle – channel should depressurize, demonstrating no blockage of channel.*

The following procedure is typically used:

1. Clamp both ends closed to seal the air channel.
2. Insert needle with pressure gauge, or other approved pressure feed device within 300 mm of one of the sealed ends, into the air channel created by the dual wedge weld (see Figure 18.3).

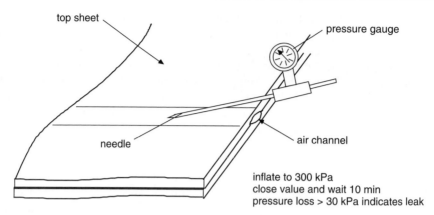

top sheet

pressure gauge

needle

air channel

inflate to 300 kPa
close value and wait 10 min
pressure loss > 30 kPa indicates leak

Figure 18.3 Schematic showing air pressure testing of double-wedge welds by inserting a needle into the central air channel as per ASTM D-5820 (or ASTM D-7177 for PVC welds). Dual hot wedge welders produce two weld tracks separated by a small unbonded channel. By sealing off both ends of this channel and pressurizing with air the integrity of the entire seam can be checked quickly and efficiently

3. Energize the air pump and pressurize the channel to 250 kPa for a 13 mm wide channel. Close the valve and sustain the pressure for a minimum of 5 min. A pressure drop of less than 10–25 kPa is allowable (depending on the liner thickness), but the air pressure is still required to stabilize for a minimum of 5 min.
4. If loss of pressure exceeds 25 kPa, or does not stabilize, locate faulty area and repair. If, in the judgement of the Construction Quality Officer (CQO), significant changes in geomembrane temperature have occurred during the test (e.g. due to cloud cover), the test should be repeated after the geomembrane temperature has stabilized.
5. Cut end of weld opposite to the pressure gauge and observe that the pressure drops. If the pressure does not drop, locate the obstruction(s) in the weld, repair and retest weld.
6. Remove needle or other approved pressure feed device and repair all holes and damage made to the air channel by extrusion welding over the damage.

Note the air pump should be placed on an adequate cushion to prevent damage to the geomembrane. A flexible hose is used to connect the pump to the air pressure device. This hose should have a quick connect on its end for disengagement after pressure is supplied to its desired value, that is, the pump is not to be attached while the air pressure is being monitored.

A hook bladed knife is recommended for trimming the geomembrane material. Straight bladed knives should not be used as they may damage the geomembrane by cutting through the material being trimmed and into the underlying geomembrane.

If the pressure does not drop below the maximum allowable value after the specified test period, open the air channel at the end away from the pressure gauge. Air should rush out and the pressure gauge should register an immediate drop in pressure, indicating that the entire length of the seam has been tested. If this does not happen, either the air channel is blocked or the equipment is faulty, and the test is not valid.

If the pressure drop is greater than the maximum allowable value after the test period, check the end seals of the air channel. Reseal these areas if a leak is noticed and then repeat the entire test.

Important note. If there are significant changes in geomembrane temperature during pressure testing (e.g. if there is cloud cover or other shading) then a variation in channel pressure may be recorded due to expansion or contraction of the air channel. If an increase or decrease in temperature is suspected of having caused a pressure variation, repeat the test after the geomembrane temperature has stabilized.

The pressures set out in Table 18.5 can be used as a guide.

The recommended air test pressures for HDPE geomembranes can also depend on the geomembrane temperature:

Table 18.5　Pressure test guide pressures for various liner thicknesses[a]

Material thickness (mm)	Minimum pressure (kPa)	Maximum pressure (kPa)
1.00	170	210
1.50	190	250
2.00	210	250
2.50	210	250

[a]Procedure: Observe and record the pressure 5 min after the initial reading. If the loss of pressure exceeds that shown in Table 18.6 below, or if the pressure does not stabilize, the faulty area should be located and repaired. The pressure differential after 5 min should be not more than that given in Table 18.6.

Table 18.6　Pressure differential commonly listed for various liner thicknesses[a]

Material thickness (mm)	Max pressure differential (kPa)
1.0	25
1.5	20
2.0	10
2.5	10

[a]Procedure: At the conclusion of all pressure tests, the other end of the air channel to where the pressure gauge is located should be cut to release the air pressure and demonstrate the continuity of the air channel by loss of pressure at the gauge. If this is not demonstrated the air channel will be considered 'blocked' and the test should be repeated from the point of blockage. If the point of blockage cannot be found, the air channel should be cut in the middle of the seam (inspection cut) and each half-treated as a separate test.

Pressurized Air Channel Testing of PVC Welds

All PVC field seams made by a dual-track fusion welding device can be tested by applying air pressure within the air channel to a sealed length of seam, and monitoring the pressure over time. The testing should be conducted in accordance with ASTM D-7177 (2005).

For the geomembrane, the initial inflation pressure needs to be equal to or greater than the minimum specified according to ASTM D-7177. The maximum allowable pressure drop over a 30 s period should be 34 kPa. A pressure gauge is inserted into the opposite end of the air channel to check for continuity in the air channel. Air channels that do not hold the minimum specified air pressure need to be further inspected to identify the location and nature of any defects or unbonded sections of seam. The seam will then be repaired and retested.

18.3.3   AIR LANCE TESTING OF FLEXIBLE LINERS

Non-bonded regions or defects in seams of flexible liners (e.g. PVC) can be detected using an air lance. This method is only really applicable to geomembranes less than 1.00 mm thick. For this test the seam needs to be clearly visible to the operator during the test. The air lance will be capable of supplying 350 kPa through a 5 mm diameter nozzle. The air stream is directed at the upper edge of the seam no more that 50 mm from the seam edge. Non-bonded areas can be observed by a 'fluttering effect' and resulting inflation at that point. Any unbonded areas or voids in the seam are then marked, repaired and retested with the air lance.

🖈 *The seams in flexible membranes can be checked for integrity using an air lance nozzle directed on the upper edge and surface to detect any loose edges or ripples indicating unbonded areas within the seam as per method ASTM D-4437.*

18.3.4   VACUUM BOX TESTING

Vacuum boxes are the most common method for checking the integrity of extrusion welds. An area with a length and width greater than the vacuum box is wetted with soapy water and then the vacuum box is placed on that area and a vacuum of at least 15 kPa is applied (note some practitioners use 20–35 kPa). Leaks are detected by the appearance of bubbles within 10–15 s of the application of the vacuum (see Figure 18.4). This is due to air entering from beneath the liner and passing through the non-bonded zone. The box is then moved to the adjoining section of seam. The test is slow to perform, and it is often difficult to make a vacuum-tight joint at the bottom of the box where it sits on the seam edges.

Only geomembrane thicknesses greater than 1.0 mm are generally tested by this technique since the thinner membranes can be sucked into the vacuum box. For thin flexible geomembranes, the bottom of the box can be fitted with a steel mesh to avoid excessive deformation. 'Vacuum boxing' is a common form of nondestructive test used for patched areas, anchor trenches and sumps where a pressurized dual seam is not possible. The method is very labour-intensive and is essentially impossible to use on sides of tanks,

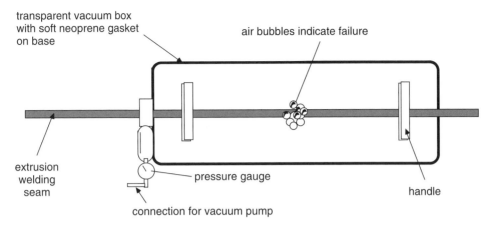

transparent vacuum box
with soft neoprene gasket
on base

air bubbles indicate failure

extrusion
welding
seam

pressure gauge

handle

connection for vacuum pump

'ensure leak tight seal for not less than 15 seconds'

Figure 18.4 Schematic of the set-up for vacuum box testing of geomembrane welds as per ASTM D-5641

Figure 18.5 Photograph of a vacuum box for testing geomembrane welds as per ASTM D-5641. A soapy solution is used to identify the location of a leak. Reproduced by permission of Silverwing Specialized Inspection Equipment, UK

since an adequate downward pressure required to make a good seal is difficult to achieve (as this is usually done by standing or leaning on top of the box).

The vacuum box is a box (typically about 0.5 m long) with a transparent viewing window, a soft neoprene gasket attached to the bottom, a port hole, a valve assembly and a vacuum gauge (see Figure 18.5). The negative pressure or vacuum is obtained within the box by using an industrial vacuum pump. In order to obtain a good seal around the neoprene gasket and to locate a leak accurately, the seam overlap must be trimmed prior to vacuum box testing. Therefore, if destructive testing is required, it should be taken prior to the removal of the seam overlap and vacuum box testing.

📌 *Vacuum box testing is useful in applications with regions of complex welding detail, or where other methods are not practical.*

👁 *AT A GLANCE – VACUUM BOX TESTING CONDITIONS for HDPE GEOMEM-BRANES*

- *Vacuum of 35 kPa.*
- *Dwell time of 15 s.*
- *Move the box over the next adjoining area with a minimum of 75 mm overlap.*

Note for fusion seams it is necessary to cut off the free flap with an approved cutter (taking care to ensure that the lower geomembrane is not damaged) prior to vacuum box testing the seam.

📌 *In vacuum box testing the weld is flooded with detergent solution and the vacuum box placed over the top. The open edge of the vacuum box incorporates a soft but hard-wearing neoprene seal so a vacuum can be generated within the box. The operator views the weld area through a clear polycarbonate observation window in the top of the box. A stream of bubbles will indicate any leaks (e.g. a puncture or pinhole) in the weld. If no bubbles appear after 15 s, the vacuum release valve is opened and the box moved over the next adjoining area with a minimum of 75 mm overlap, and the process repeated.*

Any pinholes, porosity, or non-bonded areas are detected by the appearance of soap bubbles in the vicinity of the defect. Dwell time should be between 10 and 15 s.

Lightweight and portable vacuum boxes are available where the vacuum is generated with a portable electric vacuum pump (e.g. Silverwing Specialized Inspection Equipment, UK). Specialized vacuum boxes are also available for testing geomembrane welds in corners or along tank edges (see Figure 18.6).

18.3.5 SPARK TESTING

Spark testing is suitable for geomembrane welds which are difficult to air pressure test (for instance, welds where there is no central air channel) or vacuum box test (e.g. due to presence of non-flat surfaces or irregular corners).

There are two main methods for spark testing:

- DC spark testing where a metal strip or wire is encapsulated within the weld and grounded to complete the circuit. ASTM D-6365 describes the method utilizing a conductive wire in the weld (see Figure 18.7). A high voltage potential is applied above the welded section. If a spark occurs then there is a leak. The spark will concentrate into any areas where a satisfactory welded seam has not been achieved. The dielectric constant of HDPE is quoted as approximately 600 V/0.025 mm and thus a spark will discharge through a 1 mm thick geomembrane at a voltage of 24 kV and through 2 mm gauge at 48 kV, provided there is a conductor to return the current

Figure 18.6 Photograph of a vacuum box for testing geomembrane welds at corners or along tank edge as per ASTM D-5641. A soapy solution is used to identify the location of a leak. Reproduced by permission of Silverwing Specialized Inspection Equipment, UK

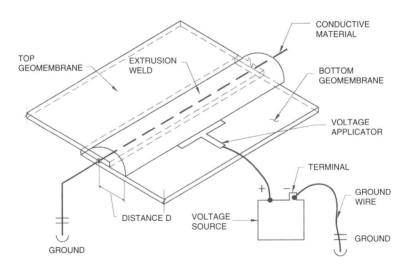

Figure 18.7 Schematic of the spark testing procedure for checking geomembrane seams as per ASTM D-6365

and complete the circuit. This method is used principally for joints made with hand welders.

- AC spark testing where an 'earth ground' in good contact with the underside of the liner such as a concrete substrate or, humidity or moisture in the soil substrate that is grounded, is required. AC spark testing can also be done with conductive foil behind the weld (such as aluminium foil). This method depends on the presence of an 'earth' under the geomembrane or otherwise no spark will be detected and possible leaks will be overlooked (Peggs, 2008).

Spark testing is applicable to welds made by the extrusion welding method, seams made by using welding tape (a strip of the same type of material as the geomembrane, that is welded over adjacent sections of geomembrane to create a seam) and welds where it is practical to insert a conductive material in the seam just prior to or during fabrication.

Spark testing is one of the oldest techniques used in the lining industry and is very reliable. It is most cost effectively applied for small projects.

Spark testing of geomembrane field seams is particularly useful in locations where other nondestructive test methods are not practical; for example in tight corners or for a circumferential seam around a pipe.

ASTM D-6365 'Standard Practice for the Nondestructive Testing of Geomembrane Seams using the Spark Test' describes the nondestructive testing of field seams in geomembranes using the spark test where a suspected defective area is indicated by the generation of a spark.

📌 *Spark testing involves inserting a conductive material (e.g. a fine copper wire) into the field seam just prior to or during fabrication. The conductive material in the seam is connected to the negative terminal of the test apparatus and a positive voltage is applied across the seam edge. A faulty area in the seam is indicated by a spark from the voltage source to the conductive material.*

Note that unless the voltages and distances prescribed in the standard ASTM D-6365 test are carefully adhered to, a 'false positive'[1] result may occur. Such a false positive occurs when the arc distance is too large for the voltage applied at the time and conditions of testing. The range of the high voltage source is typically 20 000 to 35 000 V (see Figure 18.8) which is required to cause a discharge (spark) between the positive electrode and the negative electrode wire.

Step by Step Method for Spark Testing

Spark testing involves the following eight steps (in accordance with ASTM D-6365).

Step 1. As the weld is fabricated a continuous electrically conductive wire is inserted in the lapped area of the panels at a distance 2–5 mm from the edge of the top panel of membrane.

Step 2. Prior to testing, it is necessary to connect the conductive material installed in the field seam to the negative terminal of the voltage source or a separate ground (see Figure 18.7).

---

[1] Note a false positive test is defined as a weld with a suspect area that results in no spark (perhaps because too low a voltage was used) and thus it incorrectly tests as a 'good seam'.

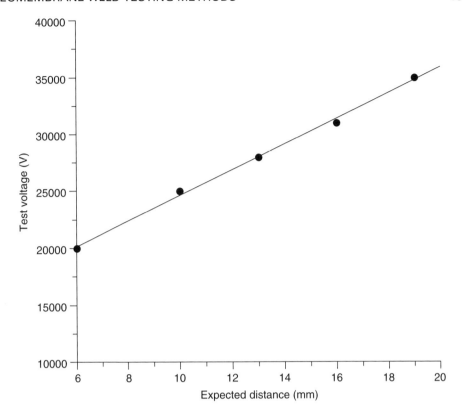

Figure 18.8 Typical test voltages for various expected distances for spark testing (ASTM D-6365)

Step 3. Set the voltage source to a voltage needed for the expected distance. Typical distances and the required voltages are shown in the graph in Figure 18.8. The user should verify that the test voltage is adequate by using a trial seam with a simulated defect prior to testing and also to verify that the test voltage will not damage the geomembrane. Test voltages that are too high (greater than 40 000 V) can cause pinholes in geomembranes.

Step 4. Check that the field seams and areas adjacent to the seams are dry prior to testing.

Step 5. Place the voltage applicator (such as a wire brush or conductive neoprene 'squeegee') connected to the positive terminal of the voltage source, in contact with the geomembrane at the seam.

Step 6. Move the voltage applicator along the seam at a uniform rate of 6–9 m/min. The voltage applicator must make intimate contact with the seam being inspected.

Step 7. As the test progresses, the generation of a spark indicates a suspect area in the seam. It is desirable that the equipment has an audible tone signal that activates when a spark is generated.

Step 8. The suspect area is marked for repair.

📌 *Bare copper wire is embedded in the seam during extrusion welding. Bare doorbell wire or alternator winding wire is generally used. One end of the wire is attached to the negative terminal of the voltage source. Any discontinuity in the weld will expose the wire and this creates a discernable spark when the wand passes over it.*

18.3.6 ULTRASONIC WELD TESTING

Geomembrane welds can also be checked using ultrasonic probes to check for inhomogeneities such as voids, air bubbles, entrapped dirt and so on. The benefits of ultrasonic weld inspection is that it is nondestructive and can also reliably measure the thickness of the welded seam. The ultrasonic method employs the pulse–echo technique whereby an ultrasonic measuring head is held in intimate contact with the weld and the delay time of the echoes from the back of the weld are recorded. The delay time of the geomembrane back echo is proportional to the weld thickness and thus any thickness reduction caused by poor welding can be detected. Furthermore, defects or anomalies in the weld such as debonded areas, air gaps and voids can be detected. Short echo delay times are indicative of the presence of weld defects.

Nondestructive testing of welds can be performed using a manual ultrasonic weld-checking device. The ultrasonic instrument can check the integrity of welds by determining the width of a good fusion weld and detecting the presence of any unbonded channels. Ultrasonic weld testing can therefore provide confirmation of weld integrity and water tightness.

The thickness of weld seams can be quickly and conveniently measured by using a portable ultrasonic thickness instrument. This is done by placing an ultrasonic measuring head on top of the geomembrane. The 'delay time' for the echo of the ultrasonic impulse to bounce back from the back side of the geomembrane to the measuring head is proportional to the overall weld thickness.

Ultrasonic thickness measurements on trial seams can assist in selecting optimum welding parameters. Before the installation of the geomembrane commences on the site, test seams (or trial welds) should be performed to adjust the welding parameters to suit the field conditions and also to verify that the welding machine is functioning correctly. It is necessary to carry out preliminary weld tests in order to define the most appropriate welding parameters (e.g. temperature, speed, pressure) for the specific climatic conditions and material type. Field ultrasonic thickness measurements have been found to be very effective in optimizing the welding of the geomembrane liner. The actual field seams (or 'production seams') should be carried out only after successful trials of seams under field conditions.

Ultrasonic thickness measurements should be performed at intervals of not more than 8 m on field welds seamed by the hot wedge welding technique. If the seam thickness reductions are greater than the allowable range then this spacing should be decreased accordingly.

Furthermore, ultrasonic techniques including the pulse–echo and 'pitch–catch' methods can be used to assess bond quality and identify the presence of internal flaws and occlusions (voids) that may initiate failure or serve as a labyrinth of channels for leaks to occur.

The ASTM Standard D-7006-03 'Standard Practice for Ultrasonic Testing of Geomembranes' describes the procedures for ultrasonic testing of geomembranes using the pulse–echo method. Ultrasonic wave propagation in the polymeric geomembranes can be correlated to physical and mechanical properties and condition of the materials. In ultrasonic testing, two wave propagation characteristics are commonly determined: velocity (based on wave travel time measurements) and attenuation (based on wave amplitude measurements). The velocity of wave propagation is used to determine thickness, density and elastic properties of materials. The attenuation of waves in solid polymeric geomembranes is used to determine microstructural properties of the materials. In addition, frequency characteristics of waves are analyzed to investigate the properties of a test material. Travel time, amplitude and frequency distribution measurements are used to assess the condition of materials to identify damage and defects in solid materials. Measurements are conducted in the time-domain (time versus amplitude) or frequency-domain (frequency versus amplitude).

18.4 DESTRUCTIVE WELD TESTS

Destructive testing is necessary to validate the strength and integrity of a weld and is part of the overall construction quality assurance programme where a sample of an installed geomembrane weld is cut out of the geomembrane and tested for shear and peel strength. Figure 18.9 shows a destructive test weld sample marked out for cutting. The resulting hole is then patched by extrusion fillet welding. Unfortunately this process can lead to flaws and weakness in the geomembrane liner, thereby increasing the likelihood of leakage in the liner. The repair method (extrusion welding) is widely recognized as being inferior to the original seaming method (which is generally dual-track hot wedge welding). Therefore it is suggested that field destructive seam samples be taken only at the start and end of the field seam as necessary.

Destructive shear and peel tests reveal little information regarding the continuity and integrity of the entire seam.

Once the recorded welding parameters are constant and within the allowable tolerance range, a section of the field weld should be cut out at the extremities (beginning and end of the seam) of each major weld run. Samples can also be cut out from completed production welds as directed by the client's engineer in accordance with contractual arrangements. Each sample is subjected to a tensile test to break in a field tensiometer (see Figure 18.10) where the force exerted is displayed, under conditions approximating those required for laboratory tensile testing. The recommended test speeds for HDPE and LDPE geomembranes are 50 mm/min and 25 mm/min, respectively (the graph of load versus displacement needs to be recorded). At the very least two specimens are tested, one in the tensile mode (shear test), and one in the peel mode; but usually more peel specimens are tested since they provide a more critical assessment of the weld integrity.

Figure 18.11 shows that the resulting bending moment of the geomembrane in a shear test, stresses to occur at the edges of the seam.

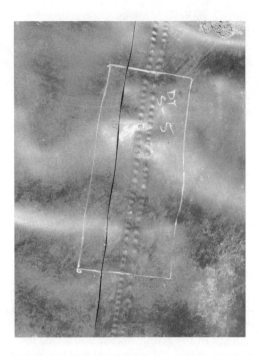

Figure 18.9 Photograph of a destructive test (DT) sample before cutting from a welded HDPE

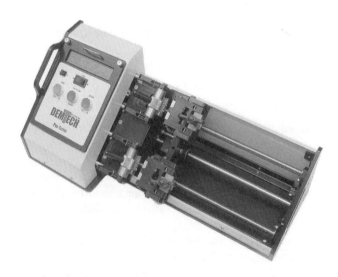

Figure 18.10 Photograph of a field tensiometer for peel, shear and tensile testing of geomembrane weld specimens in the field. Reproduced by permission of Demtech, Diamond Springs, CA, USA

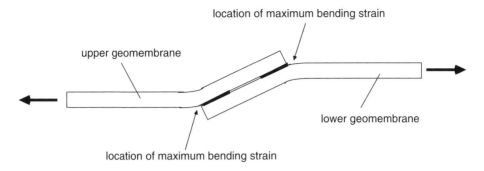

Figure 18.11 Owing to the bending moments of geomembranes in a shear test, peel stresses occur at the edges of the seam (note weld flap not shown and blending moments are exaggerated)

In each case, the sample must break in the parent liner material (achieving a satisfactory strength value) and not in the welded section. If necessary, welding conditions are adjusted and further samples taken until these criteria are achieved.

The seam is acceptable only if the parent material is stretched outside of the seam area and without the occurrence of peeling back of the weld track. The dimensions of the test pieces and evaluation of test results are provided in the standards of ASTM D-4437 and DVS-2226.

For on-site testing, generally three specimens of 25 mm width are die cut – one near each end and one near the mid-point of the trial weld sample. The two end specimens are generally tested in peel and the mid-point sample tested in shear. The contractor should supply evidence demonstrating that the tensiometer has been calibrated within the previous 12-month period.

The strength of the parent geomembrane material should be determined before each trial weld by die cutting one 25 mm wide specimen (e.g. from liner material adjacent to the trial weld). This procedure provides 'comparative' testing which eliminates the effect of temperature, tensiometer load units or variability in the material influencing the results of the trial weld.

Typical HDPE failure criteria currently used in the industry are as follows:

1. Specimen fails within the weld (i.e. the weld is weaker than the material and does not hold).
2. If peel strength is less than 65% of the parent material's tensile yield strength.
3. If it fails in shear at less than 95% of the parent material's tensile yield strength.

In the event that a trial weld fails, the entire trial weld procedure should be repeated after the appropriate adjustments to the welding device and/or improved operator skill level has been implemented.

In the case of PVC shear tests, the minimum required seam strength is typically 80% of the parent geomembrane break strength.

If the weld strength achieved with the tensiometer is acceptable, then the entire seam length should be tested by the air channel pressure test in accordance with ASTM D-5820 or its equivalent.

18.4.1   INDEPENDENT LABORATORY DESTRUCTIVE MECHANICAL TESTING

The typical external geomembrane weld testing requirement is that out of five test specimens cut from a weld sample only four must pass destructive testing which means an allowable failure rate of 20%. However, many specifications require that all five specimens should pass.

Sampling locations for destructive weld testing should be determined during welding and may be prompted by suspicion of the following:

- excess crystallinity;
- contamination;
- excess moisture/humidity;
- offset welds;
- any other potential cause of imperfect welding.

👁 *AT A GLANCE - DESTRUCTIVE WELD TESTS BY INDEPENDENT LABORATORY*
   ASTM D-6392 test method 'Determining the Integrity of Nonreinforced Geomembrane Seams Produced Using Thermo-Fusion Methods'.

   Destructive weld testing according to ASTM D-6392 requires that five specimens are tested in peel and five in shear. The precise pass requirements vary depending on the project and on the specification.

- *Approximately one DT every 150–170 m of weld.*
- *Cut out sample with minimum dimensions 30 cm × 20 cm (with the seam centred).*
- *Samples tested as alternative specimens in shear and peel.*
- *The specimen is stretched at a speed of 51 mm/min which is referred to as the crosshead speed.*
- *The deformation and failure modes of the weld sample are observed and recorded.*
- *Any failure within the seam region is a weld failure.*
- *The peel separation is the area of separation (i.e. loss of adhesion) within the weld region itself.*
- *Number of DT tests should be kept to a minimum.*

Tests should be performed on 10 specimens prepared from each field sample. From these samples 5 specimens should be tested in shear and 5 specimens should be tested in peel. The shear and peel specimens should be selected from the sample alternately, so that no two adjacent specimens are tested in the same mode (see Figure 18.12). Both tracks of double wedge welds must be tested in peel (i.e. 10 peel test results obtained).

Figure 18.13 shows the test geometry for shear and peel testing of weld specimens as per ASTM D-6392.

Figure 18.14 shows a DT weld specimen undergoing a shear test as per ASTM D-6392 while Figure 18.15 shows a peel test in progress.

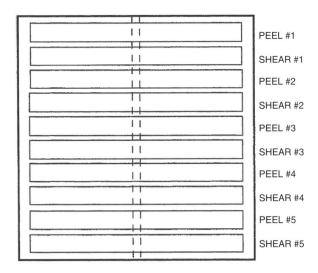

Figure 18.12 Weld specimens for peel and shear testing should be selected alternately from the weld sample and labelled as shown as per ASTM D-6392. Reprinted, with permission, from ASTM D6392-99 Standard Test Method for Determining the Integrity of Nonreinforced Geomembrane Seams Produced using Thermo-Fusion Methods, copyright ASTM International, 100 Barr Harbor Drive, West Conshohocken, PA 19428

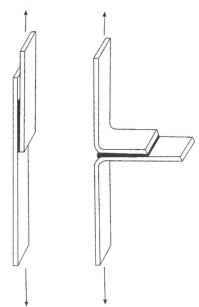

Figure 18.13 Test geometry for shear and peel testing of weld specimens as per ASTM D-6392. Reprinted, with permission, from ASTM D6392-99 Standard Test Method for Determining the Integrity of Nonreinforced Geomembrane Seams Produced Using Thermo-Fusion Methods, copyright ASTM International, 100 Barr Harbor Drive, West Conshohocken, PA 19428, USA

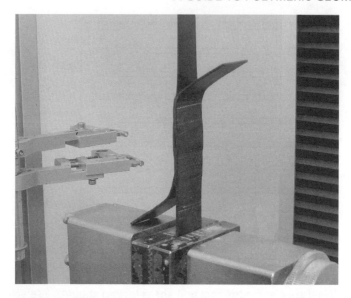

Figure 18.14 Photograph of a destructive test weld coupon (25 mm wide) undergoing a shear test as per ASTM D-6392

Figure 18.15 Photograph of a weld peel test on a white/black geomembrane

Shear tests on weld coupons can verify if there is proper elongation of the material immediately adjacent to the weld. This is useful to check if the material has been adversely affected by excessive heat, scoring or over-grinding. In contrast, the peel test verifies the degree of proper bonding of the weld.

Figure 18.16 shows examples of DT weld specimens that have failed on the weld boundary in a brittle fashion without displaying any obvious elongation. Figure 18.17 is a schematic of a dual wedge weld showing the typical failure location in case of weld overheating, excessive thickness reduction or scorelines.

The weld shear strength should be, at a minimum, equal to 95% of the rated yield tensile strength of the geomembrane sheet, as specified in the minimum property values

Figure 18.16 Photograph of weld coupons breaking at low elongation at scorelines during a shear elongation weld test (note that the breaks are practically brittle with very low elongation and essentially no ductile behaviour)

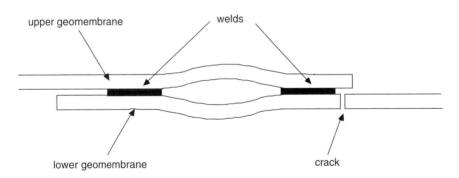

Figure 18.17 Schematic of a dual-wedge weld showing a typical cracking location. Low elongation cracking in this area can be related to one or a combination of overheating of the weld track, recrystallization of the polymer, excessive thinning and scorelines

provided by the geomembrane manufacturer. Shear strength measures the continuity of tensile strength through the seam and into the parent material.

The weld peel strength should be equal to or greater than 65% of the rated tensile yield strength of the geomembrane, as specified in the minimum property values provided by the geomembrane manufacturer in accordance with the project specification. Peel strength is a measure of the 'weld opening force'.

For both peel and shear tests the specimen should fail in the geomembrane sheet, not in the weld area, referred to as a 'film tear bond' (FTB). The specimen should be defined as failing in the weld if any portion of the weld exhibits separation across more than 10–20% (or as defined by project specifications) of the weld track width. Each track of a dual-track fusion weld should be considered to be a separate weld for the purpose of calculating the percentage separation. If more than one specimen fails, the entire weld destructive test sample should be considered as failing.

ASTM D-4437 classifies breaks for fillet extrusion and double hot wedge welds and classifies an acceptable break as where the break occurs in the sheet rather than in the weld (referred to as a 'film tear bond' (FTB) in the past).

Note. The weld shear test specimens must also exhibit clear measurable elongation since low strains can indicate the following:

- *the geomembrane is brittle;*
- *a localized decrease in thickness near the seam due to possible overheating during seaming;*
- *surface damage such as deep score lines or grinding gouges parallel to the seam direction.*

The main seam peel and shear destructive test procedure for geomembrane field welds is ASTM D-6392 (which replaced ASTM D-4437). Note although ASTM D-4437 is still widely quoted in project specifications it was actually withdrawn in 1998. ASTM D-6392 requires the measurement of shear elongation and peel separation in addition to peel and shear strength values.

ASTM D-6392 also uses 'locus-of-break' codes to classify the various rupture modes of welds. Figure 18.18 shows the locus-of-break codes for dual-wedge welds. AD (total adhesion failure) and AD-BRK (partial adhesion failure) are undesirable break codes while a break through the top or bottom sheet (denoted as BRK, SE1 and SE2) is acceptable only when certain minimum specification values for strength and elongation at break are met.

Figure 18.19 shows the locus-of-break codes for extrusion fillet welds. Under this classification scheme adhesion failures (AD, AD1 and AD2) or break in the weld (AD-BRK) are unfavorable while a break through the fillet (denoted as AD-WLD) is acceptable only when certain minimum specification values for strength and elongation at break are met. ASTM D-6392 does not specifically refer to FTB although it is still widely referred to in project specifications (a legacy of ASTM D-4437)

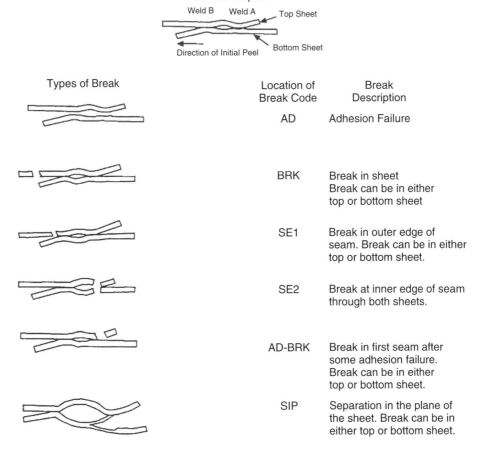

Figure 18.18 'Locus-of-Break' codes for dual hot wedge welds in unreinforced geomembranes tested for seam strength in shear and peel modes specimens as per ASTM D-6392. Reprinted, with permission, from ASTM D6392-99 Standard Test Method for Determining the Integrity of Nonreinforced Geomembrane Seams Produced using Thermo-Fusion Methods, copyright ASTM International, 100 Barr Harbor Drive, West Conshohocken, PA 19428

Due to the fact that the term FTB is somewhat ambiguous, some geosynthetic experts are proposing that FTB no longer be used and instead recommend that the locus-of-break descriptors be applied.

The following are unacceptable break codes *as per their description in ASTM D-6392. Hot Wedge AD and AD-BRK (i.e. partial adhesive failure where adhesion loss is >10% of the weld front).*

Figure 18.20 shows black/white HDPE geomembrane weld specimens exhibiting total adhesion failure (AD).

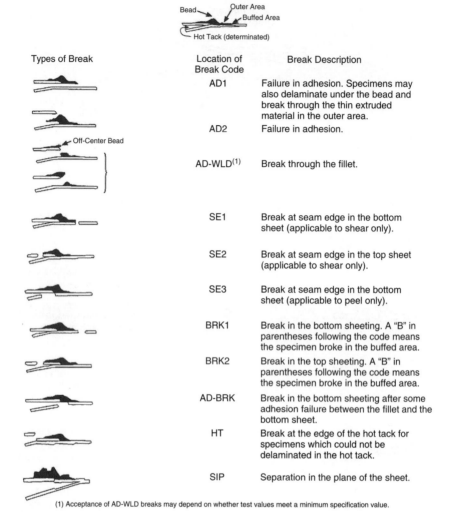

Types of Break	Location of Break Code	Break Description
	AD1	Failure in adhesion. Specimens may also delaminate under the bead and break through the thin extruded material in the outer area.
	AD2	Failure in adhesion.
	AD-WLD[1]	Break through the fillet.
	SE1	Break at seam edge in the bottom sheet (applicable to shear only).
	SE2	Break at seam edge in the top sheet (applicable to shear only).
	SE3	Break at seam edge in the bottom sheet (applicable to peel only).
	BRK1	Break in the bottom sheeting. A "B" in parentheses following the code means the specimen broke in the buffed area.
	BRK2	Break in the top sheeting. A "B" in parentheses following the code means the specimen broke in the buffed area.
	AD-BRK	Break in the bottom sheeting after some adhesion failure between the fillet and the bottom sheet.
	HT	Break at the edge of the hot tack for specimens which could not be delaminated in the hot tack.
	SIP	Separation in the plane of the sheet.

(1) Acceptance of AD-WLD breaks may depend on whether test values meet a minimum specification value.

Figure 18.19 'Locus-of-Break' codes for fillet extrusion welds in unreinforced geomembranes tested for seam strength in shear and peel modes specimens as per ASTM D-6392. Reprinted, with permission, from ASTM D6392-99 Standard Test Method for Determining the Integrity of Nonreinforced Geomembrane Seams Produced using Thermo-Fusion Methods, copyright ASTM International, 100 Barr Harbor Drive, West Conshohocken, PA 19428

Traditionally the specifications for HDPE welds were based only on the shear and peel strength values from destructive weld tests. However at present the four main criteria for seam integrity are:

- shear strength
- shear elongation

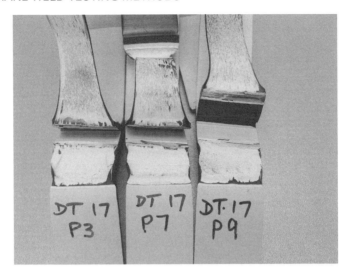

Figure 18.20 Photograph of a welded HDPE black/white geomembrane where peel specimens show AD failure (total adhesion failure) of the secondary weld track. The white zone directly above the writing on the specimens is where delamination has occurred

- peel strength
- peel separation

18.4.2 WELD STRENGTH CRITERIA

The ASTM Standards (D-6392 and D-4437) are only useful standards for producing relative values of weld peel and weld shear strengths with respect to the base material strength. That is, they are not performance standards and they give no basis for quantitative assessment as to what are 'good' strength values.

Minimum weld strength criteria for HDPE GMLs of varying thicknesses are published in the GRI GM-19 specification (see Table 18.10 below).

Usually however the minimum requirements are included in the technical specification set by the geotechnical engineering company for the particular job. The specification would include a table of geomembrane properties that the base material needs to comply with, together with laboratory weld sample destructive testing requirements.

For instance, for a 2 mm thick HDPE geomembrane, the geomembrane property table would include the main physical property requirements, including typical tensile values of:

- Tensile strength >53 N/mm
- Break elongation: >700%
- Yield strength: >29 N/mm
- Yield elongation: >12%

The shear/peel strength is the force in newtons divided by the width of the specimen and is expressed in N/mm.

Figure 18.21 Photograph of weld coupons breaking at low elongation at scorelines during a shear elongation weld test (note that the breaks are practically brittle with very low elongation)

Shear elongation is a very important parameter since it can determine whether the welding/installation process has resulted in a loss of ductility of the polymer, which can affect the longevity of the weld. Shear elongation is typically expected to be greater than 200% for smooth HDPE (and 'structured', i.e. embossed HDPE) and greater than 50% for randomly textured HDPE.

Figure 18.21 shows a weld specimen breaking at low elongation at scorelines during a shear weld test. Figure 18.22 shows an example of good elongation and poor elongation of an HDPE weld specimen during weld shear testing of an extrusion fillet weld.

Peel separation of a weld is another important parameter since this can introduce a 'notch effect' which serves as a stress concentration and in the case of HDPE could initiate stress cracking of the weld.

It is now being recognized that shear elongation and peel separation provide the most useful information on weld integrity.

Peel separation is an important consideration, particularly for HDPE, since it can result in the introduction of crazes and stress cracks into the separated surfaces which in turn can lead to a reduction in the stress cracking resistance of the remaining seam. Thus even partial peel separation can adversely affect the long-term durability of the welded area.

Figure 18.23 shows weld specimens exhibiting partial adhesion failure (AD-BRK) due to porosity in the weld region.

Figure 18.24 shows an extrusion weld exhibiting total adhesion failure due to an insufficient welding temperature which resulted in poor fusion and hence a peelable weld.

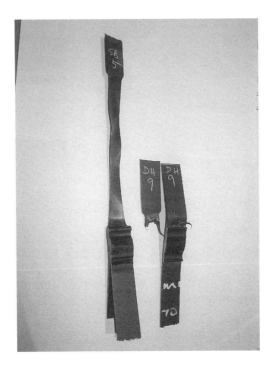

Figure 18.22 Photograph of HDPE extrusion fillet welds after testing. The extrusion weld on the left shows good elongation while the weld specimen on the right exhibits low ductility

Figure 18.23 Photograph of wedge weld coupons in an HDPE geomembrane showing a partial adhesion failure (AD-BRK) due to porosity in the weld region

Figure 18.24   Photograph of an HDPE extrusion weld showing total delamination due to an insufficient welding temperature

The area of peel separation can occur as a number of non-uniform patterns across the seam width and/or depth. The estimated dimensions of this separated area is assessed visually and since this is somewhat subjective it must be done carefully. There are a number of ways to assign a value to peel separation – a ratio based on the extent of peel back as a function of: (1) the weld front (i.e. weld width or specimen width); (2) the weld depth (i.e. extent of incursion into the weld); (3) overall weld track area. Importantly the estimated area must not include 'squeeze-out' which is a normal part of the welding process. ASTM D-6392 recommends that the assessment of the extent of peel separation be based on the linear measurement of incursion depth. The GRI Test Method GM19 specification on the other hand is based on incursion area which is regarded to be more representative of the behaviour of peel separation.

In ASTM D-6392 it is necessary to estimate the amount of seam separation in the peel test by determining the distance of separation as a percentage of the original width of the bonded seam (not the width of the test specimen). It is important to recognize that if there is an angled peel profile with no separation on one side of the 25 mm wide test specimen and complete separation on the other side, then the amount of separation is still considered 100%. Even though the separation area may vary from say 10% to 50% with an angled peel profile the distance separation is 100% in each case (see Figure 18.25). This is because a leak of any size (due to a 100% separation distance) is a potential 100% failure of the seam (Peggs and Allen, 2001).

The degree of acceptable peel separation is often quoted as less than 10% of the intended bonded area. This specification was set approximately 20 years ago when it was difficult to achieve a zero percentage peel separation on a regular basis. With today's equipment used by experience operators it is now routinely possible to achieve a zero percentage peel separation. Given this, zero percentage peel separation should be the target.

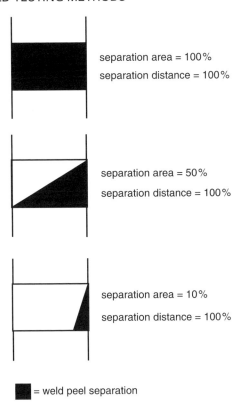

separation area = 100%
separation distance = 100%

separation area = 50%
separation distance = 100%

separation area = 10%
separation distance = 100%

■ = weld peel separation

Figure 18.25 Schematic showing that during peel testing of welds according to ASTM D-6392 the peel distance is more significant than the peel separation area. While the samples show very different peel separation areas they all exhibit 100% peel separation distances

Peel separation should be effectively zero since any degree of peel separation can produce a notch effect (i.e. stress concentration) that can reduce the stress cracking resistance of HDPE and lead to a peelable weld in the case of PVC.

When testing HDPE welds, a surface colour/texture change usually occurs when the interface separates during an incursion into the weld track. This can be used to assess the extent of weld separation. Note however that any surface colour/texture change caused by the 'squeeze-out' at the edge of fusion welds should be disregarded. A fracture face that is slightly out of line with the seam edge and greyish in appearance is characteristic of peel separation in HDPE.

When conducting peel tests they need to be performed on both tracks of the weld. If the overlapping flap is not long enough to hold in the tensile tester grips then it is necessary to cut one side of the air channel and peel test the specimen from the air channel side.

AT A GLANCE. See Table 18.7 for important weld testing facts.

Table 18.7 Important weld testing facts (adapted from Peggs [http://www.geosynthetica.net/ tech_docs/hdpe_sc_paper.asp])

Important weld testing facts	Comments
Shear strength and peel strength values alone do not give sufficient information about the quality of the weld	Other important parameters to measure in weld testing are shear elongation and extent/nature of peel separation
Shear elongation should exceed 100% of the distance between the edge of the weld and the nearer grip (i.e. this approximates to 80–100% for break elongation)	This is to ensure and validate that the welding procedure has not adversely affected the ductility of the adjacent geomembrane due to overheating or changing the polymer structure
It is not adequate that a weld achieves the required shear strength with location of the failure being outside the weld region	Such a failure must also be ductile, accompanied by obvious elongation
There should be no (or negligible) peel separation for both tracks of a double-track weld	Incipient peel separation can act as a stress concentration for stress cracking or further crack propagation under bending strains

Table 18.8 Typical weld specifications (according to ASTM D-6392)

Weld parameter	Geomembrane/weld type	Specification
Shear strength	All types	>90% of the yield strength[a]
Shear elongation	Smooth geomembrane Structured geomembrane 'Textured on smooth' (secondary process)	>100%
Shear elongation	Randomly textured Textured during manufacture (primary process)	>50%
Peel strength	Extrusion welds	>60% of the yield strength[a]
Peel strength	Fusion welds	>70% of the yield strength[a]
Peel separation	All types	0% or sometimes 10% maximum

[a]It is important to note that the weld strength values are based on the *nominal* yield strength (i.e. which is specified by the manufacturer) rather than the actual measured strength of the geomembrane liner.

Typical weld specifications (according to ASTM D-6392) are shown in Table 18.8. Shear elongation specifications (according to ASTM D-6392) can be summarized as follows:

>100% smooth membranes
>50% textured (primary process)
>100% textured (secondary process)
>100% (structured, embossed profiles)

Table 18.9 Typical geomembrane weld pass requirements

Pass requirements	Test failure rate (%)
All ten specimens need to pass	0
One specimen is allowed to fail	10
Two specimens are allowed to fail (but only one peel specimen and one shear specimen but not two of the same)	20

The typical pass requirements are shown in Table 18.9.

For critical installations the project engineer would specify that all ten specimens need to pass as 10–20% of a weld failing when subject to peel forces is cause for concern, particularly since destructive testing is only on a fraction of the overall installation.

Note that soft and flexible geomembranes such as PVC, EPDM and nitrile rubber show different weld test behaviour to HDPE. For instance, PVC and EPDM welds generally peel off and the peel strength is recorded.

Specification Values for HDPE Welds

The weld integrity parameters for hot wedge welded smooth and textured HDPE geomembranes as per the GRI GM-19 specifications are shown in Table 18.10.

The weld integrity parameters for extrusion fillet welded smooth and textured HDPE geomembranes as per the GRI GM-19 specifications are shown in Table 18.11.

Specification Values for LLDPE Geomembrane Welds

The weld integrity parameters for hot wedge welded smooth and textured LLDPE geomembranes as per the GRI GM-19 specifications are shown in Table 18.12.

The weld integrity parameters for extrusion fillet welded smooth and textured LLDPE geomembranes as per the GRI GM-19 specifications are shown in Table 18.13.

Table 18.10 Weld integrity parameters for hot wedge welded smooth and textured HDPE geomembranes (also applies for hot air and ultrasonic welded seams) (GRI Test Method GM-19)

Weld integrity parameter	0.75 mm	1.00 mm	1.25 mm	1.50 mm	2.00 mm	2.50 mm	3.0 mm
Shear strength (N/25 mm)[a]	250	350	438	525	701	876	1050
Shear elongation at break (%)	50	50	50	50	50	50	50
Peel strength (N/25 mm)[a]	197	263	333	398	530	661	793
Peel separation (%)	25	25	25	25	25	25	25

[a]Note that the values listed for shear and peel strengths are for a minimum of four out of five specimens. The 5th specimen can be as low as 80% of the listed values.

Table 18.11 Weld integrity parameters for extrusion fillet welded smooth and textured HDPE geomembranes (GRI Test Method GM-19)

Weld integrity parameter	0.75 mm	1.00 mm	1.25 mm	1.50 mm	2.00 mm	2.50 mm	3.0 mm
Shear strength (N/25 mm)[a]	250	350	438	525	701	876	1050
Shear elongation at break (%)	50	50	50	50	50	50	50
Peel strength (N/25 mm)[a]	170	225	285	340	455	570	680
Peel separation (%)	25	25	25	25	25	25	25

[a]Note that the values listed for shear and peel strengths are for four out of five specimens. The 5th specimen can be as low as 80% of the listed values.

Table 18.12 Weld integrity parameters for hot wedge welded smooth and textured LLDPE geomembranes (also applies for hot air and ultrasonic welded seams) (GRI Test Method GM-19)

Weld integrity parameter	0.75 mm	1.00 mm	1.25 mm	1.50 mm	2.00 mm	2.50 mm	3.0 mm
Shear strength (N/25 mm)[a]	197	263	328	394	525	657	788
Shear elongation at break (%)	50	50	50	50	50	50	50
Peel strength (N/25 mm)[a]	166	219	276	328	438	547	657
Peel separation (%)	25	25	25	25	25	25	25

[a]Note that the values listed for shear and peel strengths are for four out of five specimens. The 5th specimen can be as low as 80% of the listed values.

Table 18.13 Weld integrity parameters for extrusion fillet welded smooth and textured LLDPE geomembranes (GRI Test Method GM-19)

Weld integrity parameter	0.75 mm	1.00 mm	1.25 mm	1.50 mm	2.00 mm	2.50 mm	3.0 mm
Shear strength (N/25 mm)[a]	197	263	328	394	525	657	788
Shear elongation at break (%)	50	50	50	50	50	50	50
Peel strength (N/25 mm)[a]	150	190	250	290	385	500	595
Peel separation (%)	25	25	25	25	25	25	25

[a]Note that the values listed for shear and peel strengths are for four out of five specimens. The 5th specimen can be as low as 80% of the listed values.

fPP Geomembrane Weld Testing

During peel testing, fPP welds often show a colour change (a light grey discoloration due to stress whitening) that appears to indicate a peel separation. Note however that this is an artifact of the shallow break angle and does not necessarily mean that peel separation is occurring.

fPP exhibits a characteristic failure phenomenon during weld testing. In fPP the break often generally occurs at a shallow angle (less than 45 degrees) to the central plane of the geomembrane weld. This is in contrast to breaks in HDPE welds which occur perpendicular to the plane of the geomembrane.

For fPP welds the force that is required to cause peel separation is generally similar to that which is necessary to break the parent geomembrane sheet and thus this is an acceptable mode of break. This mode of break has been referred to as a 'Z-break'.

Specification Values for fPP and fPP-R Welds

The weld integrity parameters for hot wedge welded non-reinforced and reinforced flexible polypropylene geomembranes as per GRI GM-19 are shown in Table 18.14.

The weld integrity parameters for extrusion fillet welded non-reinforced and reinforced flexible polypropylene geomembranes as per GRI GM-19 are shown in Table 18.15.

PVC Weld Integrity

There are two differing specifications for PVC welds depending on whether the seams are thermal fusion welds or chemical welds.

In contrast to HDPE, PVC welds are allowed to separate during peel testing provided the minimum peel force exceeds a specified value.

Chemical welds will typically separate during peel testing at a lower force value than thermal fusion welds (which may still peel but at a much higher force).

Table 18.14 Weld integrity parameters for hot wedge welded non-reinforced and reinforced flexible polypropylene geomembranes (also applies for hot air and ultrasonic welded seams) (GRI Test Method GM-19)[a]

Weld integrity parameter	0.75 mm (NR)	1.00 mm (NR)	0.91 mm (R)	1.14 mm (R)
Shear strength (N/25 mm)[b]	110	130	890	890
Shear elongation at break (%)	50	50	n/a	n/a
Peel strength (N/25 mm)[b]	85	110	90	90
Peel separation (%)	25	25	n/a	n/a

[a]NR, non-reinforced (i.e. no tensile scrim support is present); R, reinforced (i.e. the fPP is supported on a tensile scrim reinforcement).
[b]Note that the values listed for shear and peel strengths are for four out of five specimens. The 5th specimen can be as low as 80% of the listed values.

Table 18.15 Weld integrity parameters for extrusion fillet welded non-reinforced and reinforced flexible polypropylene geomembranes (GRI Test Method GM-19)[a]

Weld integrity parameter	0.75 mm (NR)	1.00 mm (NR)	0.91 mm (R)	1.14 mm (R)
Shear strength (N/25 mm)[b]	110	130	890	890
Shear elongation at break (%)	50	50	n/a	n/a
Peel strength (N/25 mm)[b]	85	110	90	90
Peel separation (%)	25	25	n/a	n/a

[a]NR, non-reinforced (i.e. no tensile scrim support is present); R, reinforced (i.e. the fPP is supported on a tensile scrim reinforcement).
[b]Note that the values listed for shear and peel strengths are for four out of five specimens. The 5th specimen can be as low as 80% of the listed values.

PVC seams are tested for both peel adhesion (ASTM D-413) and breaking strength (ASTM D-3083). The seam requirement for peel adhesion is a 25 mm wide strip that must pass ten pounds per inch width. The breaking strength requirement is that 25 mm wide strips be tested and meet 80% of the specified sheet strength.

18.4.3 CRITICISM OF DESTRUCTIVE TESTING

Destructive testing of geomembrane welds is an important part of conventional geomembrane CQA measures. However, some of the criticisms that have been levelled against destructive testing of geomembrane welds are shown in Table 18.16.

Industry experience has shown that geomembrane welds seldom fail, and that destructive testing requires the test area to be repaired using a much longer length of inferior extrusion fillet welding. A more significant problem is construction/installation damage caused while placing protective drainage material on the geomembrane and in practice, little or no CQA resources are used to prevent or detect such damage to the geomembrane. For instance, electrical leak location surveys can locate and eliminate construction damage (Darilek and Laine, 2001).

18.5 FORENSIC WELD EXAMINATION

Weld specimens can be cross-sectioned and then thin slices cut with a microtome. Microscopic evaluation of such thin geomembrane sections enables the identification of sites that can lead to initiation of stress cracking in seams.

18.5.1 FRACTURE SURFACE MORPHOLOGY

The fracture surface of a cracked HDPE geomembrane weld was studied microscopically to correlate the type of fracture surface morphology to the different applied stress levels (Halse et al., 1990). It was found that a qualitative relationship existed between the observed fracture surface morphology and the magnitude of the applied global stress.

Table 18.16 Critical comments regarding destructive testing of geomembrane welds (Darilek and Laine, 2001)

Criticisms	Comments
Destructive testing can only check a fraction of a percent of the length of the welds	DT only samples ∼0.2% of the installed seams which equate to about 0.0015% of the total area of the liner
It is expensive	The cost of sample dispatch (particularly from remote locations) and DT tests is an additional installation overhead
It can delay the project	The turnaround is typically days
It requires the area that is tested to be repaired using much longer lengths	The patch to be repaired depends on an inferior manual extrusion fillet welding technique and is some four times the size as the original weld that is sampled
The majority of geomembrane welds pass the destructive test	Test results seldom fail the minimum shear/peel strengths. Rather, where samples fail frequently concerns peel separation and sometimes shear elongation – note however that both of these failure modes can be picked up during DT testing
A failure of a destructive test does not necessarily mean the geomembrane would have failed in service	Welds in the field are rarely subjected to the same magnitude of forces as in DT. Thus DT testing entails a considerable safety factor
Welding technologies and hence weld integrity have improved over time	The advent of double-wedge welding (which has replaced the older technology) has the advantages of providing a double weld, a weld of superior strength with less required operator skill and finally provides an intermediate channel that can be easily air pressure tested for leaks through the welds

The following patterns were observed:

- long fibrous (long ductile fibrils) (moderate stress);
- short fibrous (short ductile fibrils) (low stress);
- flake (flaky appearance) (high stress);
- 'hackle' (raised ridges) (appeared before the plastic failure as the cross-sectional area of the specimen decreased);
- lamellar (flat, thin) (a combination of stress cracking with local plastic failure).

18.6 LEAK LOCATION TESTING

The electric field test, currently called the electric leak location (ELL, also known as the Leak Location Survey, LLS) method, was developed in the early 1980s. It utilizes a liquid-covered geomembrane to contain an electric field. The base of the liner system

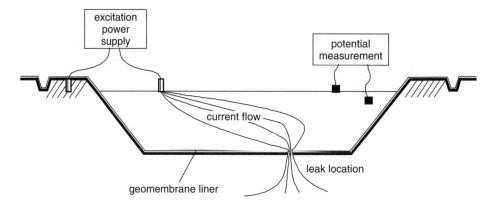

Figure 18.26 Schematic illustrating the principle of electrical leak location surveys

should be covered with water; however the depth can be nominal and even a water film is adequate. It can test the entire installation by applying a voltage across the geomembrane. Electrical potentials are measured to detect the points where electrical current flows through the geomembrane (see Figure 18.26).

Potentials measured on the surface can be used to locate the source of the leak. These potentials are measured by 'walking' a probe across the surface. The operator walks on a predetermined grid layout and marks where anomalies exist. These can be rechecked after the survey is completed by other more localized methods, such as the vacuum box.

In the case of covered geomembrane installations, the covering of the geomembrane with sand or aggregates utilizing heavy earth moving equipment is a major cause of damage.

The Leak Location Survey using a hand-held electrical probe is specified in ASTM Standard D-7007 and is a very efficient method to locate the holes in a geomembrane suspected of being damaged, or for quality control purposes. Electric leak location testing does not however give any information on the strength of the welds.

When a DC current is initiated across the geomembrane the Leak Location Survey hand-held probe will measure the potential gradient on top of the cover layer. It is necessary that the cover layer has moisture present for optimum results. An important advantage of this method is that it provides the installer with the opportunity to test the integrity of the geomembrane liner after the liner has been in contact with the loading of aggregate surface materials.

ASTM Standards for leak location surveys of exposed and covered geomembrane installations are listed in Table 18.17.

In a Leak Location Survey (LLS), a voltage is introduced across the liner system being tested between two electrodes. One electrode is placed in a conductive medium beneath the liner being tested and the second electrode is placed in the material above the liner being tested (or within the water source for exposed liner testing). If there is a leak in the

Table 18.17 Test methods for leak detection

Method	Title and comments
ASTM D-6747	Leak Location Survey – Standard Testing Guide
ASTM D-7002	Leak Location Survey – Exposed Geomembranes
ASTM D-7007	Leak Location Survey – Covered Geomembranes
ASTM D-7240	Spark Testing – Conductive Geomembranes

liner a current will flow through the leak and specialized voltage meters are utilized to detect the location of the electrical potential through the leak.

The following conditions must be met in order to perform a successful Leak Location Survey or Liner Integrity Survey. There must be a conductive medium above the geomembrane, there must be a conductive medium in the holes through the geomembrane and there must be a conductive medium beneath the geomembrane. Additionally, there must not be any contact with the medium above the geomembrane and below the geomembrane other than through the holes. The goal of the survey is for current to flow only through leaks in the geomembrane.

There are four general Leak Location Survey methods:

- wading survey utilizing a submersible hand held probe;
- deep water survey utilizing a Leak Location remote probe;
- or a Leak Location water lance survey performed on exposed geomembranes and,
- a soil covered geomembrane survey utilizing a soil hand held Leak Location probe.

Leak testing using electrical methods has now become a well-established method of CQA for geomembrane liner installations and is particularly popular in the USA, Canada and Europe. In the case of landfills, this method is used after the protective drainage material is placed over the geomembrane. Therefore the geomembrane is tested under a real load, and importantly, after the main potential for damage of the geomembrane has occurred. For proper leak detection, a conductive medium such as the prepared subgrade, a geosynthetic clay liner or a water-flooded leak detection zone must be under the geomembrane for proper operation. The accuracy of the leak location is typically to within a distance smaller than the depth of cover (Darilek and Laine, 2001).

Electrical leak detection was introduced in the early 1980s for liner leak detection, uncovered liners, liquid impoundments and soil and waste-covered geomembrane. This technique relies on having a conductive medium immediately below the geomembrane and a reasonably homogenous electrically conductive medium above the geomembrane. Unfortunately extraneous current flows can reduce the sensitivity of leak location since the electrical current can sometimes flow through batten bolts, concrete pads, pipe penetrations or soils at the edge of the geomembrane being tested.

When applying electrical leak location methods for testing geomembranes in ponds and tanks filled with water, a probe can be scanned over the geomembrane while the operator is wading in the water, or a probe can be towed across the liner if the water is deep or hazardous. A simplified version of this method can be used on exposed (i.e. uncovered) geomembranes whereby a 'squeegee', sponge or water jet/stream probe which is connected to an electrical power supply is used to scan the liner as it wets the dry geomembrane (Rollin *et al.*, 1999). While the water from the probe completes the circuit with earth ground, an electrical current flows. However, as a leak is located, there will be

Table 18.18   Number of leaks detected by electrical leak location studies

Type and location	Leaks/ha	Reference
Landfill liners (Italy)	15.3	Colucci and Lavagnolo (1995)
Pond liners (USA)	22.5	Laine and Darilek (1993)
Bare geomembranes (Canada/France)	2.03	Rollin *et al.* (1999)

Table 18.19   Electrical non-destructive weld tests

Test	Description	Comments
Electric spark testing	A high-voltage (15–30 kV) current is applied to the geomembrane seam and any current leakage to earth will cause sparking to a metal support plate	Can also detect pinholes or flaws in the body of the liner
Embedded copper wire (ASTM D-6365)	A copper wire is placed between the overlapped geomembrane sheets and it actually gets embedded into the final weld. After welding, a 20 000 V charged probe is connected to the end of the wire and passed over the weld. In the event of a flaw detection in the weld, an electrical discharge occurs which is registered by a detection unit	The method is extremely sensitive and reliable when understood and properly applied. Critical installations checked with this technique typically have a very small incidence of weld failure. It does not test the main liner body away from the welds
Electric leak location (ELL) (ASTM D-6747)	A current is imposed across the boundary of the liner using a high-energy power supply. When a current is applied between the source and the remote current return electrodes, current flows either around the entire site (if no leak is present) or bypasses the longer travel path through the leak itself (if a leak is present)	The bottom of the lined area needs to be covered with a layer of water; however even a thin layer is sufficient. The operator then surveys the area by walking on a pre-marked grid and locates where the electrical anomalies occur

an abrupt increase in the current indicating that the water from the probe has located a leak source.

Colucci and Lavagnolo (1995) found an average of 15.3 leaks per hectare for 25 landfill liners in Italy. Significantly for the leaks larger than 100 mm^2, 305 out of 320 of the leaks were holes or tears were attributed to construction related damage [Colucci and Lavagnolo, 1995].

The number of leaks detected by electrical leak location are listed in Table 18.18.

Leak location surveys have shown that approximately 25% of leaks are caused during installation, mostly at welds. The bulk of the damage however occurs when the liner is covered by soil or over burden.

Table 18.19 gives a summary of the electrical nondestructive tests for geomembrane welds.

Warning. The electrical methods used for geomembrane leak location use high voltage and low current power supplies, resulting in the potential for electrical shock or electrocution. This hazard might be increased because operations may be conducted in or near water. In particular, a high voltage could exist between the water or earth material and earth ground, or any grounded conductor. These procedures are potentially VERY DANGEROUS, and can result in personal injury or death. Hence, electrical test methods should be attempted only by qualified and experienced personnel. Appropriate safety measures must be taken to protect the leak location operators as well as other people at the site.

REFERENCES

ASTM Test Method D-7177, 'Standard Test Method for Air-Channel Testing of Field PVC Geomembrane Seams' (2005).

Colucci, P. and Lavagnolo, M. C., Three Years Field Experience in Electrical Control of Synthetic Landfill Liners, in *Proceedings of the Fifth International Landfill Symposium*, Sardinia, Italy, pp. 437–451 (October, 1995).

Darilek, G. T. and Laine, D. L., Costs and Benefits of Geomembrane Liner Installation CQA, in *Proceedings of the Geosynthetics 2001 Conference*, Portland, OR, USA, p. 65 (October, 2001).

GRI Test Method GM-19, 'The Standard Specification for Seam Strength and Related Properties of Thermally Bonded Polyolefin Geomembranes' (www.geosynthetic-institute.org).

Halse, Y. H., Koerner, R. M. and Lord Jr, A. E., Stress cracking morphology of PE geomembrane seams, ASTM Special Technical Publication 1076, in R. M. Koerner (Ed.), *Geosynthetics, Microstructure and Performance*, pp. 78–79 (1990).

Laine, D. L. and Darilek, G. T., Locating Leaks in Geomembrane Liners of Landfills Covered With a Protective Soil, in *Proceedings of the Geosynthetics '93 Conference*, Vol. 3, IFAI, Vancouver, BC, Canada, pp. 1403–1412 (March/April, 1993).

Peggs, I. D. [http://www.geosynthetica.net/tech_docs/hdpe_sc_paper.asp].

Peggs, I. D. [http://www.geosynthetica.net/tech_docs/CostBenInstall.pdf].

Peggs, I. D. [http://www.geosynthetica.net/specifications/geomembrane_seam_specs. asp].

Peggs I. D, 'Liner seam/weld spark testing' [http://www.geosynthetica.net/tech_docs/Liner_Seam_Weld_Spark_Testing.asp] (2008).

Peggs, I. D. and Allen, S., Geomembrane Seal Peel Separation: How and Why?, *GFR Magazine*, **19**(3), 1 (2001).

Poly-flex, Geomembrane Literature [http://www.poly-flex.com/sbqqgi11.html] (2008).

Rollin, A., Marcotte, M., Jacquelin, T. and Chaput, L. Leak Location in Exposed Geomembrane Liners Using an Electrical Leak Detection Technique, in *Proceedings of the Geosynthetics '93 Conference*, Vol. 2, IFAI, Boston, MA, USA, pp. 615–626 (April, 1999).

19

Geomembrane Installation Factors

19.1 INTRODUCTION

Geomembranes, if poorly installed, may conceal latent defects or weaknesses that can lead to post-installation failure and leakage of the geomembrane.

The key phases of a successful geomembrane installation are:

- design;
- site preparation;
- deployment;
- welding;
- CQC;
- final construction.

Damage to geomembrane liners can occur during the entire life cycle of the geomembrane such as:

- storage and handling at the geomembrane manufacturing site;
- transportation from the manufacturing factory to the installation site;
- offloading at the installation site;
- storage at the installation site;
- deployment at the installation site;
- positioning into the final welding location;
- welding of seams (e.g. overheating, burn through, contamination);
- extremes of weather (e.g. wind gusting, snow, extreme temperatures, severe temperature cycling, high humidity);
- animal damage (e.g. kangaroos cattle);
- traffic over the membrane;
- placement of the cover material, drainage layers or soil backfill on the completed geomembrane;
- in-service damage (e.g. inappropriate loading, cleaning).

Figure 19.1 shows the types of installation damage that can occur to polymeric geomembranes.

A Guide to Polymeric Geomembranes: A Practical Approach   J. Scheirs
© 2009 John Wiley & Sons, Ltd

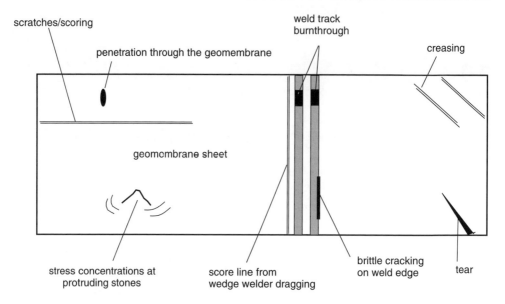

Figure 19.1   Schematic showing various types of installation damage to polymeric geomembranes

Damage from the above elements can be mitigated and largely prevented though adherence to rigorous specifications, attention to detail and a comprehensive construction quality assurance (CQA) program. In addition attentive third-party CQA personnel are also required to ensure compliance to the CQA program.

The greatest risks of damage to geomembranes during their lifetime actually occurs during the installation period. The risks can be mitigated by:

- proper storage of the geomembrane;
- proper transport of the geomembrane;
- using best practice deployment and laying techniques;
- the use of protective layers (e.g. geotextile cushioning layers);
- the use of an independent quality assurance organization;
- having comprehensive project specifications.

Damage to rolls of geomembranes can occur during storage and transport as shown in Figures 19.2 and 19.3. Wide rolls of geomembranes need to be carefully moved using the slings provided as shown in Figure 19.4.

For a single liner system with drainage layers, the geomembrane will be 'travelled' on at least three times: (i) to install the liner, (ii) to install the protective geotextile and (iii) to place the drainage layer. All of these events increase the risk of holes or tears arising from installation damage.

Even after installation, reservoir cleaning operations over its service lifetime can impose damaging stresses on a geomembrane liner. Typical damage causes and defects in geomembranes during their lifetime are summarized in Table 19.1.

Figure 19.2 Photograph of rolls of HDPE GMLs showing 'bubbling' and rippling effects caused by lateral constraint and compression of the membrane between the roll strapping points. Note the extreme rippling in the lower roll

Figure 19.3 Photograph of damage to a roll of 2 mm HDPE geomembrane by careless removal from a shipping container

19.2 DESIGN CONSIDERATIONS

Geomembranes have to withstand stresses and/or displacements imposed on them, but they are not expected to contribute to the mechanical integrity of the containment application. In service, geomembranes can be subjected to gravity stresses and shear forces on slopes, particularly at the crest of the slope (imposed stresses) as well as to differential settlements (imposed displacements).

Figure 19.4 Photograph showing how 7.5 m wide rolls of HDPE geomembranes are carried using fitted slings. Reproduced by permission of Sotrafa

An important aspect of geomembrane design is that the geomembrane (especially in the case of an HDPE geomembrane or liner) should serve only as a barrier and is not intended to serve any load-bearing or structural function. An overriding design consideration therefore is minimizing stresses on a geomembrane.

The major design considerations for geomembranes include:

- the thickness of the geomembrane required to prevent tensile failure and thinning;
- the use of a scrim reinforcement to improve/provide resistance to tearing;
- the thickness of protective layers to protect the geomembrane against puncture;
- the thickness of the protective cover required to protect the geomembrane against wheel and tracked vehicle loads;
- the hydraulic head over the geomembrane;
- the friction angle of the various layers to provide slope stability.

The liner system design components comprise:

- Subgrade/foundation (foundation settlement and maximum strain);
- Underliner (maximum particle size, interfacial friction);
- Geomembrane liner (thickness, acceptable strain);
- Overliner (maximum particle size, interfacial friction).

Basic geomembrane installation considerations include:

- Using a design to accommodate the fact that geomembranes cannot deal well with sharp corners.
- Allowing sufficient angles at corners in order to allow the geosynthetic to fit the geometry without wrinkles of folds.
- Allowing adequate distance between the top of the slope and the anchor trench, particularly in the case of long slopes.

Table 19.1 Typical defects in geomembranes and possible causes (adapted from McQuade and Needham, 1999)

Stage in geomembrane life	Types of defects	Possible cause/comments
Manufacture	Pinholes, excessive thickness changes, poor stress crack resistance, carbon agglomerates	Manufacturing defects and poor material selection. Unusual now for processes with good quality control and good resin selection
Delivery	Scuffing, cuts, brittle cracks, tears, punctures	Unloading with unsuitable equipment or lifting equipment. Impact. Poorly prepared storage areas. The geomembrane rolls should not be stacked more than five rolls high
Deployment and placement	Scratches, cuts, holes, tears	Dragging sheet along ground, trimming of panels, rough and angular subgrade, use of equipment on top of sheet without a protection layer, wind damage, large wrinkles, folds, damage by lifting bars, animal damage[a]
Welding	Cuts, overheating, scoring, poor adhesion, crimping, excessive warpage	Careless edge trimming, incorrect welding speed or welding temperatures, excessive grinding, dirt or dampness in the weld area, excessive roller pressure, poor surface preparation
Cover placement	Tears, cuts and scratches, holes, folds, stress in the membrane	Action of earthmoving equipment, insufficient cover during placement, careless probing of cover depth, contraction of sheet due to ambient temperature changes (some 73% of the leaks occur while covering the geomembrane)
Post-installation	Holes, tearing, slits, cracks	Puncture from drainage materials, puncture by sharp items in the waste, opening of scorelines under stress (see Figure 19.5), pulling apart of poor quality welds, downdrag stresses caused by settling waste, differential settlement

[a]Scratches and punctures from animals claws and hoofs.

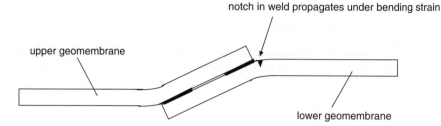

Figure 19.5   Schematic showing how a score line notch adjacent to a weld track propagates under bending strain

Table 19.2   Geomembrane design elements and typical safety factors (Koerner, 2005)

Design element	Liner stress mode	Typical safety factor
Liner self-weight	Tensile	≥ 10
Weight of cover soil	Tensile	0.5–2.0
Impact during construction	Impact	0.1–5.0
Weight of landfill	Compression	≥ 10
Puncture	Puncture	0.5–3.0
Anchorage	Tensile	0.7–5.0
Settlement of subgrade	Shear	≥ 10
Subsidence under landfill	Tensile	0.3–10

Factors which have to be taken into account for a particular geomembrane application are:

- the geology of the subgrade;
- the physical properties of the geosynthetics used;
- the nature of the product/s to be contained;
- the level of exposure to sunlight and other natural weather conditions;
- the magnitude of local and global stresses and the need to minimize these stresses.

Table 19.2 lists the major geomembrane design elements, the type of stress that it exerts on the geomembrane and typical safety factors that are associated with them.

19.3   INSTALLATION AND DAMAGE CONSIDERATIONS

19.3.1   STRESSES ON GEOMEMBRANES

Geomembranes can be subjected to a wide range of stresses during installation and operation in the end-use application (e.g. a landfill). The geomembrane must be able to support its own weight (self-weight) during and after installation due to the angle of the slope of a cell and bottom contours required for drainage. These internal stresses are due to the relatively low coefficients of friction in most geomembranes when placed on significant grades (e.g. high slope angles). Furthermore, tension is generated in the geomembrane during the placement of waste and by the final waste load.

If the subgrade underneath the landfill is not adequately prepared or if there is introduction of water under the liner then localized settlement (i.e. differential settlement) can occur to which the geomembrane liner must conform in order to remain intact. Bridging of the geomembrane may occur from the localized subsidence in the subgrade and this can present damaging stresses.

Stresses on the geomembrane can be especially important in the design consideration of side slopes and the basal liner of landfills. In the case of side slopes, the mass of the geomembrane itself and the overlying waste can drag down the liner and impose significant tensile strains on the geomembrane. The strain level can be calculated from the geomembrane liner weight, its thickness, the friction angle and the geomembrane liner's yield stress.

HDPE geomembranes exhibit a well-defined yield point at about 12% strain. Up to this point HDPE behaves elastically without damage to the polymer's microstructure structure. At greater strains the material yields and the behaviour is plastic (i.e. permanent deformation) to the failure point.

The design basis for geomembranes such as HDPE generally dictates that strains in service should not exceed about 3% in order to reduce the likelihood of creep rupture and environmental stress cracking. This level of strain however can easily occur as a result of temperature-induced stresses, basal settlements and indentation by underlying or overlying materials. In practice therefore, HDPE often performs in service at the limit of acceptable strain. Geomembranes such as LLDPE and fPP on the other hand can operate safely at much greater strains without risk of failure.

19.3.2 SUBGRADE PREPARATION

Surfaces to be lined (i.e. the subgrade) should be smooth and free of all rocks and stones greater than 12 mm diameter, sticks, sharp objects or debris of any kind so that initimate contact between the geomembrane and the subgrade is achieved (see Figure 19.6). The surface should provide a smooth, flat, firm, unyielding foundation for the membrane with no sudden, sharp or abrupt changes or discontinuity in grade. Importantly no geomembrane material should be placed on a subgrade that has become softened by water, or overly dried, until it has been properly reconditioned and/or recompacted.

Angular stones or gravel in the subgrade can be particularly damaging due to highly concentrated point stresses that can cause localized elongation and thickness reduction of the geomembrane (see Figure 19.7).

Geomembranes are designed to take up minor strains; however liners are installed to perform under confining pressures. Although some movement can be tolerated, the confinement of the liner allows for almost no movement. Therefore the liner can fail under subgrade deformation.

HDPE is a suitable geomembrane (all things being equal) where a well-compacted smooth subgrade is present. On account of its susceptibility to stress cracking, HDPE should not be placed directly on rough subgrades, especially those with angular protrusions.

Figure 19.6 It is important to achieve 'intimate contact' between the geomembrane and the subgrade (ie. substrate). Note how flat and well prepared the subgrade is. Reproduced by permission of NAUE

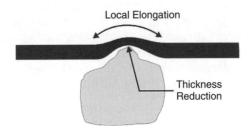

Figure 19.7 A localized thickness reduction of the geomembrane can occur in the vicinity of an angular particle, potentially puncturing the liner and reducing the durability of the geomembrane liner. If such particles are present then an appropriate protective cushion is required such as a geotextile. Reprinted with permission from the Proceedings of 56th Canadian Geotechnical Conference, Selection of Protective Cushions for Geomembrane Puncture Protection by Eric Blond, Martin Bouthot, Oliver Vermeersch and Jacek Mlynarek Copyright (2003) Eric Blond

Before placement of a geomembrane, any erosion or desiccation cracks in the subgrade need to be rectified. The subgrade then should also be rolled smooth and all protrusions and debris removed. Subgrade preparation is required to remove sharp stones, debris and vegetation. The subgrade must be compacted to provide a firm unyielding support for the geomembrane liner.

Geosynthetic installations and products are reliant on the civil construction to perform optimally and to specification. Most geosynthetic products rely on adequate subgrade preparation. Therefore a geomembrane could be adversely affected when the subgrade is not properly prepared.

👍 *The general rule is that 'the flatter the base/substrate, the better the geomembrane will perform'. Hence there is a lot of emphasis on uniform compaction. It is therefore good practice to specify very precisely the quality of the laying surface.*

19.3.3 INTIMATE CONTACT

For a composite liner system to function optimally the geomembrane is required to be in intimate contact with the underlying geosynthetic clay liner (GCL) or compacted clay liner (CCL). Due to the relatively high coefficient of thermal expansion of many geomembranes this intimate contact is difficult to achieve in practice.

The German 'Fixing Berm Construction Method' (discussed in greater detail later in this chapter) involves deploying the geomembrane in the morning and allowing it to equilibrate to the ambient temperature. The geomembrane is then welded in the warmest part of the day after which the protective geotextile is put in place. Still during the warmest part of the day fixing berms of soil are then placed along the geomembrane edges. Then during the cooler temperatures in the evening the liner lays flat on the surface. At this point the drainage or protective layers can be put in place.

If the placement of protective layers are restricted to the cooler times of the day then the likelihood of entrapment of wrinkles is very much reduced. The membrane should be somewhat loose when the temperature is low, and should be tight when the temperature is high.

LLDPE and fPP geomembranes are good choices for secondary liners as they lay flat and it is very easy to achieve intimate contact with the subgrade.

19.3.4 FOLDS IN THE SUPPLIED GEOMEMBRANE

In the case of HDPE in particular, it is necessary to watch that the geomembrane is delivered in rolls rather than folded. A folded geomembrane is not acceptable because the highly crystalline structure of the geomembrane will be damaged if it is folded. Any evidence of folding (other than from the manufacturing process) or other shipping damage can be cause for rejection of the material.

In order to avoid geomembrane damage during deployment:

- ensure that there are no stones, construction debris or other items beneath the geomembrane that could cause damage to the geomembrane;
- the geomembrane should not be dragged across a surface that could damage the material;
- appropriate lifting equipment and/or skilled manual labour needs to be available on site to deal with proper material deployment.

In the event that the geomembrane is dragged across an unprotected surface, the geomembrane must be inspected for scratches and repaired or rejected. Scratches that penetrate >0.1 mm or extend more than 20% of the geomembrane thickness are potentially damaging (Figure 19.8).

Geomembrane Damage - Notch Defects

• scratches that penetrate > 0.1 mm

• scratches that extend more than 20% of the GM thickness

• scorelines from the welder or the generator base

Figure 19.8 Notch defects in a geomembrane can occur from scratches that penetrate >0.1 mm or 20% of the liner thickness. Such scratches can arise from dragging the liner across rough subgrades or from dragging the welder or generator on the liner

Permanent crease marks or severely folded and crimped geomembrane areas are generally undesirable due to the localized stresses they create. Such areas are deemed as 'damaged' and should be cut out and patched. Wind crimping (or 'wind damage crimps') of geomembranes can cause thinning of the geomembrane cross-section.

Quality control and quality assurance are of utmost importance during geomembrane installation and fabrication in order to guarantee satisfactory long-term performance of geomembranes.

19.3.5 WATER UNDER THE GEOMEMBRANE

Care must be taken to ensure that flow or accumulation of condensate under the geomembrane does not damage the clay liner, that is, cause localized erosion or impair full-surface contact.

No standing water or excessive moisture should be allowed to accumulate on the clay liner prior to geomembrane installation. The geomembrane should not be placed onto an area that has become softened by water.

For fusion welding, a movable protective layer of plastic should be placed directly below each overlap of geomembrane that is to be welded. This is to prevent any moisture build-up between the sheets to be welded.

19.3.6 EROSION UNDER GEOMEMBRANES

It is necessary to install proper stormwater control measures prior to installation of the geomembrane to prevent erosion of the soil underneath the geomembrane. Formation of an erosion channel under the geomembrane (see Figure 19.9) can lead to regions where there is insufficient support of the geomembrane (i.e. bridging) causing elevated stresses which are potentially damaging to the geomembrane (especially HDPE geomembranes where the global strain should be kept under 3%). Furthermore, water flowing under the geomembrane can cause erosion of the protective sand layer and expose the geomembrane to sharp stone point loading (see Figure 19.10).

19.3.7 DEPLOYMENT DAMAGE

Installation of a geomembrane can result in scratching or scoring which can affect the geomembrane's ability to stretch or conform. The method used to unroll and deploy the panels shall not score, scratch or crimp the geomembrane.

Figure 19.9 Photograph of 'channeling' under a HDPE liner due to water and soil erosion (note subsidence of the liner). Hence the need to have a good compacted clay base as the subgrade for the liner

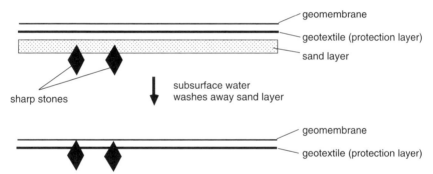

Figure 19.10 Schematic showing how erosion of the protective layer can expose the geomembrane to sharp stone point loading

19.3.8 SLOPE STABILITY

On steep slopes, the low friction characteristics of the smooth geomembranes (such as HDPE) with adjacent layers may lead to slope instability. Textured geomembranes are essential to prevent failure on steep slopes by preventing sliding. This is essential to provide structural stability given the number of layers in these multi-layer structures comprising subgrade, the geomembrane liner, geotextile protective layer and drainage layer. Hence textured geomembranes are recommended to increase the cap side slope stability.

When smooth geomembranes are pushed against other geosynthetics or soil, a degree of friction can be measured. Hard geomembranes such as HDPE have little friction; however softer materials such as LLDPE, fPP, EPDM, and PVC have greater friction under the application of an applied normal stress.

19.3.9   DIFFERENTIAL SETTLEMENT

If there is a high potential for significant differential settlement, LLDPE or fPP geomembranes are recommended on the basis of their excellent elongation and flexibility characteristics.

19.3.10   APPURTENANCES/PENETRATIONS

Appurtenances are penetrating structures in the geomembrane such as pipes and drain sumps. Penetrations through the geomembrane are sites where many of the observed leaks can occur as it is difficult to achieve good welds in these areas. Therefore special attention should be given to these areas. Pipe work and structures may be accommodated through the geomembrane by a combination of extrusion fillet welding and physical fixing. Detail work around appurtenances or sumps is generally performed by extrusion fillet welding.

Panels around piping penetrations or other projections through the panel should be cut with rounded corners to prevent tear propagation. Square corners or angular corners can act as stress concentrations where tearing or fracture can develop.

All geomembrane boots and shrouds should be of the same material and thickness as the geomembrane specified for the project. All pipe boots should fit snugly without wrinkles or 'fishmouths'. A geomembrane under the pipe boot must remain in contact with the subsurface (see Figure 19.11).

19.3.11   ATTACHING GEOMEMBRANES TO STEEL SUBSTRUCTURES

Geomembranes are typically attached and sealed to steel structures using stainless-steel batten strips and neoprene gaskets.

An improved alternative is to fix an extruded profiled strip made of the same resin as the geomembrane, onto the steel tank, to which the geomembrane can be welded. HDPE GMLs can be welded to a fixed HDPE anchor strip. However, since extrusion welds are subject to mechanical grinding and oxidation during welding it is important to design the adjacent liner/subgrade geometry to minimize the stress in the liner at the weld.

Connection of geomembranes to various structures can induce localized stresses which are beyond the design limits of the material and can lead to deterioration in the long term.

It is good practice to avoid abrupt transitions in geomembranes especially when fixing them to other structures. This is to avoid damaging stress concentrations. For example, bending a semi-rigid HDPE liner through 90° and then securing it to the side of a concrete support structure is not recommended. This is especially true if there is settlement of the subgrade which can induce deformation and strains on the geomembrane over a corner

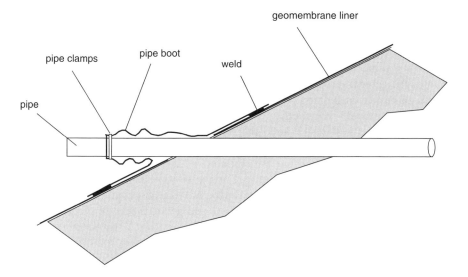

Figure 19.11 Schematic showing installation of a 'pipe boot' on a pipe penetrating a geomembrane

of a concrete pedestal or support. A heavy geotextile[1] should also be used to protect the geomembrane.

19.3.12 PROTECTION LAYERS

A protective layer is generally employed between the geomembrane and the drainage layer (e.g. 16–32 mm gravel). The purpose of the protective layer is to protect the geomembrane from puncture during construction and during the subsequent operation of the landfill. The thickness of the protective layer must be adequate to withstand the maximum load and temperature expected in operation. An additional important consideration is that the materials and design configuration used must accommodate the required shear strength for static stability on slopes under in-service loading. Failure to cater for this can lead to slope instability and sliding of the respective layers.

Drainage layers commonly used in landfills comprise at least 0.3 m of coarse aggregate or the equivalent performance with a geosynthetic drainage material. This ensures that leachate is contained within the drainage layer and thus minimizes the potential for clogging of the drainage layer. A geotextile filter should be placed over the drainage layer to protect it from clogging as a result of solids transport. Importantly leachate collection pipes must not be wrapped in a filter geotextiles as it has been demonstrated that these layers rapidly clog, thus rendering the collection pipes ineffective.

---

[1] Nonwoven geotextiles are widely used in combination with geomembranes as a cushioning material for the geomembrane, a drainage medium and a gas-venting medium.

Geotextile Cushioning

When geomembranes are placed on the top or under soils containing relatively large and/or angular stones that are protruding from the surface (or even simply resting on the surface), a protective geotextile cushioning layer is highly recommended to avoid puncture to the geomembrane. It is important to note that if the subgrade is compacted clay then a geotextile cushioning layer cannot be employed and any protruding or isolated stones must be physically removed.

Note that when positioning a textured geomembrane it is especially important to use protective geotextiles, as the action of dragging the roughened liner can dislodge protruding angular stones lying near the surface (see Figure 19.12).

There is a growing trend, especially in Europe, toward using heavier geotextiles as protection layers for geomembranes. For instance a 300 gsm geotextile can be relatively easily penetrated by a point load whereas a 1200 gsm geotextile gives superior protection against point pressure damage (i.e. point loading damage from a sharp stone). In Europe, site owners are willing to pay more for the protection layer than for the geomembrane. The effectiveness of protection layers can be quantified using the Cylinder Test where a soft metal plate is used in a compression test to imprint point loads. Since HDPE is to a degree elastic, microdeformations will relax and recover. For this reason a lead or aluminium sheet is used to record the deformations (see Figure 19.13).

There are three levels of protection that can be used for protecting geomembranes.

LEVEL I – LESS THAN 0.25% GEOMEMBRANE DEFORMATION
This protection level is sometimes referred to as the 'German Protection Philosophy'. The German Protection Philosophy requires the use of a performance puncture test to determine the mass per unit area of a protection material that is required to ensure that the localized geomembrane deformation (strain), due to the underlying, or overlying soil particles, is less than 0.25%. There can be no doubt that Level I protection is best for geomembranes; however, the cost of this level of protection can be very high. For example, for stone with a nominal particle size of 5 to 25 mm and 60 m of overlying soil, the

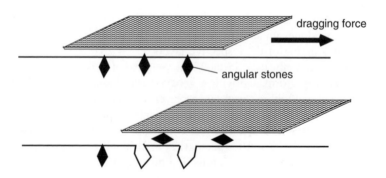

Figure 19.12   Schematic showing how dragging a textured geomembrane liner on a subgrade with protruding near-surface angular stones can cause the stones to dislodge

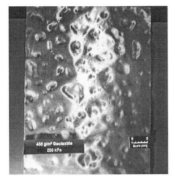
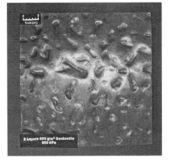

a) Geotextile1 (435 g/m²) @250 kPa   b) Geotextile2 (1200 g/m²) @ 900 kPa

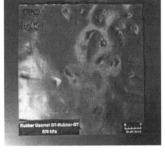

c) Sand-Filled Geocushion @ 900 kPa   d) Geomat 1(GT-Rubber-GT) @ 600 kPa   e) Geomat 2 (GT-Rubber-Grid) @ 600 kPa

Figure 19.13 Effect of different cushioning materials on stone indentations of HDPE geomembranes. Reprinted with permission from Journal of Geotechnical Engineering, Geomembrane Strain Observed in Large-Scale Testing of Protection Layers by Andrew R. Tognon, R. Kerry Rowe and Ian D. Moore, 126, 12, 1194–1208 Copyright (2000) ASCE

required mass per unit area of a nonwoven needle-punched (NW-NP) geotextile can be as high as 2000 g/m² (Narejo, 1995).

LEVEL II – INTERMEDIATE PROTECTION LEVEL
The 'Intermediate Protection Level' lies between Level I protection and the yield of HDPE geomembranes. The yield of HDPE geomembranes in the puncture mode is considered as failure for the Level II protection. A protection material that ensures that the geomembrane will not yield during the lifetime of the structure is considered satisfactory. The recommended design procedure for Level II protection requires that the geomembrane be subject to 1.5 to 2 times the design normal load for at least 100 h using a performance puncture test. At the end of the test the geomembrane test specimen is visually inspected for any signs of yield. If the geomembrane shows any indication of yield at the end of 100 h, then either the mass per unit area of the protection material must be increased, or a different type of protection material must be used (Narejo, 1995).

Table 19.3 Geomembrane application protection level selection guidelines
(Narejo, 1995)

Geomembrane application	Suitable protection level
Hazardous waste landfills	Level I
Hazardous waste surface impoundments	Level I
Municipal solid waste landfill liners	Level II
Municipal waste surface impoundment liners	Level II
Municipal waste landfill covers	Level III
Heap leach pads	Level II or Level III
Dams	Level II or Level III

LEVEL III – NONCRITICAL OR TEMPORARY PROTECTION

There are many geomembrane applications which are either noncritical or temporary. The limited loss of contained liquid through the geomembrane for such applications is neither harmful to the environment, nor otherwise unsuitable. For such applications the yield of the geomembrane may be allowed to take place. Considerably lower mass per unit area protection materials can be used for such applications (Narejo, 1995).

Geomembrane application protection level selection guidelines are summarized in Table 19.3.

In the test method GRI GM3, the geomembrane is loaded with manufactured protrusions (i.e. truncated cones) until holes in the geomembrane are formed. Appropriate safety factors are then applied to convert the truncated cone failure pressures to site-specific allowable pressures. This procedure can also be used to design for Level I and Level II geomembrane protection if the appropriate factors of safety to achieve the required level of protection are known (Narejo, 1995). The interested reader is referred to Wilson-Fahmy *et al.* (1994) for a discussion of this approach to determine the required geomembrane protection materials.

Problems can arise with using geotextiles in contact with geomembrane liners with respect to slope stability (where the liners will slide relative to each other). For 2:1 (ie. 30 degree) slopes it can be difficult to use geotextile protection layers due to slope stability considerations.

Sand Protection Layers

A good drainage layer should have little or no sand on it and so it requires coarser stones which exert higher point loading on the underlying geomembrane.

If sand is used as a protection layer for geomembranes it should always be screened to remove damaging coarse and angular stones. It is often quoted that 60% of holes in geomembranes arise from stone damage (with the next major cause being faulty welds). Note that a 2 cm layer of sand will generally protect geomembranes from stones, to less than 0.25% deformation (as is required by German regulations) (see Figure 19.14).

Note that the friction angle between the bottom face of the liner and the subgrade (or subsoil) must be larger than the friction angle between the top side of the liner and the geotextile/sand layer. In order to fulfil these requirements GM liners are textured or structured to give higher friction angles.

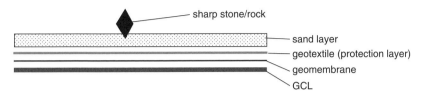

Figure 19.14 Schematic of typical protection layers for polymeric geomembranes

19.3.13 THERMAL EXPANSION

When black geomembranes are heated by exposure to direct sunlight and then experience cool temperatures during the night they can undergo significant expansion and contraction. These dimensional changes can occur both during and after installation. One of the consequences of excessive sheet stress during cold nights is that the contraction can cause the geomembrane to 'draw tight'. This stress has been implicated as a contributory factor in stress cracking of HDPE, especially at critical welds.

Example Calculation of Thermal Expansion

The variation in ambient temperature can be predicted using the following equation to calculate the maximum liner thermal movement:

$$\Delta L = L \times C \times \Delta T$$

where:

$$\Delta L = \text{liner expansion/contraction (m)}$$

$$L = \text{original liner length (m, e.g. 50 m)}$$

$$C = \text{linear coefficient of thermal expansion (m/m } ^\circ\text{C)}$$

$$\Delta T = \text{liner temperature variation } (^\circ\text{C})$$

Example: Calculate the thermal expansion of a 50 m section of an HDPE liner with an expected variation in the temperature of the ambient temperature from 10 to 40 °C.

$$\Delta T = 40 - 10 = 30\,^\circ\text{C}$$

$$\Delta L = L \times C \times \Delta T$$

$$\Delta L = 50 \times 2 \times 10^{-4} \times 30 = 0.30 \text{ m}$$

$$\Delta L = 300 \text{ mm}$$

During installation the thermal expansion can lead to wrinkles and fishmouths that can make placement and welding difficult. Furthermore such thermally induced wrinkles do not disappear when buried but instead can fold over leading to stress concentrations that act as sites for potential stress cracking to occur.

19.3.14 WRINKLES

Wrinkles generally occur in geomembranes during installation due to temperature variations or during deployment of the cover layer of soil or gravel (see Figure 19.15). Wrinkles are undesirable as they increase the incidence of construction damage, adversely affect the durability of the geomembrane (local regions of stress concentration), increase the infiltration beneath the geomembrane due to a lack of intimate contact between the geomembrane and clay substrate and also in some cases may interfere with leachate drainage. Often a compromise needs to be sought between conflicting requirements, i.e. minimizing wrinkles but also avoiding 'trampolining' or 'bridging'.

The geomembrane should be installed without undergoing buckling, wrinkling or tensioning. In particular, care shall be taken during installation of the geomembrane to ensure that the surface of the geomembrane after installation is free as possible from buckles, wrinkles, ripples, creases and folds.

On slopes that have not been covered, wrinkles can migrate down slopes and gather near the toe of the slope creating 'mega' wrinkles that can fold over flat, generating damaging localized stresses (see Figures 19.16 and 19.17). For this reason it is recommended that the tie-in-seams are welded after a few days once the wrinkles have migrated to the bottom of the slope (see Figure 19.18). This allows the geomembrane to go through a few expansion/contraction cycles before seaming the closure weld. In HDPE, although wrinkles tend to 'flow' down slopes, they do not move back up slopes at lower temperatures.

Figure 19.15 Photograph of wrinkles in an HDPE geomembrane due to thermal expansion (*expansion wrinkles*). These can lead to clay desiccation beneath the geomembrane under the influence of solar heating. Reproduced by permission of NAUE

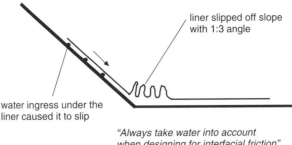

liner slipped off slope
with 1:3 angle

water ingress under the
liner caused it to slip

*"Always take water into account
when designing for interfacial friction"*

Figure 19.16   Schematic showing how geomembrane slippage can occur on slopes when there is water under the liner

Figure 19.17   On slopes, large 'waves' can form in the geomembrane. If these are still present first thing the next morning then they should be cut out and covered by a large patch, or otherwise the wave can fold over and create a crack in the geomembrane. Reproduced by permission of NAUE

Figure 19.19 illustrates the detrimental effect that wrinkles in geomembranes on a clay slope can have. Evaporated moisture from the clay can condense in the wrinkles on the underside of the liner leading to potential problems such as erosion, liner slippage and dessication of the clay.

Due to the difficulty in eliminating thermally induced wrinkles in HDPE geomembrane liners, it is almost inevitable for the geomembrane to be folded over when it is

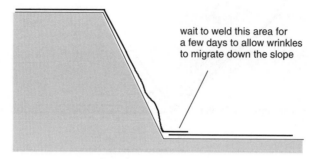

Figure 19.18 Schematic showing that geomembrane wrinkles (and waves) can migrate down a slope over time. Therefore it is good practice to wait until the wrinkles have migrated down before welding this area

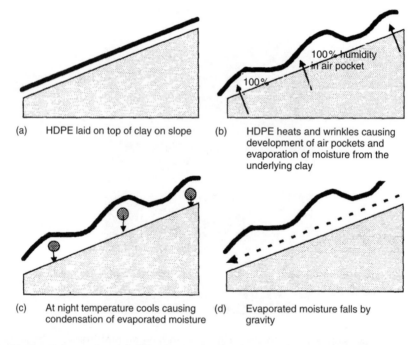

(a)   HDPE laid on top of clay on slope

(b)   HDPE heats and wrinkles causing development of air pockets and evaporation of moisture from the underlying clay

(c)   At night temperature cools causing condensation of evaporated moisture

(d)   Evaporated moisture falls by gravity

Figure 19.19 Schematic of the detrimental effect that wrinkles in an HDPE geomembrane on a clay slope can have. Evaporated moisture from the clay can condense in the wrinkles on the underside of the liner. Reproduced by permission of NAUE

covered – leading to points of high stress. Fortunately stress relaxation occurs in HDPE due to its viscous nature, so that the stresses reduce over time. Nevertheless, in these areas where the geomembrane is stressed it has a higher potential for failure. Therefore for critical applications such as landfill liners it is now standard practice to use double-lining systems (DLSs) with a leak detection system (LDS) between the liners. Any leakage between the liners is collected in a recovery bore so that no hydraulic head develops on the secondary liner. Since no leachates or liquors are allowed to accumulate between the liners, no leakage occurs into the subgrade.

It is not uncommon for bulldozers spreading the soil cover to shear off the tops of wrinkles leading to tears and hence major leaks. The common consensus amongst geotechnical engineers is that around 25% of leaks are related to installation of the geomembrane and approximately 75% are introduced during covering of the geomembrane with soil, gravel, drainage layers and so on.

Wrinkles and waves can also interfere with the operation of the impoundment, for instance, wrinkles can create local damming and pooling of leachates and interrupt drainage channels. In another example, wrinkles on the base can interfere with routine pond cleaning operations (e.g. in fish hatchery ponds).

Strategies to mitigate and minimize wrinkle formation are:

- Deployment of the geomembrane during the warmer part of the day and then covering the liner when it is cooler.
- Applying a slight pretension to geomembranes at the time of cover soil placement to eliminate wrinkles (as per the German BAM guidelines) (BAM, 1999).
- Reducing surface temperatures of the liner though the use of white or light coloured surface layers which reduce the tendency for thermally generated wrinkling by as much as 50% compared to black liners.
- Utilizing geomembrane materials such as fPP which have substantially less propensity for thermal wrinkling than HDPE. The wrinkles that form in fPP are lower in height (lower amplitude) and more frequent than the larger wrinkles that form in the more rigid HDPE.

It is necessary to ensure that the method used to deploy the sheet minimizes wrinkles but does not finish up causing bridging. It is also important to ensure that no more panels are deployed than can be welded on the same day. Wrinkles leading to folds and sharp creases are not permitted in waste containment applications. Wrinkles at the seam overlaps should be cut along the ridge of the wrinkle to achieve a flat overlap. The overlap should be welded and any portion where the overlap is inadequate should then be patched with an oval or round patch of the same geomembrane extending a minimum of 150 mm beyond the cut in all directions. Wrinkles must be 'walked-out' or removed as much as possible prior to field seaming. Any wrinkles which can folded over must be repaired either by cutting out excess material or, if possible, by allowing the liner to contract by temperature reduction. In no case can material be placed over the geomembrane which could result in the geomembrane folding.

In the case of HDPE geomembranes, there should be persons 'walking out' the wrinkles in the geomembrane liner before the advancing front of soil cover.

Giroud *et al*. (1995) found that exposed geomembranes with wrinkles were uplifted earlier and higher by wind than geomembranes without wrinkles.

Waves

Waves are a smooth undulation of the surface of a geomembrane caused either by thermal expansion or the installation method. The requirement for a smooth installation of the geomembrane is driven by the need to prevent 'waviness' or folds arising from waves

that would undergo excessive deformation under the weight of the protective-, drainage-
and topsoil layers.

The limit for permissible elongation of HDPE geomembranes under multiaxial loading
at 40 °C is 3%, while the limit under multiaxial loading at 20 °C is 6% (BAM, 1999).

It should be noted that waves in HDPE geomembranes cannot be smoothed out by
simply loading the geomembrane from above. Even heavy loading is ineffective since
HDPE is practically incompressible (Poisson ratio: 0.49). Various shaped waves can form
in geomembranes during installation and after they have been covered. Such waves do
not disappear when covered but instead can change configuration to produce 'fold overs'
or 'mushroom' shaped waves (see Figure 19.20). Soong and Koerner found that weighing
down long, high waves transforms them to small but very steep waves (known as standing
waves), causing high flexural strain in the geomembrane (Soong and Koerner, (1998,
1999)).

The German Federal Institute for Materials Research and Testing (BAM) have published
a specification with regard to the allowable wrinkle height. It states: 'Greatest clearance
between geomembrane and level supporting surface over a length of 10 m when rolled
out over a length of 12 m such that all individual measurements must be ≤50 mm'. Thus
no individual wave should be greater than 5 cm in height (BAM, 1999).

The German Geomembrane Specifications are the most stringent in the world in regard
to eliminating waves/wrinkles during geomembrane installation. They were written around
HDPE which has a tendency for stress cracking on folds.

Elimination of Wrinkles and Waves

German regulations require installation of a practically flat geomembrane over the
entire area of the installation. Thermally generated wrinkles in the geomembranes are

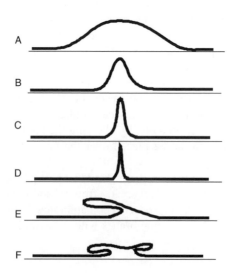

Figure 19.20   Schematic of various types of waves that can form in geomembranes.
Waves D–F are the most damaging in terms of the strains on the outer surfaces of the
membranes

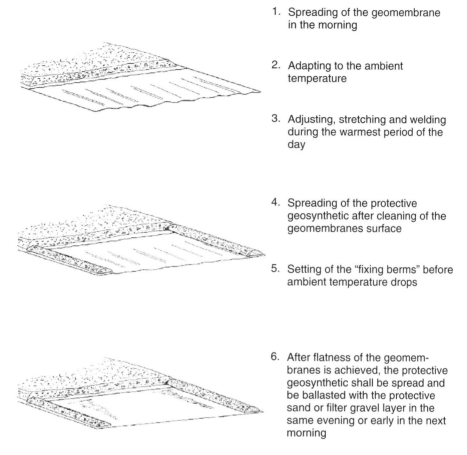

1. Spreading of the geomembrane in the morning

2. Adapting to the ambient temperature

3. Adjusting, stretching and welding during the warmest period of the day

4. Spreading of the protective geosynthetic after cleaning of the geomembranes surface

5. Setting of the "fixing berms" before ambient temperature drops

6. After flatness of the geomembranes is achieved, the protective geosynthetic shall be spread and be ballasted with the protective sand or filter gravel layer in the same evening or early in the next morning

Figure 19.21 Schematic showing the German 'Fixing Berm Construction Method' to mitigate thermal expansion wrinkles in order to achieve a flat surface and intimate contact with the subgrade before putting loads on the geomembrane surface. Reprinted from Recommendations for new installation procedures of geomembranes in landfill composite sealing systems by the Riegelbauweise by H. Schicketanz, presented at EuroGeo 2000 (Bologna) Copyright (2000) H. Schicketan

effectively eliminated using the 'Fixing Berm Construction Method' (FBCM)) (TA Abfall, 1991/1993; Seeger *et al*., 1996) (see Figure 19.21).

The FBCM method can be summarized as follows:

1. The geomembrane is deployed in the morning on the prepared subgrade.
2. After about 1 h (to allow the geomembrane to equilibrate to the ambient temperature) the geomembrane is adjusted and stretched a little to eliminate the waves in the geomembrane generated by delivery and unrolling.
3. The geomembrane is then fixed with sandbags to prevent wind uplifting, etc. After welding and testing of the seams and cleaning of the geomembrane surface, the geosynthetic protection layer is spread out.

4. Before the cooler temperatures occur in the evening, the cross-'fixing berms', consisting of protective sand layer, drainage gravel or sandbags, are placed on the geosynthetic protecting layer at the ends of the geomembranes and at special locations (e.g. toes, hollows, grooves) to fix the geomembrane and keep it in position. (Note the fixing berms should not exceed a distance of more than 30 to 50 m in order to avoid transverse contraction of the geomembranes.)
5. In the evening the lower temperature causes the geomembrane to contract and become flat and subsequently the whole area can be covered over with the protecting or drainage mineral layer either during the same night or in the early morning of the next day to prevent the formation of new waves and wrinkles.

Example: Stress Load Analysis on Thermally Contracted Liner

The following exercise illustrates a typical stress load calculation for a restrained HDPE geomembrane subjected to decreasing ambient temperature.

*   measured surface temperature of HDPE geomembranes of 2.5 mm under direct sun exposure after placement of fixing berms = 40 °C;
*   temperature of the flat geomembrane in the evening = 15 °C.

Temperature difference is therefore $\Delta T = 25\,°C$. Coefficient of linear thermal expansion of HDPE is 1.7×10^{-4} cm/cm/°C. Therefore 1.7×10^{-4} cm/cm/°C $\times 25\,°C \times 100$ cm $= 0.425$ cm and thus strain $= 0.425/100 \times 100 = 0.425\%$.

*   The geomembrane will hence undergo a strain of about 0.425%.

Using the relationship stress/strain = modulus we see that: stress = modulus $\times$ strain and therefore stress = 600 MPa $\times$ 0.425% = 2.55 MPa.

Given that HDPE (or more precisely MDPE[2]) has a modulus of elasticity of about 600 MPa the induced stress is thus approximately 2.55 MPa.

This is about 15% of the yield stress (about 17.5 MPa) of a common MDPE geomembrane.

Of course this very low contraction stress level will be reduced by stress relaxation of the geomembrane while it is in service. So, on the basis of this calculation the HDPE geomembrane is in no danger of stress cracking.

19.3.15 SLACK IN GEOMEMBRANES

During deployment and installation it is essential to include sufficient extra material in a geomembrane containment to accommodate thermal contraction. This is because plastic geomembrane materials expand and contract as the temperature changes and thermal contraction can cause extensive damage to geomembranes if it is not taken into account in the design.

---

[2] Strictly speaking HDPE geomembranes are manufactured from MDPE base resin and it is addition of the carbon black that raises the density into the density classification for HDPE.

Thus geomembranes require sufficient slack at installation to be able to contract to the lowest temperatures expected in service without damage. In order to ensure that there is enough slack at low temperatures (i.e. in the winter nights), the installer must adjust the amount of slack installed at the warmer temperatures of installation (i.e. in the summer).

If a geomembrane material is installed without sufficient slack to attain the design low temperature, then the material will tighten and possibly be stressed to failure. Environmental stress cracking failures in HDPE are commonly initiated by the stress of thermal contraction. Occasionally the anchor trench will pull out to relieve stress, or the corners of the liner will 'bridge' or 'trampoline' at cold temperatures. These stresses, if not corrected, can cause premature failure of the liner. The basic calculation for placing slack in a liner is as follows:

$$\text{Coeff}_{\text{Thermal Expansion}} \times (T_{\text{install}} - T_{\text{design}}) \times \text{Distance} \times \text{FS}$$

i.e. multiplying the coefficient of thermal expansion/contraction by the difference between the installation temperature and the design low temperature, and then multiplying by the distance between the points where the slack is being measured, multiplied by a factor of safety (FS).

Unfortunately some installers install lining materials with a complete disregard for thermal slack. Not compensating for slack in a liner may be manageable in temperate climates, but in northern hemisphere climates it is essential. It is recommended that a factor of safety of about 1.5 be used.

In considering the amount of slack, it is also important to understand the operating conditions of the lining material. For instance, if a liner is backfilled or has a permanent liquid level, the amount of thermal slack required can be reduced. The simple rule is, the more of the liner that is exposed, the more thermal slack must be incorporated. In most cases a backfilled liner requires less slack. The flexibility of a lining material also affects the amount of slack used. Flexible, unsupported materials such as PVC and PP require less slack than stiff materials such as HDPE or supported materials. This is because the flexible, unsupported materials can stretch locally (up to 100%) to accommodate thermal contraction without damage, while very low extensions (as low as 2% at low temperatures) will damage both HDPE or supported materials. (Layfield, 2008).

19.3.16 FLAWS ON THE GEOMEMBRANE EDGE

HDPE is quite notch sensitive and therefore particular care must be taken against mechanical damage while installing any HDPE geomembrane. The edges of the liner should be straight and free from nicks or cuts. Geomembrane rolls damaged in transport and/or handling should be rejected. Scorelines in HDPE geomembranes can seriously affect the elongation potential of the material under a tensile load (see Figure 19.22).

Figure 19.22 Photograph of weld test specimens demonstrating how the scorelines caused by the wedge welder are responsible for very low elongations in a shear weld test

19.3.17 WELDING-RELATED FACTORS

The welding of geomembranes is extremely sensitive to temperature differences, moisture and dust. Therefore, it is standard QA/QC procedure during an installation to do destructive on-site testing, at instances where there may be changes in ambient temperature or moisture as well as between crew breaks (i.e. interruptions to welding operation).

Prior to welding, the weld area should be clean and free of moisture, dust, dirt, debris, markings and any foreign material. It is very difficult to remove dust from the edges of textured geomembranes. For this reason it is necessary to specify that the geomembrane has a non-textured flat edge to facilitate welding.

Where abrading or surface grinding is required, the process should be completed within 1 h of the welding operation and in a way that does not damage the geomembrane. Abrasion of the geomembrane surface should not extend outside the weld bead or weld surface area.

Factors which make field geomembrane welding difficult are summarized in Table 19.4.

Overheating of the geomembrane during welding can lead to heat-affected zones that are potential sites for stress cracking and brittle failure of HDPE (see Figure 19.24).

Cross Slope Welds (Horizontal Seams)

Cross slope welds (or horizontally oriented welds) should as far as possible be avoided, especially in exposed waste containment applications, due to the potential for damaging stress concentration on weaker seams. All factory and field seams should be positioned parallel to the slope to eliminate the potential for stress concentrations at the welds. In general, geomembrane panels placed on the slopes should have no cross-slope welds

Table 19.4 Factors which make field geomembrane welding difficult

Factors	Comments
Nonconformance of the geomembrane sheets to the subsurface	Can lead to bridging, trampolining and localized stresses
Wind blown dirt and dust in seam area	Can lead poor seam integrity due to particulate matter contamination in the weld zone
Moisture on the upper surface of the geomembrane	Can lead to porosity and poorly bonded areas (see Figure 19.23)
Direct sunlight and warm temperatures	Can lead to expansion of sheets during welding
Ambient temperature variations	Can lead to inadequate fusion due to insufficient heat dwell or penetration into the geomembrane
Wind fluttering the edges of the sheets	Can lead to poor alignment and creases
Textured liners without smooth edges	Can lead to dirt entrapment since textured edges are difficult to clean[a]

[a]Some geomembranes have thin plastic tear strips (15 cm wide) on the edges of the rolls which are removed immediately before welding in order to present clean and dry edges for welding.

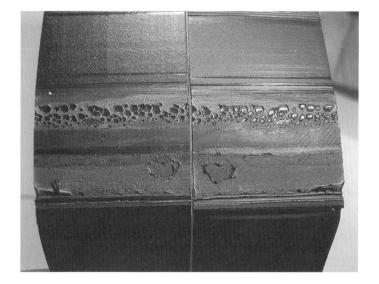

Figure 19.23 Photograph showing the effect of water on the quality of the weld in HDPE geomembranes (note internal voiding due to water vapour)

unless these can be shown to have strength of at least 95% of the parent material, and that the seam behaves in a ductile manner, being defined as a minimum extension of 150% prior to break.

In the case of long slopes where a cross-slope seam may not be avoided, the weld should be made at a 45 degree angle and be placed closer to the base of the slope (no higher than a third of the slope).

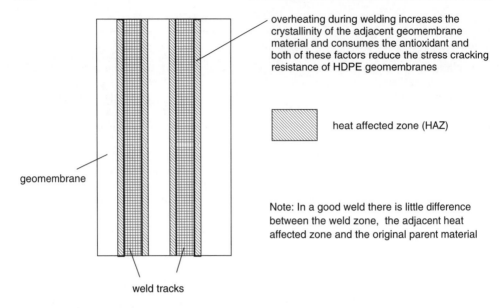

overheating during welding increases the
crystallinity of the adjacent geomembrane
material and consumes the antioxidant and
both of these factors reduce the stress cracking
resistance of HDPE geomembranes

heat affected zone (HAZ)

geomembrane

Note: In a good weld there is little difference
between the weld zone, the adjacent heat
affected zone and the original parent material

weld tracks

Figure 19.24   Schematic showing the weld track and the adjacent heat affected zones
which are susceptible to stress cracking

Side Slopes

Textured geomembranes should be used on side slopes (particularly for grades greater
than or equal to 7H:1V). It is important that seams on side slopes (slopes steeper than
or equal to 7H:1V) are oriented parallel to the slip direction, and the textured material
extends a minimum of approximately 1.5 m out past the side slope.

Corners

In corners the number of welds should be minimized. For instance, the geomembrane in
the corners of pond slopes should not be filleted to merge into single multi-panel joins.
Corners should be formed by staggered seams to satisfy the requirement of a maximum
of three panels joining at one location. All seams along the toe of an embankment and
excavation slopes should be at least 1.5 m from the toe.

'T' Seams

'T-seams' or 'T-welds' are defined as a point in the weld where three layers of geomembrane overlap with each other. They occur at the point that a dual-track field weld crosses
another seam, usually at a 90 degree angle. They are also sometimes referred to as *butt
seams* or *end seams*. The potential exists for very tiny holes to occur at the junction of
the three layers of material at each 'T'. 'T-welds' can also be susceptible to oxidation and
cracking due to the extra heating (additional thermal history) that this area is exposed to.

'T' joints produced as a result of cross/butt seams should be extrusion fillet welded to
ensure proper sealing. The overlap on each 'leg' of the 'T' joint needs to be trimmed

back 15 cm and then ground to 7.5 cm minimum on each of the three legs of the 'T'. Extrusion welds then need to be placed on all of the area prepared by surface grinding.

 All panel intersections ('T' seams) should be extrusion welded to ensure a proper seal.

In extrusion welding, a welding rod comprised of the parent geomembrane material is fed into a hand-held gun and heated to its melting point. The molten bead is laid onto an overlapped seam area that has been pre-tacked, ground to remove oxidized surface material and preheated by a hot air nozzle attached to the extrusion gun. The resulting butt seam at a 'T' intersection ensures integrity of the seam at an area where the wedge welder is not able to produce a uniform seam.

Closure Welds

Closure welds are seams between the large panels of welded geomembrane that are located in the basin of the impoundment, cell or lagoon. These closure welds are generally the last welds to be performed to 'close up' the liner system. They require careful attention both with regard to their location and the increased likelihood of dirt ingress in the welds.

Tie-in Seams

Geomembrane tie-in seams (those which are perpendicular to the slope) can be wrinkled and potentially troublesome. Dirt is known to be a problem with seam quality at tie-in seams. Often tie-in seams will have a lower peel strength or exhibit a higher peelable area due to dirt contamination. Tie-in seams should not be located within 1.5 m from the toe of the slope. Slack can be incorporated in the longitudinal direction at toe tie-in seams, or in a centre tie-in seam. Slack can also be released from the anchor trench prior to backfilling. It is important to prepare tie-in seams, or to backfill anchor trenches, at the coolest part of the day.

Detail Weldability

Detail weldability is related to the method of welding and the equipment required to perform the weld. Solvent welds and glues are the most versatile (but only for liners that will solvate), followed by hot air weldable and then extrusion weldable materials. Multiple pipe penetrations are best lined with materials that have excellent detail weldability.

Patching

Patching consists of applying a new piece of geomembrane sheet over and at least 100 mm beyond the limits of a defect. The patch is extrusion seamed to the underlying geomembrane. This method is used to repair large holes, tears, undispersed raw materials and contamination by foreign matter.

All patches should be formed with rounded edges and ends, to prevent stress concentrations in the geomembrane. All cuts should be removed or be ended off with a round cutout.

Chemical welded patches and repairs should be made with the same material with a minimum of a six-inch overlap over the damaged area. The patch should have rounded corners. Apply chemical fusion between the damaged membrane and the patch. Then apply pressure to the two surfaces to achieve intimate contact between the liners.

Spot Seaming

Spot seaming refers to applying a bead of extrudate, maximum length of 150 mm, over a defect. Spot seaming should be used only to repair dents, pinholes, air pressure test holes or other minor, localized flaws.

Capping

Capping refers to applying a new strip of geomembrane over a faulty seam. The cap strip generally extends at least 150 mm beyond the limit of the seam and the edges will be extrusion seamed to the underlying geomembrane. This method is used to repair faulty lengths of both extrusion or fusion seams.

Removal of Overhanging Flap

The overhanging flap along the edge of a wedge weld should not be ripped off as shown in Figure 19.25 as this can damage the weld. The flap should be carefully removed with a hooked blade.

19.3.18 ANCHORING OF GEOMEMBRANES

Anchoring of geomembranes can be done by a number of different methods. The most common method is by using anchor trenches. Anchor trenches are designed to suspend the geomembranes from the walls and prevent them from creeping down slopes. Smooth

Figure 19.25 Photograph of 'ripping off' the overhanging flap along the edge of a weld track in a 1 mm HDPE geomembrane

Figure 19.26 Photograph of a typical anchor trench to secure a HDPE geomembrane liner. Reproduced by permission of Sotrafa

geomembranes rely mostly on a horizontal crest and friction inside the trench to prevent them from moving (see Figure 19.26). Other anchoring techniques include run-out lengths that rely on a sufficient length to be placed at the slope top, which is subsequently covered to provide a normal load to mobilize the shear strength between the underlying surface and the geomembrane. Another method uses batten strips which are lengths of metal that hold the geomembrane secure. This method is generally used where the geomembrane is installed to a concrete beam or structure.

Adequate temporary anchorage must also be provided during the installation process of the geomembrane in order to prevent movement of the liner by forces such as wind lifting or cyclic expansion/contraction. This is particularly critical in anchor trenches at the tops of slopes since it is extremely difficult to pull back a geomembrane panel that has slipped down the slope.

The inside of the crest of the anchor trench should be rounded off to facilitate the placement of the geomembrane sheet over the edge of the trench wall. Similarly the crest and toe of the embankment slope should be rounded to support the geomembrane. Anchorage of the geomembrane should be carried out when the geomembrane is cool, to prevent bridging of the geomembrane at change of grade of the subgrade due to changes in temperature. Once the geomembrane is anchored, the geomembrane at the toe of the slope should be ballasted to prevent lateral creep over the floor of the geomembrane, which may then result in bridging (see Figure 19.27).

Slightly rounded corners should be provided in the trench where the geomembrane adjoins the trench so as to avoid sharp bends in the geomembrane. No loose soil or rocks should be allowed to underlie the geomembrane in the anchor trench. Leading edges of the anchor trench should be smooth and even.

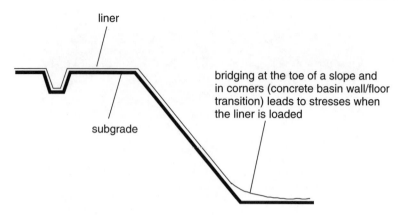

Figure 19.27 Schematic showing that 'bridging' at the toe of the slope can lead to stresses on the liner when it is loaded from above

An anchor trench is commonly used along the perimeter of the geomembrane impoundment to secure it and prevent it from sloughing or slipping down the inside side slopes during construction or service. However if the forces on the geomembrane in the anchor trench exceed the tensile strength of the geomembrane, then the liner could tear. For this reason the geomembrane should be allowed to give a little by slipping after construction to relax these forces.

Geomembranes installed on slopes are required to be fixed in anchor trenches. Usually the anchor trench is backfilled with the excavated material which is then compacted. The dimensions of the anchor trench vary depending on the length of the slope. For instance, if the slope length is 10 m then an anchor trench depth of 50 cm, a width of 40 cm and a distance of 50 cm from the anchor trench to the slope break point is required. Table 19.5 shows the anchor trench dimensions for various slope lengths.

19.3.19   TENSIONING DUE TO DOWNSLOPE DRAG

'Down drag' or 'downslope drag' is a phenomenon where the liner moves down a slope due to increased loading on the top surface. Often geomembrane ponds containing tailings can exhibit tensioning of the exposed geomembranes on the batters due to down drag where the tailings exert tension on the liner due to the downward movement of the liner under the load of the tailings. These stresses can lead to failure at any weak or thinned

Table 19.5   Anchor trench dimensions for various slope lengths

Slope length (m)	Trench width (m)	Crown width (m)
<10	>0.5	>0.5
10–40	>0.8	>0.6
>40	>1.0	>0.8

area such as the edge of welds. Downslope movement of the geomembrane is increased in cases where there is a low interface friction angle between the geomembrane liner and the wet underlying clay base.

Down drag on geomembranes can lead to significant stresses. One form of down drag can be caused by large waste settlements. Geomembranes can be subjected to multiaxial stress states from plane strain loading, such as drag down along lined slopes. For this reason the multiaxial tension test is superior to the uniaxial tension test for cases in which the geomembrane is expected to undergo multiaxial stress conditions resulting from plane strain drag down alongside slopes or from out-of-plane loading due to subsidence beneath liner and cover systems. The ability of the geomembrane to withstand downdrag during and after waste placement is an important design parameter as is the suitability of the anchorage configuration for the geomembrane and the ability of the geomembrane to support its own weight on the side slopes. The down drag force generated by waste accumulation and sliding of overlying waste material can be dissipated to a large extent through the elongation of the protecting geosynthetic overlying the geomembrane and therefore is not transferred to the geomembrane.

In mine tailing ponds singly textured geomembranes are often used to grip the geomembrane onto the compacted clay slope to prevent down drag by the tailings (Figure 19.28).

19.3.20 GEOMEMBRANE BRIDGING

As already mentioned, geomembrane bridging or trampolining needs to be minimized as far as possible. Changes of grade should be effectively ballasted to keep the geomembrane in contact with the subgrade and reduce the risk of creep due to temperature variation of the geomembrane. Sufficient slack should be provided in the geomembrane on the floor and slopes to prevent the geomembrane from being pulled away from the toe and

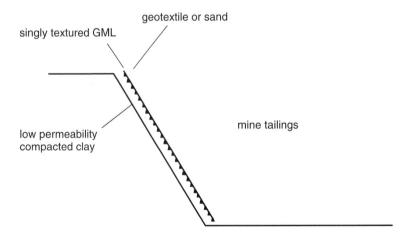

Figure 19.28 Schematic of the use of a singly textured geomembrane for lining mine tailings where the texturing is required to prevent down drag by the mine tailing

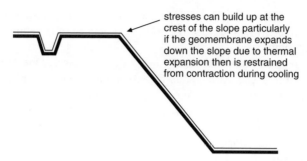

stresses can build up at the
crest of the slope particularly
if the geomembrane expands
down the slope due to thermal
expansion then is restrained
from contraction during cooling

Figure 19.29 Schematic showing that stresses on the geomembrane can concentrate at the crest of the slope. These stresses can be significant if the geomembrane expands down the slope due to thermal expansion and then is restrained from contracting during cooling

valleys of the slope, due to a drop in the temperature and subsequent contraction of the geomembrane.

During the cool temperatures of the night, the geomembrane will contract and if not enough slack is present, then trampolining at the toe and tightness at the crest can occur leading to the application of undesirable stress on the geomembrane (Figure 19.29).

There should be sufficient slack in the deployed geomembrane to prevent tensile stresses from occurring when the temperature drops, causing the geomembrane to shrink. The 'trampoline effect' occurs when a membrane shrinks and pulls away from the subgrade at corners. Inspect for the trampoline effect in the early morning when temperatures are lowest. The trampoline effect often occurs along the bottom edge of liner systems.

19.3.21 WIND DAMAGE

Deployed geomembranes are very susceptible to wind uplift if not adequately ballasted (i.e. weighted down). Wind damage can be so extreme as to pull the liner out of anchor trenches or tear it at the anchor trench. Excessive wind is that which can lift and move the geomembrane panels.

Any geomembrane panels exposed to high winds should be carefully inspected for damage such as:

- abrasion or thinning;
- crease marks;
- star creases;
- severely crimp and folded locations;
- areas of permanent deformation;
- tears, rips.

Excessively deformed or creased areas are not permitted (especially for HDPE) and must be removed and patched or the entire damaged area replaced.

Ensure there is sufficient ballasting to counteract wind uplift of a temporarily exposed liner.

Table 19.6 Mass per unit area values and minimum wind speeds to uplift various geomembrane types (after Giroud *et al.*, 1995)

Geomembrane yype	Thickness (mm)	Mass per unit area (kg/m^2)	Minimum wind velocity for uplift (km/h)
PVC	0.5	0.63	11.1
	1.0	1.25	15.7
HDPE	1.0	0.94	13.6
	1.5	1.41	16.7
	2.0	1.88	19.2
	2.5	2.35	21.5
CSPE-R	0.75	0.9	13.3
	0.90	1.15	15.0
	1.15	1.50	17.2
EIA-R	0.75	1.0	14.0
	1.0	1.30	16.0
Bituminous geomembranes	3.0	3.5	26.2
	5.0	6.0	34.3

HDPE is particularly difficult to install under windy conditions, as this makes it difficult to lay the material and can cause the sheets to move during welding.

Table 19.6 shows the mass per unit area of various geomembrane types and the minimum wind speed required to lift up the geomembrane under its own weight (Giroud *et al.*, 1995). Obviously the thinner gauge membranes require less wind speed to be uplifted. The heavier gauge geomembranes and particularly the bituminous geomembranes require higher wind speeds (as high as 34 km/h) for uplifting to occur.

Giroud *et al.* (1995) also found that exposed geomembranes with wrinkles were uplifted easier and higher by wind than geomembranes without wrinkles.

Preventing Wind Damage

In areas where high winds are prevalent the installation of the lining should begin on the upwind side of the project and proceed downwind. The leading edge of the liner needs to be secured at all times with an adequate number of sandbags of sufficient weight to hold it down during high winds and to prevent uplift by wind. The leading edges of the liner material left exposed at the end of the day's work needs to be anchored to prevent displacement and damage due to the wind. In cases of high wind, continuous loading is recommended along edges of panels to minimize risk of wind flow under the panels.

During installation the geomembrane must be secured with sandbags to protect it from wind damage or displacement.

To evaluate the forces acting on the geomembrane and the anchor trench due to wind uplift events, the design approach by Giroud *et al.* (1995) can be used. Wind uplift is considered the largest design consideration for an exposed geomembrane cover application. Using this design method, the only viable hold-down mechanism to keep the cover from being blown away due to a significant wind velocity is to use fully backfilled anchor trenches or soil berms on the geomembrane. To reduce the probability of failure, a conservative design approach needs to be used for a maximum assumed wind velocity of 40 m/s (144 km/h).

19.3.22   GAS PRESSURE UNDER THE LINER

A problem in some cases is the formation of 'whales' or 'whale backs' where the geomembrane lifts up from the substrate due to the build-up of pressure underneath. The pressure may be due to methane or sub-surface gases.

19.3.23   INSTALLATION DAMAGE BY MACHINERY

Installation damage refers to mechanical damage to the geomembrane during the construction and installation of the liner. Installation damage generally manifests itself as punctures, tears or rips in the liner causing it to leak or affecting its durability. It is difficult to detect installation damage of geomembranes after placement of soil or drainage layers on the geomembrane.

Heavy machinery should not be allowed to drive on the geomembrane or the drainage layer. The lighter the machine, the less damage will be done to the geomembrane layer by track imprints for example. Hydraulic excavators with articulated or telescopic booms are suitable. Bulldozers are therefore not very suitable because of their high machine weight and their shearing force can damage the geomembrane or geotextile. There are special bulldozers with plastic tracks known as 'bogdozers'. A competitive alternative to hydraulic excavators are 'snowcats'. These machines, originally designed for winter sports, have a low dead weight (2–6 t) and broad tracks (2 × 1.5 m), so that only low shearing forces and ground pressures (0.040–0.050 kg/cm^2) are transmitted to the geotextile and geomembrane and no waviness in the geotextile or geomembrane will arise due to the installation of the protective layer. By way of comparison: a bulldozer exerts a ten times higher specific ground pressure than a snowcat (Averesch and Schicketanz, 2000).

Construction equipment driving on the geomembrane during liner installation can impose severe stresses on the geomembrane.

Personnel on site need to ensure that low ground pressure tracked dozers are used and that these always operate on sufficient thicknesses of soil cover.

To avoid damage from tracked equipment there must be a minimum of 250 mm of compacted soil between the tracks and the geomembrane liner. Shallower depths of soil cover or fill directly under the dozer tracks can produce geomembrane damage even if low ground pressure (LGP) dozers are used (i.e. <58 kPa). It has been shown that a 0.46 m thick layer of shredded tire chips with an average size of 7.6 cm, placed over a 543 g/m^2 geotextile installed over a geomembrane liner using low-ground-pressure (<58 kPa) equipment provides adequate protection to the geomembrane liner during construction (Reddy and Siachek, 1998).

Check that machinery used on the geomembrane during installation is of a low ground pressure and is always operating on a sufficient thickness of cover soil.

Figure 19.30 Photograph of a field crew installing an HDPE geomembrane. Note the lightweight vehicle on the liner (not the white tray truck). Reproduced by permission of Layfield, Canada

Lightweight vehicles (e.g. 'Quad Bikes') can be used on deployed geomembranes, as shown in Figure 19.30.

19.3.24 FILL PLACEMENT ON TOP OF GEOMEMBRANES

Fill placement on geomembranes is highly risky and arguably the most common cause of geomembrane damage. During placement, the fill placement operator generally would not know if the geomembrane was damaged during the operation. Geomembrane damage during fill placement is not only highly reliant on the way the fill is placed, but also the type of fill. It is important that the fill should be free of angular material and any sharp objects.

The process of covering geomembranes with soil after installation can itself lead to damage of the geomembrane. Often light-ground-pressure bulldozers are used to push out the sand or gravel over the liner. However the operators should not push a large pile of soil or sand forward in a continuous manner since this can cause localized wrinkles to form in the liner and overturn (i.e. wrinkles that may fold over) in the direction of travel (see Figure 19.31). Such wrinkles are termed 'overturned wrinkles'. Overturned wrinkles can create sharp crease-lines and knife-edge folds in which damaging stresses can develop. In particular with HDPE geomembranes these stresses can lead to stress cracking and premature failure of the liner. Instead the recommended procedure is to continually place small amounts of soil or drainage material in an outward manner over the toe of the previously placed material. The soil should be spread in a manner that wrinkles in the geomembrane are smoothed out and dispersed and not folded over. Although this procedure is slow and labour-intensive it is better in the long run since finding and repairing leaks at this stage can be very time consuming and expensive. Overturned wrinkles should be cut out and seamed flat.

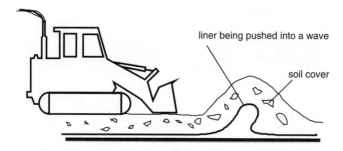

Figure 19.31 Schematic showing how a geomembrane can be pushed into a wave while deploying soil cover on a geomembrane with a bulldozer

Note the drainage and protective cover soils need to be placed on the geomembrane during the coolest part of the day when there is minimal wrinkling in the geomembrane. Also the cover soil should not contain stones, rocks or other items that might damage the geomembrane through puncturing or tearing.

Care should be taken to ensure that soil covering is not conducted in a manner that it can penetrate under a loose flap of a weld where it could exert a damaging peeling force on the weld (see Figure 19.32).

ASTM D-5818–95, 'Standard Practice for Obtaining Samples of Geosynthetics from a Test Section for Assessment of Installation Damage', outlines standardized procedures for obtaining samples of geomembranes from a test section for use in assessment of the effects of damage immediately after installation caused only by the installation techniques.

Nosko *et al*. (1996) analysed electrical leak location surveys from more than 100 sites where several thousand sites of failure were found (see Table 19.7). Of these, 73% were construction damage caused during the installation of the covering layer (i.e. emplacing earth on top of the geomembrane or liner) and 2% were accidentally caused after the covering layers were installed. It is interesting to note that only 24% of the damage found was caused during the installation of the geomembrane. The problems with the welds were

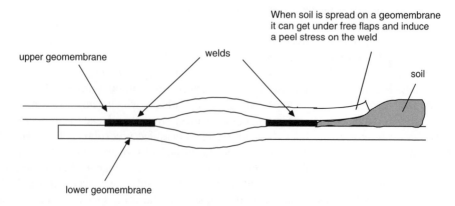

Figure 19.32 Cross-section of a dual-wedge weld showing that soil forced under the free flap during soil spreading can exert peel forces on the primary weld track

Table 19.7 Statistics of geomembrane damage (Nosko *et al.*, 1996)

	Amount of damage (%)	Detailed breakdown of causes of damage	Amount of damage (%)
When the damage occurred			
Liner installation	24	Extrusion (seaming)	61
		Melting	18
		Stone punctures	17
		Cuts	4
Covering	73	Stone punctures	68
		Heavy equipment	16
		Grade stakes	16
Post-construction	3	Heavy equipment	67
		Construction	31
		Weather, etc.	2
Where the damage occurred			
Flat floor	78	Stones	87
		Heavy equipment	13
Corner, edge	9	Stones	59
		Heavy equipment	19
			18
Under pipes	4	Stones	30
		Welds	27
		Heavy equipment	14
		Worker	15
		Cuts	14
Pipe penetrations	2	Welds	91
		Worker	8
		Cuts	1
Transport, storage, etc.	7	Heavy equipment	43
		Stones	21
		Worker	19
		Welds	17

mainly from poor extrusion welds at tees and around penetrations and corners (Nosko *et al.*, 1996).

🖈 *It was found that 73% of leaks in covered geomembranes are caused by construction damage specifically when placing earth materials on top of the geomembrane or liner.*

19.3.25 ANIMAL DAMAGE

Damage from animal traffic can be mitigated by perimeter protection (e.g. erecting fences and barriers) and by careful geomembrane material section (i.e. materials with high puncture resistance and strain capabilities). Exposed installations require fencing to prevent

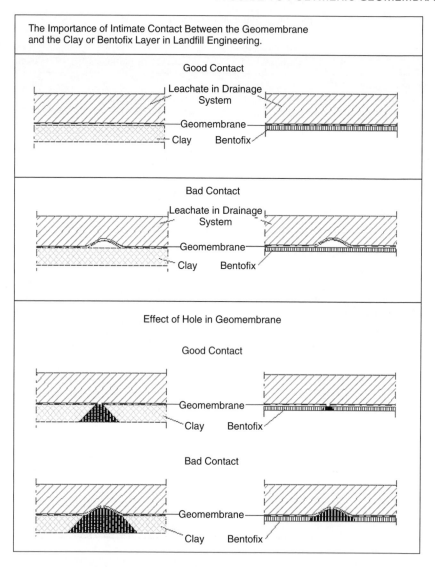

Figure 19.33 Schematic representation of the effect of 'good contact' and 'bad contact' of an HDPE geomembrane and differing subgrades. The 'black areas' in the lower parts of the figure show the leakage zones. Note that Bentofix™ is a brand of GCL. Reproduced by permission of NAUE

risk of damage by animals especially kangaroos which can damage the membrane with their claws. There have also been instances of seagulls pecking loose flaps at the edges of seams of RPP floating covers.

19.3.26   ALLOWABLE LEAKAGE RATE

It is now recognized in the geosynthetics industry that recently installed single liners will leak to some degree. Small leaks at seams and in the body of the liner or geomembrane are not unusual. The majority of singly lined geomembrane installations will have a leakage rate that is a value higher than zero. A small leakage rate is considered to be less than 100 lphd[3] while a significant leakage rate is over 3000 lphd[3].

The maximum acceptable leakage rate is generally quoted as around 200 lphd[3]. The leakage rate at which some corrective action needs to be taken is defined as the 'Action Leakage Rate' (ALR). The ALR set for US landfills is 200 lphd[3]. This is the leakage rate at which the leaks must be found and repaired (Peggs, 2003).

Relatively low leakage rates can be viewed as acceptable since the effort and cost in locating small leaks and then repairing them is often greater than the impact of the leak. Also it is not uncommon that the process of locating and repairing leaks can actually lead to more leaks or a higher leakage rate.

Figure 19.33 shows the importance of ensuring good contact between the geomembrane and the subgrade for ensuring leakage rates through holes in the liner are minimized.

19.3.27   MEASURES TO OVERCOME INSTALLATION PROBLEMS WITH GEOMEMBRANES

Factors that assist to overcome installation problems with geomembranes include:

- Ensuring that all contact surfaces are free of any protrusions and obstacles; generally anything bigger than fine gravel/coarse sand particles is unacceptable.
- Ensuring the subgrade foundation is free of sudden depressions and pinnacles.
- Ensuring that the subgrade for the geomembrane is well compacted.
- Ensuring that the geomembrane surface is dry and free of dust where welding is carried out.
- Ensuring that the geomembrane has good conformance with the subgrade.

  CQA inspection should concentrate on the following areas:

- Inspection of the surface of the geomembrane for defects (e.g. holes, tears, folds, score marks).
- Supervision of welding, weld testing and weld testing documentation (of the weather conditions during geomembrane deployment and seaming).

REFERENCES

Averesch, U. B. and Schicketanz, R. T., Recommendations for new installation procedures of geomembranes in landfill composite sealing systems by the 'Riegelbauweise',

---

[3] 1phd = litres per hectare per day.

in *Proceedings of Eurogen 2000*, Bologna, Italy; available from Düsseldorfer Consult GmbH, Düsseldorf, Germany and Ingenieurbüro Schicketanz, Aachen, Germany (2000).

BAM, 'Certification Guidelines for Plastic Geomembranes Used to Line Landfills and Contaminated Sites', Federal Institute for Materials Research and Testing (BAM) (September, 1999).

Blond, E., Bouthot, M. and Mlynarek, J., 'Selection of protective cushions for geomembrane puncture protection'; available from CTT Group/SAGEOS, St-Hyacinthe, Quebéc, Canada.

Giroud, J. P., Peltc T. and Bathurst, R. J., Uplift of Geomembranes by Wind, *Geosynthetics International*, **2**(6), 897 (1995).

Koerner, R. M., *Designing with Geosynthetics*, 5th Edition, Prentice Hall, New York, NY, USA, p. 562 (2005).

Layfield, Literature on geomembrane slack [http://www.geomembranes.com/index_ resources.cfm?copyID = 33&ID = geo&type = tech] (2008).

McQuade, S. J. and Needham, A. D., Geomembrane liner defects – causes, frequency and avoidances, *Geotechnical Engineering*, **137**, 203–213 (1999).

Narejo, D. B., Three Levels of Geomembrane Puncture Protection, *Geosynthetics International*, **2**(4), 765 (1995).

Nosko, V., Andrezal, T, Gregor, T. and Ganier, P., 1996, Sensor Damage Protection Damage Detection System (DDS) – The unique geomembrane testing method, in *Proceedings of the First European Geosynthetics Conference, Eurogeo I*, A. A. Balkema, Rotterdam, The Netherlands, pp. 743–748 (September/October, 1996).

Peggs, I. D., 'Geomembrane Liner Durability: Contributing Factors and the Status Quo' [http://www.goesynthetica.net/tech_docs/idpigsukpaper.pdf] (2003).

Reddy, K. R. and Saichek, R. E., Assessment of Damage to Geomembrane Liners by Shredded Scrap Tires, *Geotechnical Testing Journal*, **21**(4), 10 (1998).

Seeger, S. and Mueller, W. 1996. Limits of stress and strain: Design criteria for protective layers for geomembranes in landfill liner systems, in *Proceedings of the First European Geosynthetics Conference, Eurogeo I*, A. A. Balkema, Rotterdam, The Netherlands, pp. 153–157 (September/October, 1996).

Soong, T. Y. and Koerner, R. M., Laboratory study of HDPE geomembrane waves', in *Proceedings of the 6th International Conference on Geosynthetics*, Vol. 1, Industrial Fabrics Association International, Roseville, CA, USA, pp. 301–306 (1998).

Soong, T.-Y. and Koerner, R., Behavior of Waves in High Density Polyethylene Geomembranes: a Laboratory Study, *Geotextiles and Geomembranes*, **178**, 81–104 (1999).

TA Abfall, 2. Allgemeine Verwaltungsvorschrift zum Abfallgesetz. Bundesministerium für Umwelt, Naturschutz und Reaktorsicherheit, Bonn, Germany. 3. Allgemeine Verwaltungsvorschrift zum Abfallgesetz. Bundesministerium für Umwelt, Naturschutz und Reaktorsicherheit, Bonn, Germany (1991/1993).

Tognon, A. R., Rowe, K. and Moore, I. D., Geomembrane strain observed in large-scale testing of protection layers, *Geotechnical and Geoenvironmental Engineering*, **126**(12), 1194 (2000).

Wilson-Fahmy, R. F., Narejo, D. B. and Koerner, R. M., 'A Design Methodology for Puncture Protection of HDPE Geomembranes', GRI Report # 13, Philadelphia, PA, USA, p. 93 (September, 1994).

Appendix 1

Glossary

Absorption–the process by which a liquid is drawn into and tends to fill permeable pores in a porous solid body; also, the increase in mass of a porous solid body resulting from penetration of a liquid into its permeable pores.

Ageing–the process of exposing materials to an environment for an interval of time.

Anchor trench–an excavated ditch in which the edges of a geomembrane are buried in order to hold it into place.

Antioxidants–additives that protect polymer compounds from damage from oxygen, light and heat.

Antiozonants–additives that rubber compounds from cracking and deteriorating from exposure to ozone as well as providing resistance to oxygen and heat.

Appurtenances–any adjunct item necessary for proper functioning of the total geomembrane installation such as inlet and outlet piping, sumps, vents, structural support frames, etc.

Asperity–a raised area on the textured geomembrane corresponding to the peaks on a textured surface.

ASTM–American Society for Testing and Materials.

Berm–a soil ridge often forming the outer edge of a lagoon or pond.

Blooming–tendency of additives to migrate to the surface of rubber and plastic materials after processing. Blooming often manifests itself as a waxy or powdery deposit on the surfaces of rubbers and plastics. When the additive migrates to the rubber or plastic surfaces, it can be lost from the article as a result of routine abrasion or rubbing.

Breaking force–the force at failure.

Breaking load–the maximum force applied to a specimen in a tensile test carried to rupture.

Breaking strength–the ultimate tensile strength of a material per unit width.

BTX–benzene, toluene and xylene.

Butyl rubber–a synthetic rubber based on isobutylene and a minor amount of isoprene. It is vulcanizable, highly impermeable to gases and water vapour, and has good resistance to ageing, chemicals and weathering.

Cap–the barrier layer on top of a solid waste containment facility such as a landfill. The cap can serve multiple functions such as containing landfill gas (i.e. methane) and preventing rain and runoff from becoming leachate.

CCL–compacted clay liner.

Chemical resistance–the ability of a geomembrane to resist chemical attack.

CPE–chlorinated polyethylene.

Construction Quality Assurance (CQA)–relates to quality assurance (QA) performed by an inspection company working on behalf of the regulatory agency but paid by and reporting to the facility's owner and/or operator.

Construction Quality Assurance (CQA) plan–provides guidance to the site owner, environmental quality control (QC) and quality assurance (QA) personnel (or their representatives) and their contractors involved with the installation of geomembrane lining systems in existing concrete secondary containment structures.

Construction Quality Control (CQC)–relates to QC performed by the company conducting the seam fabrication.

Coupon–a portion of a material or laboratory sample from which multiple specimens can be taken for testing.

CQA Laboratory–an independent laboratory that tests material samples on behalf of the facility owner.

CQA Monitor–the facility owner's representative monitoring the quality of the material and its proper installation.

Creases–a sharp definitive line or mark produced by folding or wrinkling.

Creep–the time-dependent increase in accumulative strain (i.e. elongation) in a material resulting from an applied constant force. The slow change in length or thickness of a material under prolonged stress.

Crimpling–synonymous with wrinkling.

Critical height (CH)–the maximum exposed height of a cone or pyramid that will not cause a puncture failure of a geomembrane at a specified hydrostatic pressure for a given period of time.

Cross-machine direction (CM)–the direction in the plane of the geomembrane or scrim, perpendicular to the direction of manufacture (MD). Cross-machine direction is also referred to as the transverse direction (TD).

CSPE–chlorosulfonated polyethylene which is a synthetic rubber developed by DuPont under the tradename Hypalon® and is noted for its excellent resistance to chemicals, temperature extremes and ultraviolet light. CSPE geomembranes are manufactured by a proprietary process in which two layers of CSPE compound are calendered to fully encapsulate a polyester reinforcing scrim. CSPE geomembranes exhibit thermoplastic qualities during processing and field installation, which permits seaming by chemical or thermal fusion methods. As the membrane is exposed to UV, weather and the elements, it slowly cures forming a vulcanized membrane, with a higher tensile strength.

CVB–Chemical Vapour Barrier.

Defect–a lack of compliance of the liner with specified requirements. A defect has the potential to cause failure.

Deformation–the change in size of a material under load, from its original pre-loaded dimensions.

Density–mass per unit volume.

Design load–the load at which the geomembrane is required to withstand as it performs its intended function.

Destructive weld test–shear and peel testing of welded samples according to standards such as ASTM D-6392 and ASTM D-4437.

Differential settlement–subsidence and settlement (e.g. of waste or soil) that produces out-of-plane stresses (i.e. axisymmetric stress). For this reason landfill caps need to be made from geomembrane materials with good resistance to out-of-plane stresses, such as LLDPE.

Dimensional stability–is a reversion phenomenon of polymeric geomembranes due to 'frozen in' stresses and in-built orientation during production. Dimensional Stability of geomembranes is a measurement of the linear dimensional change resulting from exposure to temperature (also known as dimensional change).

DSC–differential scanning calorimetry. A thermal analysis technique carried out on few milligrams of material to determine its thermal properties at different temperatures. By measuring the heat flow into or out of the sample, it is possible to measure the melting point and other transitions characteristic of the material, as well as chemical changes such as oxidation.

DT–destructive testing of welds.

Dual-seam–a geomembrane seam with two parallel welded zones separated by an unwelded air space. The dual-seam itself can be made by a number of methods, the most common being the hot wedge technique. Other possible methods include hot air and ultrasonic bonding techniques.

Ductile–the deformation behaviour of a material that enables it to absorb energy by elongating under an applied load. This is often characterized by a high toughness and high elongation at fail when tested under tensile loading.

EIA–ethylene interpolymer alloy. A PVC plasticized with a non-extractable polymeric plasticizer. EIA is more commonly known by trademarked names such as XR−5™ (a Seaman Corporation product).

Elastic limit–the stress intensity at which stress and deformation of a material subjected to an increasing force cease to be proportional; the limit of stress within which a material will return to its original size and shape when the force is removed, and hence, not a permanent set.

Elasticity–the property of matter by virtue of which it tends to return to its original size and shape after removal of the stress which caused the deformation.

Elevated stress–stress and associated strain of a geomembrane material that is induced on a severely wrinkled or folded section.

Elongation at break–the elongation corresponding to the breaking load, that is, the maximum load. The percentage of elongation corresponding to the break point.

Elongation at yield–the elongation corresponding to the yield stress. The elongation at yield governs the ability of the geomembrane to withstand service-induced strains and stresses.

Embrittlement–The loss of toughness of a polymer such that if a crack is formed it will rapidly propagate under an applied load. In the absence of an applied stress the polymer will often fail due to internal stresses developed. The loss of toughness is often associated with a reduction in elongation (*see* mechanical properties).

EMMAQUA–'Equatorial Mount with Mirrors for Acceleration Plus Water Spray', is an accelerated UV ageing test for geomembrane materials to assess their degree of UV stability and resistance to weathering.

Encapsulated edge–the edge of a scrim-reinforced liner where the scrim is completely encapsulated by the polymer coating to prevent wicking.

Environmental stress cracking resistance (ESCR)–resistance to brittle cracks that occur in ductile polymer under a constant stress lower than the short-term yield or break stress of the material. Such cracking generally occurs in the case of HPDE in the presence of chemicals such as detergents, oxidizing acids and silicone fluids.

Environmental Stress Crack Resistance Test–ASTM D-5397 describes a test for evaluating the stress crack resistance of polyolefin geomembranes using a notched sample under a constant tensile load test. The samples are immersed in a solution of detergent at $50\,^{\circ}$C, subjected to a dead load to achieve 20 to 65% of the yield stress and the time to fail is measured.

EPDM–ethylene propylene diene M-class rubber is an elastomer (i.e. rubber) where the E refers to Ethylene, P to Propylene, D to diene and M refers to its classification in the ASTM standard D-1418. The 'M' class includes rubbers having a saturated chain of the polyethylene type.

EVA–ethylene vinyl acetate.

Extreme stress–stress and associated strain that approaches or exceeds the yield point of the material as induced by wrinkling resulting in creases, folding or fold over resulting in extreme creases.

Extrusion welding–this technique of extruding molten polymer resin at the edge of two overlapped geomembranes to effect a continuous bond.

Fabric reinforcement–usually an open-weaved textiled mesh also called 'scrim' that is used to add structural strength to a geomembrane. The scrim is usually enclosed between two membrane sheets that are bonded together.

Fabricator–a company that converts geomembrane rolls into seamed panels.The fabricator assembles the geomembrane into large panels.

Failure–a condition or state that prevents the liner from fulfilling its 'intended purpose'. Failure of the lining results in a leak that permits process liquids to escape in significant quantities. An arbitrary point beyond which a material ceases to be functionally capable of its intended use.

F-HPDE–fluorinated HPDE liners made by exposing regular HPDE liner to reactive fluorine gas.

Field testing–testing performed in the field under actual environmental conditions of temperature and exposure to fluids.

Fishmouths–ripples along the edges of geomembrane as it is installed leading to excessive wrinkles at the seam overlaps. Wrinkles that are perpendicular to the seam. No fishmouths should be allowed within the seam area.

Flexible polypropylene–a material having a 2% secant modulus of less than 300 MPa, as determined by ASTM Test Method D-5323, produced by polymerization of propylene with or without other alpha-olefin monomers. In particular, the Catalloy resins produced by Basell under the Astryn™ and Hifax™ trade names.

Floating cover–a geomembrane floating on a liquid containment facility (e.g. pond) that is sealed around the edges of the pond by cables, batten strips or a ring beam and that falls and rises as the liquid level changes. A floating cover is mainly used for preventing evaporative losses and contamination of potable water, preventing contamination of valuable products such as molasses, for harvesting biogas and preventing odourous emissions.

Fold over–a fold under compression due to external pressure causing sharp creases.

Folding–to double or bend over upon itself as in severe wrinkling of a liner.

Force–elongation curve–in a tensile test, a graphical representation of the relationship between the magnitude of an externally applied force and the change in length of the specimen in the direction of the applied force. Also a synonym for stress–strain curve.

Fourier-transform infrared spectroscopy–*see* FTIR spectroscopy.

fPP–flexible polypropylene.

FTB–refers to failure only in the parent material of the geomembrane and not in the weld area. It is an acceptable form of failure. Its use in discouraged nowadays as it can be ambiguous.

FTIR spectroscopy–Fourier-transform infrared spectroscopy. An analytical technique that enables the identification of the chemical groups present in polymers by measuring the amount and the energy of the radiation absorbed by the polymer in the infrared region of the electromagnetic spectrum. This method may be combined with special sampling techniques such as a microscope to enable identification of species present in small areas of a polymer such as inclusions and fracture surfaces.

GCL–geosynthetic clay liner.

Geocomposite–a product composed of two or more materials, at least one of which is a geomembrane.

Geomembrane–an essentially impermeable liner composed of one or more synthetic sheets. According to the ASTM D-4439 Standard Test Method, a geomembrane is a geosynthetic membrane or barrier with low permeability used with any geotechnical engineering related material so as to control fluid migration from a man-made project, structure or system.

Geomembrane manufacturer–manufactures the geomembrane rolls.

GML–geomembrane liner.

HALS–hindered amine light stabilizers. Additives that are incorporated into polyolefins and other polymers principally to confer UV stability but which also offer long-term thermo-oxidative stability at service temperatures by being retarders of oxidation. The polymer stabilization reactions generate reactive intermediates that are also stabilizers and are consumed only slowly in the polymer. Examples are Chimassorb 944™ and Tinuvin 622™.

HDPE–high-density polyethylene.

Heat-affected zone (HAZ)–is the region directly adjacent to a weld that may be subjected to overheating during the welding. This overheating can cause chemical (i.e. oxidation) and structural (i.e. crystallinity) changes in the geomembrane polymer which make it more susceptible to loss of ductility and brittle cracking.

Heat seaming–the process of joining two or more thermoplastic films of sheets by heating areas in contact with each other to the temperature at which fusion occurs.

Hindered amine light stabilizers–*see* HALS.

Hindered phenolic antioxidants–additives that are incorporated into polyolefins and other polymers in order to protect them against thermo-oxidative degradation during processing and in service. The additives inhibit the oxidation by scavenging the reactive radical species that lead to degradation of the polymer.

Hot air welding–this technique introduces high-temperature air or gas between two geomembrane surfaces to facilitate melting. Pressure is applied to the top or bottom geomembrane, forcing together the two surfaces to form a continuous bond.

Hot wedge (or knife) welding–this technique melts the two geomembrane surfaces to be seamed by running a hot metal wedge between them. Pressure is applied to the top or bottom geomembrane, or both, to form a continuous bond. Some seams of this kind are made with dual-bond tracks separated by a nonbonded gap. These seams are sometimes referred to as dual hot wedge seams or double-track seams.

HP-OIT–high pressure OIT. Measurement of the Oxidation Induction Time (OIT) of a polymer under the conditions of $150\,^{\circ}C$ and 3.5 MPa pressure of oxygen as described in ASTM D-5885 for HDPE.

Hydrogenated nitrile (HNBR)–is obtained by saturating the hydrocarbon chains of nitrile rubber with hydrogen. This special hydrogenation process removes many of the double bonds in the main chain of the NBR polymer and so HNBR possesses superior heat-, ozone- and chemical resistance, and mechanical characteristics over standard nitriles.

Impermeable–in geotechnical engineering, essentially 'impermeable' means that no measurable liquid flow occurs through a geomembrane.

Index test–a test procedure which may contain a known bias but which may be used to establish an order for a set of specimens with respect to the property of interest.

Initial tensile modulus–for geomembranes, the ratio of the change in force per unit width to the change in elongation of the initial portion of a force–elongation curve.

Installer–a company that installs geomembrane panels in field applications.

Laboratory sample–a portion of material taken to represent the lot sample, or the original material, and used in the laboratory as a source of test specimens.

Lapped joint–a joint made by placing one edge of a membrane partly over another surface and bonding the overlapping portions.

Leak–holes, punctures, tears, knife cuts, seam defects, cracks and similar breaches in an installed geomembrane.

Liner–a barrier layer on the floor and/or sides of a containment facility such as a landfill, reservoir, lagoon, tailings pond, etc. A geomembrane liner contains valuable product and/or protects the groundwater.

LLDPE–linear-low density polyethylene.

Lot–a single run of geomembrane material from the same production facility, where the tooling and raw materials of production have not changed during manufacturing (also can be referred to as a 'batch').

Lot sample–one or more shipping units taken at random to represent an acceptance sampling lot and used as a source of laboratory samples.

Lowest individual reading–the lowest value for each test specimen. This requirement is often used for thickness. For example, a specification may indicate a minimum average value of 1.5 mm and a lowest individual reading of 1.35 mm for liner thickness (Yazdani).

Manufacturer–a company that takes raw materials and extrudes or calenders them into geomembrane rolls.

Mechanical properties–those properties, such as strength, stiffness and toughness that have been shown to be important for the mechanical performance of a material. The testing procedure involves the subjecting of a small test sample of defined shape and size to a deformation while measuring the stress that is produced as the sample is elongated (or strained) until it fails. This measurement, if carried out in tension, produces the following parameters – yield stress, yield strain, tensile modulus, break stress and break elongation. The area under a plot of stress against strain is a measure of the toughness of a polymer.

Membrane–a continuous sheet of a material. Usually, it carries the connotation of being impermeable although this is not strictly true.

Minimum average roll value (MARV)–commonly defined as the average test results minus two standard deviations and implies that approximately 97.5% of the test results meet or exceed this value. MARV can be misinterpreted since it is sometimes confused with the minimum average value as defined below. The MARV requires a normal distribution of data, which is not the case since rolls not meeting minimum average values are not included in the data. To calculate the MARV, an independent laboratory must test a large population of rolls, making conformance testing time-consuming and expensive. Resolution of any disputes between the manufacturer's quality control certification and the conformance testing can also be time-consuming and expensive since retesting of all rolls may be necessary. For these reasons, the MARV should not be used for polyethylene geomembrane properties. The minimum average value is preferred since it is much easier to qualify one or two rolls of material without having to produce a large data bank (Yazdani).

Minimum average value–a property value representing the lowest allowable value for the average of results for the specimens tested. The lowest acceptable property value for an individual roll. For example, ASTM D-638, 'Tensile Properties of Plastics', calls for at least five specimens to be tested in machine and cross machine directions. The minimum average tensile value is the lowest average value of the five specimens for any individual roll (Yazdani).

Minimum value–property value representing the lowest individual allowable value obtained when tested according to the specified test method. This applies to individual readings, such as thickness, or where only one specimen is tested for the specified parameter.

Minimum test value–for geomembranes, the lowest sample value from documented manufacturing quality control test results for a defined population from one test method associated with one specific property.

Machine direction–the direction in the plane of the material parallel to the direction of manufacture.

Melt flow index (or melt flow rate)–a measure of the molecular weight of a polymer.

Modulus–the stiffness of a material when subjected to stress.

Modulus of elasticity–the ratio of stress (nominal) to the corresponding strain below the proportional limit of a material, expressed in force per unit area, such as megapascals (MPa).

Monomer–the reactive parent hydrocarbon from which the polymer is polymerized and which defines the chemical composition of the material, e.g. ethylene is the monomer for polyethylene.

Multiaxial tension–stress active in more than one direction.

Nitrile rubber–is a copolymer of butadiene and acrylonitrile. The acrylonitrile content varies in commercial products from 18% to 50%. As the nitrile content increases, resistance to petroleum base oils and hydrocarbon fuels increases, but the low temperature flexibility decreases. Due to its excellent resistance to petroleum products such as oil and diesel, nitrile is the most widely used elastomer in the seal industry today. Nitrile compounds do not possess good resistance to ozone, sunlight or weather unless specially formulated due to the presence of unsaturation (i.e. double bonds) in their structures.

Nominal–representative value of a measurable property determined under a set of conditions, by which a product may be described.

Nominal value–a value which is based on the historical average lot property data taken over a period of time. Nominal values are for informational purposes and should not be used in project specifications (Yazdani).

Normal stress–the component of applied stress that is perpendicular to the surface on which the force acts.

OIT–oxidation induction time. The measurement of the time taken for a sample to show heat evolution in a DSC scan due to the onset of oxidation of the sample under the applied conditions.

Ozone cracking–is characterized by the formation of small cracks or fissures on the surface of rubber which run perpendicular to the direction of strain. Ozone cracking depends both on the composition of the rubber and on the applied mechanical stress, ozone concentration and the temperature.

Ozone testing–is an accelerated test method used primarily for evaluating the resistance of rubber products to cracking during service life. Samples are fixed in rigs so that the surface is in tension and placed in the exposure chamber for some time period specified in the ASTM method. The significance of these test methods lies in the ability to differentiate between the degrees of ozone resistance under the limited and specified conditions of the accelerated tests. The degree of resistance is judged by the appearance and magnitude of the formation of cracks in the surface of the subject material.

Panel–a series of geomembrane sheets fabricated together to make a larger unit, as supplied by a fabricator, usually folded onto a pallet or folded and then rolled on a core.

Performance property–a result obtained by conducting a performance test.

Performance test–a test which simulates in the laboratory as closely as practicable selected conditions experienced in the field and the results of which can be used in design calculations. In geomembranes, a laboratory procedure which simulates selected field conditions which can be used in design. Also a synonym for 'design test'.

Permeability–the rate of flow under a differential pressure, temperature or concentration of a gas, liquid or vapour through a geomembrane material.

Permeant–a chemical species, gas, liquid or vapour that can pass through a substance.

Plasticizer–a material added to a plastic or rubber to increase its ease of working or flexibility.

Polarity–this refers to the negativity and positivity of chemical compounds. Water is a common example of a polar material. Grease is a common example of a non-polar material. These two materials do not mix. Chemical compounds with similar polarities will have an affinity for each other. *Like dissolves like* is an expression used by chemists to remember how chemical substances interact. It refers to 'polar' and 'non-polar' solvents and solutes. Since water is polar and grease is non-polar, water will not dissolve grease. Hence washing clothes with grease stains in plain water achieves no cleaning. For instance, nitrile rubber is polar and hence not affected by oil, diesel or grease which are all non-polar; however biodiesel is polar (just like nitrile rubber) and so nitrile rubber and biodiesel have a chemical affinity for each other and therefore are not compatible from a chemical resistant liner point of view.

Polymer–a macromolecule formed by the chemical combination of monomers with either the same or different chemical compositions. Plastics, rubbers and textile fibres are all high molecular weight polymers.

Polyolefins–a group of hydrocarbon-derived polymers that are based on a repeat unit of which ethylene is the parent hydrocarbon (hence polyethylene). Examples include polypropylene (PP) and many copolymers, such as linear low-density polyethylene (LLDPE). The properties of polyolefins depend on the degree of crystallinity which is controlled by the structure of the polymer backbone defined by the monomers and the type of polymerization reaction.

Polyurethane (PU)–rubber-like material made from polyols and diisocyanates that make tough, abrasion-resistant and chemical-resistant liners.

Poly(vinyl chloride) (PVC)–a synthetic thermoplastic polymer prepared by polymerization of vinyl chloride. PVC can be compounded into flexible and rigid forms through the use of plasticizers, stabilizers, fillers and other modifiers.

Polyvinylidene fluoride (PVDF)–fluoropolymer that can be used to make liners with excellent chemical resistance.

ppm–parts per million. The concentration of additives is often expressed as parts of additive per million parts of polymer. For instance, 0.1 wt% of additive is expressed as 1000 ppm.

Project engineer–the person who designs and is responsible for the functioning of the lining system.

Proportional limit–the greatest stress which a material is capable of sustaining without any deviation from the proportionality of stress to strain (Hooke's law) (also called the 'elastic Limit').

Puncture resistance–the inherent resisting mechanism of the test specimen to failure by a penetrating or puncturing object. The extent to which a geomembrane is able to withstand the penetration of an object without perforation.

Quality assurance–all those planned or systematic actions necessary to provide adequate confidence that the geomembrane material or geomembrane system, or installation, will satisfy given needs.

Quality control–the operational techniques and the activities which sustain the quality of a geomembrane material or geomembrane system, or installation; also the use of such techniques and activities – ASTM D-4354.

Rate of creep–the slope of the creep–time curve at a given time.

Reinforcement–the layer of woven or knitted textile used as a support and rip-stop in reinforced geomembranes (also known as scrim).

Resin manufacturer–manufactures the resin from which the geomembrane rolls are made.

Roll–A quantity of geomembrane rolled up to form a single package as supplied from the manufacturer.

Rubber–a polymeric material which at room temperature is capable of recovering substantially in shape and size after removal of a deforming source. Refers to both synthetic and natural rubber. Also called an 'elastomer'.

Sample–a portion of material which is taken for testing or for record purposes. Also a group of specimens used, or of observations made, which provide information that can be used for making statistical inferences about the population from which the specimens are drawn.

SCR stress crack resistance.

Scrim–a woven open-mesh reinforcing fabric made from continuous filament yarn that is laminated (or encapsulated by coating) between two geomembrane plies and used to reinforce geomembranes.

Seam–the connection of two or more pieces of geomembrane by chemical or thermal fusion methods to provide the integrity of a single piece of the geomembrane material (synonymous with weld).

Secant modulus–the ratio of stress (nominal) to corresponding strain at any specified point on the stress–strain curve. In the case of geomembranes, it is the ratio of change in force per unit width to the change in elongation between two points on a force–elongation curve.

Sheet–a part of the manufactures geomembrane material cut from the roll.

Slope–degree of deviation of a surface from the horizontal, measured as a numerical ratio, as a percentage or in degrees. Expressed as a ratio, the first number is the horizontal distance (run) and the second number is the vertical distance (rise). A 1:1 slope has a 45 degree slope angle, a 1.5:1 has a 33 degree slope angle, a 2:1 slope has a 26.6 degree slope angle, a 3:1 slope has a 18 degree slope angle and a 4:1 slope has a 14 degree slope angle.

S-OIT–Standard Oxidation Induction Time. Performance of an oxidation induction time measurement in accordance with the standard conditions described in ASTM D-3895 of 200 °C and 1 atm of oxygen for testing of HDPE.

Specific gravity–the ratio of the density of the geomembrane material to the density of a reference substance at specified conditions of temperature and pressure (related to density). The reference substance is usually distilled water.

Specification–a precise statement of a set of requirements to be satisfied by a geomembrane material or system that indicates the procedures for determining whether each of the requirements is satisfied.

Specimen–a specific portion of a material or laboratory sample upon which a test is performed or which is taken for that purpose (synonymous with test specimen).

'Squeeze-Out'–the molten polymer that is squeezed out laterally to the edge of a thermal fusion weld and indicates that sufficient temperature was applied to melt the polymer. In a good weld the 'squeeze-out' beads at each edge will be symmetrical.

Stabilizers–additives that protect polymer compounds from heat and/or UV degradation (e.g. HALS).

Stiffness–resistance to bending.

Strain–the increase in length per unit original length when a tensile stress (or flexural stress) is applied. The elongation of the polymer when expressed as a percentage increase of the original length is a measure of the strain. The change in length per unit of length in a given direction (synonymous with elongation).

Stress–the force (in newtons, N) applied to a material per unit cross-sectional area of the material in a mechanical test. The units of stress are MPa (megapascals = N/mm^2).

Stress crack–an external or internal crack in a geomembrane caused by tensile stresses less than its short-time mechanical strength.

Subgrade–soil, clay or earth prepared to support a geomembrane.

Tear strength–the force required either to start or to propagate a tear in a geomembrane under specified conditions. Test results are dependent on direction of tear, specimen geometry and rate of tear.

Tensile creep rupture strength–for geomembranes, the force per unit width that will produce failure by rupture in a creep test in a given time, at a specified constant environment tensile creep strain.

Tensile strength–the maximum resistance to deformation developed for a specific material when subjected to tension by an external force.

Tensile test–a test in which a material is stretched uniaxially to determine the force–elongation characteristics, the breaking force or the breaking elongation.

Textured geomembrane–geomembranes with roughen surface textures to provide enhanced friction on slopes.

Thermoplastic–a plastic capable of being repeatedly softened by increase in temperature and hardened by decrease in temperature. Most polymeric membranes are supplied in thermoplastic form because the thermoplastic form allows for easier seaming.

Trampolining–tautness of the geomembrane caused from thermal contraction which generates excessive localized stresses and prevents intimate contact between the liner and the subgrade. Occurs especially at the 'toe' of slopes (also known as bridging).

Transverse direction (TD)–cross-direction.

'T-welds'–these are areas where two welds intersect. Since these areas have been exposed to a heat treatment twice they are said to have experienced a 'double-heat history'.

Typical value–for geomembranes, the mean value calculated from documented manufacturing quality control test results for a defined population obtained from one test method associated with one specific property.

Ultimate elongation–the elongation of a stretched specimen at the point of break. Usually reported as a percentage of the original length. Also called elongation at break.

UV degradation–the breakdown of a geomembranes polymer when exposed to sunlight.

Vacuum box–a transparent box which is placed over a geomembrane seam (previously wetted with soapy water) and to which a vacuum is applied causing bubbling to be observed in the location of a leak.

Vulcanization–is an irreversible process during which a rubber material is chemically altered by crosslinking (linking of different rubber chains into a network) and becomes less plastic and more resistant to swelling by organic liquids and elastic properties are conferred or improved. Vulcanization and 'cure' are interchangeable terms as they share the same definition. A chemical process through which a rubber compound's physical and chemical properties are improved by crosslinking.

Waves–in geomembrane applications, a smooth undulation of the surface caused either by thermal expansion or the installation method.

Whales–ballooning of a geomembrane liner or cover due to entrapped gases such as biogas (methane) being generated under the liner or cover by biological processes (e.g. anerobic decomposition of waste) (synonymous with 'whale backs').

Wicking–transport of water along scrim reinforcement fibres due to capillary action. Via this mechanism water ingress into reinforced geomembranes can occur leading to deleterious effects such as hydrolysis of the polyester scrim, blistering and delamination.

Wide strip tensile test–for geomembranes, a tensile test in which the entire width of a 200 mm wide specimen is gripped in the clamps and the gauge length is 100 mm (synonymous with wide-width strip tensile test).

Wrinkling–in geomembrane applications, a series of ridges or furrows on a surface – more severe than waves (synonymous with crimpling).

Yield point–in geomembranes, the first point on the force–elongation curve at which there is a maximum (the first maximum).

REFERENCES

ASTM D4439 'Standard Terminology for Geosynthetics'.
Yazdani, G., 'Terminology For Polyethylene Geomembrane Specifications', Polyflex [http://www.poly-flex.com/news11.html].

Appendix 2

Geomembrane Test Methods

AMERICAN SOCIETY FOR TESTING AND MATERIALS

(arranged by number)

ASTM D-5	penetration of bitumen
ASTM D-36	softening point of bitumen
ASTM D-256	Izod impact strength
ASTM D-412	tensile properties of nonreinforced rubber geomembranes
ASTM D-413	ply adhesion of rubber to fabrics (peel)
ASTM D-471	effect of liquids on rubber
ASTM D-570	water adsorption of plastics
ASTM D-638	tensile properties of plastics
ASTM D-696	coefficient of linear thermal expansion
ASTM D-746	brittleness temperature by impact (superseded by D-1790)
ASTM D-751	thickness of coated fabrics
ASTM D-751	mass/unit area of coated fabrics
ASTM D-751	tear of coated fabrics (tougue tear)
ASTM D-751	grab tensiles of coated fabrics
ASTM D-751	hydrostatic burst
ASTM D-792	specific gravity by displacement
ASTM D-814	transmission of liquids through rubber sheet
ASTM D-882	tensile testing of thin plastic sheeting
ASTM D-1004	$90°$ tear of non-reinforced geomembranes
ASTM D-1149	ozone cracking of rubbers
ASTM D-1203	volatile loss from plastics
ASTM D-1204	dimensional change (dimensional stability)
ASTM D-1238	melt flow index
ASTM D-1388	stiffness of fabrics
ASTM D-1434	gas permeability of plastic sheeting
ASTM D-1505	density by gradient column

A Guide to Polymeric Geomembranes: A Practical Approach   J. Scheirs
© 2009 John Wiley & Sons, Ltd

ASTM D-1593	thickness of flexible PVC sheeting
ASTM D-1603	carbon black content by tube furnace
ASTM D-1621	compression of rigid cellular plastics
ASTM D-1693	ESC bent strip method
ASTM D-1790	low temperature impact (brittleness temperature)
ASTM D-1822	tensile impact
ASTM D-1898	sampling of plastics
ASTM D-2136	low temperature bending of coated fabrics
ASTM D-2240	durometer hardness of rubbers
ASTM D-2261	tongue tear test for fabrics
ASTM D-2663	test methods for carbon black
ASTM D-3015	carbon black dispersion by hot plate (withdrawn)
ASTM D-3030	volatile matter in PVC resin
ASTM D-3083	resistance to soil burial (for PVC)
ASTM D-3083	PVC water extraction
ASTM D-3083	seam strength of PVC
ASTM D-3746	impact resistance of bituminous membranes
ASTM D-3776	mass per unit area of woven fabrics
ASTM D-3895	standard OIT by DSC
ASTM D-4218	carbon black content by muffle furnace
ASTM D-4354	sampling of geomembranes for testing
ASTM D-4355	xenon arc exposure of geotextiles
ASTM D-4437	shear and peel of field welds (superceded by D-6392)
ASTM D-4439	standard terminology for geosynthetics
ASTM D-4533	trapezoid tear for reinforced geomembranes
ASTM D-4545	shear and peel of factory seams
ASTM D-4595	wide width tensile method for geotextiles
ASTM D-4703	compression moulding of plastics into test specimens
ASTM D-4759	Determining the specification conformance of geosynthetics
ASTM D-4833	pin/rod puncture (index puncture)
ASTM D-4873	Identification, storage, and handling of geosynthetic rolls
ASTM D-4885	wide width tensile method for geomembranes
ASTM D-5147	sampling and testing of bituminous membranes
ASTM D-5199	thickness of smooth geomembranes
ASTM D-5261	mass/unit area of geotextiles
ASTM D-5262	tensile creep behaviour
ASTM D-5321	coefficient of friction by direct shear
ASTM D-5322	chemical resistance of geomembranes by immersion
ASTM D-5323	2% secant modulus of PE geomembranes
ASTM D-5397	NCTL stress cracking test
ASTM D-5494	pyramidal puncture of geomembranes
ASTM D-5496	Field immersion testing of geosynthetics
ASTM D-5514	hydrostatic puncture (large scale)
ASTM D-5596	carbon black dispersion by microtome method
ASTM D-5617	multiaxial tensile testing (out of plane)
ASTM D-5641	seam testing by vacuum chamber (vacuum box)
ASTM D-5721	oven ageing of polyolefin geomembranes
ASTM D-5747	chemical resistance of geomembranes to liquids
ASTM D-5818	sampling practice for assessing installation damage

ASTM D-5819	guide for evaluation of geosynthetic durability
ASTM D-5820	pressurized air testing of dual seams
ASTM D-5884	tear of reinforced geomembranes (large scale)
ASTM D-5885	high pressure OIT by DSC
ASTM D-5886	methods for fluid permeation through geomembranes
ASTM D-5994	core thickness of textured geomembranes
ASTM D-6140	asphalt retention of bituminous geomembranes
ASTM D-6214	shear and peel testing of chemical welds
ASTM D-6364	short term compression behaviour
ASTM D-6365	spark testing of geomembrane seams
ASTM D-6392	shear and peel testing of thermal fusion welds
ASTM D-6434	guide for test methods for fPP
ASTM D-6455	guide for test methods for prefabricated bituminous GM
ASTM D-6497	guide for attachment of geomembranes to penetrations/structures
ASTM D-6636	ply adhesion of a reinforced geomembrane
ASTM D-6693	tensile properties of non-reinforced and reinforced geomembranes
ASTM D-6747	guide for electrical leak path detection
ASTM D-7002	electrical leak location of exposed GMs using water puddle
ASTM D-7003	strip tensile properties of reinforced geomembranes
ASTM D-7004	grab tensile properties of reinforced geomembranes
ASTM D-7006	ultrasonic testing of geomembranes
ASTM D-7007	electrical leak location of GMs covered with water or earth
ASTM D-7056	tensile shear strength of bituminous geomembrane seams
ASTM D-7106	guide for test methods for EPDM geomembranes
ASTM D-7176	specification for buried, nonreinforced PVC geomembranes
ASTM D-7177	air channel evaluation of PVC dual-track seamed geomembranes
ASTM D-7240	conductive geomembrane spark testing
ASTM D-7272	integrity of seams by pre-manufactured taped methods
ASTM D-7274	mineral stabilizer content of prefabricated bituminous GMs
ASTM D-7275	tensile properties of bituminous geomembranes (BGMs)
ASTM E-96	water vapour transmission by cup method
ASTM F-904	ply adhesion of flexible laminates
ASTM G-154	UV exposure by fluorescent UV lights
ASTM G-155	UV exposure by xenon arc practice

GEOMEMBRANE (GM) RELATED TEST METHODS AND STANDARDS FROM GRI

(www.geosynthetic-institute.org/meth.htm)

GM1	Seam Evaluation by Ultrasonic Shadow Method
GM2	Embedment Depth for Anchorage Mobilization
GM3	Large Scale Hydrostatic Puncture Test (now formalized as ASTM D-5514)
GM4	Three Dimensional Geomembrane Tension Test (now formalized as ASTM D-5617)

GM5 (a)	Notched Constant Tensile Load (NCTL) Test for Polyolefin Resins or Geomembranes (1992) (now formalized as ASTM D-5397)
GM5 (b)	Single Point NCTL Test for Polyolefin Resin or Geomembranes (now formalized as ASTM D-5397, Appendix)
GM5 (c)	Seam Constant Tensile Load (SCTL) Test for Polyolefin Geomembrane Seams
GM6	Pressurized Air Channel Test for Dual Seamed Geomembranes (see also ASTM D-5820)
GM7	Accelerated Curing of Geomembrane Test Strip Seams Made by Chemical Fusion Methods
GM8	Measurement of the Core Thickness of Textured Geomembranes (now formalized as ASTM D-5994) GM9 Cold Weather Seaming of Geomembranes
GM10	Specification for the Stress Crack Resistance of HDPE Geomembrane Sheet
GM11	Accelerated Weathering of Geomembranes Using a Fluorescent UVA Device
GM12	Asperity Measurement of Textured Geomembranes using a Depth Gauge
GM13[a]	Test Methods, Properties and Testing Frequency for High Density Polyethylene (HDPE) Smooth and Textured Geomembranes
GM14	Selecting Variable Intervals for Taking Geomembrane Destructive Seam Samples Using the Method of Attributes
GM15	Determination of Ply Adhesion of Reinforced Geomembranes (now formalized as ASTM D-6636)
GM16	Observation of Surface Cracking of Geomembranes
GM17[a]	Test Methods, Properties and Testing Frequency for Linear Low Density Polyethylene (LLDPE) Smooth and Textured Geomembranes
GM18[a]	Test Methods, Properties and Testing Frequency for Flexible Polypropylene (fPP and fPP-R) Nonreinforced and Reinforced Geomembranes (temporarily suspended: May 3, 2004) (Reinforced March 20, 2009)
GM19[a]	Seam Strength and Related Properties of Thermally Bonded Polyolefin Geomembranes
GM20	Selecting Variable Intervals for Taking Geomembrane Destructive Seam Samples Using Control Charts
GM21[a]	Test Methods, Properties and Testing Frequency for Ethylene Propylene Diene Terpolymer (EPDM) Nonreinforced and Scrim Reinforced Geomembranes
GM22	Test Methods, Required Properties and Testing Frequency for Scrim Reinforced Polyethylene Geomembranes used in Exposed Temporary Applications

[a]Denotes those items that are generic specifications.

GEOMEMBRANE TESTING METHODS CONVERSION TABLE ASTM AND ISO

Method	ASTM Method	ISO Method	Comments
Thickness	ASTM D5199	ISO 9863-1	similar but not equivalent
Mass per unit area	ASTM D1593	ISO 9864	similar but not equivalent
Melt flow Index	ASTM D1238	ISO 1133	equivalent
Density	ASTM D792	ISO 1183	equivalent
Tensile properties index	ASTM D6693	ISO 527-3	similar but not equivalent
Wide width tensiles	ASTM D4885	ISO 10319	similar but not equivalent
Tear resistance	ASTM D1004	ISO 34	similar but not equivalent
Carbon black content	ASTM D1603	ISO 6964	similar but not equivalent
Carbon black dispersion	ASTM D5596	ISO 11420	similar but not equivalent
OIT testing	ASTM D3895	ISO 10837	similar but not equivalent
Chemical resistance testing	ASTM D5322	ISO 175	similar but not equivalent
Oven ageing	ASTM D5721	ISO 4577	similar but not equivalent

Index